Cellulose

Cellulose: Development, Processing, and Applications covers topics related to advanced cellulose development and processing, as well as the utilization of major agricultural and biomass waste. It discusses the utilization of cellulose from other agricultural and biomass materials, including oil palm biomass, bamboo, and other non-wood forest products in emerging areas. It covers the treatments used to improve the quality of cellulosic materials in specific applications. Following that, the book delves into the use of cellulosic materials in the application of composting science and technology.

Features:

- Delves into the specific agriculture waste/biomass waste materials used for the advanced cellulose-based production
- Outlines the potential use of the covered materials for energy production and other emerging applications
- Includes composting technology and processes using cellulosic materials
- Overviews industrial applications of cellulose from agricultural waste/biomass waste and composting technology
- Discusses the main agricultural waste/biomass in the Asian region

This book is aimed at researchers and graduate students in chemical engineering, bioprocessing, composites, and biotechnology.

Cellulose
Development, Processing, and Applications

Edited by
Abu Zahrim Yaser, Mohd Sani Sarjadi, and
Junidah Lamaming

CRC Press
Taylor & Francis Group
Boca Raton London New York

CRC Press is an imprint of the
Taylor & Francis Group, an **informa** business

Designed cover image: Junidah Lamaming

First edition published 2024
by CRC Press
6000 Broken Sound Parkway NW, Suite 300, Boca Raton, FL 33487-2742

and by CRC Press
4 Park Square, Milton Park, Abingdon, Oxon, OX14 4RN

CRC Press is an imprint of Taylor & Francis Group, LLC

© 2024 selection and editorial matter, Abu Zahrim Yaser, Mohd Sani Sarjadi, and Junidah Lamaming; individual chapters, the contributors

ISBN: 9781032414386 (hbk)
ISBN: 9781032414393 (pbk)
ISBN: 9781003358084 (ebk)

DOI: 10.1201/9781003358084

Typeset in Times
by codeMantra

We dedicate this book to our family, friends, and researchers who are always passionate about sharing, crowdsourcing, and gaining knowledge to build a sustainable future.

Contents

Preface

Cellulose, as a next-generation material, will evolve in terms of its research and developments, spurred on by technological innovation. Cellulose is used in a wide range of applications, including paper, textiles, construction materials, food additives, bulking agents, and pharmaceuticals. It is also a promising renewable resource for the production of biofuels, bioplastics, compost, and other sustainable materials. The adoption of the Sustainable Development Goals as a vision has established a framework toward a more sustainable future. Given that it can be efficiently recovered not just from wood but also from agricultural waste or lignocellulosic materials, it looks promising as a sustainable resource with a low environmental impact. This eco-friendly material has the potential to address resource depletion experienced by many sectors. However, recent advances in cellulose chemistry and processing have opened up new opportunities for its use in emerging technologies, such as biocomposites, energy storage, and the biomedical field.

The book *Cellulose: Development, Processing, and Applications* aims to provide an in-depth overview of the current knowledge of cellulose and its many promising applications. This book covers a wide range of issues related to cellulose, from its molecular structure and synthesis through its processing into various forms, such as all-cellulose and nanocellulose, and its regeneration, as well as its diverse applications in different industries. Future research directions are also outlined, along with a discussion of current issues and potential solutions associated with the development of cellulose-based products. It also features a variety of processing methods, including mechanical and thermal conversion, as well as chemical and enzymatic modification. The book also delves into the various applications of cellulose, including its use in paper, textiles, construction materials, automotive applications, leather processing, and absorption materials in wastewater. It explores the emerging trends in cellulose research, including new methods for cellulose regeneration. It highlights emerging cellulose-relevant research such as new cellulose production methods, the development of innovative cellulose-based products, and the implementation of cellulose into new technologies.

The chapters in this book have been written by experts in the field of cellulose research, including scientists from academia, industry, and government research organizations. The contributors have been invited based on their expertise and their contributions to the field of cellulose research. It will be of interest to researchers working on cellulose-based materials and their applications, as well as its potential for new and innovative applications. The book will also be a valuable resource for engineers, technologists, and students in materials science, chemistry, biology, and related fields.

We hope that this book will not only contribute to the corpus of cellulose research and societal knowledge but also benefit future generations who aspire to enhance and improve the way they live through sustainable living. We extend our gratitude to the reviewers: AA Rushdan, AH Abdul Hair, AZM Asa'ari, A Ahmad, A Embrandiri, JG Boon, CK Saurabh, HK Abdul Razak, MH Mohd Roslim, J Lamaming, MN Islam, CH Ng, NI Saharudin, R Alkarimiah, SF Mhd Ramle, S Saallah, T Simioni, TNH Tuan Ismail, and ZA Abdul Hamid for their valuable comments and suggestions for the chapters in this book.

Editors
Abu Zahrim Yaser
Mohd Sani Sarjadi
Junidah Lamaming

About the Editors

Abu Zahrim Yaser, PhD is an Associate Professor in Waste Processing Technology at the Faculty of Engineering, Universiti Malaysia Sabah, Malaysia. He obtained his PhD from Swansea University. He has published 12 books and more than 100 other publications. His inventions in the composting systems have been adopted by Tongod and Nabawan districts in Sabah. He is the co-chair for Kundasang's Community Composting Site, a site dedicated for vegetable waste composting. He is a visiting scientist at the University of Hull.

Mohd Sani Sarjadi, PhD is currently an Associate Professor in Industrial Chemistry in the Faculty of Science and Natural Resources, Universiti Malaysia Sabah. His PhD was awarded in 2015 and obtained from the University of Sheffield, United Kingdom. His research interests include organic synthesis, polymer chemistry, and all aspects of industrial chemistry. He had published numerous articles in local and international refereed journals and conference proceedings. He is an Associate Member of the Royal Society of Chemistry, United Kingdom (Chartered Chemist); Associate Member of the Malaysian Institute of Chemistry; and Professional Technologist (Material Science Technology), Malaysia Board of Technologists.

Junidah Lamaming, PhD is currently a fellow researcher at the Faculty of Engineering, Universiti Malaysia Sabah. She obtained her PhD in 2016 in the field of nanocellulose and nanofiber science and technology from Universiti Sains Malaysia, Malaysia. She was rewarded an IAAM Scientist Award from International Association of Advanced Materials. Her research interests include nanocellulose fiber and applications, polymer blends, reinforced/filled polymer composites, characterization and production of lignocellulose-based composites and fire-retardants, bioadhesives, and biopolymers using sustainable methods. Over the years, she has published articles in a significant number of high impact journals, conference papers, and a book in her related research areas. She also contributes her expertise and knowledge by reviewing international and national publications. She is a member of the Malaysia Board of Technologists.

List of Contributors

Nurul Hidayah Adenan
Faculty of Applied Science
Universiti Teknologi MARA Cawangan Negeri
    Sembilan
Kuala Pilah, Malaysia

Nurul Ain Mat Akil
Faculty of Applied Science
Universiti Teknologi MARA Cawangan Negeri
    Sembilan
Kuala Pilah, Malaysia

Mohd Hazim Mohamad Amini
Faculty of Bio-Engineering and Technology
Universiti Malaysia Kelantan
Jeli, Malaysia

Nur Amira Zainul Armir
Bioresources and Biorefinery Laboratory,
    Department of Applied Physics, Faculty of
    Science and Technology
Universiti Kebangsaan Malaysia
Bangi, Malaysia

Mohd Fahmi Awalludin
Forest Products Division
Forest Research Institute Malaysia
Kepong, Malaysia

Siti Ayu Aziz
Faculty of Science and Natural Resources
Universiti Malaysia Sabah
Kota Kinabalu, Malaysia

Sumate Chaiprapat
Faculty of Engineering, Department of Civil
    and Environmental Engineering, PSU
    Energy Systems Research Institute (PERIN)
Prince of Songkla University
Songkhla, Thailand

Khim Phin Chong
Faculty of Science and Natural Resources
Universiti Malaysia Sabah
Kota Kinabalu, Malaysia

Atanu Kumar Das
Department of Forest Biomaterials and
    Technology
Swedish University of Agricultural Sciences
Umeå, Sweden

Mohd Fadhil MD Din
Centre for Environmental Sustainability and
    Water Security (IPASA), School of Civil
    Engineering, Faculty of Engineering
Universiti Teknologi Malaysia
Skudai, Malaysia

Md Omar Faruk
Forestry and Wood Technology Discipline
Khulna University
Khulna, Bangladesh

Mariliz Gutterres
Chemical Engineerin Department, Laboratory
    for Leather and Environmental Studies
Federal University of Rio Grande do Sul
Porto Alegre, Brazil

Nadya Hajar
Faculty of Applied Science
Universiti Teknologi MARA Cawangan Negeri
    Sembilan
Kuala Pilah, Malaysia

Muhammed Aidiel Asyraff Mohmad Hatta
Bioprocess Technology Division, School of
    Industrial Technology, Universiti Sains
    Malaysia

Renewable Biomass Transformation Cluster,
    School of Industrial Technology, Universiti
    Sains Malaysia
Penang, Malaysia

Juferi Idris
School of Chemical Engineering, College of
    Engineering
Universiti Teknologi MARA (UiTM)
Kota Samarahan, Malaysia

Eti Indarti
Agricultural Product Technology Department,
    Faculty of Agriculture
Universitas Syiah Kuala
Banda Aceh, Indonesia

Md Nazrul Islam
Forestry and Wood Technology Discipline
Khulna University
Khulna, Bangladesh

Shahrul Ismail
Institute of Tropical Aquaculture and Fisheries
University Malaysia Universiti Malaysia
    Terengganu
Kuala Nerus, Malaysia

Gobi Kanadasan
Department of Petrochemical Engineering,
    Faculty of Engineering and Green
    Technology
Universiti Tunku Abdul Rahman
Kampar, Malaysia

Mohd Asyraf Kassim
Bioprocess Technology Division, School of
    Industrial Technology, Universiti Sains
    Malaysia

Renewable Biomass Transformation Cluster,
    School of Industrial Technology, Universiti
    Sains Malaysia
Penang, Malaysia

Victória Vieira Kopp
Chemical Engineering Department, Laboratory
    for Leather and Environmental Studies
Federal University of Rio Grande do Sul
Porto Alegre, Brazil

Santhana Krishnan
Faculty of Engineering, Department of Civil
    and Environmental Engineering, PSU
    Energy Systems Research Institute (PERIN)
Prince of Songkla University
Songkhla, Thailand

Junidah Lamaming
Chemical Engineering Programme, Faculty of
    Engineering
Universiti Malaysia Sabah
Kota Kinabalu, Malaysia

Sofie Zarina Lamaming
Bioresource Technology Division, School of
    Industrial Technology
Universiti Sains Malaysia
Penang, Malaysia

Afroza Akter Liza
Jiangsu Co-Innovation Center for Efficient
    Processing and Utilization of Forest
    Resources and Joint International Research
    Lab of Lignocellulosic Functional Materials
Nanjing Forestry University
Nanjing, China

Mohd Hafiz Abd Majid
Faculty of Science and Natural Resources
Universiti Malaysia Sabah
Kota Kinabalu, Malaysia

Muaz Mohd Zaini Makhtar
Bioprocess Technology Division, School of
    Industrial Technology, Universiti Sains
    Malaysia

Renewable Biomass Transformation Cluster,
    School of Industrial Technology, Universiti
    Sains Malaysia

Centre of Innovation and Consultation,
    Universiti Sains Malaysia
Penang, Malaysia

Nor Haniah A. Malik
Faculty of Applied Science
Universiti Teknologi MARA Cawangan Negeri
    Sembilan
Kuala Pilah, Malaysia

Marwan Mas
Chemical Engineering Department, Faculty of
    Engineering
Universitas Syiah Kuala
Banda Aceh, Indonesia

Dianah Mazlan
School of Civil Enginnering
Universiti Sains Malaysia,
    Engineering Campus
Nibong Tebal Penang, Malaysia

Tan Kean Meng
Bioprocess Technology Division, School of
    Industrial Technology, Universiti Sains
    Malaysia
Penang, Malaysia

Renewable Biomass Transformation Cluster,
    School of Industrial Technology, Universiti
    Sains Malaysia
Penang, Malaysia

Zaim Hadi Meskam
Usaha Strategik Sdn. Bhd.
Puchong, Malaysia

Sumaya Haq Mim
Forestry and Wood Technology Discipline
Khulna University
Khulna, Bangladesh

Asniza Mustapha
Forest Products Division
Forest Research Institute Malaysia
Selangor, Malaysia

Wan Noor Aidawati Wan Nadhari
Malaysian Institute of Chemical and
    Bioengineering Technology
Universiti Kuala Lumpur
Melaka, Malaysia

Mohd Nasrullah
Faculty of Civil Engineering Technology
Universiti Malaysia Pahang (UMP)
Kuantan, Malaysia

Noor Afeefah Nordin
Institute of Power Engineering
Universiti Tenaga Nasional
Kajang, Malaysia

Swarna Devi Palanivelu
Department of Biological Sciences and
    Biotechnology, Faculty of Science and
    Technology
Universiti Kebangsaan Malaysia
Bangi, Malaysia

Ahsan Rajib Promie
Department of European Biorefinery
Université de Technologie de Troyes
Troyes, France

Vânia Queiroza
Chemical Engineering Department, Laboratory
    for Leather and Environmental Studies
Federal University of Rio Grande do Sul
Porto Alegre, Brazil

Md Lutfor Rahman
Faculty of Science and Natural Resources
Universiti Malaysia Sabah
Kota Kinabalu, Malaysia

Nur Syazwani Abd Rahman
Bioresource Technology Division, School of
    Industrial Technology
Universiti Sains Malaysia
Penang, Malaysia

Rozelyn Ignesia Raymond
Faculty of Science and Natural Resources
Universiti Malaysia Sabah
Kota Kinabalu, Malaysia

Zalniati Fonna Rozali
Agricultural Product Technology Department,
    Faculty of Agriculture, Universitas Syiah
    Kuala

Master Program of Agriculture Industrial
    Technology, Faculty of Agriculture,
    Universitas Syiah Kuala
Banda Aceh, Indonesia

Norhafizah Saari
Bioresource Technology Division, School of
    Industrial Technology
Universiti Sains Malaysia
Penang, Malaysia

Nur Izzaati Saharudin
Bioresource Technology Division, School of
    Industrial Technology
Universiti Sains Malaysia
Minden, Malaysia

Nurjannah Salim
Faculty of Industrial Sciences and Technology
Universiti Malaysia Pahang
Kuantan, Malaysia

Kushairi Mohd Salleh
Bioresource Technology Division, School of
    Industrial Technology, Universiti Sains
    Malaysia

Renewable Biomass Transformation Cluster,
    School of Industrial Technology, Universiti
    Sains Malaysia
Penang, Malaysia

João Henrique Zimnoch dos Santos
Chemistry Institute
Federal University of Rio Grande do Sul
Porto Alegre, Brazil

Mohd Sani Sarjadi
Faculty of Science and Natural Resources
Universiti Malaysia Sabah
Kota Kinabalu, Malaysia

Kallol Sarker
Forestry and Wood Technology Discipline
Khulna University
Khulna, Bangladesh

Siti Noorbaini Sarmin
Department of Wood Industry, Faculty of
    Applied Sciences
Universiti Teknologi MARA
Pusat Jengka, Malaysia

Sabrina Soloi
Faculty of Science and Natural Resources
Universiti Malaysia Sabah
Kota Kinabalu, Malaysia

Laila Sonia
Master Program of Agriculture Industrial
    Technology, Faculty of Agriculture
Universitas Syiah Kuala
Banda Aceh, Indonesia

Agency for Regional Development
Province of Aceh, Indonesia

Nurul Syuhada Sulaiman
Bioresource Technology Division, School of
    Industrial Technology
Universiti Sains Malaysia
Penang, Malaysia

Supachok Tanpichai
Learning Institute, King Mongkut's University
    of Technology Thonburi

Cellulose and Bio-based Nanomaterials
    Research Group, King Mongkut's University
    of Technology Thonburi
Bangkok, Thailand

Noor Azrimi Umor
Faculty of Applied Science
Universiti Teknologi MARA Cawangan Negeri
    Sembilan
Kuala Pilah, Malaysia

Vel Murugan Vadivelu
School of Chemical Engineering
Universiti Sains Malaysia
Nibong Tebal, Malaysia

Mohammad Harris M. Yahya
Faculty of Applied Science
Universiti Teknologi MARA Cawangan Negeri
    Sembilan
Kuala Pilah, Malaysia

Abu Zahrim Yaser
Chemical Engineering Programme, Faculty of
    Engineering
Universiti Malaysia Sabah
Kota Kinabalu, Malaysia

Dewi Yunita
Agricultural Product Technology Department,
Faculty of Agriculture, Universitas Syiah
Kuala

Master Program of Agriculture Industrial
Technology, Faculty of Agriculture,
Universitas Syiah Kuala
Banda Aceh, Indonesia

Sarani Zakaria
Bioresources and Biorefinery Laboratory,
Department of Applied Physics, Faculty of
Science and Technology
Universiti Kebangsaan Malaysia
Bangi, Malaysia

Amalia Zulkifli
Bioresources and Biorefinery Laboratory,
Department of Applied Physics, Faculty of
Science and Technology
Universiti Kebangsaan Malaysia
Bangi, Malaysia

1 Introduction

Abu Zahrim Yaser and Junidah Lamaming
Universiti Malaysia Sabah

CONTENT

Cellulose, as the main biopolymers derived from either wood or lignocellulosic materials, can be utilized in diversified products in various applications, including the energy, food and beverage industries, pharmaceutical and biomedical fields, pulp and paper industries, electric and electronic industries, as well as construction. In a report on the cellulose market, Pulidindi and Prakash of Future Business Insights (2022) estimated that the market was worth USD 219.53 billion in 2018 and that it would expand by 4.2% between 2018 and 2026 to reach USD 305.08 billion. Growing consumer demand and the development of cellulose usage and its derivatives boost increased production. Technology and engineering evolution have prompted the advanced development of new methods, inventions as well as new products which are greener and more sustainable. For example, food and beverage applications are being driven by rising demand for processed foods such as ready-to-eat meals and bakery products. Furthermore, a shift in consumer preference for plant-based ingredients in personal care and cosmetics is spurring market product development. This has led to the introduction and usage of sustainable non-wood-based alternatives made from agricultural waste. This lignocellulosic biomass includes oil palm fibers, rice husks and stalks, banana stems, bamboo, corn stalks, walnut shells, kenaf, bagasse, hemp fibers, and others.

In this book, a few lignocellulosic biomasses development, processing method, and application of cellulose have been highlighted. Among them are bamboo, oil palm wastes, including empty fruit bunches (EFB) and palm oil mill effluent (POME), allium (onion) peels, as well as sugarcane bagasse. This cellulose-rich material can be processed further using a variety of processing methods to produce desired end products. Thermochemical, biochemical, chemical, and bioconversion processes can convert lignocellulosic materials into bioenergy and biofertilizer. The isolation process of lignocellulosic materials into cellulose and nanocellulose includes pretreatment, chemomechanical, mechanical, chemical, liquefaction, and enzymatic processes. The properties of the produced cellulose and nanocellulose can be tailored to specific applications, such as biomaterials manufacturing, leather production, paper production, gels, biomedical applications, and bioadsorbent for wastewater treatment and animal waste.

With the right processing technologies and improvements, the potential of bamboo is unlimited (Lamaming et al., 2022). Bamboo lignocellulosic biomass has a lot of potential as a fossil fuel substitute. Bamboo biomass can be converted in a variety of methods (thermoelectric or biological conversion) to provide a variety of energy products (charcoal, syngas, and biofuels) that can be used as alternatives to currently available fossil fuels. Comparing bamboo biomass to other renewable resources reveals that it has both benefits and limitations. Compared to other biomass feedstocks, it has improved fuel properties and is appropriate for both thermodynamic and biochemical routes. Bamboo biomass has limitations related to the establishment, logistics, and land occupation. If poorly managed, it can also have detrimental effects on the environment, so choosing bamboo as an energy-specific feedstock requires careful consideration to avoid or reduce any potential concerns. The energy requirements cannot be met entirely by biomass from bamboo. To fully realize

DOI: 10.1201/9781003358084-1

their potential and deliver a sustainable energy supply, they must be combined with other sources. Chapter 2 explored more on the suitability of bamboo biomass as an energy source as well as process of recovering energy in bamboo biomass.

Oil palm is an essential crop commodity in Malaysia, and one of the world's major manufacturers of items derived from oil palm. The by-products of oil palm trees include a wealth of biomass resources, which enables them to be put to a variety of further productive uses. These include the production of biodiesel, palm composite, pulp, and paper. Oil palm biomass is an intriguing option for high biorefinery production efficiency because of its high cellulose content. In order to successfully execute a circular economy over the long term, biorefineries are a necessary component. They must be created extensively and reclaimed as building blocks from items that have been converted because they are dependent on renewable resources. The utilization of lignocellulosic biomass as a feedstock results in significant value addition and is an essential component of a bio-based economy. The separation of the substrate specificity of biomass enables the production of distinct product flows that were previously conceivable. Even though biorefineries investigate biochemical, morphological, and physiological processes to perform fractionation, hydrolysis, and fermentation, the amount of fresh water that is used raises worries about the quality of the water as well as economic costs. In order for biorefineries to become financially and environmentally sustainable systems, it is vital for these facilities to implement technologies that make use of non-potable resources for biomass. In order to reduce the amount of fresh water needed, efforts are being undertaken to switch to using salt water instead. Therefore, Chapter 3 delves into a comprehensive analysis of biorefineries that are supported by salt water, with an emphasis on the transformation of lignocellulosic biomass into biofuel and other value-added products.

Production of biofuel from renewable resources has gained great attention as a potential candidate to replace fossil fuels partially or completely as transportation fuels. Due to its sugary structure, cellulose has the potential to be used for biofuel production, which requires the depolymerization and isolation of smaller sugar units that could then be transformed into fuel. Production of biofuels such as biogas, bioethanol, biodiesel, and biogas can be carried out through different pathways, namely biochemical and thermochemical. Currently, both technologies are commercially available for producing biofuel, and additional research and development are being conducted to reduce the delivered cost of biofuels. To ensure the feasibility of biofuel production, selecting the most suitable technology that is eco-friendly, has less energy consumption, and is cost-effective is among the issues that need consideration. Furthermore, the biofuel production process through biochemical and thermochemical pathways can also be influenced by the type of feedstock used for the processes. Thus, Chapter 4 presents the technologies involved in the production of biofuel through current thermochemical (pyrolysis, gasification, and pyrolysis) and biochemical technologies. This chapter also describes the technological development of biofuels (bioethanol, biobutanol, liquid fuel, solid biofuel, and syngas) from different types of feedstocks.

Apart from an energy source, the enormous amount of biomass produced by the oil palm fields and mills, where it produces a large amount of lignocellulosic biomass such as oil palm trunks, oil palm fronds, EFB, palm-pressed fibers, palm shells, and palm oil mill effluent (POME), can be used as biofertilizer through composting. Composting is a widely adopted way to transform agricultural waste into organic fertilizer. This material contains a high concentration of cellulose, hemicellulose, and lignin, and its degradation affects composting efficiency (Liu et al., 2022). The oil palm biomass also contains a high concentration of nutrients, and the nutrient composition could be used as biocompost and organic fertilizer, assisting in soil conditioning, and reducing the use of inorganic fertilizer in agriculture sectors while also reducing environmental impact. The high amount of biomass generated in oil palm fields and mills is a major concern because it leads to the bioconversion of biomass into fertilizer as part of their waste management strategy, which helps reduce waste discharge into rivers while restoring nutrients to the plant nutrition cycle. Chapter 5 provides an extensive review of the latest updates on the conversion of different types of waste from the palm oil industry into fertilizer in Malaysia.

The cellulose serves as the primary source of energy for the biological transformations, as well as the resulting temperature rise and chemical changes associated with composting (Hubbe et al. 2010). The main carbon sources are cellulose and hemicellulose, which account for the majority of the carbon dioxide and heat. Cellulose degradation in the compost is important for providing carbon during the composting process. Chapter 6 reviews recent research on the use of selected wastewater composting, with a focus on using wastewater content as a nutritional enhancer and moisturizer that can be used directly and recycled, as well as the possibility of reducing load in wastewater treatment plants. The wastewater from the industries of oil mills, olive mills, alcohol or molasses distilleries, swine, composting facilities, and monosodium glutamate are among those mentioned in the review. Cellulosic material used as bulking agents can help to reduce nitrogen loss during composting and achieve an appropriate C:N ratio. In the composting process, wastewater was incorporated, mixed, and absorbed by bulking agents such as green waste, grass clippings, mushroom waste, and rice husk into the formulation of wastewater composting, which was included to increase the compost porosity, which in turn enhanced the oxygen availability, which then accelerated the microbial activities. The wastewater can help accelerate the mineralization of organic matter by microorganisms. The wastewater and the bulking agents could provide better MC, adequate nutrients for microorganism growth, degrade the compost materials, and accelerate the rate of compost, thereby obtaining a good quality fertilizer as a final product. The increasing nutrients found in this effluent vary from 1.05% to 6.1% (N), 0.06% to 2.8% (P), and 0.06% to 12.43% (K), proving their potential as nutrient enhancers.

Cellulose fibers have relatively high strength, high stiffness, and low density. For a range of applications, biocomposites are gaining popularity as a more environmentally friendly alternative to synthetic composites. This is due to the increase in environmental awareness, especially in the issues of pollution and global warming, which has diverted researchers' focus to eco-friendly biocomposites. Natural fibers such as hemp, kenaf, jute, and coir can be incorporated with bio-based polymers to form a composite. This natural fiber-reinforced composite (biocomposite) has the potential to be used in different applications such as packaging, textiles, and various structural applications such as decking and paneling. Modern society is very interested in natural biocomposite materials because of their noble mechanical qualities, light density, superior life cycle, biodegradability, and cost efficiency. In Chapter 7, the background of biocomposites, their manufacturing process, properties, challenges, and the uses of biocomposites specifically as structural components in various applications, such as in automotive, building and construction, and the furniture industry, have been summarized.

The key feature of a biocomposite in automotive applications is its lightweight material, which performs similar to conventional materials. Additionally, it also exhibits non-brittle fracture upon impact testing, which is one of the main requirements for automotive components. Chapter 8 discusses the fundamentals of natural fiber composites, its processing techniques, and the potential applications in the automotive sector, including the demands and future perspectives. The outcome from this chapter would benefit the researchers in this particular area as well as the industrial players in reducing waste disposal, greenhouse gas emissions, and life cycle considerations.

As described in Chapter 9, recently, the development of the construction industry has gone toward producing cement mortar that is high in strength and environmentally friendly. However, developing a cement mortar that is as strong as concrete is challenging because no coarse aggregate is used in the mortar mix. Thus, the development of new technologies and materials that can improve the strength of mortar without the use of coarse aggregates was studied. These days, additives to strengthen cement mortar by using natural resources have gained interest among researchers. Still, a limited study had been conducted to study the outcome of natural-based additives in cement mortar as a strengthening agent. This chapter deals with the importance of the utilization of cellulose nanocrystals (CNCs) as a natural-based additive in cement mortar as a strengthening agent. The focus of this chapter is to identify important performance criteria and parameters of research that has been done and compare them with current research. This chapter then discusses the science and different approaches to the utilization of CNCs and their ability to enhance the properties of materials. The differing performance of CNCs as admixture evaluation methods is discussed by looking at the different admixtures that each researcher reported.

In Chapter 10, Tanpichai details the production of all-cellulose composites (ACCs). The interfacial interaction between the cellulose particles and polymer matrix is a major concern for composite fabrication. ACCs have been prepared from various cellulose sources using impregnation or partial surface dissolution approaches with a specific solvent. These approaches can provide great interfacial adhesion between the reinforcement phase and polymer matrix that are both made from cellulose, affording promising improvements in the mechanical properties of ACCs compared with those of petroleum-based polymers and biopolymers. Herein, an overview of ACCs as well as factors that affect their physical properties (such as transparency and crystallinity) and mechanical properties has been provided, as well as examples of the applications of ACCs.

Nanocellulose is a strong reinforcement component in biocomposites, due to its enhanced dispersibility compared to microfibers and the tunability of its surface chemistry. Chapter 11 evaluates the potential of using the cellulose-based bioadhesives for wood, the application in wood-based composite. Polyhydroxyalkanoate (PHA) can be used as feedstock for the production of thermoplastics and can be combined with other compounds to generate blends and composites, which is another one of its many strengths. Due to the green and biobased character of PHA-derived materials, their potential applications are vast. The PHAs can be blended with nanocellulose and other polymers as a matrix. The PHA could be produced using palm oil mill effluent (POME). The POME is a thick, brownish liquid released from the palm oil mill during the oil extraction process. This liquid is highly enriched with organic contents, which need rigorous treatment before they can be released into the bodies of water. As the POME is released continuously, it provides the basis for its sustainable usage. POME could be valorized to produce the PHA. Ceasing the anaerobic treatment at the acidogenesis stage produces volatile fatty acids (VFA), which would be the precursor for PHA synthesis. PHA such as polyhydroxybutyrate and poly(3-hydroxybutyrate-co-hydroxyvalerate) has been used to produce biodegradable plastic, but it is hampered by non-sustainable resources. Pure culture microorganisms such as *Bacillus cereus*, *Cupriavidus necator*, and mixed culture microorganisms have been utilized to store the PHA within its cellular structure by supplying the feedstock with an excess carbon source but limited nutrients. Chapter 12 presented and discussed the characteristics of POME, the usage of bacterial strains to produce PHA, and the latest trend in the production of PHA from POME.

Lignocellulosic material from agricultural waste is readily available and has great potential as a cheap and sustainable raw material for nanocellulose production. Nanosized cellulosic materials have received much attention due to their excellent properties, high strength, low cost, biodegradability, abundance, and renewable properties. The advantage of using nanosized cellulosic materials is not only because of these properties, but also due to their dimensions, at the nanometer scale, which opens many possible properties that have yet to be discovered. In principle, nanoscale cellulosic materials can be extracted from a variety of cellulosic sources, including plants, bacteria, algae, and animals, using various procedures. However, the main challenge is to find a process to obtain a higher yield of nanocellulose with efficient isolation from lignocellulosic sources. It is known that the combination of two or more treatment methods can increase the yield and rate of fibrillation as well as reduce energy consumption during processing. Appropriate isolation techniques also promote the excellent properties of nanocellulose. Chapter 13 addresses the cellulose classification, a general overview of nanocellulose from various lignocellulosic sources, as well as the available factors in the extraction of nanocellulose, followed by introducing and comprehensively discussing various nanocellulose isolation techniques.

Meanwhile, Chapter 14 details the isolation and properties of nanocellulose in the form of cellulose nanocrystals (CNCs) derived from sugarcane bagasse. The chapter presented the yield and characteristics of CNC from bagasse, including the morphology and crystallinity index isolated from the various methods, including acid hydrolysis, catalyst TEMPO (2,2,6,6-tetramethylpiperidin-1-oxy) oxidation, ultrasonication, and high-pressure homogenization. At 33% sulfuric acid concentration and 30 min hydrolysis time, the acid hydrolysis method produced the highest yield (58%). The acid hydrolysis process yielded the highest crystallinity CNC (89%), with a sulfuric acid concentration of 65% in 45 min. The morphology of CNC was observed by TEM analysis. The highest axial ratio was 64 with dimensions of 2 nm width and 255 nm length, which was obtained from acid hydrolysis using a concentration of 33% sulfuric acid and a hydrolysis time of 30 min. The yield, morphology, and crystallinity of CNC obtained by each method differ depending on the concentration of chemicals used, processing time, and mechanical treatment such as tools and pressure.

Cellulose that has been liquefied becomes formable into gels that can be regenerated into complex structures and applications. In Chapter 15, regenerated cellulose in the forms of hydrogels, aerogels, cryogels, and xerogels has been thoroughly reviewed. The morphological structure, appearance, and properties of the regenerated cellulose products differ due to the preparation method, particularly how they are dried from the wet hydrogel. As a result, the various properties will lead to a variety of applications tailored to the fabricated products. This smart biopolymer has wide application in pharmaceutical and biomedical applications, particularly in wound care, therapeutics, cosmetology, cardiovascular diseases, oncology, ophthalmology, urology, drug delivery systems, tissue engineering, and tissue regeneration. It is thanks to their inherent physical, mechanical, and biological properties. In this regard, Chapter 16 discussed and highlighted the recent development of nanocellulose-based hydrogel materials in the biomedical field, particularly in drug delivery, tissue engineering, wound dressing, and wound healing.

Due to the extensive use of cellulose in the production of paper, the application of cellulose in paper production has seen a growth of over 3% CAGR through 2026. It helps shorten the paper's drying period. It raises the printing quality and makes the paper better by making it less transparent and porous. Paper made from cellulose also takes less energy and raw materials. Cellulose will also provide lightweight, enhancing transportation energy efficiency (Pulidindi and Prakash, 2022). Chapter 17 discusses the use of cellulose fiber in pulp and papermaking processes. The sources of cellulose, which cover various cellulose fiber sources including sources from wood, non-wood, recycled fiber, and rags, were explained. The advantages and disadvantages of cellulose fiber derived from various sources and their suitability for producing many grades of paper products were also discussed. The following section covers the basic processes of cellulose fiber conversion into pulp before the papermaking process. This includes types of pulping methods, bleaching methods, and pulp blending processes. The beating process, the refining process, and details on additives used in papermaking were described under the pulp blending subtopic. The basic processes involved in papermaking processes, such as forming, dewatering, pressing, drying, calendaring, reeling, and winding, were included. These basic processes were described based on the basic processes of the Fourdrinier machine. Meanwhile, the last section covers the papermaking processes used with the cellulose fiber-based pulp before it is sent to customers for further application in paper-based products.

The harvesting and processing of allium generate various waste streams. Allium peels as a raw material for papermaking are an innovative solution to global environmental problems such as deforestation and global warming. Allium peels contain cellulose and hemicellulose with low lignin content compared to commercial pulp materials such as hardwood and canola straw. However, allium peels alone cannot serve as a sole substitute to produce good-quality paper, as the technology of papermaking using allium peels is still in the development stage. As a result, Chapter 18 gives an overview of the current state of development of Allium peels as a papermaking pulp. In addition, the challenges and prospects of allium pulp for papermaking are also presented.

Leather has a significant impact on the world economy. The use of cellulose in the production of leather was covered in Chapter 19. Leather manufacturing involves a chemical process applied to a biological matrix that uses a huge quantity of chemicals and generates large amounts of residue. Cellulose, a sustainable and renewable product, can be used during leather production in three different steps: tanning, retanning, and finishing. In tanning, cellulose helps the penetration of metals into the hide matrix, contributing to the fibers' stability and improving the tanning performance. In retanning, the use of cellulose improved the physical and aesthetic properties of leather. In leather finishing, cellulose was mixed with other chemicals to obtain better leather, without compromising its aesthetic properties. During the process of adjusting the thickness of the leather or lowering it, solid waste is generated. Cellulose can be added to these wastes to obtain value-added products. They are used in different areas, such as agriculture and packaging.

Chapter 20 provides an overview of the utilization of cellulose as a main material for wastewater treatment. Because of their abundance of natural polymers and their promising combination of efficiency, low cost, and chemical freedom, cellulose-based materials have been used in wastewater treatment. Depending on the cellulose type and wastewater type, cellulose absorbent can be manufactured in powder or granule form, as fine particles, in fiber form, as gel, as a film, or in membrane form. Furthermore, the discussion in this chapter focused on the various types of cellulose modification and the most common effluent removal method using modified cellulose. The challenge and concern of using cellulose in wastewater treatment are also among the aspects highlighted in this chapter.

The demand for biodegradable cat litter adsorbent is increasing due to increased awareness of environmental and health impacts among cat owners and cat lovers. The use of clay and bentonite could be harmful to the environment if not properly managed. The EFB produced by oil palm industries have the potential to be developed into cat litter adsorbents. The hydrophilic functional groups of cellulose, hemicellulose, and lignin in the EFB structure allow it to absorb water. Furthermore, cellulose shares a structure and ion exchange with organic materials such as resin, both of which are suitable as absorbents (Saueprasearsit et al., 2010). Chapter 21 studies the comparison of imported cat litter with locally produced EFB pellets for performance. Four types of biodegradable cat litter, namely pine, EFB pellet, dried tofu, and a mixture of EFB pellet and pine (4:1), were evaluated for their physical properties, water absorption, and odor. The length and diameter of the samples varied due to the production specifications, but in terms of bulk density, dried tofu had the highest density, followed by EFB pellets. The pH of the samples was mostly in the neutral range, except for the dried tofu, which had an acidic value of 4.53. The sensory odor test showed that dried tofu caused better adsorption of the softener odor, as it was barely perceived by the respondents, followed by wood and EFB pellets. The EFB pellet-wood sample, on the other hand, was easier to smell when tested. In summary, EFB pellets have comparable characteristics to imported cat litter, which is advantageous as it is locally produced.

Chapter 22 presents some of the challenges and future perspectives of the cellulose-based products that have been encountered and some recommendations. As a versatile material, cellulose offers more research areas that can be explored in parallel with the advancement of technology that can offer sustainable products.

REFERENCES

Hubbe, M.A., Nazhad, M., & Sanchez, C. (2010). Composting as a way to convert cellulosic biomass and organic waste into high-value soil amendments: A review. *Bioresources*, 5(4), 2808–2854.

Lamaming, J., Saalah, S., Rajin, M., Ismail, N.M., & Yaser, A.Z. (2022). A review on bamboo based adsorbent for removal of contaminants in wastewater. *International Journal of Chemical Engineering*, 4, 1–14.

Liu, Q., He, X., Luo, G., Wang, K., & Li, D. (2022). Deciphering the dominant components and functions of bacterial communities for lignocellulose degradation at the composting thermophilic phase. *Bioresource Technology*, 348, 126808.

Pulidindi, K. & Prakash, A. (2022). *Cellulose Market*. Global Market Insights. Accessed on 25 October 2022 at https://www.gminsights.com/industry-analysis/cellulose-market.

Saueprasearsit, P., Nuanjarae, M., & Chinlapa, M. (2010). Biosorption of lead (Pb2+) by Luffa cylindrical fibre. *Environmental Research Journal*, 4(1), 157–166.

2 A Brief Overview of the Use of Bamboo Biomass in the Asian Region's Energy Production

Siti Ayu Aziz and Mohd Sani Sarjadi
Universiti Malaysia Sabah

CONTENTS

INTRODUCTION

Cellulose makes up between 30% and 50% of lignocellulosic biomass, whereas hemicellulose makes up between 25% and 35%, and lignin accounts for between 5% and 30% of this type of biomass. An appealing feedstock for the production of fuel alcohol is lignocellulosic biomass, which includes sources such as agricultural waste residues, wood, grass, forestry, and municipal solid wastes. This is because lignocellulosic biomass sources are abundant, and their costs are relatively cheap (Roque et al., 2012). The breakdown of cellulose into monomeric sugars and the oligosaccharides cellotetraose, cellotriose, and cellobiose results in the creation of biomass. Cellulose is the primary

DOI: 10.1201/9781003358084-2

component of all biomass (glucose and fructose). Hemicellulose is the second most common type of polymer found in biomass, and it is the polymer that is responsible for the formation of pentoses (xylose and arabinose) and hexoses (galactose, glucose, and mannose). Sugars found in hemicellulose have a lower molecular weight than those found in cellulose, and they are distinguished by having short helical chains that are easily hydrolyzed (Agbor et al., 2011). The modification of the cellulose structure, with the end goal of increasing its availability to enzymes and/or chemicals, is the primary objective of the pre-treatment of lignocellulosic biomass (Holm and Lassi, 2011).

Biorefineries are conducting extensive research into lignocellulosic biomass resources, such as bamboo, as potential replacements for the generation of industrial fuels that are derived from fossil fuels (Wang et al., 2020). The majority of the lignocellulose in bamboo species is composed of cellulose (38%–50%), hemicellulose (23%–32%), and lignin (15%–25%) (Zhang et al., 2021). The quantities of lignin found in woody and herbaceous plants are very different from one another in terms of their respective chemical makeup. Lignin content ranges from 8% to 15% in herbaceous plants to approximately 20%–38% in woody plant species (Poveda-Giraldo et al., 2021). Despite being a species of herb, bamboo has a high lignin content that is equivalent to that of woody plants despite its herbaceous nature (Jianfei et al., 2020). These components are generated by the process of photosynthesis, which makes use of carbon dioxide from the atmosphere (Wang et al., 2021). Cellulose has emerged as a critical component for use in biorefineries over the past few years (Kumar and Verma, 2021).

During the subsequent fermentation stage, which involves a great variety of microbial species, the glucose that was produced as a by-product of the hydrolysis of cellulose in a biorefinery is converted into a variety of white compounds. These substances have the potential to take the place of those that are produced by oil refineries (Islam et al., 2020). The utilization of lignocellulosic biomass results in the production of bioethanol, which is then sold on the market (Patel and Shah, 2021). Bioethanol is a well-known product of biorefineries. The production of greenhouse gases like carbon dioxide can be reduced thanks to the usage of bioethanol, which has the potential to lessen the severity of the adverse effects of climate change (Safieddin et al., 2020). Fuels for transportation that are created from petroleum can be replaced by bioethanol, which can also replace these fuels. However, the cost of producing bioethanol using present technologies from lignocellulosic biomass is not yet comparable with the cost of producing gasoline (refining oil) (Adewuyi, 2020).

Bamboo is the common name given to members of the Bambusoideae subfamily of the *Andropogoneae/Poaceae* family of grasses. This taxonomic group of gigantic woody grasses is known as the Bambusoideae. There are 1,250 species of bamboo, the majority of which are relatively fast-growing and can achieve stand maturity in 5 years or less (Scurlock et al., 2008). Bamboos are classified into 75 different genera, and there are a total of 75 different genera. The tropics are where bamboos are most commonly found; however, they can be found growing naturally on every continent, both in tropical and temperate climates. Asia has a total land area of more than 180,000 km and is home to around 1,000 different species. A significant portion of this land is made up of natural stands of native species rather than plantations or introduced species. On its own, China is home to more than 300 species, which are organized into 44 genera and spread across 33,000 km^2 (about 3% of the country's total forest land). Another country that produces a lot of bamboo is India, which has 130 different kinds and occupies 96,000 km^2 (or roughly 13% of the total wooded area). Other countries besides China and Japan that use bamboo extensively in their construction include Bangladesh, Indonesia, and Thailand. Bamboo is native to regions that range from subtropical to temperate, with the exception of Europe.

Since the beginning of human history, bamboo has been cultivated and utilized by people in a variety of contexts. Building material can be derived from the bamboo stem because of its qualities of being strong, lightweight, and flexible. From the fibers of bamboo, products such as paper, textiles, and boards can be made. Bamboo shoots of many different types are a popular source of nutrition in many Asian countries. In recent years, in an effort to find alternative energy sources to replace dwindling supplies of fossil fuels, a new method of harnessing bamboo's potential as a

TABLE 2.1

Bamboo Biomass Pilot Plant Locations in Malaysia, Indonesia, Vietnam, and Japan

Country	Location	Technology	Capacity (kW)	Launch Date	References
Malaysia	Gurun, Kedah	Gasification	4,000	2022	Pakar (2020)
Indonesia	Mentawai, West Sumatera	Gasification	700	2018	CPI (2021)
Japan	Nankan town, Kumamoto	Cogeneration	995 (power) 6,795 (thermal)	2023	Nakanishi (2019)
Vietnam	Lam Dong	Not available	Not available	Not available	Truong and Le (2014)

source of energy has been added to a list. The biomass of bamboo is used as a raw material for the production of other kinds of energy, such as electricity and biofuels.

Bamboo is a perennial plant that may thrive in climates ranging from tropical to subtropical to more temperate. In common parlance, bamboo is referred to be the "poor man's tree," but in recent years, it has gained popularity as an innovative, modern, and inexpensive material that can serve as a substitute for wood. According to Scurlock et al. (2008), India is the second most prosperous country in Asia in terms of bamboo genetic resources. China is the most prosperous country in this regard.

The countries that come together to form the ASEAN region have only recently started to grasp the potential for the development of biomass energy from bamboo's resources. Table 2.1 provides an overview of the current initiatives being developed to create biomass bamboo pilot plants in the countries of Malaysia, Indonesia, Vietnam, and Japan.

THE BENEFITS OF BAMBOO AS A POTENTIAL SOURCE OF BIOMASS

As a result of the quickening pace of economic development in these nations, bamboo has emerged as a significant biomass resource around the world over the past few decades (Scurlock et al., 2008). In addition to its usage in traditional forms of energy and a variety of other uses, bamboo biomass has more recently been reinvented to make use of gasification for the purpose of electric valorization (Kerlero and de Bussy, 2012). In addition, the incorporation of bamboo into volunteer carbon payment systems has contributed to an increase in its desirability as a plantation species. In recent years, similar developments have led to an increase in the number of bamboo plantations in Thailand. Despite the fact that specific knowledge with biomass production and energy and alternative utilization possibilities is still limited, these plantations have been able to flourish.

According to Scurlock et al. (2008), bamboo is most likely to flourish in a climate that is warm and humid (with an annual average temperature of 15°C–20°C and yearly precipitation of 1,000–1,500 mm). On three different continents, including Asia, Africa, and South America, there exist wild bamboo forests as well as planted bamboo forests. It was estimated that more than 36 million hectares were covered with bamboo forests around the world. The monsoon region of East Asia, specifically India and China, has the highest incidence of the disease worldwide (11.4 and 5.4 million ha, respectively). As shown in Figure 2.1, the total area covered by bamboo in Asia has increased by 10% over the course of the past 15 years (Lobovikov et al., 2007). This increase is mostly attributable to the massive planting of bamboo in China and India. Based on data by Lobovikov et al. (2007), the inventory of the world's bamboo resources can be found in Table 2.2.

Bamboo has a high potential to serve as a bioresource for lignocellulosic biomasses for a number of different reasons. These reasons include its rapid clonal growth, which results in the accumulation of abundant lignocellulosic biomasses in a short period of time (He et al., 2013; Ma et al., 2018), high fiber contents (Das et al., 2005), lack of agricultural land requirement, more extended harvest

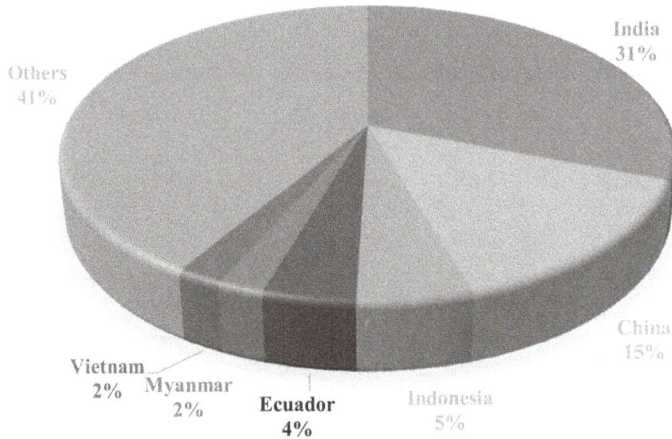

FIGURE 2.1   Countries with the most abundant supply of bamboo.

TABLE 2.2
Bamboo Forest Area

Region	Bamboo-Covered Land Area (1,000 ha)		
	2005	2000	1990
Africa	2,758	2,758	2,758
Latin America	10,399	10,399	-
Asia	23,620	22,499	21,230
Total	36,777	35,656	23,988

period, and the existence of a large, yet underutilized genetic pool (Das et al., 2008). In addition, they have the appropriate features for use as a fuel, such as a high caloric value, low levels of ash, chlorine, and moisture content, and a lowered temperature at which ash is formed (Engler et al., 2012; Fang and Jia, 2012). Cellulose, hemicellulose, and lignin are the primary structural components in plants that take up carbon (Tang et al., 2017). As a consequence of this, LCBs obtained from bamboo provide a significant potential as a biomass stock for the conversion of energy in a number of Asian countries, including Malaysia (Chin et al., 2017), Indonesia (Engler et al., 2012), India (Sharma et al., 2018; Singh et al., 2017; Hauchhum and Singson, 2019), and China (Biswal et al., 2021). Table 2.3 provides information on the fuel qualities of a number of different bamboo species (Scurlock et al., 2008; Singh et al., 2017).

AN ANALYSIS OF THE VIABILITY OF BAMBOO BIOMASS AS A SOURCE OF ENERGY

In comparison with conventional fossil fuels, traditional fossil fuels, such as oil and products derived from oil refineries, natural gas, and coal, are used extensively because they possess several desirable qualities in fuel, including the fact that they are very stable and produce a significant amount of energy, as well as the fact that they are very convenient, as shown in Table 2.4 (Scurlock et al., 2008; Singh et al., 2017). Other traditional fossil fuels, such as natural gas and products derived from natural gas refineries, are also used extensively. Fossil fuels are a potential contender for usage as portable forms of energy due to the ease with which they may be utilized, stored, and carried. They

TABLE 2.3
Fuel Properties of Certain Species of Bamboo

Species of Bamboo	Bamboo Properties				
	Higher Heating Value (kJ/kg)	Volatile Matter (%)	Ash (%)	Fixed Carbon (%)	Moisture (%)
Bambusa beecheyana	15.700	63.10	3.70	18.90	14.30
Dendrocalamus asper	17.585	71.70	2.70	19.80	5.80
Phyllostachys nigra	19.27	72.27	0.41	13.7	13.62
Phyllostachys bambusoides	19.49	75.55	0.53	14.38	9.54
Phyllostachys bissetii	19.51	64.99	0.9	12.14	21.97

TABLE 2.4
Comparison of the Thermal Conductivity of Bamboo Biomass versus Fossil Fuels

Fuel	Reduced Thermal Conductivity (MJ/kg)	Increased Thermal Conductivity (MJ/kg)
Bamboo biomass (Phyllostachys bissetii)	N/A	19.51
Natural gas	47.141	52.225
Coal	22.732	23.968
Gasoline	43.448	46.536

are the form of fuel that is most easily combustible in comparison with other types of fuel, such as biofuel or wood fuel, due to the fact that they contain a high concentration of energy. As a consequence of this, a substantial amount of energy is produced by them. Fossil fuels have the highest calorific value and are therefore the most efficient source for the production of energy (Astier, 2013).

Because biomass has a poorer thermal conductivity and a higher water content, a greater volume or mass of the material is required to produce the same amount of energy. This is due to the fact that biomass contains more water. The storing and transporting of biomass will be made more difficult as a direct result of this factor.

In addition, the grade of the fossil fuels that may be taken from each of the different sites is, for the most part, comparable. However, there can be a large amount of variation in both the quality of biomass and biofuels. It is challenging to produce biofuel from biomass that is compatible with modern engines due to the fact that these engines were built to run only on fossil fuel. When we switch to biofuel, either these engines will need to be redesigned from the ground up or the quality of the fuel itself will need to be improved in order for it to satisfy the requirements. In today's culture, however, the use of fossil fuels raises a few questions and concerns. The production of energy from these sources has two major drawbacks: First, there is a finite amount of resources available, and second, there is an increase in pollution due to the burning of fossil fuels.

Fossil fuels are non-renewable sources, which means that they are unable to replace themselves, and the rate at which they are being depleted is extremely concerning. In 40 years, it is projected that global oil production will reach its highest point, signaling the beginning of a gradual fall in oil production. It is anticipated that coal production will not begin for another 220 years, while natural gas production will not begin for another 60 years (Poppens et al., 2013). Because there are less and fewer fossil fuels available, their prices will continue to rise indefinitely. In addition to this, they are responsible for the emission of a sizeable quantity of carbon dioxide (CO_2) into the atmosphere, which is a crucial component of both global warming and climate change. This quantity of carbon dioxide was taken up by ancient plants over a period of millions of years; however, it is now being added to the atmosphere over a period of time that is significantly shorter. There is a good chance that the world will not react and adjust quickly to such a massive upheaval. Because of the destructive impacts it has, the environment and all living things, including people, will be negatively impacted.

As a direct result of these issues, a rising number of individuals are looking for alternative energy sources to minimize our reliance on fossil fuels, and biomass is increasingly being regarded as a viable choice among these alternatives. Because bamboo biomass can be burned almost instantly when it is in its dry state, it is an excellent choice for cooking and heating in rural areas and among people with poor incomes. In order to turn biomass into electricity, the cogeneration facility has been designed specifically for that purpose. It is possible to process biomass in a variety of ways, which can result in the production of char, flammable gas, and biofuel. All three of these products have properties that are comparable to those of fossil fuel. In the not-too-distant future, biomass will likely be able to completely supplant fossil fuel as an essential component of the worldwide energy grid.

When compared to fossil fuels, biomass has several significant advantages, the most important of which are its sustainability and its lower carbon dioxide emissions. Bamboo is a source of renewable energy, which means that the biomass it produces may be replaced at a rate that is compatible with its use as an extraction resource. Carbon dioxide is released during the processing of biomass, which includes both thermal conversion and biochemical conversion (CO_2). On the other hand, this does not contribute in any way to an increase in the concentration of greenhouse gases in the atmosphere. The carbon dioxide that is released as a result of these actions is the same type of carbon dioxide that is fixed into the atmosphere by bamboo plants during the process of photosynthesis.

Another aspect that needs to be mentioned is the cost of the product. At the moment, the price of electricity generated from biomass is lower than the price of power generated from fossil fuels. However, in the future there will be a shift because there won't be enough fossil fuels. When this occurs, the transition will take place naturally, and the cost of fuel derived from biomass will be lower than that of fossil fuel. The criteria were compiled and compared in Table 2.5, which compared bamboo biomass to fossil fuel (Nakanishi, 2019; Truong and Le, 2014).

IN COMPARISON WITH OTHER FORMS OF RENEWABLE ENERGY

Hydropower, wind power, and solar power are some of the other forms of renewable energy that are readily available in addition to biomass. Since the dawn of time, people have been harnessing

TABLE 2.5
Summary of the Properties of Fossil Fuels and Biomass

Criteria	Bamboo Biomass	Fossil Fuel
Accessibility	Must plant and harvest after a 3- to 4-year time frame	Drawn straight from an existing resource and used immediately
Generated energy (per same mass)	Significantly smaller	Much bigger
Logistic (storage and transportation)	More complicated (need more prominent space for transportation and storage)	Transportable and storable
Quality	Diverge	Incorporated
Sustainability	Source of renewable energy	Non-renewable source
CO_2 emission	Not raise CO_2 concentrations in the atmosphere	Increase CO_2 levels in the atmosphere

TABLE 2.6
Comparison of the Efficiencies of Various
Power Generation Methods

Technology	Range of Efficiency (%)
Biomass	16–43
Wind	23–45
Hydro	<90
Photovoltaic	4–22

the power of all of these natural resources to generate energy (hydro, wind) and heat (solar and biomass). At the current level of energy demand and consumption, the replacement of fossil fuels should be the primary focus of efforts to create renewable resource sources. It is possible to generate heat (in cement or steel industries), power (in combustion engines used in industry and transportation), and electricity by burning fossil fuels. Hydroelectric, wind, and solar power all have the potential to offer grid-connected energy, but they are not able to make up for the heat and power that fossil fuels generate. On the other hand, biomass has the potential to solve all of these problems. In a thermal power plant, it can be converted into biofuels, which can then be used to fuel combustion engines. Additionally, it can be utilized to generate electricity and heat in linked facilities. Products derived from biorefineries have the potential to meet the demand for chemicals in a wide variety of additional industries currently served by oil refineries. As a result, in this section, we only compare biomass to other forms of renewable energy in terms of its capacity to create electricity, as shown in Table 2.6 (Poppens et al., 2013). In the following section, several sustainability criteria will be applied to evaluate various forms of renewable energy in terms of power production.

EMISSIONS OF GREENHOUSE GASES

Wind, biomass residual, and water all generate relatively low levels of emissions, with averages of 25, 30, and 41 gCO_2e/kWh, respectively. The production of wind turbines is responsible for the vast bulk of the emissions produced by wind power. Emissions of biomass residue might occur as a result of the collection and transportation of fuel with a low energy density. The majority of greenhouse gases produced by hydro come from the construction of dams (methane in most cases). The average amount of emissions produced by photovoltaics and biomass energy crops (including the biomass produced by bamboo) is low to moderate. In the same way that emissions are produced by wind power, solar emissions are produced when photovoltaic panels are manufactured. The cultivation of biomass energy crops, the use of fertilizer, the collecting of biomass, and the transportation of biomass are all activities that contribute to emissions (Poppens et al., 2013).

WATER CONSUMPTION

Consumption and withdrawal are the two distinct types of water use that can be distinguished. Consumption refers to the amount of water that is lost because it either evaporates or is lost from the system and cannot be returned to the source. The term "withdrawal" refers to the total quantity required to keep the technology operational, which includes the amount of water that is still accessible for recycling. Because it does not rely on water for its operation, wind power has an extremely low water footprint (1 g/kWh). When properly cleaned, solar panels can use as little as 10 g of water per kilowatt-hour of energy production. Since the flow of water is necessary for the generation of electricity, hydropower has the greatest water intake rate (13,600 kg/kWh). However, this water is recycled back into the system after it has been used. The production of hydropower results in a water loss of 11 kg/kWh due to evaporation. Even while biomass residue consumes a significant amount of water (3.2 kg/kWh), this figure is still ten times lower than that of specialist energy crops (34 kg/kWh). This is because plants require a significant amount of water in order to survive. As a direct consequence of this, bioenergy crops such as bamboo biomass have the lowest potential for long-term water conservation (Talukder et al., 2017).

ACCESSIBILITY

Sunlight and wind can be found in any location; however, the geological and topological characteristics of a given region define the maximum permissible wind speed and solar radiation for that region. As a direct consequence of this, particular areas do not have access to an appropriate amount of wind and sunshine. For instance, the wind speed must be between 5 and 25 m/s for wind turbines to operate correctly. Another limitation of hydropower is that we are limited in the number of dams we can construct along rivers since doing so would have a large and detrimental impact on the ecosystem in the surrounding area. Because we can get our hands on sources of biomass in any country and on any continent, it demonstrates the greatest degree of availability of any resource. But before we go into that, let us take a deeper look at the particular instance of bamboo biomass. As we will see, access to it is only possible in regions that have environmental conditions that are suitable for the growth of bamboo (Evans et al., 2010).

LAND USAGE

In order for technology to function properly, a certain amount of land is necessary. Because the technology is inflicting such a significant amount of damage, it does not appear that this is discussed in the book. The acreage that has been utilized for the various technologies is outlined in Table 2.7, which may be found (Evans et al., 2010).

The information shown in Table 2.7 demonstrates that the land taken up by biomass residue is negligible. Solar power and wind power both require a significant quantity of available land. Nevertheless, both of these land-use patterns are sustainable (does not change the land quality significantly). In addition, this accounts for the total amount of land that is being utilized by the solar plant and the wind farm. If solar panels are mounted on the facade of the building as well as the roof, this space will be reduced. Only one-tenth to one-fifth of the total area represented in the table is comprised of the actual land that is occupied by wind turbines in a wind farm. The remaining portion is excellent for grazing, agricultural use, and recreational use. Because of the reservoir, hydropower accounts for the second greatest percentage of land occupancy. The quantity of land needed for specialized energy crops is tremendous compared to the amount of land needed for hydropower, and the cultivation of these crops has an effect on the quality of the soil.

TABLE 2.7
Land Utilized by Various Technologies

Technology	The Amount of Land Used (m²/kWh)
Energy derived from biomass crop	0.533
Biomass waste	0.001
Photovoltaic	0.045
Wind	0.072
Hydro	0.152

RESTRICTION IMPOSED BY TECHNOLOGY

When injected into the system, wind energy and solar energy have limitations due to the intermittent nature of their sources. This results in a rise in the cost of the power that is generated by these systems. On the other hand, hydro and biomass have the potential to collect and store fuel until there is a sufficient volume available to continue operating continuously. Therefore, the monitoring and management of the energy that is provided by biomass and hydropower plants is a great deal less complicated.

IN COMPARISON WITH OTHER FORMS OF ENERGY CROPS

Because it has a higher heating value than most other types of biomass, bamboo is an ideal choice for direct burning (e.g., co-combustion in thermal power plant). Bamboo has a lower percentage of moisture than rice husks and rice straw, but bagasse and maize stalk both have a higher percentage of moisture than bamboo does. The reduced moisture content of the biomass results in a reduction in the amount of energy that is required to dry the biomass, which increases the overall efficiency of the operation. Table 2.8 presents some of the fuel properties of various biomass feedstocks (Singh et al., 2017).

Nevertheless, there are a few drawbacks associated with using bamboo as an energy crop for the production of biomass on a big scale (Poppens et al., 2013). Only ripe shoots of bamboo should be harvested during harvesting, which makes bamboo harvesting difficult to automate.

- Other applications have a market that is more geared toward extraction.
- Because bamboo must be produced from seeds rather than from seedlings, major plantings can be rather expensive.
- The production of a stand can take several years.
- The grade of the thermal conversion is lower than that of wood.

METHODS FOR EXTRACTING ENERGY FROM BAMBOO'S BIOMASS

There are a number of different methods for extracting energy from bamboo biomass, and each of these methods produces a unique product that can be utilized in a variety of settings. The production of energy from bamboo biomass can be accomplished in one of two major ways: either through thermochemical conversion or biochemical transformation. In the past, heat was the primary method

TABLE 2.8
Characteristics of Typical Biomass Sources

Biomass Types	Higher Heating Capacity (kJ/kg)	Fixed Carbon (%)	Volatile Matter (%)	Moisture (%)	Ash (%)
Bamboo (*Bambusa beecheyana*)	15.70	18.90	68.10	14.30	3.70
Bamboo (*Dendrocalamus asper*)	17.59	19.80	71.70	5.80	2.70
Bagasse	9.67	5.86	41.99	50.76	1.75
Palm shell	18.45	16.30	68.31	12.12	3.66
Corncob	11.20	13.68	45.55	40.11	0.95
Corn stalk	11.63	8.14	46.98	41.69	3.80
Rice husk	14.64	18.88	56.98	12.05	12.73
Rice straw	13.28	18.80	60.87	10.12	10.42

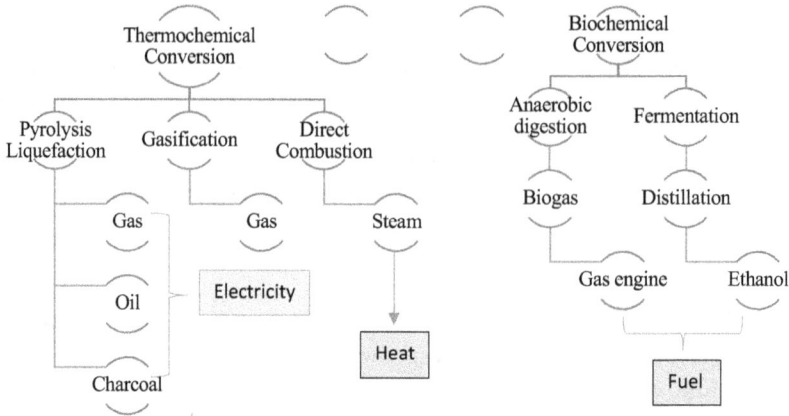

FIGURE 2.2   Principal routes of energy conversion from biomaterials.

for transforming the biomaterials included in bamboo biomass, primarily cellulose, into a variety of end products. The process of converting biomass into biogas or biofuel involves the activity of microorganisms and is referred to as biochemical conversion. The conversion of bioenergy is depicted in the Figure 2.2 (Talukder et al., 2017).

THERMOCHEMICAL CONVERSION

Pyrolysis

Pyrolysis can be described as the thermal ("pyro") decomposition ("lysis") of organic compounds when performed in an environment devoid of oxygen and at an appropriate temperature (350°C–600°C). The pyrolysis process results in the production of charcoal (which is in the solid phase), condensable pyrolysis oils (which are heavy aromatic and hydrocarbons), and tars (which are in the liquid phase), as well as con-condensable gases or syngas (gaseous phase). As a supplementary fuel, the charcoal will be used in a manner comparable to that of coal. In a gas boiler or an internal combustion engine, syngas, which is a combination of carbon monoxide, hydrogen, and methane, can be burned to generate power. In the same way that conventional crude oil is refined further in biorefineries to produce biofuels and other significant chemical compounds, pyrolysis oils may also be refined further in these facilities. The pyrolysis procedure is illustrated in Figure 2.3.

The conditions of the operation have a direct bearing on the quantities of pyrolysis products (temperature and residence time). For instance, the generation of condensable oils can be aided by a high temperature (between 500°C and 600°C) and a short residence period (sometimes referred to as flash pyrolysis). On the other hand, a carbonization process that involves a low temperature (between 350°C and 400°C) and an extended period of residence time would result in the most efficient production of charcoal and syngas. The typical distribution of the products obtained from the two separate types of pyrolysis is laid out in Table 2.9 (Truong and Le, 2014).

Gasification

The standard example of the gasification process is depicted in Figure 2.4. The term "gasification" refers to the transformation of a solid fuel into a gaseous fuel. At high temperatures and with a restricted supply of air, it consists of a complex thermal and chemical conversion of organic material that is difficult to anticipate. This happens when there is a lack of oxygen. There are two stages involved in the gasification process: The first stage is pyrolysis, and the second stage is partial combustion. It takes place at extremely high temperatures, often ranging from 750°C to 1,200°C, and there is very little or no oxygen present during the process.

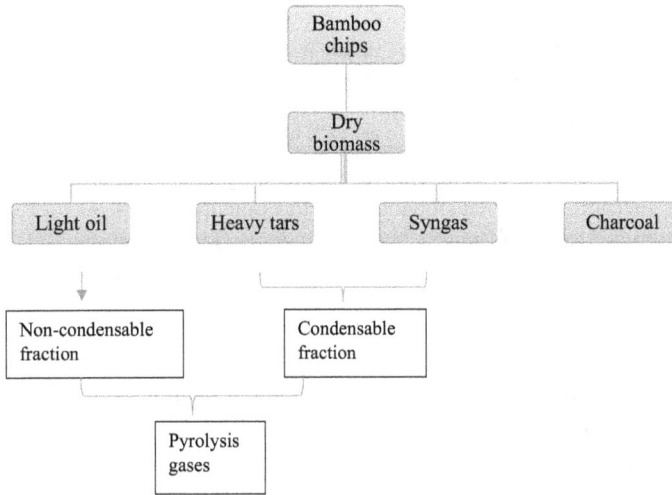

FIGURE 2.3   Reactions and products of pyrolysis.

TABLE 2.9
Types of Pyrolysis and Their Related Products

Ton of Dry Material (kg)	Carbonization	Flash Pyrolysis
Oil	190	730
Gas	380	110
Charcoal	430	160

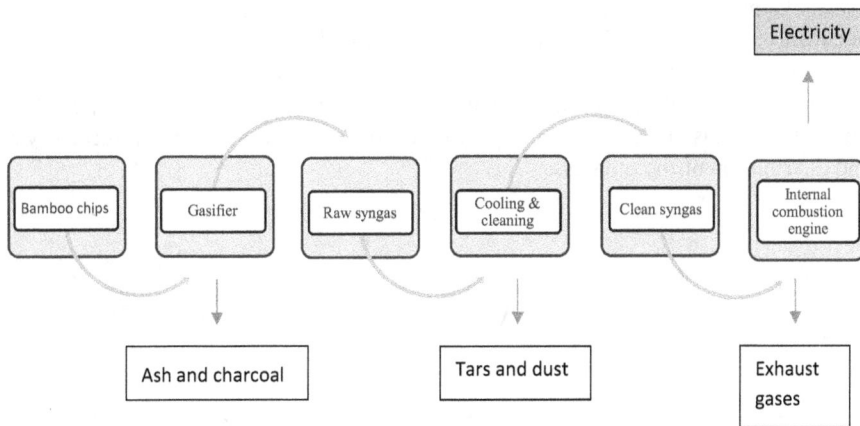

FIGURE 2.4   The flow chart of a typical gasification process.

The gasification process results in several by-products, two of which being syngas and ash. A mixture of combustible gases (such carbon monoxide, hydrogen, and methane) and non-combustible gases (like nitrogen and oxygen) is what makes up syngas (carbon dioxide, nitrogen, and other gases). It is anticipated that combustible gases make up around 40% of the total volume of syngas, which can be used to generate either electricity or heat. This syngas can also be used as a fuel source. The amount of heat that can be extracted from syngas is directly proportional to the oxygen supply mechanism. When air is used, the calorific value of the syngas that is produced is relatively low, ranging from 4 to 7 MJ/m^3. On the other hand, when oxygen-enriched air is used, the calorific value of the syngas that is produced can reach values as high as 10–15 MJ/m^3. In point of fact, because the technique for oxygen enrichment is so expensive, most of the time another gas is used in its place (NL Agency, 2012).

In comparison with combustion, gasification results in a greater energy return from the fuel and significantly lower thermal losses. Gasification has the potential to be 95% effective in terms of converting fuel to another form (dry mass), but only under perfect conditions (NL Agency, 2012). In actual reality, heat losses and other reactions bring the efficiency down to between 70% and 80% of the energy recovered in the gases created from the biomass.

Direct Combustion

Home fuel for cooking, warming, and boiling can be made from bamboo biomass that has been dried out. When there is no access to electricity, it can serve as an important source of energy for rural communities. Burning bamboo biomass directly can also be used on an industrial scale, for instance in the form of cogeneration to create heat and power in thermal power plants for the production of electricity or other plants such as cement or steel. This can be done in order to create heat and power in these plants. Through the use of cogeneration, these facilities can reduce the amount of fossil fuel that they consume. The diagrammatic representation of the combustion process can be seen in Figure 2.5.

It is necessary to have a solid understanding of the technical idea of combustion. It involves the carefully managed burning of any fuel containing carbon and hydrogen. Carbon dioxide (CO_2) and water (H_2O) are the by-products of the combustion process (CO_2). In most cases, combustion takes place within a chamber, and is then followed by a heat exchanger, which is where the heat from the hot gas stream is transferred to another fluid (water or air). The subsequent generation of power in an engine or turbine can be attributable to the utilization of this fluid. A boiler is the name given to the heat exchanger that is used when combustion is used to heat water (Darabant et al., 2014). Water boilers are utilized for the production of steam at medium and high pressures (more than 20 bar) on a big scale. Controlling the combustion process is essential to achieving maximum efficiency because it ensures that all of the biomass is completely burned, which in turn maximizes the amount of energy that is recovered while also preventing the generation and emission of non-oxidized gases like carbon monoxide (CO) and volatile organic compounds (VOC) (Darabant et al., 2014). The combustion of biomass is affected by the supply of air, the regulation of temperature, as well as the quality and distribution of the biomass.

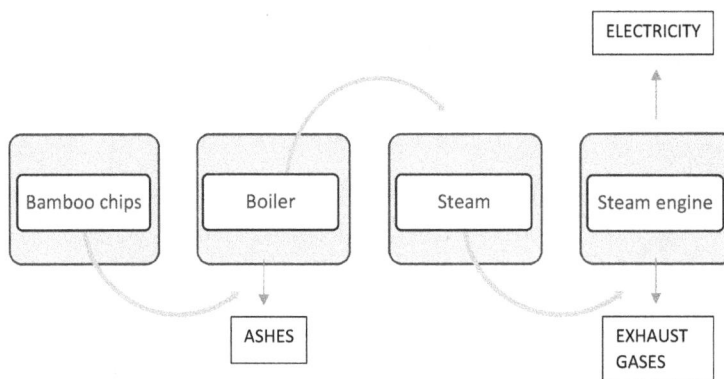

FIGURE 2.5   The combustion process from bamboo feedstock to energy production.

BIOCHEMICAL CONVERSION

During the biochemical conversion process, numerous strains of microbes are utilized to produce a variety of different biofuel products. The fermentation of sugar or other compounds that can be found in biomass by microbes results in the production of ethanol, methane, and a variety of other fuels, chemicals, and heat. This is the underlying premise of biochemical conversion. There are two different approaches that can be taken to accomplish bioconversion:

1. Anaerobic digestion, also known as anaerobic respiration, is the process by which micro-organisms, known as anaerobic bacteria, break down organic molecules in biomass in the absence of oxygen. This process is depicted in Figure 2.6. During this process, both biogas (methane) (60%) and carbon dioxide (40%) are generated (Darabant et al., 2014).
2. Microorganisms, such as yeasts and bacteria, are responsible for the breakdown of starch and sugar during the fermentation process, which results in the production of ethanol.
3. The utilization of bamboo biomass can be increased to its full potential by using an integrated process that combines the generation of ethanol with the recovery of silica and lignin. A sequential two-stage pre-treatment method consisting of autohydrolysis, and alkaline extraction was carried out to achieve the goals of reducing the amount of chemical charge required for separating silica and lignin from bamboo and increasing the digestibility of the substrate that was produced as a result. An improvement was made to a two-stage treatment to increase enzymatic hydrolysis and recovery of silica and lignin. The treatment consisted of autohydrolysis at 180°C for 90 min, followed by alkaline extraction at 100°C with 6% NaOH (based on pre-treatment chips), for 120 min. It was possible to recover around 93.7% of the silica and 75.7% of the lignin that were originally present in bamboo. Following enzymatic hydrolysis of carbohydrates and fermentation of these carbohydrates, an overall sugar yield of 88.6% of the initial sugar content was obtained, as was an ethanol recovery of 0.467% g/g sugar. Both results were reached. Silica in the pre-treatment solids prevents enzymatic hydrolysis by interacting with the enzyme cellulase,

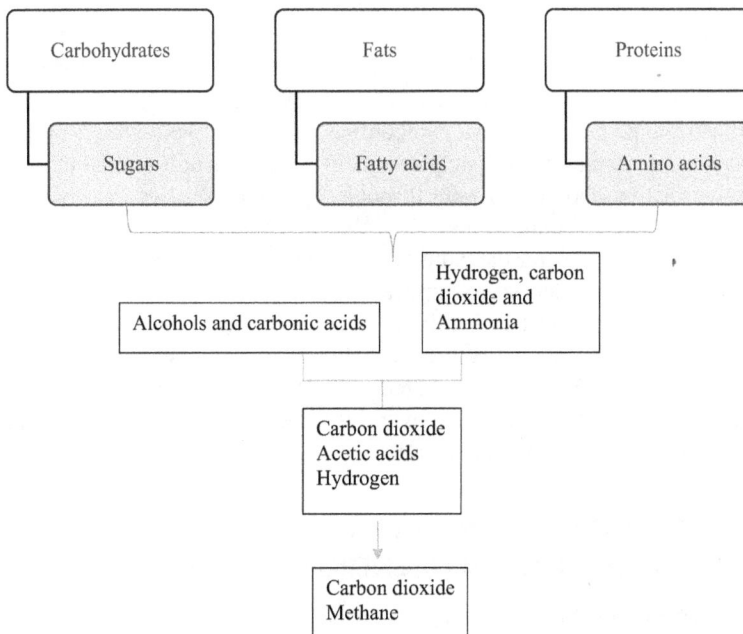

FIGURE 2.6 Pathways for anaerobic digestion.

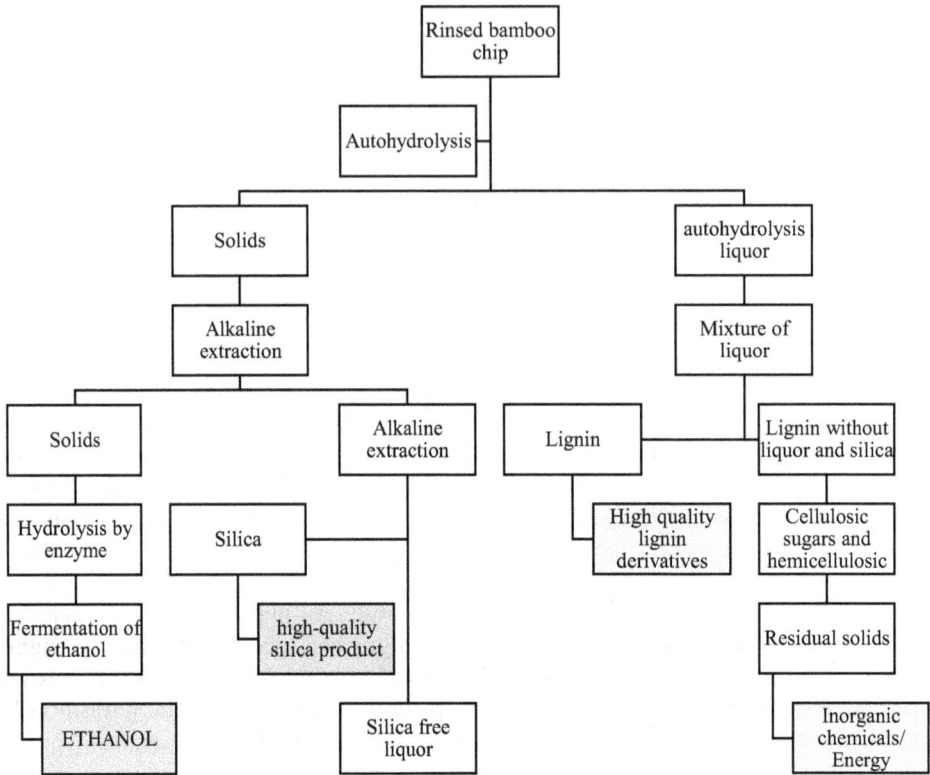

FIGURE 2.7   Depicts the proposed system biorefinery process for converting bamboo into ethanol, silica, and lignin.

which is necessary to produce bioethanol from bamboo (NL Agency, 2012). Therefore, to maximize the effectiveness of the synthesis of bioethanol from bamboo, it is necessary to reduce the challenges posed by silica. To eliminate the silica problem, it is recommended to remove silica from bamboo in advance of the enzymatic hydrolysis process. Figures 2.7 and 2.8 illustrate the procedure that is recommended for extracting high-purity silica and lignin from bamboo, which ultimately results in the production of ethanol. In the beginning, commercial bamboo chips went through a procedure called autohydrolysis, which involved the removal of hemicelluloses and the increased availability of lignin and silica to chemical researchers. After the bamboo chips had been processed, an alkaline solution was used to extract the material to remove silica and lignin. The method was successful in recovering the dissolved silica, lignin, and carbohydrates (hemicellulose and cellulose), which were recovered from the two pre-extraction liquors (autohydrolysis liquor and alkaline extraction liquor). While the recovered carbohydrates were reintroduced into the cellulose substrate to facilitate enzymatic hydrolysis and the production of ethanol, the silica and lignin were stored in a separate location.

THE PRESENT SITUATION REGARDING THE USE OF BAMBOO BIOMASS FOR THE GENERATION OF ENERGY

Energy derived from bamboo biomass on a global scale

The utilization of bamboo as a biomass resource for the generation of electricity is a more recent development. Bamboo's potential as a source of renewable energy has been the subject of a

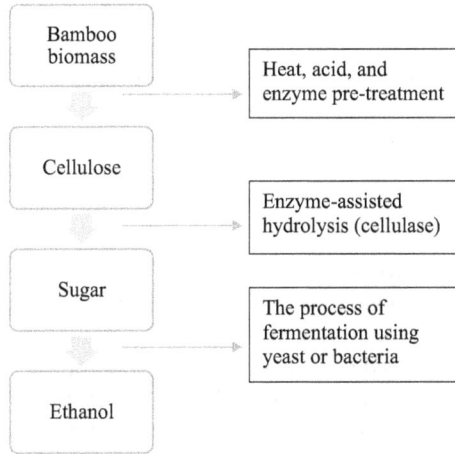

FIGURE 2.8 Simple methodology of ethanol production from bamboo biomass.

significant amount of investigation in recent years. Numerous nations, primarily those located in regions with abundant bamboo resources, such as Thailand, Indonesia, China, and India, have conducted research on this topic. Bamboo has been brought to the attention of a number of scholars as a potential alternative biomass resource. Nevertheless, national research on the potential of bamboo are few in a number of countries, including Vietnam and Thailand, both of which have considerable bamboo resources. However, there is a growing interest in bamboo in these countries, and more research and initiatives, some of which include bamboo, are now being developed.

Across the world, there are now various bamboo energy projects that are either operational or in the process of being implemented. The primary purpose of bamboo biomass projects in Africa is to provide an alternative to the use of firewood or to produce charcoal for domestic consumption. For instance, in the countries of Ethiopia and Ghana, a project with the working title "Bamboo as sustainable biomass energy" is currently being carried out with the goal of supplying bamboo charcoal to the locals so that they can meet their energy needs in an environmentally responsible manner, generate revenue, and relieve pressure on other forest resources. The effort is funded by the European Commission, and the International Network for Bamboo and Rattan is in charge of carrying it out (NBAR, 2019). Because India contains the world's largest amount of bamboo forest, the number of biomass projects utilizing bamboo is quickly growing in this country. In contrast to the initiatives that are being carried out in Africa, the processes that are being utilized in this region to extract biomass from bamboo are quite mature and diverse. The generation of electricity can be accomplished by the use of cogeneration, gasification, or pyrolysis processes with bamboo biomass. The combination of these research and implementation projects will provide a clearer picture of the potential for bamboo biomass to be used in the future as a source of renewable energy.

MALAYSIA

Pakar B2E Sdn, a partnership firm, was responsible for the construction of a bamboo biomass power plant in the Malaysian state of Kedah. Bhd. It is anticipated that the power plant will have an installed capacity of 4 megawatts (MW), with a feed-in tariff rate of 0.091 US dollars per kilowatt-hour (kWh) permitted by Malaysia's Sustainable Energy Development Authority (SEDA), and an effective lifespan of 21 years (Pakar, 2020). The construction of the power station is currently underway, and it is anticipated that it will begin operations in March of 2022. This first plant in Malaysia dedicated to the production of electricity from biomass will be beneficial to thousands of people, particularly those who live in rural areas. Bamboo farmers and growers in the surrounding area will benefit from this facility in the form of expanded employment opportunities and assistance. Around 4,000 metric tons of bamboo will be used each month at the facility. This bamboo will be provided by local farmers and citizens who harvest unharvested wild bamboo from the surrounding area. The establishment of this biomass bamboo power plant is meant to assist in the rural development of Kedah (Shah et al., 2021).

INDONESIA

Indonesia is the only country in the Asia Pacific region that has been successful in constructing a biomass power plant based on bamboo. Madobag Village, Matotoan Village in South Siberut District, and Saliguma Village in Central Siberut District are all included in the scope of the project, which can be found on Siberut Island in Mentawai, West Sumatra. The three biomass power plants were supposed to provide a combined total of 700 kW of power, with six units of 100 kW and two units of 50 kW being enough to supply 1,250 individual households. In addition to this, it will result in the creation of 450 jobs in plant operations and maintenance. PLN will purchase all of the power generated by the project from the developer at a regulated price of 0.15 USD per kilowatt-hour during the course of a 20-year Power Purchase Agreement (PPA). After that, PLN will sell the community the power generated by the project at the National Electricity Rate (TDL), which is currently set at 0.031 USD/kWh (CPI, 2021; CPI & CIFOR, 2019). This power plant has become a source of cash for the local population and has helped to strengthen the economy of the surrounding area thanks to an investment of USD 12 million made by an outside donor from another country. It is anticipated that this new planting of bamboo will additionally remove 3,000 t CO2e/y of carbon dioxide equivalents by preventing emissions and capturing carbon. The power plant, which has a cost efficiency of 70%, can assist Indonesia in increasing its national electrification rate and accomplishing its goal of incorporating at least 23% renewable energy into its energy mix by the year 2025.

It has been projected by Yoesgiantoro et al. (2019) that the entire net cash flow provided by the biomass bamboo power plant in Mentawai will be around 10 million USD in the 30th year, with a payback time of 16 years. The Indonesian government, recognizing the potential for beneficial advantages of bamboo resources, has set a goal of increasing the amount of energy that can be produced from bamboo biomass to 500 MW (GGGI, 2019). After being operational for a few years, the plants' performance has been showing signs of deterioration. The plants were unable to run reliably due to the recurring problems that plagued them (Febrianti, 2021). At the moment, the power plants are only producing a fraction of the amount of electricity that is theoretically possible for them. The natural resources that originated from bamboo were frequently replaced with alternatives that were less environmentally friendly, such as fossil fuels. According to statements made by several bamboo suppliers, the current selling price is unsustainable (Jacques, 2021).

THAILAND

Production of bamboo biomass and the following characteristics of bamboo feedstock used in Thailand's power plants:

The site productivity and the management of the bamboo plants in two different locations in eastern Thailand resulted in a considerable difference in the amount of biomass produced. Both *Bambusa beecheyana* and *Dendrocalamus membranaceus* had similar levels of biomass production, but *B. beecheyana* was more resistant to drought. The *beecheyana* culms had a moisture content that was noticeably higher than that of *D. membranaceus* culms. Although the two species were comparable in terms of their overall characteristics, *D. membranaceus* had a calorific content that was much higher than that of *B. beecheyana*. Internodes of the letter D. The moisture content of membranaceus was found to be higher than that of nodes. Moisture content was found to be decreasing with increasing culm height for both species; however, the gradient was more pronounced with *D. membranaceus*. The amount of moisture contained in B. The amount of energy stored in beecheyana culms was shown to decrease with increasing culm age, indicating that older culms are better suited for energy use. Although the upper parts of both species and the nodes of older culms have potential as a bioenergy fuel source, outputs from plantations that are planted on marginal land or without proper care are likely to be lower (Talukder et al., 2017).

There is a significant amount of variation in the amount of biomass produced by bamboo plantations in the eastern region of Thailand. This difference is caused by the different levels of site productivity. Plantations developed on marginal sites without proper plantation management will produce extremely low yields, whereas intensive management on high-productivity sites can result in very large biomass yields. To be more specific, the lower culm sections and internodes, in addition to the juvenile culms, all have exceptionally high moisture contents, which inhibits their ability to use energy. Although the outputs of biomass from different species are equivalent, *D. membranaceus* has a lower percentage of moisture and a higher calorific value than *B. Beecheyana*. If grown in the right conditions, *D. membranaceus* could represent a more desirable option as a potential source of energy. It is important to investigate potential substitute applications for lower culm sections and internodes, in addition to juvenile culms (Darabant et al., 2014).

VIETNAM

In Vietnam, bamboo has been most commonly used in the construction of houses, the production of handicrafts, and the preparation of food at the village level for the regional market, and at the industrial level for the production of floors for export markets. According to the Vietnam Trade Promotion Agency, bamboo is one of the ten industries that are contributing the most to Vietnam's overall export growth. There are an estimated 800,000 hectares of bamboo plantations in Vietnam, with an average annual output of 10 to 13 tons per hectare (NL Agency, 2012). Additionally, there are 600,000 hectares of mixed forests in Vietnam, with up to 70% bamboo in those forests. There are 6.2% bamboo farms and 16% mixed bamboo forest in the province of Lam Dong, which is located in Southern Vietnam. Bamboo is the dominating plant in this region. The four provinces of Tuyen Quang, Son La, Bac Kan, and Yen Bai in Northern Vietnam are where the majority of the country's bamboo is grown for commercial purposes. These five provinces are responsible for 7% of the country's bamboo planting and 43% of the country's mixed forest (NL Agency, 2012).

The production of bioenergy from bamboo is a unique idea that has not garnered much attention from either the corporate sector or the government. At the moment, the production of energy from biomass is mostly comprised of biogas produced from municipal waste and animal waste, the synthesis of ethanol from cassava and molasses, and the co-generation of bagasse.

The six bioethanol facilities in Vietnam that are based on cassava are having trouble functioning due to financial issues and problems with cash flow. National regulations specified that all vehicles in the United States must use E5 fuel (gasoline containing 5% ethanol) by June 2012, which led to the investment in these plants. This regulation has not yet been codified, and there is no timetable available to indicate when it will go into force. As a result of this limitation, it is not possible to invest in bamboo-derived bioethanol in Vietnam at the present time.

Despite this, there is still a market for bamboo biomass, which includes the commercialization of commodities resulting from the thermal conversion of bamboo biomass as well as the use of bamboo biomass in co-generation plants to reduce the amount of fossil fuel that is used (charcoal, syngas, and oil). This heat process works well with biomass that is high in cellulose, but it is less effective with agricultural residue (such rice husk or straw) and municipal solid trash. Hardwood is less efficient than bamboo since it has a significantly longer period during which it can be harvested. As a result, the processes of gasification and pyrolysis are best carried out using bamboo as the source of biomass.

Multiple countries are currently working on the development of a wide variety of bamboo biomass energy plants, ranging in scale from the domestic to the industrial. There is a huge amount of bamboo biomass potential in Vietnam. In spite of this, the utilization of bamboo for the production of energy is still in its infancy. Additional research is required in order to evaluate the capabilities and sustainability of utilizing this potential resource (NL Agency, 2012).

JAPAN

Bamboo Energy Co., Ltd. successfully finished construction of Japan's first biomass bamboo power plant, which began operations in 2019 in Nankan, Kumamoto Prefecture, and is owned by the Japanese government. The New Energy and Industrial Technology Development Organization is in charge of directing the progress of this initiative (NEDO). This plant was also the first of its kind to implement a cogeneration technique known as an organic Rankine cycle (ORC) in order to generate heat and power from bamboo resources. It is expected to be available to consumers in 2023 (Nakanishi, 2019). In order to produce 995 kW of power and 6,795 kW of heat output, the system consumes 8,750 t/y of bamboo (NEDO, 2019). It is anticipated that this plant will increase the value of bamboo resources by enabling a more holistic utilization of those resources. As a consequence of this, the problem of bamboo groves that are unmanaged and have become overgrown should be resolved. The project also intends to make the most of the region's biomass resources and establish production of biomass energy, both of which would enhance the economic conditions of the local community.

CONCLUSION

Since bamboo biomass is a renewable resource and can be processed to create various fuels, it can replace fossil fuel. Bamboo biomass can be converted using a variety of processes, such as pyrolysis, gasification, and thermal conversion. Charcoal, syngas, oil, and ethanol are the processes' commercially viable by-products. Bamboo biomass offers greater fuel properties than most of energy crops. It may grow on damaged ground, requiring less maintenance and less competition for land from food crops. Bamboo biomass has two drawbacks: It consumes water and occupies land. Bamboo biomass has a lot of potential for use in energy production in Vietnam.

REFERENCES

Adewuyi, A. (2020). Challenges and prospects of renewable energy in Nigeria: A case of bioethanol and biodiesel production. *Energy Reports*, 6, 77–88.

Agbor, V.B., Cicek, N., Sparling, R., Berlin, A., & Levin, D.B. (2011). Biomass pretreatment: Fundamentals towards application. *Biotechnology Advances*, 29, 675–685.

Astier, S. (2013). New technologies of energy. *Energy Challenges*, 1–10.

Biswal, D., Shinkhede, S., & Mandavgane, S.A. (2021). Bamboo valorization by fermentation and enzyme treatment, in: H. Thatoi, S. Mohapatra, S.K. Das (Eds.), *Bioprospecting of Enzymes in Industry, Healthcare and Sustainable Environment*, Springer, Singapore, 87–101.

Chin, K.L., Ibrahim, S., Hakeem, K.R., San P., H'ng, S.H.L., & Lila, M.A.M. (2017). Bioenergy production from bamboo: Potential source from Malaysia's perspective, *Bioresources*, 12(3), 6844–6867.

CPI. (2021). *700 kWp Biomass Gasifier in Mentawai (Indonesia), Clean Power Indonesia, Mentawai, Indonesia.* CPI. https://www.ruralelec.org/project-case-studies/clean-power-indonesia-700-kwp-biomass-gasifier-mentawai-indonesia, accessed 25.05.2021.

CPI, CIFOR. (2019). *Powering the Indonesian Archipelago, GLF, Luxembourg, Luxembourg.* http://www.globallandscapesforum.org/publication/powering-the-indonesian-archipelago-white-paper/, accessed 25.05.2021.

Darabant, A., Haruthaithanasan, M., Atkla, W., & Phudphong, T. (2014). Bamboo biomass yield and feedstock characteristics of energy plantations in Thailand. *Energy Procedia*, 59, 134–141.

Das, M., Bhattacharya, S., & Pal, A. (2005). Generation and characterization of SCARs by cloning and sequencing of RAPD products: A strategy for species-specific marker development in bamboo. *Annals of Botany*, 95 (5), 835–841.

Das, M., Bhattacharya, S., Singh, P., Filgueiras, T.S., & Pal, A. (2008). Bamboo taxonomy and diversity in the era of molecular markers. *Advances in Botanical Research*, 225–268.

Engler, B., Schoenherr, S., Zhong, Z., & Becker, G. (2012). Suitability of bamboo as an energy resource: Analysis of bamboo combustion values dependent on the culm's age, *International Journal of Engineering*, 23(2) 114–121.

Evans, A., Strezov, V., & Evans, T. (2010). Comparing the sustainability parameters of renewable, nuclear and fossil fuel electricity generation technologies. *World Energy Conference*, Montreal, Canada.

Fang, X., & Jia L. (2012). Experimental study on ash fusion characteristics of biomass, *Bioresource Technology*, 104, 769–774.

Febrianti. (2021). *Bamboo-Powered Plants Gone Offline, Indonesia: Tempo Inti Media TBK*. Febrianti. https://www.rainforestjournalismfund.org/stories/bamboo-powered-plants-gone-offline, accessed 15.07.2021.

GGGI. (2019). *GGGI engaged on community-based bamboo biomass to energy in Indonesia at the Global Landscape Forum Luxembourg*. https://www.gggi.org/gggi-engaged-on-community-based-bamboo-biomass-to-energy-in-indonesia-at-the-global-landscape-forum-luxembourg-2019/, accessed 24.05.2021.

Hauchhum, R. & Singson, M. (2019). Assessment of aboveground biomass and carbon storage in bamboo species in sub-tropical bamboo forests of Mizoram, north-east India, *Indian Journal of Ecology*, 46, 358–362.

He, C., Cui, K., Zhang, J. A., & Zeng, D.Y. (2013), Next-generation sequencing-based mRNA and microRNA expression profiling analysis revealed pathways involved in the rapid growth of developing culms in Moso bamboo. *BMC Plant Biology*, 13(1), 119.

Holm, J. & Lassi, U. (2011). Ionic liquids in the pretreatment of lignocellulosic biomass, in: Alexander Kokorin (Ed.), *Ionic Liquids: Applications and Perspectives*, 545– 560, ISBN: 978-953-307-248-7.

Islam, M. K., Wang, H, Shazia, R., Chengyu, D., Hsu, H.Y., Lin, Ki, C.S., & Leu, S.Y. (2020). Sustainability metrics of pretreatment processes in a waste derived lignocellulosic biomass biorefinery. *Bioresource Technology*, 298, 122558.

Jacques H. (2021). *FEATURE-Betting on Bamboo: Indonesian Villages Struggle to Source Safe, Green Power*. Thomson Reuters Foundation. https://www.reuters.com/article/indonesia-energy-climate-bamboo-idUSL8N2LU4I6, accessed 26.06.2021.

Jianfei, Y., Zixing, F., Liangmeng, N., Qi, G., & Zhijia, L. (2020). Combustion characteristics of bamboo lignin from kraft pulping: Influence of washing process. *Renewable Energy*, 162, 525–534.

Kerlero, de Rosbo & de Bussy, J. (2012). *Electrical Valorization of Bamboo in Africa*. ENEA Consulting, Paris.

Kumar, B, & Verma, P. (2021). Biomass-based biorefineries: An important architype towards a circular economy. *Fuel*, 288, 119622.

Lobovikov, M., Paudel, S., Piazza, M., Ren, H., & Wu, J. (2007). *World Bamboo Resource: A Thematic Study Prepared in the Framework of the Global Forest Resources Assessment 2005*. Food and Agriculture Organization of the United Nation, Rome, Italy. Report No. 18.

Ma, X., Zhao, H., Xu, W., You, Q., Yan, H., Gao, Z., & Su. Z. (2018). Co-expression gene network analysis and functional module identification in bamboo growth and development. *Frontiers in Genetics*, 9, 574.

Nakanishi, M. (2019). *Bamboo Power: Japanese Plant Fires up for Trial Runs, Nikkei Asia, Osaka, Japan*. https://asia.nikkei.com/Spotlight/Environment/Bamboo-power-Japanese-plant-fires-up-for-trial-runs, accessed 14.07.2021.

NBAR. (2019). *Vietnam: Key Facts, International Bamboo and Rattan Organisation*. https://www.inbar.int/country/viet-nam/, accessed 25.05.2021.

NEDO. (2019). *Ceremony Held to Commemorate Completion of Japan's First ORC Cogeneration Facility for Heat and Electric Power Generation Using Bamboo Biomass*, New Energy and Industrial Technology Development Organization, Kumamoto, Japan. https://www.nedo.go.jp/english/news/whatsnew_00176.html, accessed 27.06.2021.

NL Agency. (2012). Biomass business opportunities Viet Nam. *SVN, Netherland Development Organisation Vietnam*, NL Agency, Ministry of Economic Affairs, Utrecht, 1–85.

Pakar, B.S. (2020). *Wealth Creation through Bamboo Renewable Power and Biochar*. Pambudi, Kedah, Malaysia.

Patel, A. & Shah, A.R. (2021). Integrated lignocellulosic biorefinery: Gateway for production of second generation ethanol and value added products. *Journal of Bioresources and Bioproducts*, 6(2), 108–128.

Poppens, R., van Dam, J., & Elbersen, W. (2013). *Bamboo: Analyzing the Potential of Bamboo Feedstock for the Biobased Economy*. NL Agency, Ministry of Economic Affairs, Utrecht.

Poveda-Giraldo, J.A., Solarte-Toro, J.C., & Cardona Alzate, C.A. (2021). The potential use of lignin as a platform product in biorefineries: A review. *Renewable and Sustainable Energy Reviews*, 138, 110688.

Roque, R.M.N., Baig, M.N., Leeke, G.A., Bowra, S., & Santos, R.C.D. (2012). Study on sub- critical water mediated hydrolysis of Miscanthus a lignocellulosic biomass. *Resources, Conservation & Recycling*, 59, 43–46.

Safieddin Ardebili, SM, Solmaz, H, 'Ipci, D, Calam, A, & Mostafaei M. (2020). A review on higher alcohol of fuel oil as a renewable fuel for internal combustion engines: Applications, challenges, and global potential. *Fuel*, 279, 118516.

Scurlock, J.M.O., Dayton, D.C., & Hames, B. (2008). Bamboo: An overlooked biomass resource? *Biomass Energy*, 19, 229–244.

Shah, K.N.A.K.A., Yusop, M.Z.M., Rohani, J.M., Fadil, N.A., Manaf, N.A., Hartono, B., Tuyen, N.D., Masaki, T., Ahmad, A.S., & Ramli, A. (2021). Feasibility study on biomass bamboo renewable energy in Malaysia, Indonesia, Vietnam and Japan. *Chemical Engineering Transactions*, 89, 127–132.

Sharma, R., Wahono, J. & Baral. H. (2018). Bamboo as an alternative bioenergy crop and powerful ally for land restoration in Indonesia, *Sustainability*, 10 (12), 4367.

Singh, S., Adak, A., Saritha, M., Sharma, S., Tiwari, R., Rana, S., Arora, A., & Nain, L. (2017). Bioethanol production scenario in India: Potential and policy perspective, in: A.K. Chandel, R.K. Sukumaran (Eds.), *Sustainable Biofuels Development in India*, Springer International Publishing, Cham, 21–37.

Talukder, M.M.R., Goh, H.Y., & Puah, S.M. (2017). Interaction of silica with cellulase and minimization of its inhibitory effect on cellulose hydrolysis. *Biochemical Engineering Journal*, 118, 91–96.

Tang, X., Xia, M., Pérez-Cruzado, C., Guan, F., & Fan, S. (2017). Spatial distribution of soil organic carbon stock in Moso bamboo forests in subtropical China, *Scientific Reports*, 7(1), 42640.

Truong, A.H. & Le, T.M. (2014). *Overview of Bamboo Biomass for Energy Production*. Hanoi, Vietnam. https://shs.hal.science/halshs-01100209/document, accessed 20.05.2021.

Wang, F., Ouyang, D., Zhou, Z., Page, S.J., Liu, D., & Zhao, X. (2021) Lignocellulosic biomass as sustainable feedstock and materials for power generation and energy storage. *Journal of Energy Chemistry*, 57, 247–280.

Wang, H., Yang, B., Zhang, Q., & Zhu, W. (2020). Catalytic routes for the conversion of lignocellulosic biomass to aviation fuel range hydrocarbons. *Renewable and Sustainable Energy Reviews*, 120, 109612.

Yoesgiantoro, D., Panunggul, D.A., Corneles, D.E., & Yudha, N.F. (2019). The effectiveness of development bamboo biomass power plant (Case Study: Siberut Island, the district of Mentawai Islands). *IOP Conference Series: Earth and Environmental Science*, 265, 1977.

Zhang, H., Han, L., & Dong, H. (2021). An insight to pretreatment, enzyme adsorption and enzymatic hydrolysis of lignocellulosic biomass: Experimental and modeling studies. *Renewable and Sustainable Energy Reviews*, 140, 110758.

3 Pre-Treatment of Oil Palm Empty Fruit Bunches with Sea Water Improves the Qualities of Lignocellulose Biomass

Siti Ayu Aziz, Sabrina Soloi, and Mohd Hafiz Abd Majid
Universiti Malaysia Sabah

Juferi Idris
Universiti Teknologi MARA (UiTM)

Md Lutfor Rahman and Mohd Sani Sarjadi
Universiti Malaysia Sabah

CONTENTS

INTRODUCTION

One of the most abundant sources of lignocellulosic materials in Malaysia is the biomass that is derived from oil palms. It has a great deal of potential as a form of renewable energy that has the potential to lessen the nation's reliance on non-renewable fossil fuels while simultaneously meeting the nation's growing requirements for energy (Onoja et al., 2018). Palm oil farming and processing is currently one of Malaysia's most important economic sectors, and as a result, the country produces around 311 million tons of biowaste every year because of these activities. Figure 3.1 depicts the amount of biomass that Malaysian oil palm will produce in the year 2020. Even though it has the potential to be changed into fuel as well as many other high-quality items, oil palm biomass is not being exploited to its full potential at the present time (Awalludin et al., 2015). Table 3.1 provides a rundown of the most cutting-edge applications for oil palm biomass.

Oil palm planted area as of December 2020 (hectares), Malaysia Palm Oil Board (MPOB), (2021); fresh fruit bunch (FFB) processed by mill for the month of December 2020, Malaysia Palm Oil Board (MPOB, 2021; Hamzah et al., 2019). The amount of oil palm biomass is estimated using the regular biomass to FFB extraction ratio. The data for this ratio comes from the online publications of the MPOB. Using a combination of biological, chemical, and thermochemical processes, biomass derived from oil palms can be turned into viable biofuels. The thermochemical method is

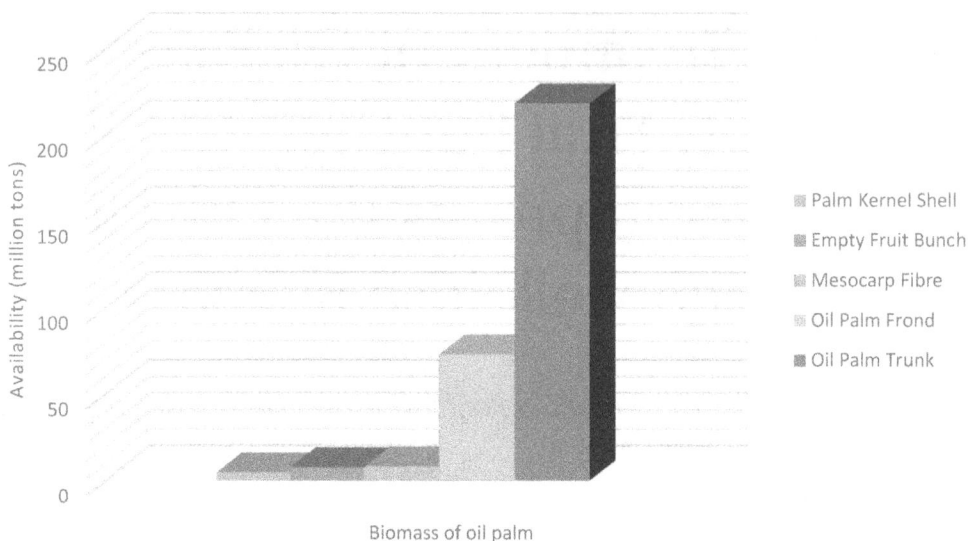

FIGURE 3.1   Malaysian oil palm biomass production in 2020.

TABLE 3.1
Lists of the Most Recent Uses for Oil Palm Biomass

Biomass of Oil Palm	Products	References
Palm Kernel Shell (PKS)	Wastewater treatment adsorbents, boiler oil	Hamzah et al. (2019), Abnisa et al. (2013), Shuit et al. (2009)
Empty Fruit Bunch (EFB)	Furnace fuel, fertilizer from ash, and source of carbon for bioplastic manufacture	Hamzah et al. (2019), Abnisa et al. (2013), Chang (2014), Shuit et al. (2009)
Mesocarp Fiber (MF)	Boiler fuel, thermoplastic infill, and board manufacturing	Onoja et al. (2018), Hamzah et al. (2019), Abnisa et al. (2013), Shuit et al. (2009)
Oil Palm Frond (OPF)	Panels of fiberboard, pulp, roof tiles, and ruminant forage	Onoja et al. (2018) Hamzah et al. (2019), Hashim et al (2012)
Oil Palm Trunk (OPT)	Gasification of syngas, plywood, and particleboard	Hamzah et al. (2019), Zhang et al. (2018), Abdul Khalil et al. (2010), Hashim et al. (2012), Nipattummakul et al. (2012)

more adaptable than the chemical and biochemical processes in terms of the feedstocks that can be used, the labor patterns that can be employed, and the product flow that can be achieved (Lee et al., 2019).

Pyrolysis would be the most viable option for the production of bio-oil from oil palm biomass due to the fact that it has a high degree of adaptability in terms of feedstock sampling (regardless of the type, material body, and physicochemical properties of the feedstock), that it is functional across a wide temperature and atmospheric condition range, and that it can produce three distinct types of products (solid, liquid, and gas). The system parameters of the pyrolysis process are what define the distribution of the process's products, which can be either solid, liquid, or gaseous. Bio-oil, a liquid byproduct of pyrolysis (such as bio-ethanol, bio-oil, and bio-diesel), will be one of the most in-demand types of biofuels in the future. This information was provided by the Malaysian Palm Oil Board (MPOB) in their report titled "Fresh fruit bunch (FFB) processed by mill for the month of December 2020."

When compared to biogas and charcoal, bio-oil is simpler to both store and move due to its compatibility with the vast majority of the equipment, pumping systems, and safety requirements that are now in use (Lee et al., 2019; Abnisa et al., 2013). The production of bio-oil through thermal decomposition and co-decomposition of lignocellulosic biomass has been the subject of numerous reviews that have been written and published. These publications concentrate on contemporary pyrolysis and bio-oil conversion systems (Bridgwater, 2012; Yaman, 2004; Krutof and Hawboldt, 2018), as well as parametric analyses for bio-oil creation during pyrolysis (Akhtar and Saidina, 2012; Guedes et al., 2018). Abnisa et al. (2014) and Zhang et al. (2016) extended the parametric analysis in order to add catalyzed and non-catalytic of biomass with polymers. Zhang et al. (2018) investigated the influence of catalysts on thermal breakdown and co-pyrolysis, in addition to focusing on the optimization of product distribution. Although these studies delve deeply into the processes of biomass decomposition and co-decomposition for the production of bio-oil, they do not concentrate on biomass derived from oil palms but rather advocate for an increase in the total amount of biomass present in their respective geographic regions. Other studies have either concentrated entirely on the pyrolysis of oil palm biomass (Abdullah et al., 2013) or have been restricted to a particular kind of oil palm biomass, such as palm kernel shell (PKS) (Qureshi et al., 2019) or empty fruit bunch (EFB) (Chang, 2014; Kasim et al., 2018). Other studies have either focused solely on the pyrolysis.

Two of the carbohydrates that make up lignocellulose are called cellulose and hemicellulose. These two carbohydrates are responsible for the production of bioenergy as well as reduced organic molecules such as methanol, ethanol, acetic acid, formic acid, and 5-hydroxymethyl furfural, amongst others. Both cellulose and hemicellulose are components of the substance known as lignocellulose. The macromolecular component of refractory lignin is a material that is employed in the production of synthetic aliphatic polymers. Polyimide, thermoplastics, thermosets, and composite films are a few examples of the types of polymers that fall under this category.

The earlier inefficient thermochemical and biochemical processes have been replaced by an effective technique for greener technology known as biomass cascade exploitation, which is a practical strategy that may methodically create a variety of bioproducts. During this step, a number of pre-treatment activities are carried out, after which the various components of the biomass are separated. It has been discovered that the most effective method for breaking down and hydrolyzing lignocellulosic materials is a series of complicated steps and processes. Techniques such as ultrasonic irradiation, extraction of supercritical fluids (such as carbon dioxide, water, and ethanol), hydrodynamic cavitation, electromagnetic disturbance, and pre-treatment of base fluid (IL) are among those that are described in this book. The production of high-value bioproducts from lignocellulosic biomass is illustrated in Figure 3.2.

OIL PALM BIOMASS: LIGNOCELLULOSIC MATERIAL CHARACTERISTICS

The use of oil palm biomass as a feedstock for pyrolysis-based bio-oil synthesis is advantageous for several different reasons. These reasons include the abundant availability of the material and the extensive amount of research that has been conducted on the topic. Bio-oil can be produced from a variety of components of oil palm biomass, including EFB, PKS, MF, OPT, and OPF, as well as oil palm leaves (Abdullah et al., 2013). The concentration of cellulose in oil palm biomass can range from 20 to 59 wt % of n-CFs, whereas the concentrations of hemicellulose and lignin can range from 20 to 40 and 18 to 50 wt %, respectively. In comparison with fossil fuels, the oxygen content of oil palm biomass is higher (40%–50%), while the nitrogen concentration is lower (1%), and the sulfur content is lower (0.2%). During the pyrolysis process, the quality of the bio-oil that is produced will be affected by the physicochemical properties of the feedstock.

During the process of gasification, the decomposition of oil palm biomass results in the production of extra volatiles, which can subsequently be compressed into bio-oil (Guedes et al., 2018). Because it contains a high percentage of cellulose fibers, biomass derived from oil palms presents an intriguing prospect for the development of high-efficiency bio-oil production. The

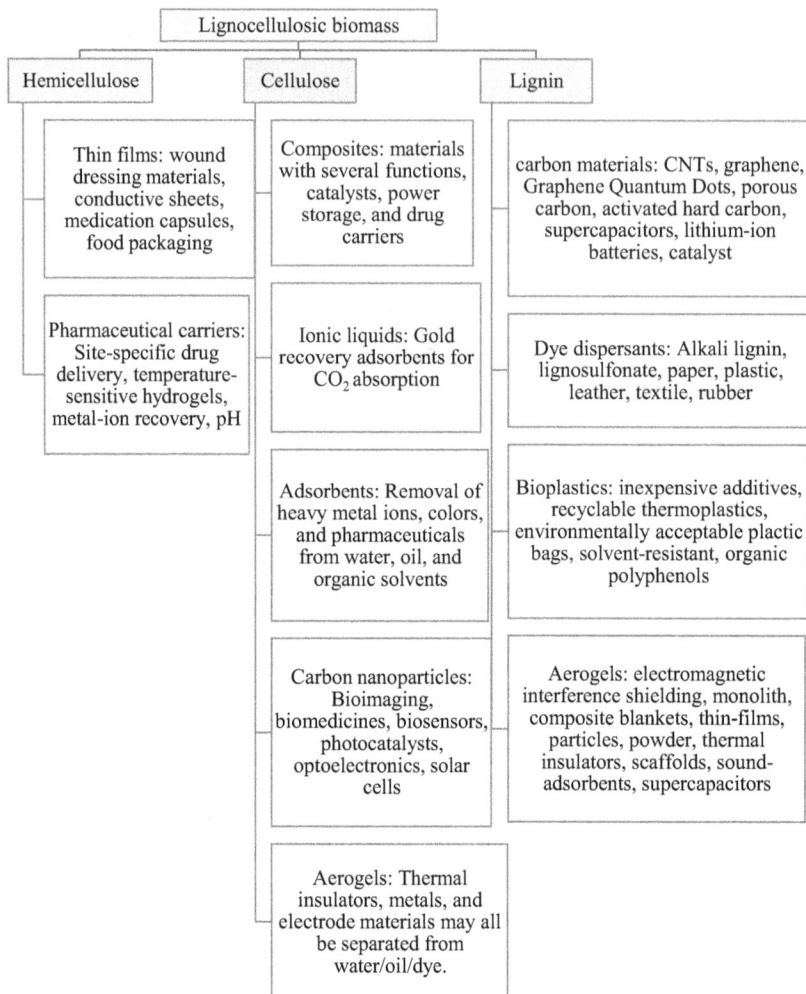

FIGURE 3.2 Bioproducts of high value derived from lignocellulosic biomass.

percentage of cellulose found in PKS is the lowest of any oil palm biomass, coming in at just 30.40 wt %. This is followed by the percentages found in MF (40 wt %), OPT (34.50 wt %), and OPF (30.70 wt %). The cellulose concentration of EFB, which weighs in at 59.70 wt %, is the greatest of any oil palm biomass. During pyrolysis at 500°C, OPF produces 45.99 wt % more bio-oil than PKS does (36 wt %), as stated by Palamanit et al. (2019). This is because OPF contains a higher percentage of cellulose.

In their research, Kim et al. (2013) and Bensidhom et al. (2018) investigated the thermal degrading properties of oil palm biomass as well as other forms of lignocellulosic biomass. It is possible that the pyrolysis of EFB at 478°C will produce more bio-oil (53.97 wt %) than the pyrolysis of PKS (51 wt %) or Jatropha seed shell cake. After carrying out their research at a temperature of 478°C, Kim et al. (2013) came to this conclusion (47.78 wt %). Bensidhom et al. (2018) carried out a subsequent investigation to check up on the situation. Pyrolysis of EFB (26 wt %) and date palm rachis (26 wt %) with a high cellulose content produced more bio-oil than pyrolysis of Date Palm Glaich (25 wt %) and Date Palm Leaflets (25 wt %), according to what they found (17.03 wt %).

The high quantity of moisture that exists in feedstock just prior to being subjected to pyrolysis may result in an increase in the amount of power that is required to complete the processing step of removing water from the biomass. According to the findings of studies Abnisa et al. (2014) and Bridgwater (2012), the manufacture of biofuel should have less than 10% moisture. The oil palm biomass has a moisture content that is lower than the maximum percentage of 10% that is allowed by weight (Table 3.1). It is possible that the high moisture levels in biomass are related to the high water content of bio-oil, which results in the bio-oil having a lower energy content (i.e., heating value).

During the plant's natural life cycle, the biomass composition of oil palms, also referred to as the mineral composition, is formed (i.e., the soil and fertilizer, which together provide the essential plant nutrients). During the production of bio-oil, it is undesirable to have a significant concentration of ash. According to Patwardhan (2010), the presence of NaCl and KCl in feedstock increased the creation of glycolaldehyde from the breakdown of cellulose while decreasing the generation of levoglucosan. This was the result of a net change in the concentration of the two salts. On the other hand, the presence of MgCl2 and CaCl2 during the dehydration processes that occur during the process of biomass degradation increased the synthesis of furans and levoglucosan. Minerals, on the other hand, have no effect whatsoever on the breakdown of lignin.

Researchers Yakub et al. (2015) discovered that the thermal decomposition of OPF at 600°C produced 34.23 wt % bio-oil, while the thermal decomposition of EFB and OPT both produced 30 wt % bio-oil (29.56 wt %). This is because OPF has a lower percentage of ash (3.61% by weight) compared to EFB (4.76% by weight) and OPT. They also investigated the ash composition of OPT samples collected at a variety of altitudes and found that materials with a lower ash concentration create more bio-oil than those with a higher ash level. The thermal breakdown of EFB (2.1 wt %), rice husk (31.2 wt %), rice straw (14.9 wt %), rubberwood (6.1 wt %), eucalyptus wood (2.4 wt %), and Teng wood (0.01% of the total mass) were all researched by Fukuda (2015). As a result of their lower ash content, EFB (55.2 wt %), eucalyptus wood (61.2 wt %), and Teng wood (58.2 wt %) produced more bio-oil than rubber wood (51.6 wt %), rice husk (49.4 wt %), and rice straw (49.1 wt %) (Scapini et al., 2021).

In addition to this, an investigation was conducted to determine the effects that pre-treatment has on the characteristics of biomass derived from oil palms: The ash content of EFB that was treated with 10% and 20% wt % HNO$_3$ was reduced to 1.27 and 1.27 wt %, respectively, compared to the ash content of untreated EFB, which was 8.24 wt %. The generation of bio-oil increased from 43.25 to 69.01 and 66.46 wt %, respectively, when acid-treated EFB was pyrolyzed at 500°C with 10 and 20% wt % HNO$_3$ (Park et al., 2019). This is because there are lower quantities of inorganic elements such as potassium, salt, and magnesium, all of which stimulate the synthesis of biochar and biogas as opposed to bio-oil (Park et al., 2019; Abdullah et al., 2013). In addition, a pre-treatment with deionized water was utilized in order to bring the ash content of EFB down from 5.19 to 2.48 wt % (Abdullah et al., 2013). The production of bio-oil was boosted with the use of EFB that had been pre-treated with H$_2$SO$_4$ and distilled water (Sukiran et al., 2018). After treating EFB with acid water and deionized water, Lim et al. (2014) impregnated it with red mud, red mud extract, acidified rinsed red mud, acidified rinsed red mud extract, and FeSO4. This was done in order to improve the properties of the EFB. As a result of the modifications and pre-treatment, the amount of ash that is contained in EFB has been brought down from 7 to 1.5 wt%.

In biorefineries, the use of fresh water as a replacement for salt water is possible at several stages (Figure 3.3), and the benefits of doing so extend beyond a reduced dependence on fresh water (Scapini et al., 2021). The placement of biorefineries in coastal areas, where salt water rather than fresh water can be used in place of it, has the potential to have a significant impact on the natural world. Studies on the life cycle assessment of a saltwater biorefinery for the production of sugar beet ethanol have shown that this alternative has a significant influence on the loss of water, a decrease in emissions of 31.2%, and reductions in the impacts of climate change and fossil fuels (Zaky et al., 2021). These findings were presented in a paper titled "Life Cycle Assessment of a Saltwater Biorefinery for the Production of Sugar Beet Ethanol." It is possible that it can increase productivity by catalyzing the separation of biomass and by serving as an alternative source of nutrients in

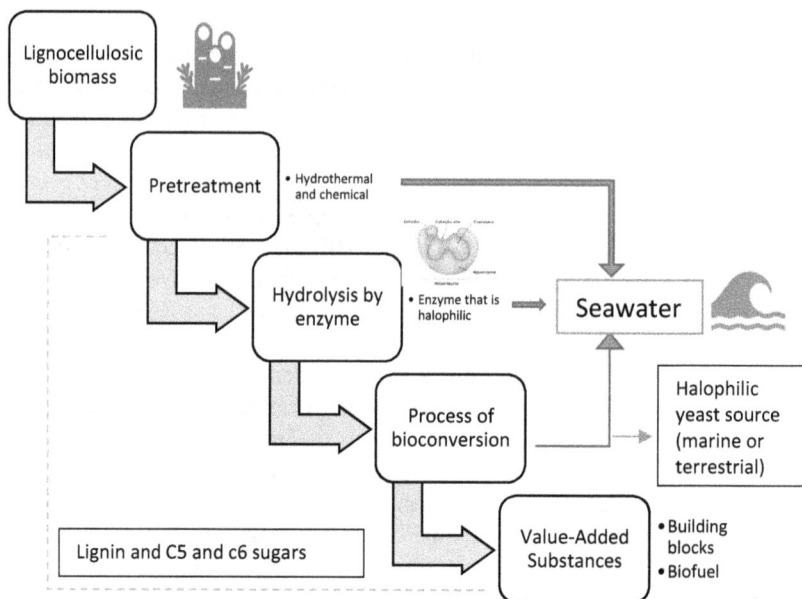

FIGURE 3.3 Overview of biorefinery routes, including salt water for lignocellulose conversion.

bioconversion processes that make use of chloride, salt, sulfate, and magnesium ions (Bonatto et al., 2021; Fang et al., 2017; Zhang et al., 2020).

Significant progress in research pertaining to the pre-treatment of biomass prior to the biological conversion process is still required before biorefineries that employ other water sources, such as salt water, can become economically viable in the future. It will be essential for bridging the gap between the several relevant businesses. For instance, the water-energy-food Nexus (WEF Nexus) may also empower other industries, minimize long-term impact, and reduce the process's water and carbon footprint (Hoff, 2011). The Nexus is an element of the Sustainable Development Goals established by the United Nations. It offers an all-encompassing and multidisciplinary framework for analyzing the interdependencies that exist across a number of different industries, with a particular emphasis on water, energy, and food security. This article discusses research that looked at alternatives to the use of fresh water in lignocellulosic-based wide product bioenergy. The purpose of the essay is to highlight the novelty and importance of the situation by examining the research (Scapini et al., 2021).

BIOMASS FRACTIONATION IN SEA WATER

Despite the fact that lignocellulosic biomasses have value in industrial, agricultural, and forestry applications, their value as a raw material is dictated by the mechanism by which waste is generated and the structural makeup of the waste itself. Because cellulose, hemicellulose, and lignin are so resistant, certain fractionation processes are necessary. These procedures must disrupt the connections between the fibers, raise the surface area, lower the crystallinity, improve the hydrolase transit, and enable agricultural activity. The bulk of systematic approaches to use organic wastes in bio-based products involve polysaccharide structures (pentoses and hexoses) and substrates based on lignin and products produced from it (Chandel et al., 2018; Moodley et al., 2019; Islam et al., 2020; Patel et al., 2021).

The pre-treatment processes that have been utilized, including those that are chemical, physical, physicochemical, hydrothermal, and biological, have all been documented in the relevant research.

In addition to lowering cellulose crystallinity and polymerization level, the method needs to take into account the proportion of biomass involved, the things that are being targeted, the formation of sugar degradation products, the digestibility of enzymes, the economic feasibility of the process, and the protection of the environment (Chandel et al., 2018; Patel et al., 2021; Fang et al., 2015; Gomes et al., 2021; Nakasu et al., 2021; Medina et al., 2018). Even while research on biomass separation was carried out in a unified fashion and produced a single product, very few people looked at ways to cut down on the amount of water that was wasted by finding replacement water resources for fresh water. If it does not have a comprehensive understanding of the demand in the system, a plan to purchase a product exclusively on the basis of its economic benefits may lose its potential in the market.

In order to recover fractions in the least amount of unit operations possible, elements need to be separated at the biomass fractionation stage in biorefineries (Chandel et al., 2018). In addition, recent studies Chen et al. (2018) and de Araújo Padilha et al. (2019) have revealed viable options for biorefineries that involve the use of various pre-treatment procedures to fractionate biomass under gentler conditions and with greater product recoveries. Since the introduction of biorefineries and the ongoing research into the best combinations to utilize, there has been an increase in the amount of ethanol that is produced using sea water. In hydrothermal procedures during pre-treatment, fresh water has been established as the reaction mixture. This is the case despite the paucity of data on seawater-based fractionation processes. Zhang et al. (2020) and Fang et al. (2015) verified the demonstrated promise for biocompatibility in pre-treatments with ionic liquids replacing fresh water (Das et al., 2021). In these approaches, the ions in the water act as catalysts, which efficiently eliminates hemicellulose.

Inorganic ions found in salt water, such as potassium chloride (KCl), sodium chloride (NaCl), calcium chloride ($CaCl_2$), magnesium chloride ($MgCl_2$), and, most importantly, iron (III) chloride ($FeCl_3$), accelerate the breakdown of cellulose and xylan during a variety of hydrothermal pretreatments (Chen et al., 2015; Moodley and Kana, 2017; Yang et al., 2018; Zhang et al., 2019; Zhang et al., 2018). There are traces of a few different chemical elements that are soluble in aqueous solvents in sea water. Some of these elements are Fe^{2+}, Mg^{2+}, and Ca^{2+}. It is possible that it could act in the same way that Lewis' acids do in the medium. These Lewis' acids have been shown to be capable of disrupting glycosidic connections and thereby facilitating xylan breakdown (Fang et al., 2017; Moodley and Kana, 2017; Zhang et al., 2019; Loow et al., 2015). In oligomeric hemicellulose, the presence of Cl- ions has the ability to disrupt the chemical bonds. As a direct consequence of this, a relatively high concentration of NaCl is essential, because high quantities of NaCl generate a higher release of acids from the component of hemicellulose.

According to Jiang et al. (2018), these acids have the potential to depolymerize lignocellulosic materials. It is possible that the theorized sea salt action mechanisms in lignocellulosic biomass structures are accountable for unintended repercussions spanning from the denaturing process to its later routes (hydrolysis and fermentation). Because of this, there is a possibility that structural disintegration will lead to the production of ethanol and the release of toxins (such as xylooligosaccharides—XOS).

The use of NaCl for the depolymerization of XOS in xylose under microwave hydrolytic conditions decreased the reaction time from 60 (only H_2O) to 10 min (H_2O/NaCl), and the replacement of H_2O with sea water increased the concentration of Cl- ions and improved the performance of the process (Jiang et al., 2018). It was revealed that adding sodium chloride had a beneficial effect when microwaves were used to cut down on the amount of sugar in sugarcane leaf waste. According to Moodley and Kana (2017), the energy consumption of the microwave equipment drops as the salt concentration rises. In a similar manner, the addition of NaCl to the sugarcane bagasse that had been subjected to hydrothermal pre-treatment and artificial sea water led to an increase in the yield of XOS. A shorter pre-treatment residence period was found to have an effect on the growth of XOS at high temperatures (Zhang et al., 2020). This was another finding made by the researchers.

The study of the interaction between hydrothermal processes and salt water is advancing at a rapid pace, and it is necessary to research the synergistic impacts of the various process parameters. It is well known that factors such as temperature and salinity can influence the effectiveness of pre-treatments that use sea water. The breakdown of cellulose, hemicellulose, and lignin is stimulated when the temperature is raised, which results in improved glucose absorption. However, breakdown chemicals such as acetic acid, furfural, 5-hydroxymethylfurfural (HMF), and phenolic compounds can be produced, which results in a direct decrease in bioconversion yield (Fang et al., 2017; Ko et al., 2015). There is also the possibility that the procedures described above will raise the salinity of the reaction mixture, which would have an effect. It is possible that it will have a negative impact on the recovery of cellulose from solids when paired with an increase in the temperature of the pre-treatment (Fang et al., 2017).

When compared to a pre-treatment with fresh water, the structural changes in plantain leaflet biomass that were discovered by Fang et al. (2015) after hydrothermal pre-treatment with synthetic salt water imply the elimination of hemicellulose, a reduction in the thermodynamic properties of cellulose, and a reduction in the amount of cellulose solids that are absorbed. The breakdown and elimination of xylan, glucan, and lignin are thought to have been influenced by the ion composition of salt water, which includes alkaline and alkaline-earth metal chlorides, according to experts. On the other hand, the authors Fang et al. (2015) found that there were no significant differences in either the pre-treatment processes or the ethanol yields. The increase in the Cl- ion complex, which, in addition to depolymerizing hemicellulose, also breaks the glycosidic connections of the cellulose, resulting in a reduction in the parameters of its thermodynamics (Fang et al., 2017). Due to the fact that it boosts enzymatic digestion, this procedure may be delightful; nonetheless, it needs to be properly managed in order to prevent a considerable loss of sugar.

HALOPHILIC ENZYMES CHALLENGE SEAWATER-BASED BIOCATALYSIS IN BIOREFINERIES

Pre-treatment biomass hydrolysis is necessary for biorefineries that use lignocellulose monomer conversion methods. These activities are primarily responsible for selecting the optimal blend for each biomass, pre-treatment method, and end product (Adsul et al., 2020; Indira et al., 2016; Scapini et al., 2020). Hydrolases (cellulases and xylanases) that function on the geometrical bonds of cellulose and hemicellulose are primarily responsible for these activities.

Cellulases are enzymes that work on cellulose by cleaving bonds-1,4 and releasing units of lower chain sugars, primarily glucose (Adsul et al., 2020; Jayasekara and Ratnayake, 2019). This process is described in Adsul et al. (2020) and Jayasekara and Ratnayake (2019). In addition to cellulase, the breakdown of lignocellulose and the cleaving of specific structures requires the presence of a number of other enzymes, such as xylanases and amylases. There are a number of different enzymatic cocktails available on the market right now, each of which has a specificity for a certain kind of lignocellulosic biomass.

In seawater-based biorefineries, the presence of salts such as NaCl may increase the effectiveness of hydrolase enzymes engaged in the biological process of monosaccharide conversion. This process is typically accompanied with high osmotic pressure (Fang et al., 2017; Grande and de María, 2012). In order to reduce the amount of water that is used by biorefineries, scientists are searching for halotolerant enzymes and microorganisms that are resistant to high osmotic conditions (Alves et al., 2019). In comparison with terrestrial ecosystems, halophilic ecosystems are typically given less attention from researchers (like soil and plants). As a direct consequence of this, unfavorable conditions such as high salt concentrations, low oxygen levels, or high pH may serve as a suitable habitat for bacteria that possess biotechnologically significant characteristics. The ability of microorganisms to adapt to harsh environments is dependent on a metabolic pathway that is site-tolerant as well as the evolution of enzymes that are capable of stimulating reactions in these settings (Adsul et al., 2020; Alves et al., 2019). In this context, saline biodiversity is a valuable technique for discovering new species of extremophilic organisms capable of thriving in extremely salty environments and producing effective halotolerant proteins (Ruginescu et al., 2020; Vogler et al., 2020). In this context, saline biodiversity is a valuable technique for discovering new species of extremophilic organisms capable of thriving in extremely salty environments and producing halotolerant proteins.

In the experiment, the results showed an increase in the concentration of salts in the reaction fluid, which is evidence that halophilic enzymes are biocatalysts that can catalyze processes. They have a high threshold for the effects of salt. The presence of salt in the medium has the effect of altering the temperature, osmotic pressure, and pH of the medium, so producing an unfavorable environment in which only halophilic extremophilic enzymes are able to function (Dayakar et al., 2021). Salt-tolerant enzymes are produced by microorganisms that reside in salty environments or harsh habitats with fluctuating salt concentrations, such as NaCl, KCl, $MgSO_4$, $CaSO_4$, and $MgCl_2$, or by insect microbiotas (Alves et al., 2019; Daoud et al., 2020; Villanova et al., 2021).

Both marine and terrestrial sources of salt are a possibility in locations that are salty. Any isolated system in which the withdrawal of water causes the environment to become saturated can be referred to as a "oceanic source." Examples of such systems include oceans, swamps, and even ponds. On the other hand, salts in terrestrial habitats come from rocks and soil. Like in marine environments, these salts are dissolved and concentrated by rainwater runoff in the same way that they occur in marine ecosystems. In this particular instance, though, each of the salty sides has its own unique chemical profile. Halophilic bacteria can also be found in locations that process salty materials, such as aquaculture, underground saline lakes, and oil resources (Daoud et al., 2020; Villanova et al., 2021).

Halophilic bacteria have a lot of potential for use in industrial and biotechnological applications since they can create enzymes such amylases, proteases, xylanases, and cellulases. There are microbial sources of halophilia that can be found in every element of living things. They can be found in prokaryotes (Archaea and Bacteria) the majority of the time and are categorized according to their ability to live in salty environments. Through the utilization of cultural media design, this position contributes to the identification of prospective strains. These microorganisms are divided into three classes based on the levels of salt concentration required for growth. The mildly halophilic microbes account for 2%–5% of salt, the moderately halophilic bacteria account for 5%–20% of salt, and the halophilic extremes account for 20%–30% of salt (Daoud et al., 2020).

A suitable strain for obtaining halophilic enzymes is derived based on actual bioprospecting processes in halophilic habitats, such as collecting an adequate habitat sample, isolating microorganisms in saline culture conditions, and extracting enzymes (Daoud et al., 2020). These processes include collecting an adequate habitat sample, isolating microorganisms in saline culture conditions; extracting enzymes. Halophilic bacteria utilize chemical, physiological, structural, and enzymatic strategies (Dumorne et al., 2017) to overcome the osmotic pressure that is caused by high concentrations of salt in their environment.

The negatively charged amino acid structure of halophilic enzymes allows them to distinguish themselves in both aqueous and saltwater environments which results in enzymes that are durable and coiled. The chemical structure of the protein becomes attached to ions (such as sulfate, phosphate, and ammonium), which in turn causes the enzyme structures to gather and fold. The majority of the interactions that take place on enzyme contact surfaces involve hydrogen bonds and hydrophobic interactions. Because halophilic enzymes are utilized in hostile settings, the structural formation of these enzymes needs to be flexible enough to allow for catalytic activity while remaining rigid enough to prevent denaturation (Mevarech et al., 2000). The solubility of these enzymes is controlled by the quantity of acids (Dalmaso et al., 2015; Donato et al., 2018) that are created by folded layers in which surplus water is covalently coupled to cationic ions to give durability (Li et al., 2017; Sekar and Kim, 2020).

Halophilic bacteria are major sources of potential halophilic enzymes because of the saline ions that they store inside for the purpose of osmotic control. Because of this, the enzymes that were released have been altered in such a way that they can keep their activity and stability even in the presence of high salt concentrations. These enzymes are characterized by a negatively charged nature, which makes it unlikely for them to precipitate in aqueous environments. Alternately, due to the quick dissociation that occurs when enzymes are exposed to low levels of salt, it is possible that they are irreversibly destroyed (Marhuenda-Egea et al., 2002).

The activity of non-halophilic enzymes is reduced when exposed to high salt concentrations because this disrupts the hydrophobic contacts and surface hydrogen bonds that are present on the enzyme surface. Negative charges on the surface of halophilic enzymes can interact with water and salt ions to hydrate the surface and inhibit protein aggregation due to electrostatic repulsion (Contreras et al., 2020). This is accomplished while maintaining protein properties such as stability, flexibility, stiffness, and solubility (Figure 3.4).

Karan and colleagues (2020) used X-ray crystallographic and quantum chemical models to explain the galactosidase from *Halorubrum lacusprofundi*. This was done in order to gain a better understanding of the mechanisms by which enzymes adapt to environments that are high in salt content. The writers found that the aforementioned paths are associated to stability. Furthermore, halophilic enzymes have a decreased hydrophobicity, which reduces clumping and maintains high specific strength at high levels of salt according to the researchers. Diverse microorganisms are utilized to isolate halophilic enzymes, which are then utilized in a variety of sectors due to their unique features. Halophilic enzymes' negatively charged surface makes them beneficial in a variety of circumstances, including seawater-based biorefineries and ionic liquids prior to salt-catalyzed pre-treatments.

Indira et al. (2016) evaluated the viability of saccharifying lignocellulosic biomass with a halo-tolerant cellulase to create 2G ethanol. Because the enzyme was more effective in salt water than in fresh water, it is possible that these enzymes could be utilized in biorefineries that operate in saltwater environments. An increase in the concentration of salt, up to a maximum that varies depending on the enzyme and the place of separation, can improve the enzymatic activity of enzymes that were created by microorganisms that had been removed from habitats that included high concentrations of salt (Bano et al., 2019; Pasin et al., 2020; Zhao et al., 2021).

Despite the fact that environmental circumstances may be damaging to hydrolases that are used for lignocellulose bioconversion, hydrolase enzymes derived from *Aspergillus flavus* and *Aspergillus*

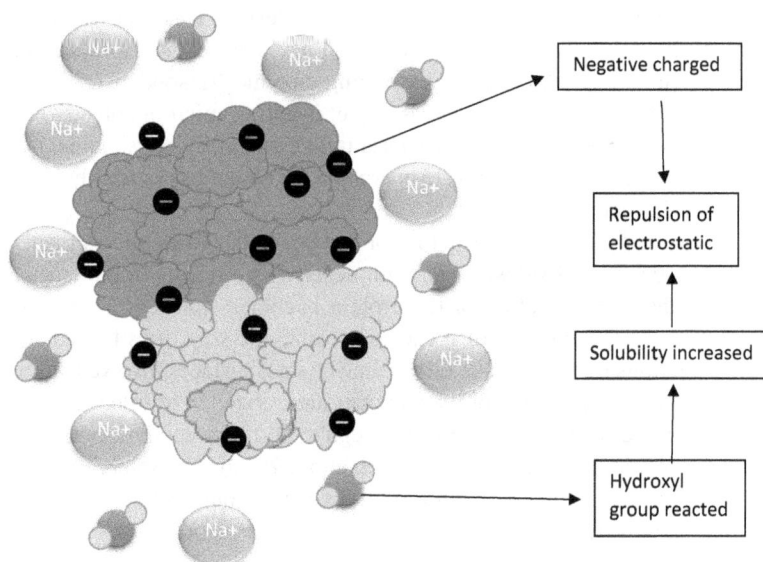

FIGURE 3.4 Adaptation techniques of halophilic enzymes in environments of high salt concentration.

penicillioides were improved by NaCl concentrations of up to 20% (Sinha et al., 2021). They were able to establish that bamboo is capable of saccharification (Zhao et al., 2021). A clavatus xylanases that were isolated from a terrestrial environment demonstrated relative activities of more than 80% in the presence of varying amounts of NaCl, ranging from 0.04 to 3 M. Furthermore, after 24 h of salt exposure, enzyme activation occurred, which may be connected to the enzymes' stability in the existence of Cl⁻ ions (Pasin et al., 2020).

When high salt concentrations are used in the pre-treatment process, using halophilic enzymes may result in decreased water consumption because the solid rinsing phase is not needed (Zhao et al., 2021; Sinha et al., 2021; Gunny et al., 2014; Sharma et al., 2019). This may be the case because the use of halophilic enzymes makes the solid rinsing phase unnecessary. In this regard, random mutagenesis studies that measure cellulase stability in ionic liquids and sea water are useful. This investigation into cellulase optimization might include Cel5A as a candidate cellulase. As a consequence of this, the utilization of genetic engineering in conjunction with computational methods is the most effective strategy for comprehending the molecular processes that are responsible for enzyme enhancement. Contreras et al. (2020), Zhao et al. (2021), and Pasin et al. (2020) revealed the efficacy of enzyme activity of halophilic fungal hydrolases in the enzymatic hydrolysis of bamboo once bamboo was treated with aqueous solutions. This reduced the need to wash the solid after pre-treatment, resulting in decreased water usage during the process.

Because of their specific features, halophilic enzymes in brine biorefineries have the potential to increase system performance. In suboptimal conditions for the biotechnological enzymes routinely utilized, halophilic enzymes are effective. Halophilic enzymes could be deployed as a single hydrolysis step in biorefineries to generate sugars for biomass conversion or even as high-value outputs in their own right.

BUILDING COMPONENTS IN LIGNOCELLULOSIC BIOREFINERIES BASED ON SEA WATER

Multiproduct fluxes are critical for lignocellulosic biomass viability (Medina et al., 2018). The mix of lignocellulosic biomass may yield a variety of products with considerable market demand, allowing biorefinery systems to be adaptable. As was mentioned earlier, lignocellulosic biomass is made up of three different components: cellulose, hemicellulose, and lignin. Each of these components can serve as an indicator of essential components that are relevant to specific areas of the economy, such as the pharmaceutical industry, the chemical industry, or the food and beverage industry. When lignocellulosic biomass is processed in a biorefinery, the production of furfuran molecules is an inevitable consequence. The synthesis of chemicals, polymers, and biofuels all include the utilization of furfural, which is obtained from hemicellulose. Even though the market for furfural is not very large, there is an expectation that interest will increase as the production process becomes more effective (Hongsiri et al., 2014).

Previous research implies that salt water could be used in the production of furfural while simultaneously reducing the amount of water required. In this situation, saltwater salts may accelerate the synthesis of furfural. The use of salt water for corncob hydrolysis boosted furfural output and selectivity at a given reaction temperature, potentially boosting existing furfural manufacturing techniques (Hongsiri et al., 2014). Using salt water and tetrahydrofuran, the cellulose modification of cornstalk to HMF yielded 55.4%, proving that Cl⁻ ions were essential in cellulose to glucose transformation, glucose phase separation, and fructose fluid loss, resulting in a selective process for HMF manufacturing (Li et al., 2018).

Xylan, another hemicellulose component, is a potential raw material for the manufacture of xylooligosaccharides (XOS) (XOS). Low-polymerization XOS may protect against disease and boost the formation of good gut flora (*bifidobacteria* and *lactobacilli*) (Zhang et al., 2017, 2018, 2020). In addition to that, the industries of food production and animal nutrition have a significant need for this chemical. This demand has encouraged the quest for additional XOS sources and acquisition methods. As a consequence of this, lignocellulosic biomass can be converted to XOS through the processes of enzymatic hydrolysis and microbial fermentation (Kumar et al., 2021). By using a hydrothermal pre-treatment with salt water, Zhang and colleagues (2020) created an environmentally friendly technique for generating XOS from sugarcane bagasse. This approach involved using hydrothermal energy.

It was established that hydrothermal pre-treatment with salt water is a reliable and profitable method for XOS synthesis and fermentation products. Cl⁻ ion interactions triple the amount of XOS that may be produced from the process. These findings highlight the relevance of improvements in seawater pre-treatment research, which have resulted in enhanced protection for biodiversity, reduced consumption of chemicals, and the development of systems for the high-quality, continuous manufacture of materials.

The pre-treatment sorghum with the yeast *Rhodosporidium toruloides* was subjected to additional research, during which it was investigated whether or not it was possible to produce prepatent bio-jet fuel by exchanging fresh water for salt water during the process. According to the findings, the processing of sorghum by salt water was on par with that of fresh water, which suggests that salt water might be used in place of fresh water, hence reducing the demand for this natural resource (Das et al., 2021).

In a microwave environment, an investigation was also conducted into the creation of feedstock from sawdust. It is anticipated that salt water would result in a lower yield than fresh water due to the fact that while it will boost phenolic chemical condensation, this will come at the expense of bioconversion. Additionally, in this investigation, additional biomasses with reduced fiber content were assessed, and the researchers found that the addition of NaCl had no influence on the quality of the biocrude (Yang et al., 2020). In order to properly evaluate the system, the authors believe that additional in-depth research is necessary. Succinic acid (Lin et al., 2011) and itaconic acid (Klement et al., 2012) are two distinct building blocks and products that are manufactured by biorefineries that make use of salt water as a resource (Lin et al., 2011).

The production of 2G ethanol is the primary focus of the vast bulk of research being conducted on saltwater biorefineries. On the other hand, the market for building blocks is still in its infant stages, particularly for products derived from lignocellulosic biorefineries that are based on sea water. Other compounds, such as xylitol, biofuels, nanomaterials, components of bioplastics, biotechnologically interesting enzymes, and others, continue to cast doubt on the viability of developing biorefineries and the concept of projects that would replace fresh water with salt water.

If the output recovery process is not hindered by the presence of sea salts, then the approach outlined above can be used to synthesize a variety of different compounds. They will use enzymes and microorganisms that are halophilic or halotolerant in order to execute efficient pre-treatment, hydrolysis, and fermentation operations. This holds true for the glycerol, xylitol, flavoring, and aromatic chemicals that are produced by halotolerant or aquatic yeasts. Some of these compounds can be produced if there is an increase in the concentration of salt (Chi et al., 2016; Musa et al., 2018; Capusoni et al., 2019).

Millerozyma farinosa, a well-known halophilic yeast, has been shown to be able to survive concentrations of up to 3 M NaCl (Lages et al., 1999), which is five times greater than the quantity of salt found in sea water. When M. farinose was inoculated into a culture environment that contained waste coffee grounds extracts and sea water (Hirono-Hara et al., 2021), the researchers found that the organism produced a substantial amount of glutathione. The pharmaceutical and food sectors make extensive use of the tripeptide glutathione, which is an antioxidant. M. farinose produced glutathione (50 mg/L) in their lignocellulosic-seawater-based medium after 48 h of culture, which corresponds to roughly half of the amount found for S. cerevisiae (Hirono-Hara et al., 2021; Li et al., 2004).

As was mentioned earlier, the high performance of halophilic bioprocesses satisfies the requirements for a biorefinery that uses salt water as its source of water. However, additional public regulations and government incentives to reduce water use are required to promote studies in this field in order to enhance and diversify bioproduct manufacturing. These regulations and incentives are required to promote studies in this field in order to improve and diversify bioproduct manufacturing.

CONCLUSION

In Malaysia, a country that is one of the main manufacturers of oil palm items around the world, oil palm is an essential crop commodity. Waste from oil palm trees is an abundant source of biomass, which permits the manufacture of a wide variety of valuable goods such as bio-diesel, palm composite, pulp, and paper, amongst others. It is possible to run it in a wide range of temperatures and weather conditions, and it provides a great deal of versatility in terms of the feedstock that can be sampled. Lignocellulose is made up of many different components, including cellulose and hemicellulose. The macromolecular component known as refractory lignin is utilized in the process of artificially producing aliphatic polymers. When compared to fossil fuels, the oxygen concentration of oil palm biomass is significantly higher (40%–50%), while the nitrogen level is significantly lower (1%), and the sulfur content is significantly lower (0.2%). For saltwater biorefineries, it can be useful to increase the efficiency of the enzymes that are produced by microorganisms. Because the solid rinsing step is not required when utilizing halophilic enzymes, it is possible to reduce the amount of water used. In lignocellulosic biomass-based biorefineries, furfuran molecules are an essential byproduct that must be produced. The production of chemicals, polymers, and biofuels all start with furfural, which can be obtained from hemitriglyceride.

REFERENCES

Abdul Khalil, H.P.S., Nurul Fazita, M.R., Bhat, A.H., Jawaid M., & Nik Fuad, N.A. (2010). Development and material properties of new hybrid plywood from oil palm biomass. *Materials & Design*, 31, 417–424.

Abdullah, N., Sulaiman, F. & Aliasak, Z. (2013). A case study of pyrolysis of oil palm wastes in Malaysia. *AIP Conference Proceedings*, 1528, 331–336.

Abnisa, F., Arami-Niya, A., Wan Daud, W.M.A., Sahu, J.N. & Noor, I.M. (2013). Utilization of oil palm tree residues to produce bio-oil and bio-char via pyrolysis. *Energy Conversion Management*, 76, 1073–1082.

Abnisa, F., Wan Daud, W.M.A., Arami-Niya, A., Ali, B. S. & Sahu, J. N. (2014). Recovery of liquid fuel from the aqueous phase of pyrolysis oil using catalytic conversion. *Energy and Fuels*, 28(5), 3074–3085.

Adsul, M., Sandhu, S.K., Singhania, R.R., Gupta, R., Puri, S.K., & Mathur, A. (2020) Designing a cellulolytic enzyme cocktail for the efficient and economical conversion of Lignocellulosic biomass to biofuels. *Enzyme and Microbial Technology*, 133, 109442.

Akhtar, J. & Saidina N. A. (2012). A review on operating parameters for optimum liquid oil yield in biomass pyrolysis. *Renewable and Sustainable Energy Reviews*, 16, 5101–5109.

Alves, S.L., Müller, C., Bonatto, C., Scapini, T., Camargo, A.F., Fongaro, G., Treichel, H. (2019). Bioprospection of enzymes and microorganisms in insects to improve second- generation ethanol production. *Industrial Biotechnology*, 15(6), 336–349.

de Araújo Padilha, C. E., da Costa Nogueira, C., de Santana Souza, D. F., de Oliveira, J. A., & dos Santos, E. S. (2019). Valorization of green coconut fibre: Use of the black liquor of organolsolv pretreatment for ethanol production and the washing water for production of rhamnolipids by *Pseudomonas aeruginosa* ATCC 27583. *Industrial Crops and Products*, 140, 111604.

Awalludin, M.F., Sulaiman, O., Hashim, R., & Nadhari, W.N.A.W. (2015). An overview of the oil palm industry in Malaysia and its waste utilization through thermochemical conversion specifically via liquefaction. *Renewable and Sustainable Energy Reviews*, 50, 1469–1484.

Bano, A., Chen, X., Prasongsuk, S., Akbar, A., Lotrakul, P., Punnapayak, H., Anwar, M., Sajid, S., & Ali, I. (2019). Purification and characterization of cellulase from obligate halophilic Aspergillus flavus (TISTR 3637) and its prospects for bioethanol production. *Applied Biochemistry and Biotechnology*, 189(4), 1327–1337.

Bensidhom, G., Hassen-Trabelsi, A. B., Alper, K., Sghairoun, M., Zaafouri, K., & Trabelsi, A. (2018). Pyrolysis of date palm waste in a fixed-bed reactor: characterization of pyrolytic products. *Bioresource Technology*, 247, 363–369.

Bonatto, C., Scapini, T., Zanivan, J., Dalastra, C., Bazoti, S.F., Alves, S., Fongaro, G., de Oliveira, D., & Treichel, H. (2021). Utilization of seawater and wastewater from shrimp production in the fermentation of papaya residues to ethanol. *Bioresource Technology*, 321, 124501.

Bridgwater, A.V. (2012). Review of fast pyrolysis of biomass and product upgrading. *Biomass Bioenergy*, 38, 68–94.

Capusoni, C., Arioli, S., Donzella, S., Guidi, B., Serra, I., & Compagno, C. (2019). Hyper- osmotic stress elicits membrane depolarization and decreased permeability in halotolerant marine Debaryomyces hansenii strains and in Saccharomyces cerevisiae. *Frontiers in Microbiology*, 10 (64).

Chandel, A.K., Garlapati, V.K., Singh, A.K., Antunes, F.A.F., & Silva, S.S. (2018). The path forward for lignocellulose biorefineries: bottlenecks, solutions, and perspective on commercialization. *Bioresource Technology*, 264, 370–38.

Chang, S.H. (2014). An overview of empty fruit bunch from oil palm as feedstock for bio- oil production. *Biomass Bioenergy*, 62, 174–181.

Chi, Z., Liu, G.-L., Lu, Y., Jiang, H., & Chi, Z.-M. (2016). Bio-products produced by marine yeasts and their potential applications. *Bioresource Technology*, 202, 244–252.

Chen, L., Chen, R., & Fu, S. (2015). FeCl$_3$ pre-treatment of three lignocellulosic biomass for ethanol production. *Sustainable Chemistry & Engineering*, 3(8), 1794–1800.

Chen, X., Li, H., Sun, S., Cao, X., & Sun, R. (2018). Co-production of oligosaccharides and fermentable sugar from wheat straw by hydrothermal pre-treatment combined with alkaline ethanol extraction. *Industrial Crops and Production*, 111, 78–85.

Contreras, F., Pramanik, S., Rozhkova, A.M., Zorov, I.N., Korotkova, O., Sinitsyn, A.P., Schwaneberg, U., & Davari, M.D. (2020). Engineering robust cellulases for tailored lignocellulosic degradation cocktails. *International Journal of Molecular Sciences*, 21(5), 1589.

Dalmaso, G., Dalmaso, G., Ferreira, D., & Vermelho, A. (2015). Marine extremophiles: a source of hydrolases for biotechnological applications. *Marine Drugs*, 13, 1925–1965.

Daoud, L. & Ben Ali, M. (2020). *Physiological and Biotechnological Aspects of Extremophiles*. Elsevier, Amsterdam, 51–64.

Das, L., Geiselman, G.M., Rodriguez, A., Magurudeniya, H.D., Kirby, J., Simmons, B.A., & Gladden, J.M. (2021.) Seawater-based one-pot ionic liquid pre-treatment of sorghum for jet fuel production. *Bioresource Technology Reports*, 13, 100622.

Dayakar, B., Xavier, K.A.M., Das, O., Porayil, L., Balange, A.K., & Nayak, B.B. (2021). Application of extreme halophilic archaea as biocatalyst for chitin isolation from shrimp shell waste. *Carbohydrate Polymer Technologies and Applications*, 2, 100093.

Di Donato, P., Buono, A., Poli, A., Finore, I., Abbamondi, G., Nicolaus, B., & Lama, L. (2018). Exploring marine environments for the identification of extremophiles and their enzymes for sustainable and green bioprocesses. *Sustainability*, 11, 149.

Dumorne, K., Cordova, D.C., Astorga-Elo, M., & Renganathan, P. (2017). Extremozymes: a potential source for industrial applications. *Journal of Microbiology and Biotechnology*, 27(4), 649–659.

Fang, C., Thomsen, M.H., Brudecki, G.P., Cybulska, I., Frankaer, C.G., Bastidas- Oyanedel, J.-R., & Schmidt, J.E. (2015). Seawater as alternative to freshwater in pre-treatment of date palm residues for bioethanol production in coastal and/or arid areas. *ChemSusChem*, 8(22), 3823–3831.

Fang, C., Thomsen, M.H., Frankær, C.G., Bastidas-Oyanedel, J.-R., Brudecki, G.P., & Schmidt, J.E. (2017). Factors affecting seawater-based pre-treatment of lignocellulosic date palm residues. *Bioresource Technology*, 245, 540–548.

Fukuda, S. (2015). Pyrolysis investigation for bio-oil production from various biomass feedstocks in Thailand. *International Journal of Green Energy*, 12, 215–224.

Gomes, D.G., Michelin, M., Romaní, A., Domingues, L., & Teixeira, J.A. (2021). Co- production of biofuels and value-added compounds from industrial Eucalyptus globulus bark residues using hydro. *Fuel*, 285, 119265.

Grande, P.M. & de María, P.D. (2012). Enzymatic hydrolysis of microcrystalline cellulose in concentrated seawater. *Bioresource Technology*, 104, 799–802.

Guedes, R.E., Luna, A.S., & Torres, A.R. (2018). Operating parameters for bio-oil production in biomass pyrolysis: a review. *Journal of Analytical and Applied Pyrolysis*, 129, 134–149.

Gunny, A.A.N., Arbain, D., Edwin Gumba, R., Jong, B.C., & Jamal, P. (2014). Potential halophilic cellulases for in situ enzymatic saccharification of ionic liquids pretreated lignocelluloses. *Bioresource Technology*, 155, 177–181.

Hamzah, N., Tokimatsu, K. & Yoshikawa, K. (2019). Solid fuel from oil palm biomass residues and municipal solid waste by hydrothermal treatment for electrical power generation in Malaysia: a review, *Sustainability*, 11(4), 1060.

Hashim, R., Wan Nadhari, W.N.A., Nadhari, A., Sulaiman, O., Sato, M., Hiziroglu, S., Kawamura, F., Sugimoto, T., Tay, G. S., & Tanaka, R. (2012). Properties of binderless particleboard panels manufactured from oil palm biomass, *Bioresources*, 7(1), 1352–1365.

Hirono-Hara, Y., Mizutani, Y., Murofushi, K., Iwahara, K., Sakuragawa, S., Kikukawa, H., & Hara, K.Y. (2021). Glutathione fermentation by Millerozyma farinosa using spent coffee grounds extract and seawater. *Bioresource Technology Reports*, 15, 100777.

Hoff, H. (2011). Understanding the nexus. Background paper for the Bonn 2011 Conference: The Water, Energy and Food Security Nexus. Stockholm Environment Institute (SEI), Stockholm, Sweden.

Hongsiri, Wijittra, Danon, Bart, & de Jong, W. (2014). Kinetic study on the dilute acidic dehydration of pentoses toward furfural in seawater. *Industrial & Engineering Chemistry Research*, 53(13), 5455–5463.

Indira, D., Sharmila, D., Balasubramanian, P., Thirugnanam, A., & Jayabalan, R. (2016). Utilization of sea water-based media for the production and characterization of cellulase by Fusarium subglutinans MTCC 11891. *Biocatalysis and Agricultural Biotechnology*, 7, 187–192.

Islam, M.K., Wang, H., Rehman, S., Dong, C., Hsu, H-Y, Lin., C.S.K., & Leu, S-Y. (2020). Sustainability metrics of pre-treatment processes in a waste derived lignocellulosic biomass biorefinery. *Bioresource Technology*, 298, 122558.

Jayasekara, S. & Ratnayake, R. (2019). Microbial cellulases: an overview and applications, in: Pascual, A.R., Martin, M.E.E. (Eds.), *Cellulose*. Intech Open, London.

Jiang, Zhicheng, Budarin, Vitaliy L., Fan, Jiajun, Remón, Javier, Li, Tianzong, Hu, Changwei, & Clark, James H. (2018). Sodium chloride-assisted depolymerization of xylo-oligomers to xylose. *ACS Sustainable Chemistry & Engineering*, 6(3), 4098–4104.

Karan, R., Mathew, S., Muhammad, R., Bautista, D.B., Vogler, M., Eppinger, J., Oliva, R., Cavallo, L., Arold, S.T., & Rueping, M. (2020). Understanding high-salt and cold adaptation of a polyextremophilic enzyme. *Microorganisms*, 8(10), 1594.

Kasim, N.N., Mohamed, A.R., & Ismaili. K. (2018). An upgraded bio-oil derived from untreated and treated empty fruit bunches (EFB) by catalytic pyrolysis: a review. *Journal of Advanced Research in Engineering Knowledge*, 4, 8–16.

Kim, S.W., Koo, B.S., Ryu, J.W., Lee, J.S., Kim, C. J., Lee, D.H., Kim, G.R., & Choi. S. (2013). Bio- oil from the pyrolysis of palm and Jatropha wastes in a fluidized bed. *Fuel Processing Technology*, 108, 118–124.

Klement, T., Milker, S., Jäger, G., Grande, P.M., Domínguez de María, P., & Büchs, J. (2012). Biomass pre-treatment affects Ustilago maydis in producing itaconic acid. *Microbial Cell Factories*, 11, 43.

Ko, J., Kim, Y., Ximenes, E. L., & Michael R. (2015). Effect of liquid hot water pre-treatment severity on properties of hardwood lignin and enzymatic hydrolysis of cellulose. *Biotechnology and Bioengineering*, 112(2), 252–262.

Krutof, A. & Hawboldt, K. (2018). Upgrading of biomass sourced pyrolysis oil review: focus on co-pyrolysis and vapour upgrading during pyrolysis. *Biomass Conversion and Biorefinery*, 8.

Kumar, V., Bahuguna, A., Ramalingam, S., & Kim, M. (2021). Developing a sustainable bioprocess for the cleaner production of xylooligosaccharides: an approach towards lignocellulosic waste management. *Journal of Cleaner Production*, 316, 128332.

Lages, F., Silva-Graça, M., & Lucas, C. (1999). Active glycerol uptake is a mechanism underlying halotolerance in yeasts: a study of 42 species. *Microbiology*, 145, 2577–2585.

Lee, S.Y., Sankaran, R., Chew, K.W., Tan, C.H., Krishnamoorthy, R., Chu, D.-T., & Show, P.- L. (2019). Waste to bioenergy: a review on the recent conversion technologies. *BMC Energy*, 1, 4.

Li, P-Y, Zhang, Y. X., Zhang, B. B., Hao, Y.Q., Wang, J., Wang, Y., Li, P., Qin, C-Y, Zhang, Q-L., Su, X-Y, Shi, H-N., Zhang, M., Chen, Y.Z., Zhou, X-L., & Ning, Y. (2017). Structural and mechanistic insights into the improvement of the halotolerance of a marine microbial esterase by increasing intra- and interdomain hydrophobic interactions. *Applied and Environmental Microbiology*, 83(18), e01286-17.

Li, Xiangcheng, Zhang, Yayun, Xia, Qineng, Liu, Xiaohui, Peng, Kaihao, Yang, Sihai, & Wang, Yanqin. (2018). Acid-free conversion of cellulose to 5-(hydroxymethyl)furfural catalyzed by hot seawater. *Industrial & Engineering Chemistry Research*, 57(10), 3545–3553.

Li, Y., Wei, G., & Chen, J. (2004). Glutathione: a review on biotechnological production. *Applied Microbiology and Biotechnology*, 66(3), 233–242.

Lim, X., Sanna, A., & Andrésen, J.M. (2014). Influence of red mud impregnation on the pyrolysis of oil palm biomass-EFB. *Fuel*, 119, 259–265.

Lin, Carol S.K., Luque, Rafael, Clark, James H., Webb, Colin, & Du, Chenyu. (2011). A seawater-based bio-refining strategy for fermentative production and chemical transformations of succinic acid. *Energy & Environmental Science*, 4(4), 1471.

Loow, Yu-Loong, Wu, Ta Yeong, Tan, Khang Aik, Lim, Yung Shen, Siow, Lee Fong, Md. Jahim, Jamaliah, Mohammad, Abdul Wahab, & Teoh, Wen Hui. (2015). Recent advances in the application of inorganic salt pre-treatment for transforming lignocellulosic biomass into reducing sugars. *Journal of Agricultural and Food Chemistry*, 63(38), 8349–8363.

Malaysian Palm Oil Board (MPOB). (2021). Oil palm planted area as at December 2020 (hectares), Malaysia Palm Oil Board (MPOB). https://bepi.mpob.gov.my/images/overview/Overview2021.pdf. Accessed March 21, 2021.

Marhuenda-Egea, Frutos C, Bonete, & Mariá Jośe. (2002). Extreme halophilic enzymes in organic solvents. *Current Opinion in Biotechnology*, 13(4), 385–389.

Medina, J.D.C., Woiciechowski, A.L., Filho, A.Z., Brar, S.K., Magalhães Júnior, A.I., & Soccol, C.R. (2018). Energetic and economic analysis of ethanol, xylitol and lignin production using oil palm empty fruit bunches from a Brazilian factory. *Journal of Cleaner Production*, 195, 44–55.

Mevarech, M., Frolow, F., & Gloss, L. M. (2000). Halophilic enzymes: proteins with a grain of salt. *Biophysical Chemistry*, 86, 155–164.

Moodley, P. & Kana, E.B.G. (2017). Microwave-assisted inorganic salt pre-treatment of sugarcane leaf waste: effect on physiochemical structure and enzymatic saccharification. *Bioresource Technology*, 235, 35–42.

Moodley, P., Rorke, D.C.S., & Gueguim Kana, E.B. (2019). Development of artificial neural network tools for predicting sugar yields from inorganic salt-based pre-treatment of lignocellulosic biomass. *Bioresource Technology*, 273, 682–686.

Musa, H., Kasim, F.H., Nagoor Gunny, A.A., & Gopinath, S.C.B. (2018). Salt-adapted moulds and yeasts: potentials in industrial and environmental biotechnology. *Process Biochemistry*, 69, 33–44.

Nakasu, P.Y.S., Ienczak, J.L., Rabelo, S.C., & Costa, A.C. (2021). The water consumption of sugarcane bagasse post-washing after protic ionic liquid pre-treatment and its impact on 2G ethanol production. *Industrial Crops and Production*, 169, 113642.

Nipattummakul, N., Ahmed, I.I., Kerdsuwan, S., & Gupta, A.K. (2012). Steam gasification of oil palm trunk waste for clean syngas production, *Applied Energy*, 92, 778–782.

Onoja, E. S. Chandren, F. Razak, N. Mahat, & Wahab, R. (2018). Oil Palm (Elaeis guineensis) biomass in Malaysia: the present and future prospects. *Waste and Biomass Valorization*, 10(28), 2099–2117.

Palamanit, A., Khongphakdi, P., Tirawanichakul, Y., & Phusunti, N. (2019). Investigation of yields and qualities of pyrolysis products obtained from oil palm biomass using an agitated bed pyrolysis reactor, *Biofuel Research Journal*, 6(4), 1065–1079.

Park, J.-W., Heo, J., Ly H.V., Kim, J., Lim H., & Kim, S.-S. (2019). Fast pyrolysis of acid-washed oil palm empty fruit bunch for bio-oil production in a bubbling fluidized- bed reactor, *Energy*, 179, 517–527.

Pasin, Thiago Machado, Salgado, José Carlos Santos, Scarcella, Ana Sílvia de Almeida, de Oliveira, Tássio Brito, de Lucas, Rosymar Coutinho, Cereia, Mariana, Rosa, José César, Ward, Richard John, Buckeridge, Marcos Silveira, Polizeli, & de Moraes Polizeli, Maria de Lourdes Teixeira. (2020). A halotolerant endo-1,4-β-Xylanase from Aspergillus clavatus with potential application for agroindustrial residues saccharification. *Biotechnology and Applied Biochemistry*, 191(3), 1111–1126.

Patel, Amisha & Shah, Amita R. (2021). Integrated lignocellulosic biorefinery: gateway for production of second-generation ethanol and value added products. *Journal of Bioresources and Bioproducts*, 6(2), 108–128.

Patwardhan, P. (2010). *Understanding the Product Distribution from Biomass Fast Pyrolysis*. Iowa State University, USA.

Qureshi, S.S., Nizamuddin, S., Baloch, H.A., Siddiqui, M.T.H., Mubarak, N.M., & Griffin, G. J. (2019). An overview of OPS from oil palm industry as feedstock for bio-oil production. *Biomass Conversion and Biorefinery*, 9(4), 827–841.

Ruginescu, R., Gomoiu, I., Popescu, O., Cojoc, R., Neagu, S., Lucaci, I., Batrinescu- Moteau, C., & Enache, M. (2020). Bioprospecting for novel halophilic and halotolerant sources of hydrolytic enzymes in brackish, saline, and hypersaline lakes of Romania. *Microorganisms*, 8, 1903.

Santos, Everaldo Silvino. (2019). Fractionation of green coconut fiber using sequential hydrothermal/alkaline pre-treatments and Amberlite XAD-7HP resin. *Journal of Environmental Chemical Engineering*, 7(6), 103474.

Scapini, T., Camargo, A.F., Bonatto, C., Stefanski, F.S., Dalastra, C., Zanivan, J., Viancelli, A., Michelon, W., Fongaro, G., & Treichel, H. (2020). Sustainability of biorefineries: challenges associated with hydrolysis methods for biomass valorization, in: Verma, P. (Ed.), *Biorefineries: A Step towards Renewable and Clean Energy*, Springer, Singapore, 255–272.

Scapini, Thamarys, dos Santos, Maicon S.N., Bonatto, Charline, Wancura, João H.C., Mulinari, Jéssica, Camargo, Aline F., Klanovicz, Natalia, Zabot, Giovani L., Tres, Marcus V., Fongaro, Gislaine, Treichel, & Helen, T. (2021). Hydrothermal pre-treatment of lignocellulosic biomass for hemicellulose recovery. *Bioresource Technology*, 342, 126033.

Sekar, A. & Kim, K. (2020). Halophilic bacteria in the food industry, in: Kim, S.-K. (Ed.), *Encyclopedia of Marine Biotechnology*. John Wiley & Sons Ltd, New Jersey, United States, 2061–2070.

Sharma, Vishal, Nargotra, Parushi, & Bajaj, Bijender Kumar. (2019). Ultrasound and surfactant assisted ionic liquid pre-treatment of sugarcane bagasse for enhancing saccharification using enzymes from an ionic liquid tolerant Aspergillus assiutensis VS34. *Bioresource Technology*, 285, 121319.

Shuit, S.H., Tan, K.T., Lee, K.T., & Kamaruddin A.H. (2009). Oil palm biomass as a sustainable energy source: a Malaysian case study. *Energy*, 34, 1225–1235.

Sinha, S., Datta, K., Maithili, D., & Supratim. (2021). A glucose tolerant β-glucosidase from Thermomicrobium roseum that can hydrolyze biomass in seawater. *Green Chemistry*, 23(18), 7299–7311.

Sukiran, M.A., Loh, S.K., & Bakar, N.A. (2018). Conversion of pre-treated oil palm empty fruit bunches into bio-oil and bio-char via fast pyrolysis. *Journal of Oil Palm Research*, 30, 121–129.

Villanova, Valeria, Galasso, Christian, Fiorini, Federica, Lima, Serena, Brönstrup, Mark, Sansone, Clementina, Brunet, Christophe, Brucato, Alberto, & Scargiali, Francesca. (2021). Biological and chemical characterization of new isolated halophilic microorganisms from saltern ponds of Trapani, Sicily. *Algal Research*, 54, 102192.

Vogler, M., Karan, R., Renn, D., Vancea, A., Vielberg, M.-T., Grötzinger, S.W., DasSarma, P., DasSarma, S., Eppinger, J., Groll, M., & Rueping, M. (2020). Crystal structure and active site engineering of a halophilic γ-carbonic anhydrase. *Frontiers in Microbiology*, 11 (742).

Yakub, M.I., Abdalla1, A.Y., Feroz, K.K., Suzana, Y., Ibraheem, A., & Chin, S.A. (2000). Halophilic enzymes: proteins with a grain of salt. *Biophysical Chemistry*, 86(2–3), 155–164.

Yaman, S. (2004). Pyrolysis of biomass to produce fuels and chemical feedstocks. *Energy Conversion and Management*, 45, 651–671.

Yakub, M.I., Abdalla1, A.Y., Feroz, K.K., Suzana, Y., Ibraheem, A., & Chin, S.A. (2015). Pyrolysis of oil palm residues in a fixed bed tubular reactor. *Journal of Power and Energy Engineering*, 3, 185–193.

Yang, J., Chen, H., Liu, Q., Zhou, N., Wu, Y., & He, Q. (2020). Is it feasible to replace freshwater by seawater in hydrothermal liquefaction of biomass for biocrude production? *Fuel*, 282, 118870.

Yang, Q., Huo, D., Si, C., Fang, G., Liu, Q., Hou, Q., Chen, X., & Zhang, F. (2018). Improving enzymatic saccharification of eucalyptus with a pre-treatment process using MgCl2. *Industrial Crops and Production*, 123, 401–406.

Zaky, A.S., Carter, C.E., Meng, F., & French, C.E. (2021). A preliminary life cycle analysis of bioethanol production using seawater in a coastal biorefinery setting. *Processes*, 9, 1399.

Zhang, Weiwei, Lei, Fuhou, Li, Pengfei, Zhang, Xiankun, & Jiang, Jianxin. (2019). Co- catalysis of magnesium chloride and ferrous chloride for xylo-oligosaccharides and glucose production from sugarcane bagasse. *Bioresource Technology*, 291, 121839.

Zhang, W., You, Y., Lei, F., Li, P., & Jiang, J., 2018. Acetyl-assisted autohydrolysis of sugarcane bagasse for the production of xylo-oligosaccharides without additional chemicals. *Bioresource Technology*, 265, 387–393.

Zhang, X., Lei, H., Chen, S., & Wu, J. (2016). Catalytic co-pyrolysis of lignocellulosic biomass with polymers: a critical review, *Green Chem*, 18(15), 4145–4169.

Zhang, Xiankun, Zhang, Weiwei, Lei, Fuhou, Yang, Shujuan, & Jiang, Jianxin. (2020). Coproduction of xylooligosaccharides and fermentable sugars from sugarcane bagasse by seawater hydrothermal pretreatment. *Bioresource Technology*, 309, 123385.

Zhao, Bo, Al Rasheed, Haroon, Ali, Imran, & Hu, Shanglian. (2021). Efficient enzymatic saccharification of alkaline and ionic liquid-pretreated bamboo by highly active extremozymes produced by the co-culture of two halophilic fungi. *Bioresource Technology*, 319, 124115.

4 Biorefinery of Biofuel Production

Thermochemical and Biochemical Technologies from Renewable Resources

Tan Kean Meng, Muaz Mohd Zaini Makhtar,
Muhammed Aidiel Asyraff Mohmad Hatta,
and Mohd Asyraf Kassim
Universiti Sains Malaysia

CONTENTS

DOI: 10.1201/9781003358084-4

INTRODUCTION

Biofuel is a type of renewable energy that is derived from organic materials such as microbial, plant, and animal wastes. An increase in petroleum-based oil prices and environmental concerns has imparted a new strategy to explore clean and sustainable energy resources. Biofuel includes biodiesel, bioethanol, biobutanol, biogas, and biosynthetic gas (syngas), which are among the common biofuels that are currently being used to partially replace fossil fuel for various applications. These biofuels can be produced via many different approaches either thermochemical or biochemical via chemical reactions, fermentation, and heat. Production of biofuel through these technologies involves a complex process by converting the organic material into biofuel by breaking down the feedstock complex structure into small molecules. Production of biofuel can be carried out using a wide range of organic feedstocks (Figure 4.1).

BIOFUEL FEEDSTOCKS

Most of the traditional biofuel can be derived from different types of feedstocks such as wood, crops, and waste material (Figure 4.2). Feedstock for biofuel production can be further classified into several groups such as first-, second-, and third-generation biofuels.

First-generation biofuel is a biofuel that is derived from food crops such as sugar, soybean, corn, straw, and edible oil. Other more marginal feedstocks that are used or considered to produce first-generation bioethanol include but are not limited to whey, barley, potato wastes, and sugar beets. Second-generation biofuel is a biofuel defined as bioenergy that can be generated from non-edible resource especially by-products from the agriculture industry. Generally, the feedstock for second-generation biofuels is typically produced from lignocellulosic material which is inexpensive and easily available locally. Lignocellulosic biomass is composed of three main biopolymers: cellulose, hemicellulose, and lignin. Different types of resources have different compositions. Table 4.1 shows the lignocellulosic material component distribution from various feedstocks.

In contrast, the third-generation biofuel is the biofuel that is produced from the conversion of microalgal biomass. Production of biofuel from this type of biomass poses more advantages over first- and second-generation feedstock biofuel. Among the main advantages of biofuel production from microalgae are that it can be cultivated on non-arable land and that these microalgae are able to consume carbon dioxide (CO_2) from the atmosphere for their cell growth. On the other hand, this microorganism exhibits a higher growth rate in comparison to other higher plants. This microalgal

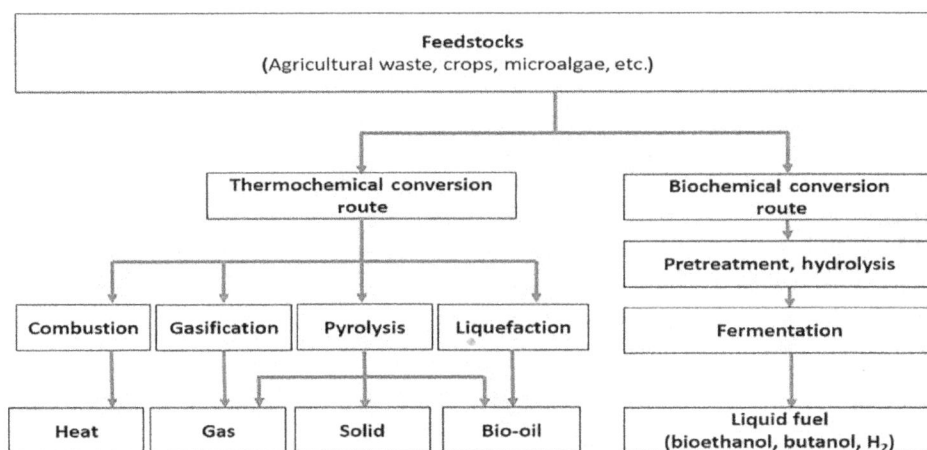

FIGURE 4.1 Biofuel production using different conversion routes.

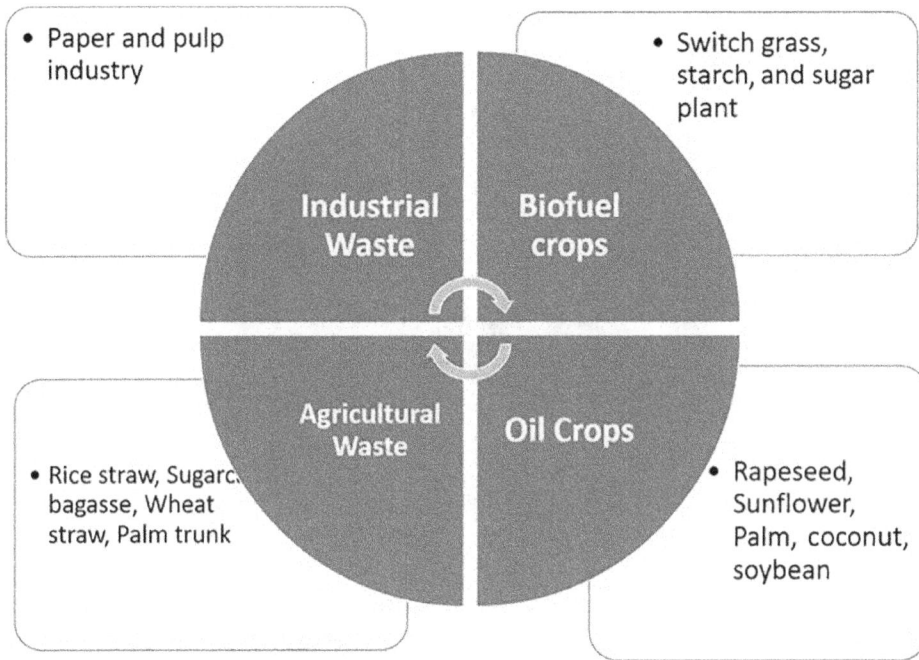

FIGURE 4.2   Different types of feedstocks for biofuel feedstock.

TABLE 4.1
Chemical Composition of Various Potential Biofuel Feedstocks

	Cellulose	Hemicellulose	Lignin	Ash	Moisture	Reference
RS	39.04	21.64	16.20	18.0	-	Singh et al. (2014)
Kenaf fiber	46.10	29.70	22.10	1.60	-	Norshahida et al. (2012)
Banana pseudostem	46.30	12.50	15.80	10.70	96.70	Jayaprabha et al. (2011)
Palm trunk	40.56	21.10	46.44	2.36	1.45	Sahari et al. (2012)
Sugarcane bagasse	30.20	56.70	13.40	1.90	8.34	El-Tayeb et al. (2012)
EFBs	54.17	29.10	15.13	2.86	-	Kassim et al. (2011)

can accumulate several biochemical compounds such as lipid, carbohydrates, and protein which could be potentially converted into various types of biofuel and bio-based products. Currently, microalgal biorefinery has been introduced to ensure the feasibility of biofuel production from this type of feedstock. Figure 4.3 shows the potential microalgal biorefinery concept for the integration of biofuel and bio-based production from single feedstock.

THERMOCHEMICAL CONVERSION

Biofuel feedstocks can directly be converted into energy using a thermochemical conversion pathway. Thermochemical gasification is a promising technology that can exploit the embedded energy in various types of biomass and convert it to valuable products suitable for different industrial applications. Bioenergy conversion technologies such as pyrolysis, gasification, combustion, and liquefaction are the most commonly used to produce bioenergy thermochemical conversion approaches. The choice of which actions are used depends upon the type and quantity of the biomass feedstock,

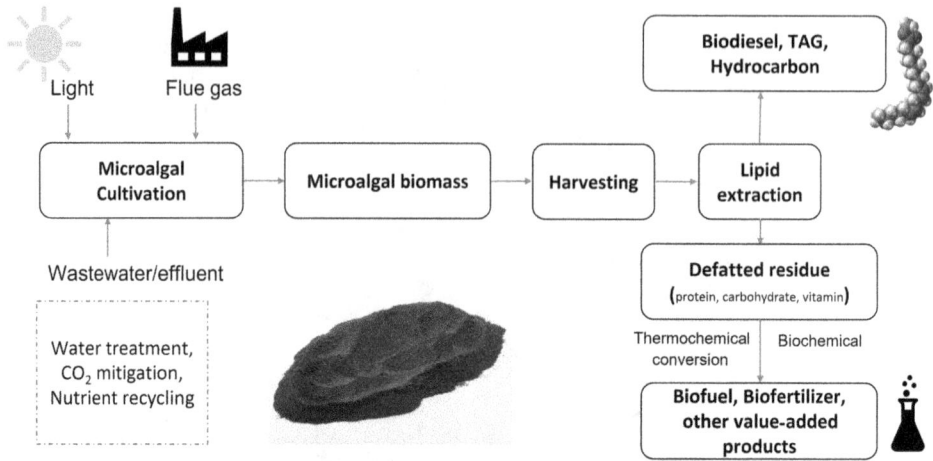

FIGURE 4.3 Microalgal biorefinery for sustainable bio-based products.

TABLE 4.2
Biofuel Production Using Different Type of Biomass and Technologies

Feedstock	Conversion Technology	Features
Dry biomass	Pyrolysis	Thermal decomposition of feedstock at high temperature 100°C–600°C
		Type of pyrolysis: slow, fast, and flash
	Combustion	Conversion of materials to produce heat
	Gasification	Thermal decomposition of feedstock using air or steam at temperature >1,000°C
		Produce syngas
Wet biomass	Hydrothermal conversion	High pressure conversion reaction for feedstock with high moisture content
	Supercritical gasification	Conversion of high moisture feedstock under supercritical condition.

the desired type of fuel, the environmental standards, the economic conditions, and the various reaction parameter condition (Canabarro et al., 2013). Different approaches were found to be able to produce different types of biofuels (Table 4.2). Typically, a combination of liquid, solid, and gas can be produced from the conversion of feedstock via pyrolysis. In addition, conversion of organic materials using liquefaction produces crude bio-oil which can be further purified to produce other chemicals such as alcohol, dimethyl ether, and methyl alcohol (Zhang et al., 2019), while biosynthetic gas that consists of carbon dioxide, carbon monoxide, hydrogen, and methane can be produced via gasification reaction. The direct combustion of feedstock is one of the dominant thermochemical conversion pathways globally (Tanger et al., 2013).

PYROLYSIS

Pyrolysis is a direct thermal conversion which involves decomposition of organic material at temperature range of 300°C–800°C in the absence of oxygen (Chowdhury et al., 2017). A variety of products such as solid, liquids, and gases can be generated from this thermal process (Figure 4.4). The pyrolysis process is a complex process that manipulation of the reaction process could significantly influence the final product. In the pyrolysis process, it involves several major steps including drying,

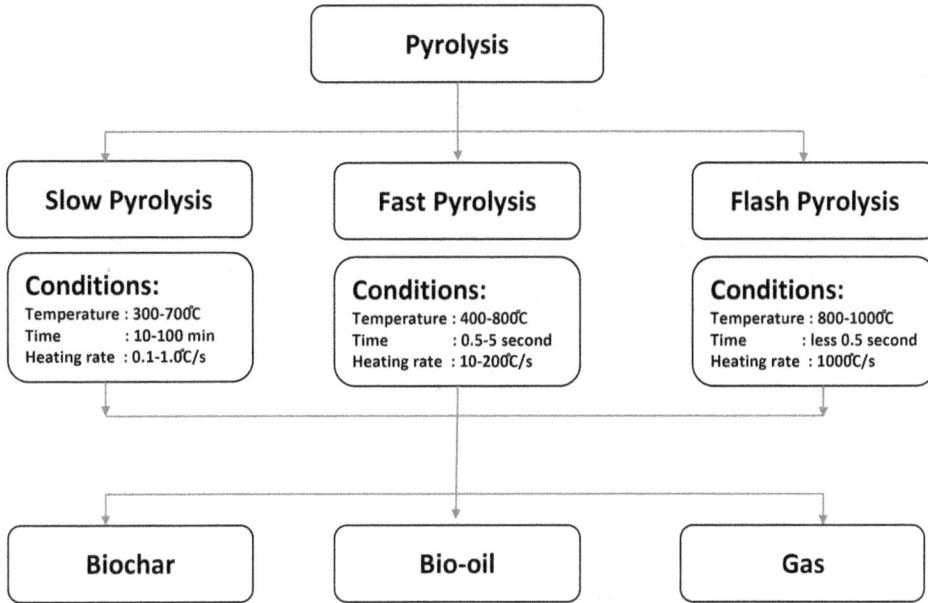

FIGURE 4.4 Type of pyrolysis and it products.

devolatilization, and char formation process. The thermal decomposition of organic materials starts at a temperature of 300°C–500°C. During this stage, the chain of carbon, oxygen, and hydrogen in the materials breaks down into smaller molecules to form gases and condensable vapors, and formation of solid charcoal can be obtained at the end of the pyrolysis. The charcoal yield decreases as the temperature increases. The yield of products resulting from biomass pyrolysis can be maximized as follows: charcoal (a low temperature, low heating rate process), liquid products (a low temperature, high heating rate, short gas residence time process), and fuel gas (a high temperature, low heating rate, long gas residence time process) (Mohan et al., 2006, Balat et al., 2009, Abella et al., 2007).

According to Tanger et al. (2013), pyrolysis at high temperature and longer residence time favors then gas production, whereas reaction at moderate temperatures ranging 400°C–600°C and short residence reaction are suitable to generate more liquid. In contrast, pyrolysis reaction at lower temperature and longer residence times is optimum for production of solid products (Williams and Besler, 1996). The pyrolysis reaction at different condition not only affect the products, but also influence the chemical composition distribution such as alcohol, ketone, aldehyde, ester, phenolic compounds, and heterocyclic derivatives (Lyu et al., 2015, Sarkar and Wang, 2020, Ben et al., 2019). The pyrolysis reaction for bioenergy from biomass could also be divided into three different reactions, namely catalytic, fast, and flash. Generally, the difference between those reactions is the process condition which is mainly contributed by residence time, heating rate, and temperature. The details of each pyrolysis reaction are discussed later.

SLOW PYROLYSIS

Slow pyrolysis is one of the thermal decomposition processes that is conducted at low temperature and low heating rate. This pyrolysis process is commonly used to produce high-quality and reliable char from various feedstocks. The temperature range for this reaction is typically between 300°C and 700°C, with heating rate of 1°C–30°C/min (Ronsse et al., 2013, Noor et al., 2019). Several studies on the slow pyrolysis process on various feedstocks have been reported in the literature. For example, pyrolysis of cellulose, *Lemna minor*, rice straw (RS), pine, and microalgae such as

Chlorella pyrenoidosa has been reported in the literature. According to Wang et al. (2019), their study on pyrolysis of six different feedstocks reported that pyrolysis products mainly gas, solid, and bio-oil can be influenced by the hydrogen-to-carbon ratio available in the feedstock. Another study on slow pyrolysis of mixture biomass and plastic in fixed-bed reactor also indicated that co-pyrolysis could positively affect the final product generated from the process. The study also indicated that difference on the feedstock structural characteristic plays important role in the production of char and bio-oil yield (Çepelioğullar and Pütün, 2014). The study showed that the biochar produced from the pyrolysis of different feedstocks such as cotton stalk, hazelnut shells, sunflower residues, and *Euphorbia rigida* were 28.01%, 30.10%, 21.69%, and 23.12%, respectively. However, slow pyrolysis process exhibited technological limitation particularly on the quality of bio-oil produced from the reaction. Moreover, low heating transfer for long residence time was found to have contributed to the extra energy input to produce desired final products (Bridgwater et al., 1999, Hilal DemirbaŞ, 2005).

Fast Pyrolysis

Fast pyrolysis is a thermal reaction of either organic or inorganic materials that commonly conducted at moderate temperature between 450°C and 600°C with rapid heating rates of more than 100°C/min. This thermal reaction generally occurs at short residence time and produces high yield of good-quality bio-oil with a minimum char and gas of 12% and 13%, respectively. The high-quality bio-oil produced from fast pyrolysis potentially substitutes fuel oil in electricity application. Fast pyrolysis of various feedstocks such as woody materials, agricultural residues, aquatic biomass, and microalgae for bio-oil production has been reported (Morgan et al., 2016, Muradov et al., 2010, Wang et al., 2013, Thangalazhy-Gopakumar and Adhikari, 2016). Most of the studies indicate that the bio-oil production from various feedstocks is significantly influenced by the type of feedstock. The main operational conditions including temperature, gas flow rate, residence time, and feedstock particle size are among the parameters that have been reported and play important role on bio-oil yield production via fast pyrolysis in fluidized bed reactor (Jahirul et al., 2012).

According to Onay (2007), their study of bio-oil production via fast pyrolysis of safflower seeds indicated that maximum bio-oil yield of 54% was obtained for pyrolysis at 600°C. Another study on the fast pyrolysis of manure sample pyrolyzed at 600°C, 800°C, and 1000°C indicated that maximum bio-oil production of 27% was achieved for the pyrolysis at 800°C (Fernandez-Lopez et al., 2017). Similar bio-oil trend has also been reported by Sukiran et al. (2016) in their investigation of fast pyrolysis of EFB, which concluded that high bio-oil yield was obtained when the pyrolysis was conducted at 500°C. The study also indicated that particle size of feedstock could play crucial role in bio-oil yield. Fast pyrolysis of EFB in fluidized reactor indicated that maximum bio-oil was observed for the pyrolysis reaction using 126–150 μ. Low bio-oil produced from pyrolysis using small particle size could be explained by the fact that small particle size will lead to un-uniform heating process, thus resulting to low pyrolysis yield (Encinar et al., 2000, Seebauer et al., 1997).

The changes of pyrolysis operational condition not only influence the bio-oil yield, but could also affect the quality of bio-oil produced. According to the previous characterization of bio-oil from feedstock, it is indicated that the common chemical compounds present are hydrocarbon, fatty acid, alcohol, ester, ether, ketone, and phenolic compound (Chukwuneke et al., 2019, Morgan et al., 2015). Furthermore, hydrogen, carbon, and oxygen ration are among the important indicators for the future potential of bio-oil application. It was found that bio-oil from pyrolysis of anaerobic sludge at high temperature contains high hydrogen-to-carbon ratio and oxygen-to-carbon ratio (Kim et al., 2012). Another study on fast pyrolysis of chestnut capsule contains low oxygen content with hydrogen-to-carbon ratio which is close to conventional fuel gas (Kar and Keles, 2013). Bio-oil with high oxygen content is highly reactive, is unstable, and is not suitable for use as fuel application due to storage limitation. Thus, due to this problem, further upgrading is required to improve the storage stability.

FLASH PYROLYSIS

Another type of pyrolysis reaction is flash pyrolysis that involves rapid heating rate >1000°C/s at high temperature between 900°C and 1300°C. Similar to other pyrolysis reaction, the main product generated from the flash pyrolysis is bio-oil. Generally, the reaction rate for this pyrolysis is less than 0.5 s.

Many investigations on flash pyrolysis of various feedstocks have reported a promising approach to produce high-quality bio-oil. Flash pyrolysis of wood fibers obtained maximum conversion with 60% liquid yield at reaction temperature range between 450°C and 550°C at residence time less than 4 s (Imran et al., 2016). The study also reported that flash pyrolysis at longer residence time could be negatively impacted by increasing the bio-oil product. Another study by Urban et al. (2017) on pyrolysis of milled soybean at temperature range 250°C–610°C using 0.2 and 0.3 s reaction produced nearly 70% of bio-oil with low gas and tar residue. This study concludes that high liquid produced from the flash pyrolysis could be attributed to low secondary reaction of oil produced due to short retention time. Another study on the flash pyrolysis of mixed wood also indicated that maximum bio-oil yield of 67% with high polycyclic aromatic hydrocarbon was obtained at high temperature more than 550°C (Horne and Williams, 1996). Further increase of temperature was found to slightly reduce bio-oil yield. Similar study has been observed on the flash pyrolysis of other types of feedstocks such as corn stalk, and wheat straw in heated laminar entrained reactor (Shuangning et al., 2005).

Although many studies have demonstrated the potential of producing bio-oil from a variety of feedstocks, this technology has some limitations, such as poor thermal stability, low viscosity and high oxygenate compounds, necessitating more in-depth investigation into bio-oil upgrading and reactor design to further improve the process.

PYROLYSIS TECHNOLOGY

Several pyrolysis technologies such as microwave pyrolysis, plasma pyrolysis, and vacuum pyrolysis have been introduced to improve production of bio-oil and biochar from biomass. Development of new pyrolysis technology makes the thermal conversion process become more rapid, selective, and efficient compared to the conventional pyrolysis system.

Microwave-assisted pyrolysis process is among new technology that has been introduced to improve pyrolysis process. The main differences of this technology compared to the conventional approach lies on the type of heating method. Fundamentally, conventional pyrolysis involves transfer of thermal energy from the surface of biomass into the depth of feedstock and partially combusts to produce bio-oil, hydrocarbon-rich gas, and carbon-rich biochar, whereas the microwave pyrolysis involves transfer of electromagnetic energy into inside the biomass and constantly accumulates inside the biomass. High energy accumulation at the core of biomass causes temperature gradient from inside to outside of the biomass and release volatile compounds of the feedstock from inside core to the biomass surface (Zhang et al., 2017). Microwave pyrolysis poses more advantages including selective, rapid, uniformly internal heating of biomass and flexible over conventional pyrolysis process (Xie et al., 2015).

Investigation on the microwave pyrolysis toward different lignocellulose biomass feedstocks has been conducted for bio-oil and biochar production. For instance, this microwave pyrolysis was used to convert RS to produce biochar at low temperature for CO_2 absorbent application. The study indicated that production of biochar from RS using microwave pyrolysis could reduce reaction time, cost, and energy consumption (Huang et al., 2015). Another investigation on the microwave pyrolysis of coconut shell for bio-oil production has also been reported. Study by Nuryana et al. (2020) indicated that bio-oil from coconut shell produced from microwave pyrolysis contained high phenolic compound which exhibits growth of *E. coli*. This indicated that the product generated from this pyrolysis process could be used as antibacterial agent. Furthermore, a microwave pyrolysis with the presence of catalyst has also been developed to improve the pyrolysis process. Microwave pyrolysis of sewage sludge showed that the presence of catalyst HZSM-5 could reduce pyrolysis temperature, and maximum bio-oil was observed when the pyrolysis reaction was performed at 550°C (Xie et al., 2014).

Although this microwave pyrolysis exhibits great potential as future thermal conversion technology, it is obvious that processing at large scale is a big challenge in comparison with conventional pyrolysis technology. Currently, several efforts involving scaling-up of microwave pyrolysis can be made by both academic institutes and industrial companies, especially involving local safety and environmental issue.

Vacuum Pyrolysis

Vacuum pyrolysis is another technology that could be the future potential for thermal conversion of biomass. This technology involves pyrolysis reaction under sub-atmospheric pressure within the reactor. The low pressure occurring in the system makes the carbon conversion to become more efficient and avoids side reaction occurring during the process. This pyrolysis has also been reported, is able to reduce decomposition temperature, and shortens organic vapor residence time (Luda, 2012).

Investigation of the products generated from vacuum pyrolysis on different biomass feedstocks such as wood (Garcìa-Pérez et al., 2007), pine sawdust (Zhang et al., 2010), coffee residue (Chen et al., 2017), and palm kernel cake (Dewayanto et al., 2013) has been reported by many researchers. To date, most of the studies are performed at the bench scale level.

Most of the studies indicated that bio-oil is the dominant product produced from vacuum pyrolysis. A study by Chen et al. (2017) on the vacuum pyrolysis of coffee residue indicated that bio-oil was the highest product with 42.29% followed by biochar and gas with 33.14% and 24.57%, respectively. Similar study has also been reported on the vacuum pyrolysis of lignocellulosic biomass such as forest pinewood waste, which indicated that bio-oil represents 70% of the final product produced followed by biochar and gas (Amutio et al., 2011). Investigation on the pyrolysis product from different biomass feedstocks such as softwood and hardwood rich in fiber also found that the main product produced from this process is bio-oil followed by biochar (Garcìa-Pérez et al., 2007).

The potential application of the bio-oil and biochar derived from vacuum pyrolysis of biomass has also been investigated by several researchers. For instance, characterization of bio-oil produced from vacuum pyrolysis of soft back biomass contains low Na^+K^+Ca with low viscosity, and high net energy can be used to operate gas turbine (Boucher et al., 2000). In addition, another study on the characterization of wood-derived bio-oil produced via vacuum pyrolysis that contains low oxygen level can be potentially used for future liquid fuel (Özbay and Özçifçi, 2021). Furthermore, biochar produced from waste palm shell (WPS) through vacuum pyrolysis can be potentially used as substrate for oyster mushroom (*Pleurotus ostreatus*) cultivation (Kaźmierczak et al., 2017). It was found that WPS-derived biochar contains high porous structure and BET surface area that could exhibit high absorption properties as biofertilizer. This type of biochar could also enhance and stimulate the growth of microbes and targeted plant.

Plasma Pyrolysis

Plasma pyrolysis technology is an emerging thermal conversion technology that could provide complete solution for solid waste management including lignocellulosic materials. This technology involves disintegration of carbonaceous material into new compound fragments under limited oxygen environment. Plasma pyrolysis uses high energy from burning of working gas such as argon (Ar), helium (He), hydrogen (H_2), and nitrogen (N_2), and provides extremely high temperature to decompose feedstock materials via chemical reaction including decomposition, evaporation, pyrolysis, and oxidation. The conversion of biomass occurred in a plasma reactor zone that contained active electron, ion, and excited molecules with high energy radiation. Exposure of biomass in this area will rapidly decompose, and volatile matter is released from the biomass. In general, the final products produced from this technology are mainly CO, H_2, and non-leachable solid hydrocarbons.

There are four different plasma generators that are commonly being developed for plasma pyrolysis including direct current, alternate current, radio frequency (RF) plasma pyrolysis, and microwave plasma pyrolysis. Numerous investigations on plasma pyrolysis for valorization of biomass into other value-added products were carried out. A study by Tu et al. (2009) indicated that application of RF plasma pyrolysis was able to produce higher hydrocarbon-rich solid product compared to gas from RS. The study indicated that the formation of pyrolysis product is significantly influenced by reaction temperature, and the suitable temperature to obtain high solid fraction from RS was obtained at 740 K. Another study on similar feedstock reported that plasma pyrolysis using pilot-scale plasma torch thermolysis reactor indicated that almost 90% of the RS was converted into CO- and H_2-rich gas product (Je-Lueng et al. 2010). Similar study on pyrolysis of glycerol and crushed wood using water steam plasma pyrolysis reactor also found that higher synthesis gas was obtained when the reaction was conducted in the presence of steam (Tamošiūnas et al., 2016). This clearly indicated that that moisture content of the feedstock could significantly affect the pyrolysis product and gas composition obtained from the process. A study by Dermawan et al. (2022) (seed waste into high-efficiency biochar by atmospheric pressure microwave plasma) indicated that biochar obtained from the microwave plasma pyrolysis of RS and golden shower has a great potential to be used for bioremediation application. Biochar produced from this pyrolysis exhibited higher methylene blue absorption capacity compared to that produced from conventional pyrolysis approach. The study indicated that high surface area and high pore volume were observed from both biochar of RS and GS pyrolyzed using microwave plasma pyrolysis technology.

HYDROTHERMAL LIQUEFACTION

Conversion of high-moisture-content feedstock for biofuel production such as microalgae, sludge, and animal waste is one of the problems in this area. Processing of wet feedstock requires a lot of energy for drying process prior to be subjected for biofuel production via thermal conversion process. Liquefaction has been introduced to be applied to overcome this limitation. Liquefaction is considered as a promising conversion process to convert biomass that is presence of water or organic solvent in the presence or without catalyst into liquid biofuel. The liquefaction is commonly conducted under high temperature in hot liquid water, and biocrude produced has a complex mixture of ketone, phenols, alcohol, aldehydes, and carboxylic acid (Shuangning et al., 2005, Reddy et al., 2016).

Analysis of HL on various feedstocks has been reported and showed that the quality of the biocrude produced is depending on the chemical composition distribution in the feedstock. Comparative analysis on biocrude production from various bark species via liquefaction showed that biocrude quality, chemical composition, and conversion efficiency varied significantly with type of feedstock (Feng et al., 2014). In addition, another study by Karagoz et al. (2005) investigated the influence of feedstock types on biocrude production via liquefaction and indicated that cellulose showed the highest conversion compared to sawdust and rice husk. Another study on hydrothermal of forest and agricultural feedstock indicated that agricultural biomass such as wheat straw and sugarcane exhibited high conversion under thermal and catalytic reaction conditions compared to forest biomass (Singh et al., 2015).

The biocrude yield obtained from hydrothermal reaction could also be contributed by the operational conditions. Numerous studies have reported that alteration of hydrothermal liquefaction condition could affect the biocrude yield and chemical distribution in the bio-oil (Guo et al., 2015). The operational conditions such as reaction temperature, residence time, and solvent-to-feedstock ratios are among the parameters that play crucial role in conversion of biomass feedstock to biocrude via hydrothermal liquefaction. Studies by Wei and Jie (2018) evaluated the interaction of liquefaction parameters on biocrude production from microalgae Spirulina sp. indicated that the maximum biocrude production of 41.5% was obtained when the reaction was conducted at temperature 357°C for

37 min with reaction condition of 10.5%. Their study also reported that operational parameters such as temperature and feedstock ratio exhibited the most influential factor on the biocrude production process. Mathanker et al. (2020) conducted the liquefaction of corn stover at different temperatures of 250°C, 300°C, 350°C, and 375°C which indicated that the maximum biocrude was observed for the liquefaction at 300°C. Further increase of temperature resulted in low biocrude production due to the hydrocracking of hydrochar and biocrude generated from the process. Similar observation has also been reported on the hydrothermal liquefaction of barley straw (Zhu et al., 2015), oil palm empty fruit bunches (EFBs), palm mesocarp fiber, and palm kernel shell.

Motavaf and Savage (2021) investigation on the biocrude oil production from food waste indicated that further increase of pressure from 10.2 to 16.9 MPa could enhance 20% of biocrude yield. In another study on liquefaction of palm frond biomass at different pressure, it was found that this parameter could be influenced by solvent used during the process (Prasetyo et al., 2020). The study indicated that the pressure of the solvent can be predicted according to its boiling temperature. The study suggested that the utilization of solvent with low boiling point should obtain maximum biocrude oil yield at low temperature.

GASIFICATION

Gasification is another thermal conversion pathway that has received attention recently (Canabarro et al., 2013, Verma et al., 2012). In this process, the feedstock is converted to a simplified biosynthetic gas or syngas that consists of carbon monoxide (CO), hydrogen (H_2), carbon dioxide (CO_2), NO_x, SO_x, tar, and slag (Sikarwar et al., 2017). Typically, the gasification reaction is performed at temperature of >800°C under optimized oxygen or water (H_2O). Usually, the gasification involved a complex reaction of organic feedstock including drying, pyrolysis, combustion, cracking oxidation, and reduction which can be carried out via following reaction (Figure 4.5).

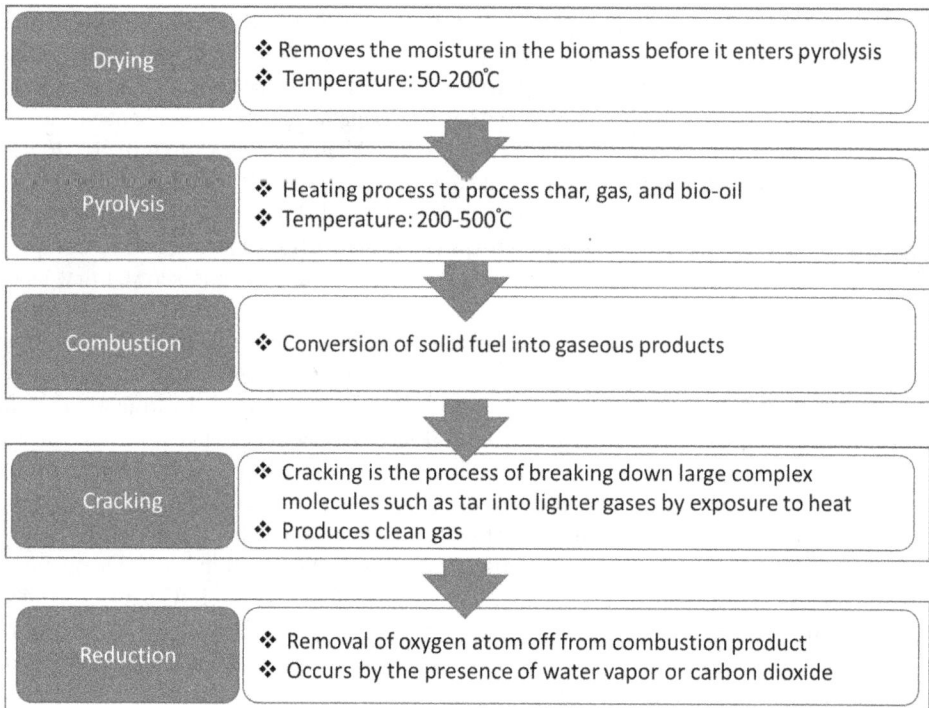

FIGURE 4.5 Gasification of feedstock reaction process.

GASIFICATION PARAMETERS

Syngas formed from gasification of biomass varies depending on the type of feedstock and reaction conditions such as particle size, pressure, and temperature.

Effect of Feedstock Types

Selection of the suitable feedstock is not considerable for better quality syngas. Previous study indicated that different biomass feedstocks used in gasification can directly link to the chemical composition of syngas (Table 4.3). Most of the suitable reported that the syngas yield is related to the chemical composition between cellulose and hemicellulose content in the feedstock, while the residues generated are correlated with the lignin content. Numerous investigations on gasification of different feedstocks employed significantly affect the thermal decomposition kinetics, reactivity, gas composition, and its calorific value (Matsumoto et al., 2009).

Effect of Temperature

Another factor that could affect the gasification reaction is temperature that will influence the syngas production yield, quality, ash, and tar that related to problem in the conversion process. Typically, the gasification of biomass feedstock is performed at temperature ranging between 400°C–1000°C. The gasification temperature can be classified into three ranges, namely low (400°C–600°C), median (600°C–900°C), and high temperatures (beyond 1000°C). Studies on the gasification temperature on various feedstocks including bamboo (Wongsiriamnuay et al., 2013), switch grass pellet (Madadian et al., 2017), microalgae (Raheem et al., 2017), and olive bagasse (Almeida et al., 2019) have been reported elsewhere. Most of the investigations indicated that increase of gasification temperature will result into increase in the proportion of H_2, CO, carbon conversion, and gas yield. It is expected that gasification at higher temperature will affect the several reaction rates during the process. According to Sadhwani et al. (2016), the increase of temperature reaction from 800°C to 950°C will lead to increase the syngas production yield from southern pine biomass. The study also indicated that gasification reaction at different temperatures could significantly affect the syngas composition. Further increase of gasification temperature was found to increase CO and H_2 concentration. Similar observation was reported on the air gasification of microalgal biomass *Chlorella vulgaris* (Raheem et al., 2017). The study reported that temperature was the most important factor that influences overall gasification of microalgal biomass sample. The maximum syngas yield was observed for the gasification conducted at 950°C, corresponding to 20% for the gasification at 700°C–950°C. A similar observation has also been reported on the influence of temperature on different types of feedstock such as switchgrass, cardboard, and hardwood (Madadian et al., 2017).

FISCHER–TROPSCH CATALYTIC REACTION

The syngas produced from gasification can be further converted into wide range of chemicals and fuel via Fischer–Tropsch (FT) reaction synthesis. Several types of chemicals such as methanol, ethanol, hydrocarbon, and oxygenate chemicals can be produced through this reaction. In general, this FT is a complex chemical reaction of hydrogen and carbon monoxide into liquid chemical that typically occurs at high temperatures of 150°C–350°C and pressure with the presence of metal catalyst. The FT reaction can be written as follows:

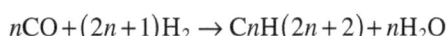

$$n\mathrm{CO} + (2n+1)\mathrm{H}_2 \rightarrow \mathrm{C}n\mathrm{H}(2n+2) + n\mathrm{H}_2\mathrm{O}$$

where CnH (2n+2) represents a range of hydrocarbons, ranging from low-molecular-weight gases (n=1, methane), by way of gasoline (n=5–12), diesel fuel (n=13–17), and as far as solid waxes (n>17).

TABLE 4.3

Syngas Production From Various Feedstocks

Feedstock	Dry Gas Composition (%)							HHV/LHV (MJ/kg)	References
	H_2	CO	CO_2	CH_4	C_2H_2	C_2H_4	C_2H_6		
Rice husk	13.6	14.9	12.9	2.3	ND	ND	ND	4.26	Yoon et al. (2012)
EFB	5.6	16.6	19.2	4.3	ND	ND	ND	5.9	Lahijani and Zainal (2011)
Palm trunk	15.09	20.97	13.19	3.65	ND	ND	ND	ND	Jalil et al. (2011)
Wood pellet	32.1	29.8	7.9	50.9	ND	ND	ND	ND	Raibhole and Sapali (2012)
Kenaf	2.0	33.1	63.1	1.7	ND	ND	ND	ND	Hasanoğlu et al. (2019)
Eucalyptus	22.5	24.2	9.7	2.1	ND	ND	ND	8.08	Borges et al. (2019)
Pine	10	14.18	16.35	2.94	ND	ND	ND	LHV-5.32	Abdoulmoumine et al. (2014)

Note: HHV, high heating value; LHV, low heating value; ND, not detected.

The hydrocarbons produced by FT process can be refined and used in place of more conventional liquid fuels derived from crude oil. Synthetic fuel can be produced by a variety of gasification methods, with gas-to-liquid, coal-to-liquid, and biomass-to-liquid being the most widespread.

Syngas produced from gasification process that contains tars, particulate, and contaminants needs to be clean prior to be subjected for FT synthesis reaction. This cleaning process is required because the presence of these contaminants will cause poison and clog at the downstream process stage. The cleaning process can be carried out either through chemical or through physical methods. To date, physical cleaning method through several techniques such as cyclone and filter is considered the most suitable technique to remove particulate or tars when this technique is economically feasible.

Production of liquid fuel and hydrocarbon from syngas can be influenced by several factors including temperature, pressure, and the presence of catalyst. Reaction pressure and different H_2/CO ratios are among the most important operating parameters that could affect final FT reaction synthesis products. Study by Sauciuc et al. (2012) on their FT synthesis of biomass at different pressures ranging from 16 to 24 bar indicated that maximum CO conversion was achieved when the reaction was conducted at 24 bar. In addition, most study on FT reaction from gasification of different biomasses indicated that the hydrocarbon produced is significantly affected by temperature reaction (Pendyala et al., 2014, Farias et al., 2007, Zamani et al., 2016). In general, the FT synthesis is conducted at temperature range of 180°C–350°C depending on the pressure and catalyst presence in the reaction. Studies indicated that increase of reaction temperature will increase the reaction rate non-linearly. According to Farias et al. (2007), it is indicated that the maximum hydrocarbon and liquid fuel selectivity can be observed for reaction that performed at low temperature 240°C. The study indicated that this temperature was favorable to produce liquid fuel from syngas. Maximum hydrocarbon can also be achieved by introducing the suitable metal catalyst. Several studies have reported that catalysts such as Fe-Mn, Rh/SiO$_2$, Cu-ZnO/Al$_2$O$_3$, Rg/MCM-41, Cu/MCM-41, and Cu-Fe-K/MCM-41 are among the common catalysts used to produce liquid fuel from syngas via FT synthesis (Okoye-Chine et al., 2022, Valero-Romero et al., 2021). Selecting the suitable catalyst is important to achieve maximum selectivity, in which the catalyst properties may affect the interaction, active site, surface hydrophobicity, and performance during the FT synthesis reaction.

BIOCHEMICAL CONVERSION

Besides thermochemical reactions, biochemical conversion is considered an environmentally friendly method, which involves the utilization of bacteria, microorganisms, and enzymes to break down biomass into liquid fuels and gases such as bioethanol, biobutanol, and biogas. The good yield and quality of the liquid fuels and gases could be used to replace non-renewable sources of fossil fuels (Yusoff et al., 2020). Generally, several steps are involved in the biochemical conversion of biomass into liquid biofuel or gas production, including biomass pretreatment, hydrolysis, and fermentation (Figure 4.6).

PRETREATMENT

The main purpose of biomass pretreatment is to disrupt the natural recalcitrance carbohydrate lignin that is located at the outer membrane of the biomass, which could limit the accessibility of enzymes to cellulose and hemicelluloses (Cheah et al., 2020). The choice of the pretreatment types is very crucial, since this could directly affect the rate of hydrolysis and quality and of the liquid fuels and gases production (Aftab et al., 2019). Therefore, a successful pretreatment method must be taken into consideration in terms of the sugar released and solid loading concentration in conjunction with the overall pretreatment process, biomass feedstock, enzymes, or organisms to be applied.

Generally, the pretreatment of the biomass can be categorized into physical, chemical, and biological. We have emphasized the different types of pretreatment methods toward the different types

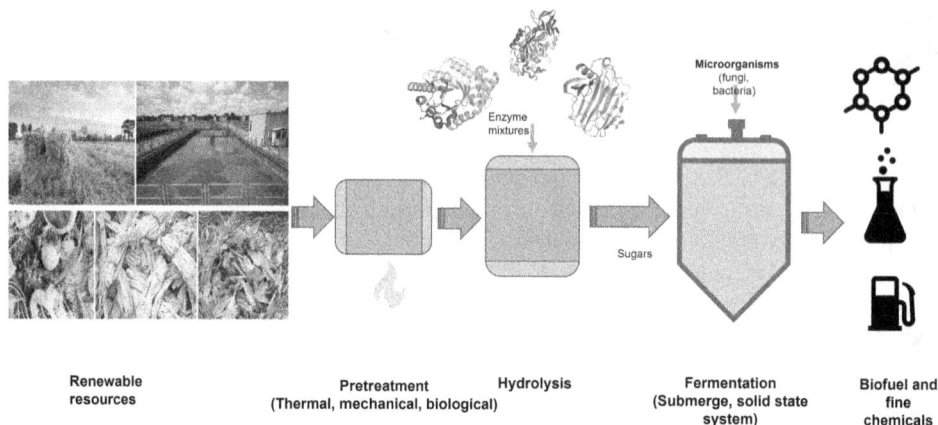

FIGURE 4.6 The formation of value-added products from lignocellulosic biomass.

of lignocellulosic biomasses, aiming to remove the recalcitrance lignin while releasing the cellulose and hemicellulose components for bioconversion into liquid biofuels (Kumar and Sharma, 2017).

Physical Pretreatment

Physical pretreatment is a common pretreatment that involves the breakdown of the outer recalcitrance membrane surrounding the lignocellulose biomass into fermentable sugars. This pretreatment method typically involves the reduction of the lignocellulose biomass size and crystallinity through increasing the temperature and pressure toward the biomass. The energy required for physical pretreatment toward the lignocellulose biomass is dependent on the final particle size and the reduction in the crystallinity of the lignocellulosic material (Brodeur et al., 2011). Hence, this reduction could improve the mass transfer characteristic from the reduction in the biomass particle size, while also improving the hydrolysis results (Maurya et al., 2015).

Physical pretreatments such as milling, extrusion, ultrasound, microwave irradiation, and grinding could be applied on lignocellulosic biomass prior to the hydrolysis process (Ariffin et al., 2008). The biomass was exposed to a high temperature of 300°C, followed by mixing and shearing. This softens the biomass fibers. The feasibility of the physical pretreatment was in agreement with a previous study, which demonstrates that the milling pretreatment on barely strew could obtain a glucose concentration of up to 7 g/L during the hydrolysis process, which was economically feasible for bioethanol production (Raud et al., 2020). Apart from that, Marta et al. (2020) also offered an ultrasonic pretreatment on the mixture of *Sida hermaphrodita* (L.) Rusby mixed with cattle manure which could enhance the production of biogas. El Achkar et al. (2018) also proved that ultrasonic pretreatment could enhance biomethane production using a sonication frequency of 50 kHz, a temperature of less than 25°C, and a residence time of 40–70 min. In order to achieve an effective lignocellulosic biomass pretreatment process, the combination of physical pretreatment with other technologies can be empirically explored.

Chemical Pretreatment

Chemical pretreatment is the most popular method on a commercial scale, which has been extensively applied for cellulosic delignification in the pulping industry (Baruah et al., 2018). Chemicals that are commonly applied toward the lignocellulosic biomass pretreatment are acid and alkali.

Generally, sulfuric acid, nitric acid, phosphoric acid, and hydrochloric acid are common acids typically used for acid pretreatment. This acid pretreatment technology can be conducted under concentrated or diluted acid toward the lignocellulosic biomass. However, using concentrated acid is less popular, is attributable to the inhibitory compound formation, and has negative impact on the environment (Maurya et al., 2015). Therefore, Baruah et al. performed diluted acid pretreatment (<10%) on switchgrass and corn stover under higher temperature (100°C–250°C) conditions. The diluted acid pretreatment was also in agreement with a previous study, which showed that using hydrogen peroxide–acetic acid toward pine wood results in 97.2% of the lignin being removed (Wi et al., 2015). This was due to the diluted acid solubilizing the hemicelluloses and lignin of the lignocellulosic biomass and releasing the cellulose for more accessibility to the enzymatic hydrolysis (Maurya et al., 2015).

Alkaline pretreatment typically involves the application of bases such as sodium, potassium, calcium, and ammonium for the pretreatment on the lignocellulosic biomass (Brodeur et al., 2011). Alkaline application causes the degradation of ester and glycosidic side chains, thus resulting in the structural alteration of lignin, cellulose swelling, and partial decrystallization of cellulose (Cheng et al., 2010). Numerous previous studies were conducted based on the alkaline pretreatment to disrupt the lignin structure of the lignocellulosic biomass to enhance the accessibility of enzymes to cellulose and hemicellulose during the subsequent hydrolysis process (Soto et al., 1994, Zhao et al., 2008). Sun et al. (1995) found out that pretreatment using 1.5% sodium hydroxide (NaOH) was effective on wheat straw, resulting in up to 60% lignin and 80% hemicellulose being released for 144 h at 20°C.

A similar study was conducted by Yuan et al. (2018), who used mild alkaline pretreatment as 8% NaOH toward the bamboo biomass feedstock, which can be used as a potential feedstock for the production of sugars and alcohols. Apart from that, the influence of the alkaline pretreatment was studied previously, whereby the low concentration of NaOH was feasible to remove the lignin and hemicellulose from banana pseudostem. This treatment has the potential for liquid biofuel production (Shimizu et al., 2018).

Biological Pretreatment

Biological pretreatment is a low-cost and eco-friendly technique to treat lignocellulosic biomass prior to the hydrolysis process. In biological pretreatment, the microorganisms, mainly fungi, were used to biologically degrade the lignin and hemicellulose, while remaining intact for the cellulose component in the lignocellulosic biomass (Sánchez, 2009, Shi et al., 2008). The fungi secrete the lignin-degrading enzyme called "ligninase" for removing the lignin (Dashtan et al., 2010). White-rot, brown-rot, and soft-rot fungi are examples of fungus used in pretreating the lignocellulosic biomass (Baruah et al., 2018). However, the white-rot fungus is more dominant compared to the other types of fungus, due to its unique ability of selectively degrading lignin in the lignocellulosic biomass, while being associated with high sugar production during enzymatic saccharification (Baruah et al., 2018). da Silva Machado and Ferraz (2017) used white-rot fungus *Ceriporiopsis subvermispora* to pretreat sugarcane bagasse, which secretes a satisfactory result for sugar production of up to 47% under 27°C±2°C for 60 days of biotreatment time. Another study also proved the feasibility of white-rot fungi as *Ceriporiopsis subvermispora* toward the lignin removal from wheat straw under solid-state fermentation, and that it was suitable for biogas production. Table 4.4 shows the summarization of the advantages and limitations for various types of lignocellulosic pretreatment methods.

Enzymatic Hydrolysis

Enzymatic hydrolysis is the central technology which operates as a biochemical conversion process. Proceeded by pretreatment and followed by a microbial fermentation process, enzymatic hydrolysis is the process that breaks down complex macromolecules such as cellulose and hemicellulose into simpler sugars (Stickel et al., 2014). In order for the hydrolysis process to be economically viable, several important parameters, such as environmental temperature, buffer pH, and enzyme-to-biomass ratio, need to be considered. This is to ensure that optimal maximum sugar conversion rates could be achieved with a minimum energy input and chemicals introduced.

TABLE 4.4

The Advantages and Limitations of Different Pretreatment Methods

Pretreatment	Advantages	Limitations
Physical	Reduce the particle size	High energy consumption
	Increase in surface area	
	Decrease degree of polymerization and crystallinity	
	Eco-friendly and seldom produce any toxic material	
Chemical	Cleavage of the intermolecular ester linkages between hemicelluloses and lignin	Recovery from chemical added
	Change of the lignocellulosic structure via cellulose swelling	Chemicals are highly toxic and corrosive
	Breakdown of the long cellulose and hemicellulose chains into sugar monomers	
	Reduction in crystallinity	Environmental issues
	Reduction in degree of polymerization	Formation of inhibitory compounds
	Increase of internal surface area	
Biological	Low cost	Require longer time
	Eco-friendly	Non-specific reaction
	No formation of inhibitor	Low efficiency
	Less energy consumption	

The typical operating temperature for enzymatic hydrolysis ranges between 40°C and 55°C (Fenila and Shastri, 2016). Below this temperature range causes inactivation enzyme activity, and exceeding this range reduces the enzyme activity due to the denaturation of the active site that is located at the enzyme proteins (Robinson, 2015). Han et al. (2012) also studied the effect of different temperatures on the cellulase activity toward wheat straw. They obtained a maximum saccharification rate of 42.74% using the cellulase produced by *Penicillium waksmanii* F10–2 under an optimum temperature of 55°C. The significance of the effect of temperature on the enzymatic activity was further studied by Rozenfelde et al. (2017), who showed both cellulose as Accellerase 1500 and cellulose as Accellerase XC have optimum conditions at 50°C for hydrolyzing complex wheat straw lignocellulose to simpler sugars. This was also in line with a previous study which found out that hydrolysis temperature was also important for third-generation biomass as a microalgae biomass. The suitable hydrolyzing temperature of 40°C played an important role for cellulase obtained from *Trichoderma reesei* ATCC 26921, reflecting the highest glucose yield up to ~68% for *Chlorococum humicola* biomass (Harun and Danquah, 2011a). The temperature below or above 40°C resulted in a lower glucose conversion yield.

Apart from the temperature, the pH buffer can also influence the enzymatic hydrolysis yield toward lignocellulosic biomass. The pH buffer served to adjust and stabilize the desired pH during the enzymatic hydrolysis assay (Bisswanger, 2014). The normal pH solution for the enzymatic hydrolysis falls in the range of 4.8–6.0 pH (Abdulsattar et al., 2020). da Silva Machado and Ferraz (2017) suggested that a pH of 4.80 was a suitable condition to pretreat sugarcane bagasse with 47% of sugar recovery using a mixture of commercial enzymes, namely cellulases and β-glucosidases. Similar pH condition was also applied by a previous study to maintain the pH conditions with the citrate buffer for performing enzymatic hydrolysis toward the sorghum husk for bioethanol production (Waghmare et al., 2018). This buffer solution successfully replaced the usage of acid, alkaline, or other organic solvents which can cause environmental pollution and have a negative impact on the biomass itself (Huang et al., 2017).

In order to enhance biofuel production from lignocellulosic biomass, high solid concentrations are applied in enzymatic hydrolysis to increase product concentration and reduce energy input.

The biomass-to-enzyme ratio also plays an important role for affecting the enzymatic hydrolysis process prior to biofuel production. Rosgaard et al. (2007b) demonstrated that the effectiveness ratio was 11 mg/g for the cellulase enzymes to dry the barley straw. This ratio was able to convert up to 81% glucose during the enzymatic hydrolysis process. Optimizing the enzyme-to-biomass ratio to obtain the maximum sugar yield was further studied by Banerjee et al. (2010), who used a ratio of 15 mg/g commercial enzyme to corn stover biomass for 48 h digestion conditions with a temperature of 50°C. They were successful to release 54% and 41% available glucose and xylose, respectively. Another study also proved that the hydrolysis yield could be affected by the combination of the enzyme and biomass ratio. Harun and Danquah (2011a) found out that a 0.02 g cellulase enzyme-to-substrate ratio was able to generate the highest glucose yield of up to ~64% compared to other ratios. This result showed that an economic feasible process can be achieved with the more fermentable sugars produced prior to the fermentation process.

However, increasing solid concentrations decreased cellulose conversion yields, which is called the "high solid effect." It was reported that an inefficient hydrolysis step was observed from the barley straw due to the high biomass loading of 10% (w/w) (Rosgaard et al., 2007a). The presence of a high content of insoluble materials could cause a substrate viscosity effect. This viscosity effect could lead to a decrease in the efficiency of the enzymes to hydrolyze the substrates during the hydrolysis step. Therefore, the short duration for the optimum release of fermentable sugars during the hydrolysis process offers many advantages, which eliminate contamination, reduce inhibition effects, and make the process economically effective for biofuel production.

BIOFUEL PRODUCTION VIA BIOCHEMICAL ROUTE

Biofuels have emerged as among the most strategic alternative fuel sources to replace conventional petroleum fuel. The biofuels are considered important progress for limiting greenhouse gas emissions and improving air quality. Generally, biofuel is a fuel that is produced from photosynthetic organisms such as photosynthetic bacteria, micro- and macro-algae, and vascular land plants. The most common examples of biofuels are bioethanol, biobutanol, and biohydrogen.

Bioethanol

Bioethanol is considered among the important renewable fuels that have partially replaced conventional petroleum-derived fuels (Kang et al., 2014). Bioethanol is commonly derived through biochemical reactions using yeast or bacteria as biocatalysts. The general chemical expression of converting glucose-based lignocellulose biomass into bioethanol is as follows (Tran et al. 2019):

$$(C_6H_{10}O_5)n + nH_2O \rightarrow n(C_6H_{12}O_6) \rightarrow 2nC_2H_5OH + 2nCO_2$$

Several studies reported bioethanol production using lignocellulosic biomass (Saini et al., 2015). Sindhu et al. (2014) demonstrated 1.76% (w/w) bioethanol production from *Saccharomyces cerevisiae* fermentation using the Indian bamboo variety (*Dendrocalamus* sp.) as a raw material. The feasibility of bioethanol production is based on renewable biomass and was further explored by Marx et al. (2014), who proved that the potential of sweet sorghum bagasse to biosynthesis was up to $0.252 \, m^3/ton^{-1}$ of bioethanol. Bioethanol production is not only limited to the first and second generation of lignocellulosic biomass, the microalgae biomass as a third generation has also been widely reported, due to the carbohydrate-rich feedstock microalgae biomass. This was due to the absolute absence or near absence of lignin, making the enzymatic hydrolysis process simple prior to bioethanol fermentation (John et al., 2011). It was reported that the *Chlorella vulgaris* FSP-E was used as a feedstock for producing an 11.7 g/L bioethanol yield (Ho et al., 2013). The feasibility of the microalgae for bioethanol production was further explored by Harun and Danquah (2011b), who showed that *Chlorococcum humicola* biomass can act as a carbon food source for *S. cerevisiae* to undergo fermentation with a biosynthesis of up to 7.20 g/L of bioethanol. The bioethanol production

TABLE 4.5

Bioethanol Yield and Fermentation Conditions from Different Biomasses

Types of Biomass	Pretreatment Method	Fermentation Conditions	Bioethanol Yield	References
Indian bamboo	Dilute H_2SO_4 acid	Temperature: 30°C Time: 72 h	1.76% (v/v)	Sindhu et al. (2014)
Sweet sorghum bagasse	Microwave pretreatment	Temperature: 32°C Time: 24 h pH: 4.8	0.252 m³/ton	Marx et al. (2014)
Oil palm fronds	Ultrasonic	Temperature 40°C Time: 5 h pH: 5.0	18.2 g/L	Ofori-Boateng and Lee (2014)
Sweet sorghum	Alkaline pretreatment (NaOH)	Temperature: 37°C Time: 48 h pH: 6.0	91.9 kg/tons	Li et al. (2013)
Olive pruning	liquid hot water	Temperature: 35°C Time: 72 h	3.7% (v/v)	Manzanares et al. (2011)
Microalgae biomass	Dilute acidic hydrolysis (1% sulfuric acid)	Temperature: 30°C pH: 6.0	11.7 g/L	Birgen et al. (2019)

based on the biological process was proven to be feasible to make the process economically attractive to replace petroleum-based fuel (Table 4.5).

Biobutanol

Butanol is a promising biofuel candidate compared to other more established biofuels such as ethanol and methanol. Butanol provides several advantages as follows (Birgen et al., 2019):

- Longer carbon chain with a higher heating value
- Lower volatility
- Lower polarity
- Lower corrosively
- Lower heat of vaporization
- Lesser ignition problems
- Without modification of diesel engines

Typically, biobutanol was synthesized through the petrochemical route or by fermentation using anaerobic bacteria, predominantly the Clostridia species by acetone-biobutanol-ethanol (ABE) fermentation (Van Hecke et al., 2018). Zhao et al. (2018) explored the feasibility of RS for biobutanol production through ABE fermentation after the mild acid pretreatment followed by enzymatic hydrolysis using cellulase. A high cell density of *Clostridium saccharoperbutylacetonicum* ATCC 13564 based on ABE fermentation resulted in a biobutanol production of 6.68±0.24 g/L. The high content of glucose in the RS was suitable for the *Clostridium* sp. to perform ABE fermentation and biosynthesis biobutanol. This was in agreement with Moradi et al. (2013), who showed 163 g glucose pretreated from the RS was able to synthesize up to 112.7 g of biobutanol over kg of pretreated RS through ABE fermentation by *Clostridium acetobutylicum*. The development of non-food crop later appeared due to its composed readily fermented carbohydrates and less lignin, which is suitable for biofuel production. Gao et al. (2016) showed the advantages of the hexane-extracted microalgae that contained more than 80% of carbohydrates in the form of starch components, which were economically feasible to undergo ABE fermentation to biosynthesis of up to 6.63 g/L of biobutanol titers.

TABLE 4.6

The Biobutanol Yield and Fermentation Conditions from Different Biomasses

Types of Biomasses	Pretreatment Method	Fermentation Conditions	Biobutanol Yield	References
RS	Dilute H_2SO_4 acid	Temperature: 30°C Time: 15 h	6.04% (v/v)	Zhao et al. (2018)
RS	Acid and alkaline pretreatment	Temperature: 37°C Time: 72 h	44.2 g/kg	Moradi et al. (2013)
RS	Dilute H_2SO_4 acid	Temperature:37°C Time: 96 h pH: 6.7	5.52 g/L	Gottumukkala et al. (2013)
Chlorella sorokiniana CY1	Methanol and microwaved	Temperature:37°C Time: 48 h pH: 6.0	3.86 g/L	Cheng et al. (2015)
Chlorella vulgaris JSC-6.	Acid and alkali	Temperature: 37°C Time: 60 h pH: 5.0	13.1 g/L	Wang et al. (2016)
Chlorella sp.	Dilute acidic hydrolysis (1M sulfuric acid)	Temperature: 37°C pH: 4.8	6.23±0.19 g/L	Onay (2018)

However, further research is still needed to enhance the biobutanol production from lignocellulosic biomass at an industrial scale, especially to achieve higher biobutanol concentrations during ABE fermentation and to improve biobutanol separation and purification technologies (Table 4.6).

Biohydrogen

Biohydrogen was considered a clean and attractive biofuel that could partially replace conventional fossil fuels and progress to a sustainable hydrogen-oriented economy. Abundant biomass from various industries could be a valuable source for biohydrogen production, in which this process could be carried out through the combination of waste treatment and energy production (Chong et al., 2009). This was suggested by Sekoai et al. (2019), who used potato waste as a carbon food source for the biohydrogen spore-forming bacteria, namely *Clostridium* sp. The result showed that *Clostridium* sp. was able to generate a biohydrogen yield of up to 68.54 mL H_2/g total volatile solids under anaerobic conditions. The biohydrogen production process from lignocellulosic biomass was further conducted by Nasirian et al. (2011), who proved that acid pretreated wheat straw could generate 1.19 mol of hydrogen/mol glucose released from the biomass. The main key for the economic feasible conversion of lignocellulosic biomass into biohydrogen depends on the glucose availability after the pretreated method and hydrolysis steps, prior to the fermentation process. The high content of glucose, galactose, mannose, xylose, and rhamnose presence in the microalgae biomass makes it valuable as a raw material for biohydrogen production (Markou et al., 2012). This finding shows that biohydrogen production was inversely proportional to sugar reduction in the *Scenedesmus obliquus* biomass (Ortigueira et al., 2015). The biohydrogen yield was 2.74 mol/mol of glucose consumption in the biomass after 24 h of incubation time. Therefore, further studies should be carried out to maximize the production of biohydrogen, which possess the advantages of a reduction in energy-related environmental issues, such as greenhouse emissions or acid rain (Table 4.7).

CONCLUSION AND FUTURE OUTLOOK

Biofuel from wide range of renewable feedstock as an alternative to fossil fuel has become one of the approaches to ensure the environmentally friendly future energy. In order to make the production

TABLE 4.7

The Biohydrogen Yield and Fermentation Conditions from Different Biomasses

Types of Biomasses	Pretreatment Method	Fermentation Conditions	Biohydrogen Yield	References
Palm oil mill effluent	-	pH: 5.5 Temperature: 60°C	6.33 L H_2/L-POME	O-Thong et al. (2013)
Cheese whey	-	pH: 6.0 Temperature: 30°C	7.89 mmoL/g lactose	Ferchichi et al. (2005)
Corn stover	Acid	pH: 7.0 Temperature: 37°C	4.17 mmoL/g sugar	Zhang et al. (2014)
Cornstalk	Alkaline	pH: 7.0 Temperature: 37°C	80 mL/g substrate	Bala-Amutha and Murugesan (2013)
Scenedesmus obliquus	Physical grinding	pH: 7.0 Temperature: 37°C	35.0 mL H_2/g volatile solid	Ortigueira et al. (2015)

successful, two potential conversion pathways have been introduced, namely (1) thermochemical and (2) biochemical conversion pathways. According to the discussion presented, it can be drawn that different conversion pathways will produce different products depending on the application. Also, wide range of chemical distribution in different feedstock could clearly affect the final biofuel product either in quality or in quantity aspect. Thus, selection of the most suitable feedstock for final application is important. Another issue that needs to be considered prior to be used as fuel is the stability of the product. Further upgrading of bio-oil and new technology is required to overcome this issue. Currently, biofuels such as ethanol and butanol are produced via biochemical conversion pathway that involves many steps to produce these final products. Further exploration on the utilization of robust microbes to convert syngas or hydrogen from thermochemical pathway could be an alternative to reduce the steps involved in the process. On the other hand, development of single-step process, which is either simultaneous hydrolysis and fermentation or consolidate fermentation process, could be beneficial to ensure biofuel production. Further research to attempt the maximum biofuel production from wide range of feedstock using low energy and cheap technology is very important to ensure the feasibility of the future biofuel production.

REFERENCES

Abdoulmoumine, N., Kulkarni, A. & Adhikari, S. (2014). Effects of temperature and equivalence ratio on pine syngas primary gases and contaminants in a bench-scale fluidized bed gasifier. *Industrial & Engineering Chemistry Research*, 53, 5767–5777.

Abdulsattar, M.O., Abdulsattar, J.O., Greenway, G.M., Welham, K.J. & Zein, S.H. (2020). Optimization of pH as a strategy to improve enzymatic saccharification of wheat straw for enhancing bioethanol production. *Journal of Analytical Science and Technology*, 11, 17.

Abella, L., Nanbu, S. & Fukuda, K. (2007). A theoretical study on levoglucosan pyrolysis reactions yielding aldehydes and a ketone in biomass. *Memoirs of the Faculty of Engineering, Kyushu University*, 67, 67–74.

Aftab, M.N., Irfana, I., Fatima, R., Ahmet, K. & Meisam, T. (2019). Different pretreatment methods of lignocellulosic biomass for use in biofuel production. In: Abd El-Fatah, A. (ed.) *Biomass for Bioenergy-Recent Trends and Future Challenges*. Rijeka: IntechOpen.

Almeida, A., Neto, P., Pereira, I., Ribeiro, A. & Pilao, R. (2019). Effect of temperature on the gasification of olive bagasse particles. *Journal of the Energy Institute*, 92, 153–160.

Amutio, M., Lopez, G., Aguado, R., Artetxe, M., Bilbao, J. & Olazar, M. (2011). Effect of vacuum on lignocellulosic biomass flash pyrolysis in a conical spouted bed reactor. *Energy & Fuels*, 25, 3950–3960.

Ariffin, A., Hassan, M.A., Umi-Kalsom, M.S., Abdullah, N. & Shirai, Y. (2008). Effect of physical, chemical and thermal pretreatments on the enzymatic hydrolysis of oil palm empty fruit bunch (OPEFB). *Journal of Tropical and Food Science*, 36(2), 1–10.

Bala-Amutha, K., & Murugesan, A. (2013). Biohydrogen production using corn stalk employing *Bacillus licheniformis* MSU AGM 2 strain. *Renewable Energy*, 50, 621–627.

Balat, M., Balat, M., Kirtay, E. & Balat, H. (2009). Main routes for the thermo-conversion of biomass into fuels and chemicals. Part 1: Pyrolysis systems. *Energy Conversion and Management*, 50, 3147–3157.

Banerjee, G., Car, S., Scott-Craig, J.S., Borusch, M.S., Aslam, N., & Walton, J.D. (2010). Synthetic enzyme mixtures for biomass deconstruction: Production and optimization of a core set. *Biotechnology and Bioengineering*, 106(5), 707–720.

Baruah, J., Nath, B.K., Sharma, R., Kumar, S., Deka, R.C., Baruah, D.C. & Kalita, E. (2018). Recent trends in the pretreatment of lignocellulosic biomass for value-added products. *Frontiers in Energy Research*, 6, 141.

Ben, H., Wu, F., Wu, Z.A.O., Han, G., Jiang, W. & Ragauskas, A.A.O. (2019). A comprehensive characterization of pyrolysis oil from softwood barks. *Polymers*, 11(9), 1387.

Birgen, C., Durre, P., Preisig, H.A., & Wentzel, A. (2019). Butanol production from lignocellulosic biomass: Revisiting fermentation performance indicators with exploratory data analysis. *Biotechnology for Biofuels*, 12(1), 1–15.

Bisswanger, H. (2014). Enzyme assays. *Perspectives in Science*, 1(1–6), 41–55.

Borges, A.C.P., Onwudili, J.A., Andrade, H.M.C., Alves, C.T., Ingram, A., Vieira de Melo, S.A.B. & Torres, E.A. (2019). Catalytic supercritical water gasification of eucalyptus wood chips in a batch reactor. *Fuel*, 255, 115804.

Boucher, M.E., Chaala, A. & Roy, C. (2000). Bio-oils obtained by vacuum pyrolysis of softwood bark as a liquid fuel for gas turbines. Part I: Properties of bio-oil and its blends with methanol and a pyrolytic aqueous phase. *Biomass and Bioenergy*, 19, 337–350.

Bridgwater, A.V., Meier, D. & Radlein, D. (1999). An overview of fast pyrolysis of biomass. *Organic Geochemistry*, 30, 1479–1493.

Brodeur, G., Yau, E., Badal, K., Collier, J., Ramachandran, K.B. & Ramakrishnan, S. (2011). Chemical and physicochemical pretreatment of lignocellulosic biomass: A review. *Enzyme Research*, 2011, 787532.

Canabarro, N., Soares, J.F., Ancieta, C.G., Kelling, C. S. & Mazutti, M. A. (2013). Thermochemical processes for biofuels production from biomass. *Sustainable Chemical Processes*, 1, 22.

Çepelioğullar, Ö. & Pütün, A.E. (2014). Products characterization study of a slow pyrolysis of biomass-plastic mixtures in a fixed-bed reactor. *Journal of Analytical and Applied Pyrolysis*, 110, 363–374.

Cheah, W.Y., Sankaran, R., Show, P.L., Ibrahim, T.N.B., Chew, K.W., Culaba, A. & Chang, J.S. (2020). Pretreatment methods for lignocellulosic biofuels production: Current advances, challenges and future prospects. *Biofuel Research Journal*, 7, 1115–1127.

Chen, N., Ren, J., Ye, Z., Xu, Q., Liu, J. & Sun, S. (2017). Study on vacuum pyrolysis of coffee industrial residue for bio-oil production. *IOP Conference Series: Earth and Environmental Science*, 59, 012065.

Cheng, H.H., Whang, L.M., Chan, K.C., Chung, M.C., Wu, S.H., Liu, C.P., Tien, S.Y., Chen, S.Y., Chang, J.S., & Lee, W.J (2015). Biological butanol production from microalgae-based biodiesel residues by *Clostridium acetobutylicum*. *Bioresource Technology*, 184, 379–385.

Cheng, Y.S., Zheng, Y., Yu, C.W., Dooley, T.M., Jenkins, B.M. & Vandergheynst, J.S. (2010). Evaluation of high solids alkaline pretreatment of rice straw. *Applied Biochemistry and Biotechnology*, 162, 1768–1784.

Chong, M.L., Sabaratnam V., Shirai, Y., & Hassan, M.A. (2009). Biohydrogen production from biomass and industrial wastes by dark fermentation. *International Journal of Hydrogen Energy*, 34(8), 3277–3287.

Chowdhury, Z.Z., Kaushik, P., Wageeh, A.Y., Suresh, S., Syed, S., Ganiyu, A., A., Emy, M., Rahman, F. R. & Rafie, B. J. (2017). Pyrolysis: A sustainable way to generate energy from waste. In: Mohamed, S. (ed.) *Pyrolysis*. Rijeka: IntechOpen, London, 1–36.

Chukwuneke, J.L., Ewulonu, M.C., Chuwujike, I.C. & Okolie, P.C. (2019). Physico-chemical analysis of pyro-lyzed bio-oil from swietenia macrophylla (mahogany) wood. *Heliyon*, 5, e01790.

da Silva Machado, A. & Ferraz, A. (2017). Biological pretreatment of sugarcane bagasse with basidiomycetes producing varied patterns of biodegradation. *Bioresource Technology*, 225, 17–22.

Dashtan, M., Schraft, H., Syed, T.A. & Qin, W. (2010). Fungal biodegradation and enzymatic modification of lignin. *International Journal of Biochemistry and Molecular Biology*, 1, 36–50.

Dermawan, D., Febrianti, A.N., Setyawati, E.E.P., Pham, M.T., Jiang, J.J., You, S.J. & Wang, Y.F. (2022). The potential of transforming rice straw (*Oryza sativa*) and golden shower (*Cassia fistula*) seed waste into high-efficiency biochar by atmospheric pressure microwave plasma. *Industrial Crops and Products*, 185, 115122.

Dewayanto, N., Isha, R., & Nordin, M.R. (2013). Catalytic pyrolysis of palm oil decanter cake using CaO and y-Al2O3 in vacuum bed reactor to produce bio-oil. *International Conference of Chemical Engineering and Industrial Biotechnology*, Kuantan, 1–10.

El-Achkar, J.H., Lendormi, T., Salameh, D., Louka, N., Maroun, R.G., Lanoiselle, J.L. & Hobaika, Z. (2018). Influence of pretreatment conditions on lignocellulosic fractions and methane production from grape pomace. *Bioresource Technology*, 247, 881–889.

El-Tayeb, T. S., Abdelhafez, A. A., Ali, S. H. & Ramadan, E. M. (2012). Effect of acid hydrolysis and fungal biotreatment on agro-industrial wastes for obtainment of free sugars for bioethanol production. *Brazilian Journal of Microbiology*, 43(4), 1523–1535.

Encinar, J.M., Gonzlez, J.F. & Gonzalez, J. (2000). Fixed-bed pyrolysis of *Cynara cardunculus* L. product yields and compositions. *Fuel Processing Technology*, 68, 209–222.

Farias, F.E.M., Silve, F.R.C., Cartaxo, S.J.M., Fernandes, F.A.N. & Sales, F.G. (2007). Effect of operating conditions on Fischer-Tropsch liquid products. *Latin America Applied Research*, 37, 283–287.

Feng, S., Yuan, Z., Leitch, M. & Xu, C.C. (2014). Hydrothermal liquefaction of barks into bio-crude – Effects of species and ash content/composition. *Fuel*, 116, 214–220.

Fenila, F. & Shastri, Y. (2016). Optimal control of enzymatic hydrolysis of lignocellulosic biomass. *Resource-Efficient Technologies*, 2, S96–S104.

Ferchichi, M., Crabbe, E., Gil, G.H., Hintz, W., & Almadidy, A. (2005) Influence of initial pH on hydrogen production from cheese whey. *Journal of Biotechnology*, 120(4), 402–409.

Fernandez-Lopez, M., Anastasakis, K., De-Jong, W., Valverde, J.L. & Sanchez-Silva, L. (2017). Temperature influence on the fast pyrolysis of manure samples: Char, bio-oil and gases production. *E3S Web Conference*, 22, 1–7.

Gao, K., Orr, V. & Rehmann, L. (2016). Butanol fermentation from microalgae-derived carbohydrates after ionic liquid extraction. *Bioresource Technology*, 206, 77–85.

Garcìa-Pérez, M., Chaala, A., Pakdel, H., Kretschmer, D. & Roy, C. (2007). Vacuum pyrolysis of softwood and hardwood biomass: Comparison between product yields and bio-oil properties. *Journal of Analytical and Applied Pyrolysis*, 78, 104–116.

Gottumukkala, L. D., Parameswaran, B., Valappil, S.K., Mathiyazhakan, K., Pandey, A., & Sukumaran, R.K. (2013). Biobutanol production from rice straw by a non acetone producing *Clostridium sporogenes* BE01. *Bioresource Technology*, 145, 182–187.

Guo, Y., Yeh, T., Song, W., Xu, D. & Wang, S. (2015). A review of bio-oil production from hydrothermal liquefaction of algae. *Renewable and Sustainable Energy Reviews*, 48, 776–790.

Han, L., Feng, J., Zhang, S., Ma, Z., Wang, Y. & Zhang, X. (2012). Alkali pretreatment of wheat straw and its enzymatic hydrolysis. *Brazilian Journal of Microbiology*, 43(1), 53–61.

Harun, R. & Danquah M.K. (2011a). Enzymatic hydrolysis of microalgal biomass for bioethanol production. *Chemical Engineering Journal*, 168(3), 1079–1084.

Harun, R. & Danquah, M.K. (2011b). Influence of acid pre-treatment on microalgal biomass for bioethanol production. *Process Biochemistry*, 46(1), 304–309.

Hasanoğlu, A., Demirci, I. & Secer, A. (2019). Hydrogen production by gasification of Kenaf under subcritical liquid–vapor phase conditions. *International Journal of Hydrogen Energy*, 44, 14127–14136.

Hilal DemirbaŞ., A. (2005). Yields and heating values of liquids and chars from spruce trunk bark pyrolysis. *Energy Sources*, 27, 1367–1373.

Ho, S.H., Huang, S.W., Chen, C.Y., Hasunuma, T., Kondo, A., & Chang, J.S. (2013). Bioethanol production using carbohydrate-rich microalgae biomass as feedstock. *Bioresource Technology*, 135, 191–198.

Hornem P.A. & Williams, P.T. (1996) Influence of temperature on the products from the flash pyrolysis of biomass. *Fuel*, 75(9), 1051–1059.

Huang, L., Ding, X., Dai, C. & Ma, H. (2017). Changes in the structure and dissociation of soybean protein isolate induced by ultrasound-assisted acid pretreatment. *Food Chemistry*, 232, 727–732.

Huang, Y.F., Chiueh, P.T., Shih, C.H., Lo, S.L., Sun, L., Zhong, Y. & Qiu, C. (2015). Microwave pyrolysis of rice straw to produce biochar as an adsorbent for CO_2 capture. *Energy*, 84, 75–82.

Imran, A., Bramer, E.A., Seshan, K. & Brem, G. (2016). Catalytic flash pyrolysis of biomass using different types of zeolite and online vapor fractionation. *Energies*, 9(3), 187.

Jahirul, M.I., Rasul, M.G., Chowdhury, A.A. & Ashwath, N. (2012). Biofuels production through biomass pyrolysis —A technological review. *Energies*, 5(12), 4952–5001.

Jalil, R., Sakanishi, K., Miyazawa, T., Nor, M.Y., Ibrahim, W., Sarif, M., Hashim, S. & Elham, P. (2011). Effects of different gasifying agents on syngas production from oil palm trunk. *Journal of Tropical Forest Science*, 23, 282–288.

Jayaprabha, J.S., Brahmakumar, M., & Manilal, V. B. (2011). Banana pseudostem characterization and its fiber property evaluation on physical and bioextraction. *Journal of Natural Fibers*, 8, 149–160.

Je-Lueng, S., Feng-Ju, T., Kae-Long, L. & Ching-Yuan, C. (2010) Bioenergy and products from thermal pyrolysis of rice straw using plasma torch. *Bioresource Technology*, 1010(2), 761–768.

John, R.P., Anisha G.S., Nampoothiri, K.M., & Pandey, A. (2011). Micro and macroalgal biomass: A renewable source for bioethanol. *Bioresource Technology*, 102(1), 186–193.

Kang, Q., Appels, L., Tan, T., & Dewil, R. (2014). Bioethanol from lignocellulosic biomass: Current findings determine research priorities. *The Scientific World Journal*, 2014, 1–13.

Kar, T. & Keles, S. (2013). Fast pyrolysis of chestnut cupulae: Yields and characterization of the bio-oil. *Energy Exploration & Exploitation*, 31, 847–858.

Karagoz, S., Bhaskar, T., Muto, A. & Sakata, Y. (2005). Comparative studies of oil compositions produced from sawdust, rice husk, lignin and cellulose by hydrothermal treatment. *Fuel*, 84, 875–884.

Kassim, M.A., Loh, S.K., Nasrin, A.B., Aziz, A. A. & Som, R. M. (2011). Bioethanol production from enzymatic saccharified empty fruit bunches hydrolysate using Saccharomyces cerevisiae *Research Journal of Environmental Sciences*, 5(6), 573–586.

Kaźmierczak, B., Nam, W.L., Su, M.H., Phang, X.Y., Chong, M.Y., Liew, R.K., Ma, N.L., Lam, S.S., Kutylowska, M., Piekarska, K., Jouhara, H. & Danielewics, J. (2017). Production of bio-fertilizer from microwave vacuum pyrolysis of waste palm shell for cultivation of oyster mushroom (Pleurotus ostreatus). *E3S Web of Conferences*, 22, 00122.

Kim, K. H., et al. (2012). "Influence of pyrolysis temperature on physicochemical properties of biochar obtained from the fast pyrolysis of pitch pine (*Pinus rigida*)." *Bioresource Technology*, 118, 158–162.

Kumar, A.K. & Sharma, S. (2017). Recent updates on different methods of pretreatment of lignocellulosic feedstocks: A review. *Bioresources and Bioprocessing*, 4, 7.

Lahijani, P. and Z. A. Zainal (2011). Gasification of palm empty fruit bunch in a bubbling fluidized bed: A performance and agglomeration study. *Bioresource Technology*, 102(2), 2068–2076.

Li, J., Li, S., Han, B., Yu, M., Li, G. & Jiang, Y. (2013). A novel cost-effective technology to convert sucrose and homocelluloses in sweet sorghum stalks into ethanol. *Biotechnology for Biofuels*, 6(1), 1–12.

Luda, M.P. (2012). Pyrolysis of WEEE plastics. In: Goodship, V. & Stevels, A. (eds.) *Waste Electrical and Electronic Equipment (WEEE) Handbook*. Woodhead Publishing Series in Electronic and Optical Materials, Elsevier, 239–263.

Lyu, G., Wu, S. & Zhang, H. (2015). Estimation and comparison of bio-oil components from different pyrolysis conditions. *Frontiers in Energy Research*, 3(28), 1–11.

Madadian, E., Orsat, V. & Lefsrud, M. (2017). Comparative study of temperature impact on air gasification of various types of biomass in a research-scale down-draft reactor. *Energy & Fuels*, 31, 4045–4053.

Manzanares, P., Negro, M.J., Oliva, J.M., Saez, F., Ballesteros, I., Ballesteros, M., Cara, C., Castro, E., & Ruiz, E. (2011). Different process configurations for bioethanol production from pretreated olive pruning biomass. *Journal of Chemical Technology & Biotechnology*, 86(6), 881–887.

Markou, G., Angelidaki, I., & Georgakakis, D. (2012). Microalgal carbohydrates: An overview of the factors influencing carbohydrates production, and of main bioconversion technologies for production of biofuels. *Applied Microbiology and Biotechnology*, 96(3), 631–645.

Marta, K., Paulina, R., Magda, D., Anna, N., Aleksandra, K., Marcin, D., Kazimierowicz, J. & Marcin, Z. (2020). Evaluation of ultrasound pretreatment for enhanced anaerobic digestion of Sida hermaphrodita. *BioEnergy Research*, 13, 824–832.

Marx, S., Ndaba, B., Chiyanzu, I., & Schabort, C. (2014). Fuel ethanol production from sweet sorghum bagasse using microwave irradiation. *Biomass and Bioenergy*, 65, 145–150.

Mathanker, A., Pudasainee, D., Kumar, A. & Gupta, R. (2020). Hydrothermal liquefaction of lignocellulosic biomass feedstock to produce biofuels: Parametric study and products characterization. *Fuel*, 271, 117534.

Matsumoto, K., Takeno, K., Ichinose, T., Ogi, T. & Nakanishi, M. (2009). Gasification reaction kinetics on biomass char obtained as a by-product of gasification in an entrained-flow gasifier with steam and oxygen at 900–1000°C. *Fuel*, 88, 519–527.

Maurya, D.P., Singla, A. & Negi, S. (2015). An overview of key pretreatment processes for biological conversion of lignocellulosic biomass to bioethanol. *3 Biotech*, 5, 597–609.

Mohan, D., Pittman, C.U. & Steele, P.H. (2006). Pyrolysis of wood/biomass for bio-oil: A critical review. *Energy & Fuels*, 20, 848–889.

Moradi, F., Amiri, H., Soleimanian-Zad, S., Ehsani, M.R., & Karimi, K. (2013). Improvement of acetone, butanol and ethanol production from rice straw by acid and alkaline pretreatments. *Fuel*, 112, 8–13.

Morgan, T.J., Turn, S.Q. & George, A. (2015). Fast pyrolysis behavior of bana grass as a function of temperature and volatiles residence time in a fluidized bed reactor. *PLOS ONE*, 10, e0136511.

Morgan, T.J., Turn, S.Q., Sun, N. & George, A. (2016). Fast pyrolysis of tropical biomass species and influence of water pretreatment on product distributions. *PLOS ONE*, 11, e0151368.

Motavaf, B. & Savage, P.E. (2021). Effect of process variables on food waste valorization via hydrothermal liquefaction. *ACS ES&T Engineering*, 1, 363–374.

Muradov, N., Fidalgo, B., Gujar, A.C. & T-raissi, A. (2010). Pyrolysis of fast-growing aquatic biomass – Lemna minor (duckweed): Characterization of pyrolysis products. *Bioresource Technology*, 101, 8424–8428.

Nasirian, N., Almassi M., Minaei, S., & Widmann, R. (2011). Development of a method for biohydrogen production from wheat straw by dark fermentation. *International Journal of Hydrogen Energy*, 36(1), 411–420.

Noor, N. M., Shariff, A., Abdullah, N. & Aziz, N.S.M. (2019). Temperature effect on biochar properties from slow pyrolysis of coconut flesh waste. *Malaysian Journal of Fundamental and Applied Sciences*, 15, 152–158.

Norshahida, S., Ismail, H., & Ahmad, Z. (2012). Effect of fiber loading on properties of thermoplastic sago starch/kenaf core fiber biocomposite. *BioResource*, 7, 4292–4306.

Nuryana, D., Alim, M.F.R., Loveyanto, R.O., Wibowo, B.A., Sulong, R.S.B.R., Hamzah, M.A.A.M., Zakaria, A.A. & Kusumaningtyas, R.D. (2020). Phenolic compound derived from microwave-assisted pyrolysis of coconut shell: Isolation and antibacterial activity testing. *E3S Web of Conferences*, 202, 10007.

Ofori-Boateng, C., & Lee, K.T. (2014). Ultrasonic-assisted simultaneous saccharification and fermentation of pretreated oil palm fronds for sustainable bioethanol production. *Fuel*, 119, 285–291.

Okoye-Chine, C.G., Moyo, M. & Hildebrandt, D. (2022). The influence of hydrophobicity on Fischer-Tropsch synthesis catalysts. *Reviews in Chemical Engineering*, 38, 477–502.

Onay, M. (2018). Investigation of biobutanol efficiency of *Chlorella* sp. cultivated in municipal wastewater. *Journal of Geoscience and Environment Protection*, 6(10), 40–50.

Onay, O. (2007). Influence of pyrolysis temperature and heating rate on the production of bio-oil and char from safflower seed by pyrolysis, using a well-swept fixed-bed reactor. *Fuel Processing Technology*, 88, 523–531.

Ortigueira, J., Alves, L., Gouveia, L., & Moura, P. (2015). Third generation biohydrogen production by *Clostridium butyricum* and adapted mixed cultures from *Scenedesmus obliquus* microalga biomass. *Fuel*, 153, 128–134.

O-Thong, S., Prasertsan, P., Intrasungkha, N., Dhamwichukorn, S. & Birkeland, N.K. (2013). Improvement of biohydrogen production and treatment efficiency on palm oil mill effluent with nutrient supplementation at thermophilic condition using an anaerobic sequencing batch reactor. *Enzyme and Microbial Technology*, 41(5), 583–590.

Özbay, G. & Özçifçi, A. (2021). Vacuum pyrolysis of woody biomass to bio-oil production. *Politeknik Dergisi*, 24, 1257–1261.

Pendyala, V.R.R., Shafer, W.D., Jacobs, G. & Davis, B.H. (2014). Fischer–Tropsch synthesis: Effect of reaction temperature for aqueous-phase synthesis over a platinum promoted Co/alumina catalyst. *Catalysis Letters*, 144, 1088–1095.

Prasetyo, J., Murti, G.W., Kismanto, A., Murti, S.D.S., Rekso, A., Rahmadi, A. & Ssaputra, H. (2020). Preliminary study on low pressure hydrothermal liquefaction processes of biomass for biofuels: Bio crude oil. *AIP Conference Proceedings*, 2248, 060002.

Raheem, A., Dupont, V., Channa, A.Q., Zhao, X., Vuppaladadiyam, A.K., Taufiq-Yap, Y.H., Zhao, M. & Harun, R. (2017). Parametric characterization of air gasification of *Chlorella vulgaris* biomass. *Energy & Fuels*, 31, 2959–2969.

Raibhole, V.N. & Sapali, S.N. (2012). Simulation and parametric analysis of cryogenic oxygen plant for biomass gasification. *Mechanical Engineering Research*, 2(2), 97–107.

Raud, M., Orupold, K., Rocha Meneses, L., Rooni, V., Trass, O. & Kikas, T. (2020). Biomass pretreatment with the Szego Mill™ for bioethanol and biogas production. *Processes*, 8, 1327.

Reddy, H.K., Muppaneni, T., Ponnusamy, S., Sudasinghe, N., Pegallapati, A., Selvaratnam, T., Seger, M., Dungan, B., Nirmalakhandan, N., Schaub, T., Holguin, F.O., Lammers, P., Voorhies, W. & Deng, S. (2016). Temperature effect on hydrothermal liquefaction of *Nannochloropsis gaditana* and *Chlorella* sp. *Applied Energy*, 165, 943–951.

Robinson, P. K. (2015). Enzymes: Principles and biotechnological applications. *Essays in Biochemistry*, 59, 1.

Ronsse, F., Van-Hecke, S., Dickson, D. & Prins, W. (2013). Production and characterization of slow pyrolysis biochar: Influence of feedstock type and pyrolysis conditions. *GCB Bioenergy*, 5, 104–115.

Rosgaard, L., Andric, P., Johansen, K.D., Pedersen, S., & Meyer, A.S. (2007a). Effects of substrate loading on enzymatic hydrolysis and viscosity of pretreated barley straw. *Applied Biochemistry and Biotechnology*, 143(1), 27–40.

Rosgaard, L., Pedersen, S., Langston, J., Akerhielm, D., Cherry, J.R. & Meyer, A.S. (2007b). Evaluation of minimal *Trichoderma reesei* cellulase mixtures on differently pretreated barley straw substrates. *Biotechnology Progress*, 23, 1270–1276.

Rozenfelde, L., Puke, M., Kruma, I., Poppele, I., Matjuskova, N., Vedernikovs, N. & Rapoport, A. (2017). Enzymatic hydrolysis of lignocellulose for bioethanol production. *Proceedings of the Latvian Academy of Sciences. Section B. Natural, Exact, and Applied Sciences*, 71(4), 275–279.

Sadhwani, N., Adhikari, S. & Eden, M.R. (2016). Biomass gasification using carbon dioxide: Effect of temperature, CO_2/C ratio, and the study of reactions influencing the process. *Industrial & Engineering Chemistry Research*, 55, 2883–2891.

Sahari, J., Sapuan, S., Zainudin, E. S. & Maleque, M. (2012). Sugar palm tree: A versatile plant and novel source for biofibres, biomatrices, and biocomposites. *Polymers from Renewable Resources*, 3(2), 61–77.

Saini, J.K., Saini, R., & Tewari, L. (2015). Lignocellulosic agriculture wastes as biomass feedstocks for second-generation bioethanol production: Concepts and recent developments. *3 Biotech*, 5(4), 337–353.

Sánchez, C. (2009). Lignocellulosic residues: Biodegradation and bioconversion by fungi. *Biotechnology Advances*, 27(2), 185–194.

Sarkar, J. K. & Wang, Q. (2020). Characterization of pyrolysis products and kinetic analysis of waste jute stick biomass. *Processes*, 8, 837–852.

Sauciuc, A., Abosteif, Z., Weber, G., Potetz, A., Rauch, R., Hofbauer, H., Schaub, G. & Dumitrecu, L. (2012). Influence of operating conditions on the performance of biomass-based Fischer–Tropsch synthesis. *Biomass Conversion and Biorefinery*, 2, 253–263.

Seebauer, V., Petek, J. & Staudinger, G. (1997). Effects of particle size, heating rate and pressure on measurement of pyrolysis kinetics by thermogravimetric analysis. *Fuel*, 76, 1277–1282.

Sekoai, P. T., Ayeni, A.O., & Daramola, M.O. (2019). Parametric optimization of biohydrogen production from potato waste and scale-up study using immobilized anaerobic mixed sludge. *Waste and Biomass Valorization*, 10(5), 1177–1189.

Shi, J., Chinn, M.S. & Sharma-Shivappa, R.R. (2008). Microbial pretreatment of cotton stalks by solid state cultivation of *Phanerochaete chrysosporium. Bioresource Technology*, 99, 6556–6564.

Shimizu, F.L., Monteiro, P.Q., Ghiraldi, P.H.C., Melati, R.B., Pagnocca, F.C., Souza, W.D., Sant'anna, C. & Brienzo, M. (2018). Acid, alkali and peroxide pretreatments increase the cellulose accessibility and glucose yield of banana pseudostem. *Industrial Crops and Products*, 115, 62–68.

Shuangning, X., Weiming, Y. & Li, B. (2005). Flash pyrolysis of agricultural residues using a plasma heated laminar entrained flow reactor. *Biomass and Bioenergy*, 29, 135–141.

Sikarwar, V.S., Zhao, M., Fennell, P.S., Shah, N. & Aanthony, E.J. (2017). Progress in biofuel production from gasification. *Progress in Energy and Combustion Science*, 61, 189–248.

Sindhu, R., Kuttiraja, M., Binod, P., Sukumaran, R.K., & Pandey, A. (2014). Bioethanol production from dilute acid pretreated Indian bamboo variety (*Dendrocalamus* sp.) by separate hydrolysis and fermentation. *Industrial Crops and Products*, 52, 169–176.

Singh, R., Prakash, A., Balagurumurthy, B., Singh, R., Saran, S. & Bhaskar, T. (2015). Hydrothermal liquefaction of agricultural and forest biomass residue: Comparative study. *Journal of Material Cycles and Waste Management*, 17, 442–452.

Singh, R., Tiwari, S., Srivastava, M. & Shukla, A. (2014). Microwave assisted alkali pretreatment of rice straw for enhancing enzymatic digestibility. *Journal of Energy*, 2014, 483813.

Soto, M.L., Dominguez, H., Nunez, M.J. & Lema, J.M. (1994). Enzymatic saccharification of alkali-treated sunflower hulls. *Bioresource Technology*, 49, 53–59.

Stickel, J.J., Elander, R.T., McMillan, J.D. & Brunecky, R. (2014). Enzymatic hydrolysis of lignocellulosic biomass. In: Bisaria, B.S., & Kondo, A. (eds) *Bioprocessing of Renewable Resources to Commodity Bioproducts*. John Wiley & Son, New Jersey, 77–103.

Sukiran, M.A., Loh, S.K., & Nasrin, A.B. (2016). Production of bio-oil from fast pyrolysis of oil palm biomass using fluidised bed reactor. *Journal of Energy Technologies and Policy*, 6, 52–62.

Sun, R., Lawther, J.M. & Banks, W.B. (1995). Influence of alkaline pre-treatments on the cell wall components of wheat straw. *Industrial Crops and Products*, 4, 127–145.

Tamošiūnas, A., Valatkevicius, P., Valincius, V. & Levinskas, R. (2016.) Biomass conversion to hydrogen-rich synthesis fuels using water steam plasma. *Comptes Rendus Chimie*, 19, 433–440.

Tanger, P., Field, J., Jahn, C., Defoort, M. & Leach, J. (2013). Biomass for thermochemical conversion: Targets and challenges. *Frontiers in Plant Science*, 4(218), 1–20.

Thangalazhy-Gopakumar, S. & Adhikari, S. (2016). Fast pyrolysis of agricultural wastes for bio-fuel and bio-char. In: Karthikeyan, O., Heimann, K. & Muthu, S. (eds) *Recycling of Solid Waste for Biofuels and Bio-chemicals. Environmental Footprints and Eco-design of Products and Processes*. Singapore: Springer, 301–332.

Tran, T.T.A., Le, T.K.P., Mai, T.P., Nguyen, D.Q. (2019). Bioethanol production from lignocellulosic biomass. In: Yun, S (ed) *Alcohol Fuels-Current Technologies and Future Prospect*. IntechOpen, London, 1–13.

Tu, W.K., Shie, J.L., Chang, C.Y., Chang, C.F., Lin, C.F., Yang, S.Y., Kuo, J.T., Shaw, D.G., You, Y.D. & Lee, D.J. (2009). Products and bioenergy from the pyrolysis of rice straw via radio frequency plasma and its kinetics. *Bioresource Technology*, 100, 2052–2061.

Urban, B., Shirazi, Y., Maddi, B., Viamajala, S. & Varanasi, S. (2017). Flash pyrolysis of oleaginous biomass in a fluidized-bed reactor. *Energy & Fuels*, 31, 8326–8334.

Valero-Romero, M.J., Rodriguez-Cano, M.Á., Palomo, J., Rodriguez-Mirasol, J. & Cordero, T. (2021). Carbon-based materials as catalyst supports for Fischer–Tropsch Synthesis: A review. *Frontiers in Materials*, 7, 1–27.

Van Hecke, W., Meyvis, E.J., Beckers, H., & Wever, H.D. (2018). Prospects & potential of biobutanol production integrated with organophilic pervaporation–A techno-economic assessment. *Applied Energy*, 228, 437–449.

Verma, M., Godbout, S., Brar, S.K., Solomatnikova, O., Lemay, S.P. & Larouche, J.P. (2012). Biofuels production from biomass by thermochemical conversion technologies. *International Journal of Chemical Engineering*, 2012, 542426.

Waghmare, P.R., Khandare, R.V., Jeon, B.H. & Govindwar, S.P. (2018). Enzymatic hydrolysis of biologically pretreated sorghum husk for bioethanol production. *Biofuel Research Journal*, 5, 846–853.

Wang, K., Brown, R.C., Homsy, S., Martinez, L. & Sidhu, S.S. (2013). Fast pyrolysis of microalgae remnants in a fluidized bed reactor for bio-oil and biochar production. *Bioresource Technology*, 127, 494–499.

Wang, M., Zhang, S.L. & Duan, P.G. (2019). Slow pyrolysis of biomass: Effects of effective hydrogen-to-carbon atomic ratio of biomass and reaction atmospheres. *Energy Sources, Part A: Recovery, Utilization, and Environmental Effects*, 45(1), 1–14.

Wang, Y., Guo, W., Cheng, C.L., Ho, S.H., Chang, J.S., & Ren, N. (2016). Enhancing bio-butanol production from biomass of *Chlorella vulgaris* JSC-6 with sequential alkali pretreatment and acid hydrolysis. *Bioresource Technology*, 200, 557–564.

Wei, X. & Jie, D. (2018). Optimization to hydrothermal liquefaction of low lipid content microalgae *Spirulina* sp. using response surface methodology. *Journal of Chemistry*, 2018, 2041812.

Wi, S.G., Cho, E. J., Lee, D.S., Lee, S. J., Lee, Y. J. & Bae, H.J. (2015). Lignocellulose conversion for biofuel: A new pretreatment greatly improves downstream biocatalytic hydrolysis of various lignocellulosic materials. *Biotechnology for Biofuels*, 8, 228.

Williams, P. T. & Besler, S. (1996). The influence of temperature and heating rate on the slow pyrolysis of biomass. *Renewable Energy*, 7, 233–250.

Wongsiriamnuay, T., Kannang, N. & Tippayawong, N. (2013). Effect of operating conditions on catalytic gasification of bamboo in a fluidized bed. *International Journal of Chemical Engineering*, 2013, 297941.

Xie, Q., Addy, M., Liu, S., Zhang, B., Cheng, Y., Wan, Y., Li, Y., Liu, Y., Lin, X., Chen, P. & Ruan, R. (2015). Fast microwave-assisted catalytic co-pyrolysis of microalgae and scum for bio-oil production. *Fuel*, 160, 577–582.

Xie, Q., Peng, P., Liu, S., Min, M., Cheng, Y., Wan, Y., Li, Y., Lin, X., Liu, Y., Chen, P. & Ruan, R. (2014). Fast microwave-assisted catalytic pyrolysis of sewage sludge for bio-oil production. *Bioresource Technology*, 172, 162–168.

Yoon, S.J., Son, Y.I., Kim, Y.-K. & Lee, J.-G. (2012). Gasification and power generation characteristics of rice husk and rice husk pellet using a downdraft fixed-bed gasifier. *Renewable Energy*, 42, 163–167.

Yuan, Z., Wen, Y. & Kapu, N.S. (2018). Ethanol production from bamboo using mild alkaline pre-extraction followed by alkaline hydrogen peroxide pretreatment. *Bioresource Technology*, 247, 242–249.

Yusoff, M.N.A.M., Zulkifli, N.W.M., Sukiman, N.L., Chyuan, O. H., Hasnul, M. H., Zulkifli, M. S. A., Abbas, M. M. & Zakaria, M. F. (2020). Sustainability of palm biodiesel in transportation: a review on biofuel standard, policy and international collaboration between Malaysia and Colombia. *BioEnergy Research*, 14, 43–60.

Zamani, Y., Rahimizadeh, M. & Seyedi, S.M. (2016). The effect of temperature on product distribution over Fe-Cu-K catalyst in Fischer-Tropsch synthesis. *Journal of Petroleum Science and Technology*, 6, 46–52.

Zhang, K., Ren, N.Q., & Wang, A.J. (2014). Enhanced biohydrogen production from corn stover hydrolyzate by pretreatment of two typical seed sludges. *International Journal of Hydrogen Energy*, 39(27), 14653–14662.

Zhang, Q., Wang, T., Wu, C., Ma, L. & Xu, Y. (2010). Fractioned preparation of bio-oil by biomass vacuum pyrolysis. *International Journal of Green Energy*, 7, 263–272.

Zhang, S., Yang, X., Zhang, H., Chu, C., Zheng, K., Ju, M. & Liu, L. (2019). Liquefaction of biomass and upgrading of bio-oil: A review. *Molecules*, 24(12), 2250.

Zhang, Y., Chen, P., Liu, S., Fan, L., Zhou, N., Min, M., Cheng, Y., Peng, P., Anderson, E., Wang, Y., Wan, Y., Liu, Y., Li, B. & Ruan, R. (2017). Microwave-assisted pyrolysis of biomass for bio-oil production. In: Samer, M (ed) *Pyrolysis*. IntechOpen, London, 129–166.

Zhao, T., Tashiro, Y., Zheng, J., Sakai, K., & Sonomoto, K. (2018). Semi-hydrolysis with low enzyme loading leads to highly effective butanol fermentation. *Bioresource Technology*, 264, 335–342.

Zhao, Y., Wang, Y., Zhu, J.Y., Ragauskas, A. & Deng, Y. (2008). Enhanced enzymatic hydrolysis of spruce by alkaline pretreatment at low temperature. *Biotechnology and Bioengineering*, 99, 1320–1328.

Zhu, Z., Rosendahl, L., Toor, S.S., Yu, D. & Chen, G. (2015). Hydrothermal liquefaction of barley straw to bio-crude oil: Effects of reaction temperature and aqueous phase recirculation. *Applied Energy*, 137, 183–192.

5 Review on the Current Updates on Palm Oil Industry and Its Biomass Recycling to Fertilizer in Malaysia

Rozelyn Ignesia Raymond and Khim Phin Chong
Universiti Malaysia Sabah

CONTENTS

INTRODUCTION

Oil palm was first cultivated in Western Africa, where it was considered an indigenous plant that could be used for a variety of uses, and the plantation has since spread throughout the world, particularly in South America and Asia (Onoja et al., 2019). In 1763, Jacquin gave the oil palm the botanical name *Elaeis guineensis Jacq.* where the *Elaeis* comes from the Greek word elaion, which means "oil," and the specific name *guineensis* indicates that Jacquin assigned it to the Guinea coast (Goh et al., 2016). *E. guineensis* is a single-stemmed, erect palm tree that grows to be 20–30 m tall with an adventitious root system where it forms a dense mat on top of the soil (35 cm); the stem is cylindrical (75 diameter) and coated with petiole bases in young palms while smooth in mature trees ranging from 10 to 12 years (Dijkstra, 2015). *E. guineensis* is a tropical, perennial crop grown primarily for vegetable oil, which consists of palm oil and kernel oil, with palm oil obtained from the mesocarp and kernel oil obtained from the endosperm, as shown in Figure 5.1 (Goh et al., 2016).

The palm oil plant is currently one type of plantation crops that occupy an important position in the agricultural sector, and the plantation sector in particular. Oil palm is considered the largest producer of oil or fats in the world (Widians et al., 2019). Palm oil is one of the most widely used vegetable oils in the world, accounting for one-quarter of worldwide consumption and roughly 60% of international vegetable oil trade with food products usage estimated around 74% globally while industrial usage accounts 24% (Ferdous Alam et al., 2015). The palm oil has become the principal cooking oil for the majority of people in Asia, Africa, and the Middle East due to its significantly higher oil output compared to other oilseeds. The oil palm produces four to seven times more oil than rapeseed and soy, and it also offers a more economical price for consumers (Pirker et al., 2016). Although there was an impact from the increased demand of bio diesel in European countries which resulted in an increase in price for other competitive vegetable oil such as rapeseed oil, soy oil, and

DOI: 10.1201/9781003358084-5

FIGURE 5.1 Cross section of the mature fruit showing two sources of vegetable oil (palm oil from mesocarp and palm kernel oil from endosperm or kernel).

palm oil, palm oil remains the most affordable oil compared to other vegetable oils still remain the cheapest vegetable oil. Palm oil is currently the "marginal oil;" in future, increasing demand for vegetable oils will predominantly be met by palm oil rather than other vegetable oils (Schmidt and Weidema, 2008; Pirker et al., 2016).

Oil palm-planted area in Malaysia has shown dramatic growth, and most of the expansion took place due to the agricultural transformation and declining availability of suitable land, which leads to rubber land converted into oil palm during 1960s and 1970s (Nambiappan et al., 2018). In the year 2021, the oil palm-planted area in Malaysia has decreased to 5.74 million ha that shows 2.2% reduction from 2020 (5.87 million ha) due to the spread of COVID-19 pandemic (Parveez et al., 2022). In countries, such as Malaysia and Indonesia, oil palm is a major cash crop with Indonesia being world's largest producer accounting for 57% followed by Malaysia (26%) and Thailand (3%). Both Malaysia and Indonesia account for more than 80% global output (Maluin et al., 2020; Aziz et al., 2021). The palm oil's demand has increased particularly in India and China as a result of its high productivity at a lower cost than other vegetable oils (Zulkifli et al., 2017).

Currently, 71% of Malaysia's cultivable land is under palm oil cultivation (Begum et al., 2019; Mohd-Azlan, 2020); considering the rapid growth of oil palm plantation has its downsides, one of which being the massive amount of biomass create (Onoja et al., 2019) that caused environmental problems with a detrimental effect on the environment (Hassan et al., 2019). The solid biomass wastes generated from the oil palm sector are mainly from the plantations and mills, and the wastes include empty fruit bunches (EFBs), palm kernel shell (PKS), mesocarp fiber (MF), palm oil mill effluent (POME), oil palm trunks (OPTs), oil palm leaves (OPLs), and oil palm fronds (OPFs) (Abdullah and Sulaiman, 2013; Awalludin et al., 2015). Palm oil biomass has been used in the production of pellets, bio-briquettes, bio-char, plywood, and other value-added products, as well as in the generation of power, that makes palm oil biomass is currently in the spotlight since the oil palm business has been identified as a vital indicator of the country's economic performance (Ng et al., 2012; Chong et al., 2017). Since expanding oil palm sector, Malaysia has led to growing concerns over the oil palm biomass waste where it should be disposed in a sustainable way, and one of the options for solving this problem is to employ oil palm biomass waste as a renewable energy source (Chee et al., 2019). Therefore, this paper aims to provide a review on the current updates on palm oil industry and its biomass recycling to fertilizer in Malaysia.

OIL PALM INDUSTRY IN MALAYSIA

The British introduced the oil palm tree to Malaysia as an ornamental plant in 1871, with the creation of the first oil plantation at Tenamaran Estate in Selangor in 1917; around a century ago, Malaysia began to cultivate oil palm plants on a large scale (Onoja et al., 2019). Palm oil is one of the most widely used vegetable oils in the world, accounting for one-quarter of worldwide consumption (Ferdous Alam et al., 2015); for the food products, usage is estimated around 74% globally while industrial uses account for 24% (Begum et al., 2019). According to Department of Statistic Malaysia (2022), Malaysia's land area was 330,524 km^2 that covers 14 states with a population density of 98 people km^2 (DOSM, 2022). The equatorial climate of Malaysia was influenced by the alternating northeast and southwest monsoons, which produce rain and dry weather, respectively (Tang and Al Qahtani, 2020). Malaysia environment is conducive to the growth of oil palm trees; without a question, it has become Malaysia's most significant agricultural crop and has played a key role in the country's economic development (Awalludin et al., 2015). Oil palm-planted area in Malaysia has shown dramatic growth from a mere 55,000 ha in 1960 to 5.74 million ha in 2016 due to the agricultural transformation and setback of rubber prices (Nambiappan et al., 2018). In comparison to 2017, Malaysia's oil palm industry had a poor year in 2018; the FFB yield, CPO production, and palm oil exports dropped, while palm oil imports rose, due to weaker vegetable oil price has dragged down the CPO price which reduced the export revenue (Kushairi et al., 2019) (Tables 5.1 and 5.2).

PALM OIL WASTE IN MALAYSIA

All organic matters or substances produced by agriculture, forests, or marine life are referred to as biomass (Awalludin et al., 2015). Agricultural waste comes from organic compounds discarded by people during agricultural operations, primarily plant waste, livestock and poultry manure, farm and side line waste management materials, and rural residential waste (Dai et al., 2018). Oil palm biomass refers to agricultural by-products produced during replanting, pruning, and milling activities by the oil palm industry, which are left to decompose in the fields in most cases (Onoja et al., 2019). These are non-product outputs of agricultural product production and processing that may contain material beneficial to humans, but whose economic values are lower than the cost of selecting, transporting, and processing agricultural products for the benefit of their intended users (Obi et al., 2016). Oil palm planting in Malaysia is one of the world's top-producing fruit crops for oil palm biomass (Hosseini et al., 2015; Parveez et al., 2021). Improper handling of this residue and bio-waste can cause environmental and health problems by contributing to eutrophication, pollution, and other types of disruptions in both aquatic and terrestrial life (Maluin et al., 2020).

The oil palm industry produces lignocellulosic biomass such as OPTs, OPFs, EFB and palm-pressed fibers (PPFs), palm shells, and POME (Abdullah and Sulaiman, 2013) as shown in Figure 5.2. The solid biomass wastes in the oil palm sector are created in two ways; first, it comes from the oil palm plantations, where trash is collected in the form of harvested trunks and fronds

TABLE 5.1

Malaysian Oil Palm Industry Performance 2020 (Parveez et al., 2021)

	2021	2020	Difference Vol./Value	%
Planted area (mil hectares)	5.74	5.87	(0.13)	(2.20)
CPO production (mil tons)	18.12	19.14	(1.02)	(5.30)
FFB yield (t ha^{-1})	15.47	16.73	−1.26	−7.50
Oil extraction rate (%)	20.01	19.92	0.09	0.50
PO exports (mil tons)	14.84	16.22	−1.38	−8.50
PO imports (mil tons)	1.18	0.95	0.23	24.30
Closing stocks (mil tons)	1.60	1.27	0.33	0.26
CPO price (RM t^{-1})	4 407.00	2, 685.50	1 721.50	64.10
Export revenue (RM billion)	108.52	73.25	35.27	48.00

Note: CPO, crude palm oil; FFBs, fresh fruit bunches; PO, palm oil; mill, million.

TABLE 5.2

Malaysian Oil Palm Industry Performance 2021 (Parveez et al., 2022)

	2020	2019	Difference Vol./Value	%
Planted area (mil hectares)	5.87	5.90	0.03	0.60
CPO production (mil tons)	19.14	19.86	(0.72)	(3.60)
FFB yield (t ha^{-1})	16.73	17.19	(0.46)	(2.70)
Oil extraction rate (%)	19.92	20.21	(0.29)	(1.40)
PO exports (mil tons)	16.22	17.43	(7.00)	16.70
PO imports (mil tons)	0.95	0.98	−0.03	(3.10)
Closing stocks (mil tons)	1.27	2.01	(0.74)	(37.00)
CPO price (RM t^{-1})	2, 685.50	2, 079.00	606.50	29.20
Export revenue (RM billion)	73.25	67.55	5.70	8.40

Note: CPO, crude palm oil; FFBs, fresh fruit bunches; PO, palm oil; mill, million.

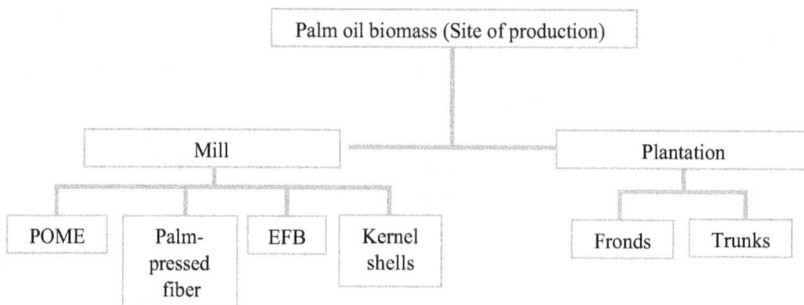

FIGURE 5.2   Oil palm biomass residue and source of generation.

that have been clipped while the second source comes from palm oil extraction mills, which include EFB, mesocarp fiber, and PKS (Awalludin et al., 2015). As one of the world's leading producers and exporters of palm oil, Malaysia's palm oil industry produces massive amounts of palm oil waste, which is discarded as effluent from EFB, medium fuel oil, oil palm shell, and palm oil mills, and is often left in the palm oil mill's vicinity to decompose naturally (Mushtaq et al., 2015). In an oil palm plantation, the extracted oil accounts for just 10% of the total, while the remaining 90% is deemed waste, resulting in an average of 50–70 tons of biomass residues, complicating conventional waste management techniques (Kurnia et al., 2016).

UTILIZATION OF PALM OIL WASTE

Satisfying the core sustainability criteria has been a challenging task, and sustainability developments in the oil palm value chain have been one of the industry's key concerns (Parveez et al., 2020). Knowing the chemical makeup of this biomass is now necessary in order to unlock its hidden potentials, which leads to correct utilization. The enormous volume of biomass generated in oil palm farms and mills is a significant source of concern (Onoja et al., 2019). The utilization of oil palm biomass could be used in various applications such as cultivation of straw mushroom, animal feed, mulching, production of compost, organic fertilizer, hard fiberboard, biomass fuel, and production of biogas fuel (Phoochinda, 2020). The potential of palm oil industry waste for different purposes is shown in Table 5.3.

The palm oil industry's prospects and ideas in terms of long-term viability and practicality should be explored where palm oil can be used in the energy industry for heat/power generation and as a biofuel in the transportation sector (Kaniapan et al., 2021). POME which is high in organic matter

TABLE 5.3
Utilization of Palm Oil Industry Waste

Materials	Benefits	References
POME	Biogas production	Chin et al. (2013); Sari et al. (2019)
	Treated POME used as liquid fertilizer	Haryati and Theeba (2021)
	Bioplastic and bioethanol production	Salihu and Alam (2012)
EFBs	Mulch and fertilizer substitute	Banjarnahor et al. (2020)
	Renewable feedstock (biofuel and bioethanol production)	Sudiyani et al. (2013); Derman et al. (2018); James Rubinsin et al. (2020)
	Cellulose nanofiber production	Supian et al. (2020)
	Bioplastic production	Yang et al. (2021)
PPF	Vermicompost and organic fertilizer	Rupani et al. (2019)
	Alternative carrier oil in emulsion	Teh et al. (2019)
	Livestock feed	Sundalian et al. (2021)
PKS	Supercapacitor electrode application	Misnon et al. (2019)
	Used as an aggregate in concrete and laterite blocks	Raju and Ramakrishnan (1972); Ikumapayi and Akinlabi (2018)
	Water purification	Edmund et al. (2014)
	Fuel and biodiesel production	Abdullah et al. (2020); Ikumapayi and Akinlabi (2018)
Decanter cake	Feedstock for biodiesel production	Maniam et al. (2013)
	Soil condition, organic fertilizer	Embrandiri et al. (2017); Embrandiri et al. (2016)
	Enzyme production	H-Kittikun et al. (2021)
	Biogas fermentation	Kanchanasuta and Pisutpaisal (2016)
OPFs and OPLs	Animal feeds	Puastuti (2017)
	Mulching and soil amendment	Pulunggono et al. (2019)

content has been used to make biogas, and the technology has been commercialized in the Asia Pacific region, while EFB is one of the most sought after oil palm biomass waste; it may be used to make bio-alcohol, solid fuel, pulp, and a variety of other high-value products (Aljuboori, 2013). The bulk of decanter cake (DC) produced is currently used as fertilizer and soil cover materials in plantation areas or for biogas production; because to its acidic composition, DC is used in conjunction with inorganic fertilizers, and it has a synergistic impact that improves crop nutrient (Embrandiri et al., 2015). Oil palm lignocellulosic fibers, such as hemicellulose, lignin, and especially cellulose, could be used in nanotechnology, the pulp fiber from the oil palm fiber is mechanically treated to form a network structure unit such as nano-sized mesh called cellulose microfibril, which is obtained through smoothing and high-pressure homogenizer processes (Dungani et al., 2018). The most abundant non-edible biomass is lignocellulosic biomass that is a source of value-added chemicals made from oil palm biomass, which might provide additional cash for the country and lead to a variety of uses for oil palm biomass, such as solid biofuels and biofertilizer (Noorshamsiana et al., 2017; Wang et al., 2017).

CONVERSION OF PALM OIL WASTE INTO FERTILIZER

Fertilization is the practice of giving plants nutrients to help them develop and produce high-quality fruit, where the nutrients that are transported during harvest can be restored by fertilizing which is the key to soil fertility (Purba et al., 2020). A total of 17 elements are necessary for the plants to reach their full potential that consists of carbon (C), hydrogen (H), and oxygen (O) that were obtained from the water and air, whereby the other nutrients are obtained from the earth (Griengo et al., 2020). Macronutrients and micronutrients have an important role in disease control and management, making them a vital component for optimal plant growth and development (Nadeem et al., 2018). Macronutrients utilize primary nutrients (nitrogen (N), phosphorus (P), and potassium (K)) in relatively significant amounts, and they are frequently supplemented with fertilizers, while secondary nutrients such as calcium (Ca), magnesium (Mg), and sulfur (S) are used in high amounts, yet are usually readily available (Griengo et al., 2020). Terms like micronutrients and trace elements were coined to describe nutrients that were only necessary in trace amounts in the physiology of the organism that are vital plant components that are only required in trace amounts consisting of iron (Fe), zinc (Zn), molybdenum (Mo), manganese (Mn), boron (B), copper (Cu), cobalt (Co), and chlorine (Cl) (Nadeem et al., 2018; Griengo et al., 2020; Brown et al., 2021).

Prior to now, palm biomass was only disposed of through open burning, incineration, or waste ponds, all of which contributed to the global climate change by emitting greenhouse gases, and today, palm biomass appears to be one promising raw material for the production of fertilizer as a win–win strategy for sustainable development. (Chong et al., 2017). Rather than discarding these materials, the business can make a full use of them as a renewable nutrient supply for a sustainable oil palm cultivation, whereby converting oil palm waste materials into compost has sparked a lot of interest in Malaysia (Yi et al., 2019). Current agricultural waste management strategies include using EFB as mulch and fertilizer for soil conditioning; direct application, on the other hand, was unable to remedy the problem of EFB excess (Wahi and Yusup, 2016). Many researches have looked into the use of solid oil palm wastes, as well as the use of EFB waste as a fertilizer alternative (Dungani et al., 2018). Both the public and private sectors have recently paid attention to the conversion of palm biomass into fertilizers since the need for input fertilizers has increased as a result of Malaysia's agriculture industry's rapid expansion (Chong et al., 2017).

Many oil palm mills in Malaysia are turning to bioconversion of EFB and POME into compost as part of their waste management strategy, with the goal of reducing waste discharge into rivers while also restoring nutrients to the plant nutrition cycle (Hoe et al., 2016). POME and EFB both are treated or used as fertilizer where it offers a full nutritional profile (Table 5.4) since it contains both macro- and micronutrients in tiny amounts, that helps to keep the soil nutrients in balance (Rangkuti et al., 2018). Studies also show that within 10 weeks, the composting process increased

TABLE 5.4

Mean Chemical Composition of OPF, EFBs, OPT, and Raw POME

Property (%)	OPF	EFB	OPT	POME
C	42.10	42.80	34.14	31.50
N	0.60	0.90	0.65	4.70
P	0.10	0.16	0.09	0.38
K	1.10	2.59	2.50	0.37
Ca	0.60	0.46	0.26	0.19
Mg	0.10	0.37	0.09	0.25

Sources: Sung (2016), Rosenani et al. (2016), Ooi et al. (2017), Agida et al. (2020), Windiastuti et al. (2022), Siang et al. (2022).

Note: Dry weight basis is in percentages.

nutrient content, particularly in the compost formed from EFB, POME, and oil palm decanter cake with a ratio of 1:3:0.2, which was the most optimum compost with a low C/N ratio of 23:64, pH 8.4, and high nutrient content, revealing it as a potential organic fertilizer source (Adam et al., 2016).

According to a recent study, temperature and a variety of organic matters play a critical role in achieving a successful composting process where the experiment reveals that without active aeration, a mixture of EFB, POME, and recycle compost (10:1:5) may achieve thermophilic temperatures and maintain them for 20 days (Alkarimiah and Suja, 2020). In another study, compost made up from EFB mixed with activated liquid organic fertilizer for 40 days had a positive result on pH 9.0; MC 52.59%; WHC 76%; C/N ratio 12:15; N 1.96%; P 0.58%; and K 0.95% that was used to grow cactus, sansevieria, and anthurium, where the planting medium consists of compost–sand–husk rice mixture with the best ratio of 1:3:1; 1:1:1; and 1:0:1, respectively (Trisakti et al., 2018). Hau and colleagues (2020) suggested that EFB with POME (1:1) combination based fertilizer and composted with various organic wastes to enhance its nutrient content where the result in studies shows that mixed composting of base fertilizer combining with fishmeal as N source, bonemeal for P source, and bunch ash as source in the presence of 50 *Eisenia Fetida* increased in N, P, and K nutrients, which the composting took about 40 days to reach maturity with a pH of 6.3 that provides a porous structure that allows for aeration, high mass yields of 88%–90%, and moisture contents ranging from 55% to 70% (Hau et al., 2020).

CONCLUSION AND FUTURE PROSPECT

Oil palm is a high-value crop that contributes to the growth of Malaysian economies. The growing demand for oil palm on the worldwide market has resulted in an increase in plantations that is up to 60% of agriculture land were used by this crop. Despite the fact that palm oil offers a variety of uses, the palm oil business has received a lot of negative attention in recent years, as a result of various environmental and social issues. Although the oil palm industry waste is associated with number of challenges, it also gives a number of possibilities in terms of products obtained from it especially being a renewable energy to help nation development. Oil palm biomass contains large amounts of nutrients, and the nutrient composition shown from previous studies has unveiled that this biomass could be used as bio-compost and organic fertilizer which helps in soil conditioning, thus reducing the usage of inorganic fertilizer in agriculture sectors and at the same time reducing the environmental impact. Recent studies also prove mixing palm oil waste with other organic wastes could accelerate the composting process and increase the macronutrients and micronutrients in the compost for growing medium especially in horticulture plants depending on the ratio needed. Therefore, in order to achieve optimal production of palm oil in a more sustainable manner, bioconversion of palm oil wastes needs to be a strong plan and policy to be applied in the future.

REFERENCES

Abdullah, N., & Sulaiman, F. (2013). The oil palm wastes in Malaysia. In Matovic, M. D. (ed) *Biomass Now - Sustainable Growth and Use*. Intech Open, London, 75–100.

Abdullah, R. F., Rashid, U., Taufiq-Yap, Y. H., Ibrahim, M. L., Ngamcharussrivichai, C., & Azam, M. (2020). Synthesis of bifunctional nanocatalyst from waste palm kernel shell and its application for biodiesel production. *RSC Advances, 10*(45), 27183–27193.

Adam, S., Syd Ahmad, S. S. N., Hamzah, N. M., & Darus, N. A. (2016). Composting of empty fruit bunch treated with palm oil mill effluent and decanter cake. In Yacob, N.A., Mohamed, M., & Hanafiah, M. A. K. M (eds) *Regional Conference on Science, Technology and Social Sciences (RCSTSS 2014)*, Springer, Singapore, 437–445.

Agida, C. A., Amaduruonye, W., Nsa, E. E., & Nathaniel, J. (2020). Serum biochemistry, haematological profile and organ proportion of broiler starter chicks fed graded levels of palm oil mill effluent (POME). *Journal of Animal Science and Veterinary Medicine, 5*(6), 202–211.

Aljuboori, A. H. R. (2013). Oil palm biomass residue in Malaysia: Availability and sustainability. *International Journal of Biomass & Renewables, 2*, 13–18.

Alkarimiah, R., & Suja, F. (2020). Composting of EFB and POME using a step-feeding strategy in a rotary drum reactor: The effect of active aeration and mixing ratio on composting performance. *Polish Journal of Environmental Studies, 29*(4), 2543–2553.

Awalludin, M. F., Sulaiman, O., Hashim, R., & Nadhari, W. N. A. W. (2015). An overview of the oil palm industry in Malaysia and its waste utilization through thermochemical conversion, specifically via liquefaction. *Renewable and Sustainable Energy Reviews, 50*, 1469–1484.

Aziz, N. F., Chamhuri, N., & Batt, P. J. (2021). Barriers and benefits arising from the adoption of sustainable certification for smallholder oil palm producers in Malaysia: A systematic review of literature. *Sustainability, 13*(18), 10009.

Banjarnahor, L. R., Rahmah, S., Damanik, M., & Zubir, M. (2020). Synthesis of Fe and Zn organic fertilizer from palm oil waste. *Indonesian Journal of Chemical Science and Technology (IJCST), 3*(2), 57.

Begum, H., Choy, E. A., Alam, A. S. A. F., Siwar, C., & Ishak, S. (2019). Sustainability of Malaysian oil palm: A critical review. *International Journal of Environment and Sustainable Development, 18*(4), 387–408.

Brown, P. H., Zhao, F. J., & Dobermann, A. (2021). What is a plant nutrient? Changing definitions to advance science and innovation in plant nutrition. *Plant and Soil, 476*(1–2), 11–23.

Chee, K. Y., Chee, W. Y., Peng, S. H. T., Sinniah, U. R., Leow, C. S., He, Y., & Ng, W. K. (2019). Oil palm biomass wastes as renewable energy sources in Malaysia: Potentials and challenges. *Global Journal of Civil and Environmental Engineering, 1*, 20–24.

Chin, M. J., Poh, P. E., Tey, B. T., Chan, E. S., & Chin, K. L. (2013). Biogas from palm oil mill effluent (POME): Opportunities and challenges from Malaysia's perspective. *Renewable and Sustainable Energy Reviews, 26*, 717–726.

Chong, M. Y., Ng, W. P. Q., Ng, D. K. S., Lam, H. L., Lim, D. L. K., & Law, K. H. (2017). A mini review of palm based fertiliser production in Malaysia. *Chemical Engineering Transactions, 61*, 1585–1590.

Dai, Y., Sun, Q., Wang, W., Lu, L., Liu, M., Li, J., Yang, S., Sun, Y., Zhang, K., Xu, J., Zheng, W., Hu, Z., Yang, Y., Gao, Y., Chen, Y., Zhang, X., Gao, F., & Zhang, Y. (2018). Utilizations of agricultural waste as adsorbent for the removal of contaminants: A review. *Chemosphere, 211*, 235–253.

Department of Statistic Malaysia (DOSM). (2022). Basic Information Statistics. https://www.dosm.gov.my/v1/index.php?r=column/cthree&menu_id=YU9jTGdWVlNGMkVJMzkwV3dTNTNxdz09. Accessed July 14, 2022. Last updated July 14, 2022.

Derman, E., Abdulla, R., Marbawi, H., & Sabullah, M. K. (2018). Oil palm empty fruit bunches as a promising feedstock for bioethanol production in Malaysia. *Renewable Energy, 129*(Part A), 285–298.

Dijkstra, A. J. (2015). Palm oil. *Encyclopedia of Food and Health*, 199–204.

Dungani, R., Aditiawati, P., Aprilia, S., Yuniarti, K., Karliati, T., Suwandhi, I., & Sumardi, I. (2018). Biomaterial from Oil Palm Waste: Properties, Characterization and Applications. In Waisundara, V. (ed) *Palm Oil*. Intech Open, London, 31–51.

Edmund, C. O., Christopher, M. S., & Pascal, D. K. (2014). Characterization of palm kernel shell for materials reinforcement and water treatment. *Journal of Chemical Engineering and Materials Science, 5*(1), 1–6.

Embrandiri, A., Quaik, S., Rupani, P. F., Srivastava, V., & Singh, P. (2015). Sustainable utilization of oil palm wastes: Opportunities and challenges. In Singh, R. P., & Sarkar, A (eds) *Waste Management: Challenges, Threats and Opportunities*, Nova Science Publishers, Inc, New York, 217–232.

Embrandiri, A., Rupani, P. F., Ismail, S. A., Singh, R. P., Ibrahim, M. H., & Kadir, M. O. b. A. (2016). The effect of oil palm decanter cake on the accumulation of nutrients and the stomatal opening of Solanum melongena (brinjal) plants. *International Journal of Recycling of Organic Waste in Agriculture*, 5(2), 141–147.

Embrandiri, A., Rupani, P. F., Shahadat, M., Singh, R. P., Ismail, S. A., Ibrahim, M. H., & Kadir, M. O. A. (2017). The phytoextraction potential of selected vegetable plants from soil amended with oil palm decanter cake. *International Journal of Recycling of Organic Waste in Agriculture*, 6(1), 37–45.

Ferdous Alam, A. S. A., Er, A. C., & Begum, H. (2015). Malaysian oil palm industry: Prospect and problem. *Journal of Food, Agriculture and Environment*, 13(2), 143–148.

Goh, K. J., Wong, C. K., & Ng, P. H. C. (2016). Oil palm. *Encyclopedia of Applied Plant Sciences*, 3(2), 382–390.

Griengo, S. G., Bandera, A. D., & Magolama, A. A. (2020). Application of different fertilizer types and levels on vegetable production: A critical review. *International European Extended Enablement in Science, Engineering & Management (IEEESEM)*, 8(10), 151–157.

H-Kittikun, A., Cheirsilp, B., Sohsomboon, N., Binmarn, D., Pathom-Aree, W., & Srinuanpan, S. (2021). Palm oil decanter cake wastes as alternative nutrient sources and biomass support particles for production of fungal whole-cell lipase and application as low-cost biocatalyst for biodiesel production. *Processes*, 9(8):1365, 1–15.

Haryati, M., & Theeba, M. (2021). Utilisation of DOBE: Palm oil mill waste as organic fertiliser. *Journal of Tropical Agriculture and Food Science*, 49(1), 1.

Hassan, N., Idris, A., & Akhtar, J. (2019). Overview on bio-refinery concept in Malaysia: Potential high value added products from palm oil biomass. *Jurnal Kejuruteraan*, 2(1), 113–124.

Hau, L. J., Shamsuddin, R., May, A. K. A., Saenong, A., Lazim, A. M., Narasimha, M., & Low, A. (2020). Mixed composting of palm oil empty fruit bunch (EFB) and palm oil mill effluent (POME) with various organics: An analysis on final macronutrient content and physical properties. *Waste and Biomass Valorization*, 11(10), 5539–5548.

Hoe, T. K., Sarmidi, M. R., Syed Alwee, S. S. R., & Zakaria, Z. A. (2016). Recycling of oil palm empty fruit bunch as potential carrier for biofertilizer formulation. *Jurnal Teknologi*, 78(2), 165–170.

Hosseini, S. E., Abdul Wahid, M., Jamil, M. M., Azli, A. A. M., & Misbah, M. F. (2015). A review on biomass-based hydrogen production for renewable energy supply. *International Journal of Energy Research*, 39(12), 1597–1615.

Ikumapayi, O. M., & Akinlabi, E. T. (2018). Composition, characteristics and socioeconomic benefits of palm kernel shell exploitation: An overview. *Journal of Environmental Science and Technology*, 11(4), 1–13.

James Rubinsin, N., Daud, W. R. W., Kamarudin, S. K., Masdar, M. S., Rosli, M. I., Samsatli, S., Tapia, J. F., Karim Ghani, W. A., & Lim, K. L. (2020). Optimization of oil palm empty fruit bunches value chain in peninsular Malaysia. *Food and Bioproducts Processing*, 119, 179–194.

Kanchanasuta, S., & Pisutpaisal, N. (2016). Waste utilization of palm oil decanter cake on biogas fermentation. *International Journal of Hydrogen Energy*, 41(35), 15661–15666.

Kaniapan, S., Hassan, S., Ya, H., Nesan, K. P., & Azeem, M. (2021). The utilisation of palm oil and oil palm residues and the related challenges as a sustainable alternative in biofuel, bioenergy, and transportation sector: A review. *Sustainability*, 13(6), 3110.

Kurnia, J. C., Jangam, S. V., Akhtar, S., Sasmito, A. P., & Mujumdar, A. S. (2016). Advances in biofuel production from oil palm and palm oil processing wastes: A review. *Biofuel Research Journal*, 3(1), 332–346.

Kushairi, A., Ong-Abdullah, M., Nambiappan, B., Hishamuddin, E., Bidin, M. N. I. Z., Ghazali, R., Subramaniam, V., Sundram, S., & Parveez, G. K. A. (2019). Oil palm economic performance in Malaysia and R&D progress in 2018. *Journal of Oil Palm Research*, 31(2), 165–194.

Maluin, F. N., Hussein, M. Z., & Idris, A. S. (2020). An overview of the oil palm industry: Challenges and some emerging opportunities for nanotechnology development. *Agronomy*, 10(3), 356.

Maniam, G. P., Hindryawati, N., Nurfitri, I., Jose, R., Mohd, M. H., Dahalan, F. A., & Yusoff, M. (2013). Decanter cake as a feedstock for biodiesel production: A first report. *Energy Conversion and Management*, 76, 527–532.

Misnon, I. I., Zain, N. K. M., & Jose, R. (2019). Conversion of oil palm kernel shell biomass to activated carbon for supercapacitor electrode application. *Waste and Biomass Valorization*, 10(6), 1731–1740.

Mohd-Azlan, A. (2020). Is there a sustainable future for wildlife in oil palm plantations in Malaysia? *Journal of Oil Palm Research*, 33(4), 732–738.

Mushtaq, F., Abdullah, T. A. T., Mat, R., & Ani, F. N. (2015). Optimization and characterization of bio-oil produced by microwave assisted pyrolysis of oil palm shell waste biomass with microwave absorber. *Bioresource Technology*, *190*, 442–450.

Nadeem, F., Hanif, M. A., Majeed, M. I., & Mushtaq, Z. (2018). Role of macronutrients and micronutrients in the growth and development of plants and prevention of deleterious plant diseases: A comprehensive review. *International Journal of Chemical and Biochemical Sciences (IJCBS)*, *14*, 1–22.

Nambiappan, B., Ismail, A., Hashim, N., Ismail, N., Shahari, D. N., Idris, N. A. N., Omar, N., Salleh, K. M., Hassan, N. A. M., & Kushairi, A. (2018). Malaysia: 100 years of resilient Palm oil economic performance. *Journal of Oil Palm Research*, *30*(1), 13–25.

Ng, W. P. Q., Lam, H. L., Ng, F. Y., Kamal, M., & Lim, J. H. E. (2012). Waste-to-wealth: Green potential from palm biomass in Malaysia. *Journal of Cleaner Production*, *34*, 57–65.

Noorshamsiana, A. W., Nur Eliyanti, A. O., Fatiha, I., & Astimar, A. A. (2017). A review on extraction processes of lignocellulosic chemicals from oil palm biomass. *Journal of Oil Palm Research*, *29*(4), 512–527.

Obi, F., Ugwuishiwu, B., & Nwakaire, J. (2016). Agricultural waste concept, generation, utilization and management. *Nigerian Journal of Technology*, *35*(4), 957.

Onoja, E., Chandren, S., Abdul Razak, F. I., Mahat, N. A., & Wahab, R. A. (2019). Oil palm (Elaeis guineensis) biomass in Malaysia: The present and future prospects. *Waste and Biomass Valorization*, *10*, 2099–2117.

Ooi, Z. X., Teoh, Y. P., Kunasundari, B., & Shuit, S. H. (2017). Oil palm frond as a sustainable and promising biomass source in Malaysia: A review. *Environmental Progress and Sustainable Energy*, *36*(2), 1864–1874.

Parveez, G. K. A., Hishamuddin, E., Loh, S. K., Ong-Abdullah, M., Salleh, K. M., Bidin, M. N. I. Z., Sundram, S., Hasan, Z. A. A., & Idris, Z. (2020). Oil palm economic performance in Malaysia and R&D progress in 2019. *Journal of Oil Palm Research*, *32*(2), 159–190.

Parveez, G. K. A., Kamil, N. N., Zawawi, N. Z., Ong-Abdullah, M., Rasuddin, R., Loh, S. K., Selvaduray, K. R., Hoong, S. S., & Idris, Z. (2022). Oil palm economic performance in Malaysia and R&D progress in 2021. *Journal of Oil Palm Research*, *34*(2), 185–218.

Parveez, G. K. A., Tarmizi, A. H. A., Sundram, S., Loh, S. K., Ong-Abdullah, M., Palam, K. D. P., Salleh, K. M., Ishak, S. M., & Idris, Z. (2021). Oil palm economic performance in Malaysia and R&D progress in 2020. *Journal of Oil Palm Research*, *33*(4), 181–214.

Phoochinda, W. (2020). Assessment of social return on investment from the utilisation of oil palm's residues. *Journal of Oil Palm Research*, *32*(1), 145–151.

Pirker, J., Mosnier, A., Kraxner, F., Havlík, P., & Obersteiner, M. (2016). What are the limits to oil palm expansion? *Global Environmental Change*, *40*, 73–81.

Puastuti, W. (2017). Pemanfaatan pelepah daun sawit sebagai pakan sumber serat: Strategi dan respon produksi pada sapi potong. *Pastura*, *5*(2), 98.

Pulunggono, H. B., Anwar, S., Mulyanto, B., & Sabiham, S. (2019). Decomposition of oil palm frond and leaflet residues. *Agrivita*, *41*(3), 524–536.

Purba, R., Purba, J., & Damanik, F. H. (2020). The Influence of solid palm oil waste and NPK fertilizer on the growth and the production of green eggplant (Solanum Melongena L.). *International Journal of Scientific Research and Management*, *8*(7), 296–303.

Raju, N. K., & Ramakrishnan, R. (1972). Properties of laterite aggregate concrete. *Matériaux et Constructions*, *5*(5), 307–314.

Rangkuti, I. U. P., Giyanto, Novayanty, R., Raja, P. M., & Zakwan. (2018). The micronutrient contents of composting from empty bunch after added palm oil mill effluent. *IOP Conference Series: Materials Science and Engineering*, *434*, 012232.

Rosenani, A. B., Rabuni, W., Cheah, P., & Noraini, J. (2016). Mass loss and release of nutrient from empty fruit bunch of oil palm applied as mulch to newly transplanted oil palm. *Soil Research*, *54*(8), 985–996.

Rupani, P. F., Alkarkhi, A. F. M., Shahadat, M., Embrandiri, A., El-Mesery, H. S., Wang, H., & Shao, W. (2019). Bio-optimization of chemical parameters and earthworm biomass for efficient vermicomposting of different palm oil mill waste mixtures. *International Journal of Environmental Research and Public Health*, *16*(12), 2092, 1–10.

Salihu, A., & Alam, Z. (2012). Palm oil mill effluent: A waste or a raw material? *Journal of Applied Sciences Research*, *8*(1), 466–473.

Sari, D. A. P., Fadiilah, D., & Azizi, A. (2019). Utilization of palm oil mill effluent (POME) for biogas power plant; Its economic value and emission reduction. *Journal of Advanced Research in Dynamical and Control Systems*, *11*(7), 465–470.

Schmidt, J. H., & Weidema, B. P. (2008). Shift in the marginal supply of vegetable oil. *International Journal of Life Cycle Assessment*, *13*(3), 235–239.

Siang, C. S., Wahid, S. A. A., & Sung, C. T. B. (2022). Standing biomass, dry-matter production, and nutrient demand of tenera oil palm. *Agronomy, 12*(2), 426.

Sudiyani, Y., Styarini, D., Triwahyuni, E., Sudiyarmanto, Sembiring, K. C., Aristiawan, Y., Abimanyu, H., & Han, M. H. (2013). Utilization of biomass waste empty fruit bunch fiber of palm oil for bioethanol production using pilot: Scale unit. *Energy Procedia, 32,* 31–38.

Sundalian, M., Larissa, D., & Suprijana, O. (2021). Contents and utilization of palm oil fruit waste. *Biointerface Research in Applied Chemistry, 11*(3), 10148–10160

Sung, C. T. B. (2016). Availability, use, and removal of oil palm biomass in Indonesia. Report prepared for the International Council on Clean Transportation. Working paper.

Supian, M. A. F., Amin, K. N. M., Jamari, S. S., & Mohamad, S. (2020). Production of cellulose nanofiber (CNF) from empty fruit bunch (EFB) via mechanical method. *Journal of Environmental Chemical Engineering, 8*(1), 103024.

Tang, K. H. D., & Al Qahtani, H. M. S. (2020). Sustainability of oil palm plantations in Malaysia. *Environment, Development and Sustainability, 22,* 4999–5023

Teh, S. S., Lau, H. L. N., & Mah, S. H. (2019). Palm-pressed mesocarp fibre oil as an alternative carrier oil in emulsion. *Journal of Oleo Science, 68*(8), 803–808.

Trisakti, B., Mhardela, P., Husaini, T., Irvan, & Daimon, H. (2018). Production of oil palm empty fruit bunch compost for ornamental plant cultivation. *IOP Conference Series: Materials Science and Engineering, 309,* 1–8.

Wahi, R., & Yusup, I. 'A. (2016). Empty fruit bunches compost and germination of Raphanus sativs L. *Borneo Journal of Resource Science and Technology, 6*(1), 10–18.

Wang, S., Dai, G., Yang, H., & Luo, Z. (2017). Lignocellulosic biomass pyrolysis mechanism: A state-of-the-art review. *Progress in Energy and Combustion Science, 62,* 33–86.

Widians, J. A., Taruk, M., Fauziah, Y., & Setyadi, H. J. (2019). Decision support system on potential land palm oil cultivation using PROMETHEE with geographical visualization. *Journal of Physics: Conference Series, 1341,* 1–9.

Windiastuti, E., Suprihatin, Bindar, Y., & Hasanudin, U. (2022). Identification of potential application of oil palm empty fruit bunches (EFB): A review. *IOP Conference Series: Earth and Environmental Science, 1063*(1), 1–11.

Yang, J., Ching, Y. C., Chuah, C. H., & Liou, N. S. (2021). Preparation and characterization of starch/empty fruit bunch-based bioplastic composites reinforced with epoxidized oils. *Polymers, 13*(1), 1–15.

Yi, L. G., Wahid, S. A. A., Tamilarasan, P., & Siang, C. S. (2019). Enhancing sustainable oil palm cultivation using compost. *Journal of Oil Palm Research, 31*(3), 412–421.

Zulkifli, Y., Norziha, A., Naqiuddin, M. H., Fadila, A. M., Nor Azwani, A. B., Suzana, M., Samsul, K. R., Ong-Abdullah, M., Singh, R., Parveez, G. K. A., & Kushairi, A. (2017). Designing the oil palm of the future. *Journal of Oil Palm Research, 29*(4), 440–455.

6 Wastewater as Nutrient Enhancer and Moisturizer for Compost Production
A Review

Abu Zahrim Yaser and
Junidah Lamaming
Universiti Malaysia Sabah

CONTENTS

INTRODUCTION

Global population expansion, climate change effects, and lifestyle changes are placing increasing demand on our vital water resources, causing significant water stress in many countries (Kılıç, 2020). The demand for water treatment management systems is also being driven by the high pace of urbanization in developing nations. The increased need for high-quality water along with a focus on water reuse and recycling via initiatives like zero liquid discharge is propelling advanced technology adoption in the wastewater sector (Research and Markets, 2022). To keep people healthy and live in the modern world, the sixth sustainable development goal is being spread so that more people know about clean water and sanitization. One of the efforts is waste management, which is commonly done by the government, water sector, or the industry sector itself.

The opportunities from exploiting wastewater as a resource are enormous. Safely managed wastewater is an affordable and sustainable source of water, energy, nutrients, and other recoverable materials. One sustainable approach is using non-toxic wastewater as a nutrient enhancer and

DOI: 10.1201/9781003358084-6

moisture conditioner via composting. Wastewater, liquid manure (LM), and sewage sludge are all excellent fertilizer sources. It is composed of a variety of chemical elements that will be released into bodies of water, posing a threat to aquatic life as well as humans and animals (Ayilara et al., 2020). Studies show a clear correlation between inorganic nutrients discharged from industrial and municipal wastewater and the degradation of natural water ecosystem balance in terms of trophic state (Preisner et al., 2022). However, industrial waste also contains macro- and micronutrients, organic and inorganic components, and trace elements that are necessary for plant development. Depending on the industry, the wastewater happens to contain nutrients such as carbon, nitrogen, sulfur, and phosphorus required for sustaining and developing the growth of the bacteria. In addition, the wastewater is also rich in water content and moisture. The adjustment of initial moisture contents has become the main approach to controlling moisture content in composting systems (Wu et al., 2015).

In the last few decades (2012–2022), a few reviews on wastewater composting have been published. Not all the wastewater can be directly used in composting as some of it needs further treatment before being used. Mohammad et al. (2012) and Alkarimiah and Rahman (2014) have reviewed the POME and EFB efficient composting process, and Chowdhury et al. (2013) reviewed the olive mill composting as well as treatment technologies for POME. In addition, olive oil mill wastewater (OMW) wastewater was also reviewed by Lee et al. (2019) and vinasse by Hoarau et al. (2018). There have been few or no review papers published to date that specifically addressed industrial wastewater as a moisture conditioner and nutrient enhancer in composting. This paper will fill the gap by focusing on industrial waste used for composting and will review the composting process using wastewater from certain industries that is turned into a nutrient enhancer and moisturizer for compost production. More specifically, it discusses the wastewater from the industries including palm oil mill or known as POME, OMW, swine wastewater (SW), alcohol or molasses distillery wastewater, trickling liquid (TL) from composting plant wastewater, and monosodium glutamate wastewater (MSGW). Accordingly, previous research on composting with industrial waste has been compiled and analyzed based on their performance as both moisturizers and nutrient enhancers, as well as phytotoxicity.

METHODOLOGY

The references and publications cited in this review paper were gathered via Scopus, Springerlink, and ScienceDirect. This review article incorporates data from a variety of sources, including review papers, journal articles, and book chapters. The keywords "domestic wastewater", "industrial wastewater", "composting", "nutrient", "effluent", and "moisture" were often used in the search for articles. Articles published between 2012 and the present are included as primary sources to ensure the authenticity of this review. From 2012, 2,582 articles have been identified using automatic search and the keyword combination throughout the database. According to VOSviewer results as shown in Figure 6.1, the article was used around 2,000 with 5,255 authors around the world. The minimum number of documents an author can have is set at 30 articles. The VOSviewer result is also used to define the research paper based on the countries. There are 2,000 articles that were published separately in 103 countries. The search was narrowed to certain methods that were combined manually by screening the titles of articles and reading the abstracts that contained the keywords "wastewater" and "composting". By reading and screening the titles and abstracts, the list was whittled down to 81 papers in total. Additional screening of the articles was conducted by reading the technique and results sections to obtain additional information about the system and to check that the wastewater used was from the industrial sector. From this point, 22 articles are identified as being highly relevant to the subject of this review. The articles were classified according to the type of industrial waste they contained: palm oil mill, olive oil mill, alcohol or molasses, composting facilities as well as monosodium glutamate production. The type of wastewater, solid substrate, composting parameters, and final compost quality are focused on in each of the reviewed articles.

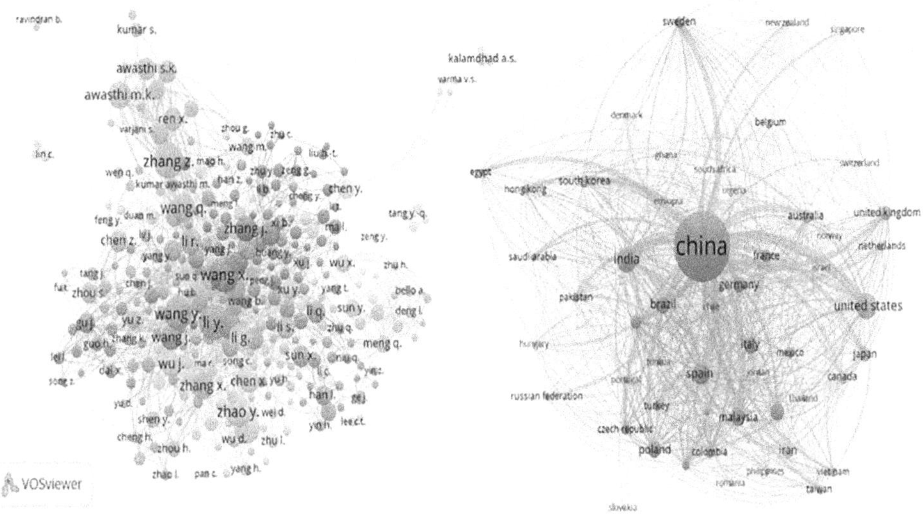

FIGURE 6.1 Documents screening using VOSviewer.

Several publications published before 2012 were also employed in the final stage of the approach to supplement the review articles with supporting evidence and arguments.

INDUSTRIAL WASTEWATER CHARACTERISTICS

Industrial wastewater usually has a lot of organic matter in it, and it can be used as a source of nutrients and moisture. Table 6.1 depicts some of the physico-chemical characteristics of the wastewater. One of the typical wastewaters is palm oil mill wastewater (POME), especially in most Asian countries that derive economic value from palm oil. A ton of crude palm oil requires 5–7.5 tons of water, half of which is released as POME and the other half as unused water that does not enter the effluent stream or the wastewater treatment system (Liew et al., 2015; Mahmod et al., 2021). The raw or partially treated POME contains an extraordinarily high concentration of degradable organic matter, which is particularly beneficial to the environment. POME is primarily made up of water (95%–96%), with 4%–5% total solids derived primarily from palm fruit mesocarp debris, and 2%–4% suspended solids. The suspended solids in POME are composed of cell walls, organelles, short fibers, various carbohydrates (hemicellulose as complex compounds and simple sugars as their monomers), nitrogenous compounds (proteins as complex compounds to amino acids as their monomers), free organic acids, and a combination of minor organic and mineral components (Ahmad et al., 2016). Hot, acidic (pH 4–5), brown-brown colloidal solution made from fresh POME has a lot of organic matter, a lot of total solids, oil and grease, and is high in chemical oxygen demand (COD), and biological oxygen demand (BOD) (Krishnan et al., 2017; Lee et al., 2019; Hau et al., 2020).

Another problematic wastewater is OMW that is derived from olive oil production, especially in Mediterranean countries. OMW produces a dark, foul-smelling waste with an acidic pH, high electrical conductivity (EC), a high carbon/nitrogen ratio, resistance to degradation (owing mostly to the waste's high concentration of phenolic compounds), high organic loads, and antimicrobial and phytotoxic characteristics (Chowdhury et al., 2013; Kefalogianni et al., 2021). Sugars, polyalcohols, pectins, lipids, and considerable amounts of aromatic compounds like tannins and polyphenols make up the OMW organic fractions. Olive oil mill discharges are a problem because of their high polyphenol content, which can range from 18 to 125 mg/g depending on the variety of olives used,

TABLE 6.1

Physicochemical Characteristics of Some Wastewater from Industries

Parameters[a]	Palm Oil Mill	Olive Mill	Alcohol/Molasses-Based Distillery	Monosodium Glutamate	Composting TL
pH	3.4–5.2	4.7–5.7	7.7	2.5–3.6	6.5–7.9
EC (mS/cm)	-	8.02	-	63.8	3.6–8.1
Color	Blackish brown	Dark	Dark brown	Dark brown	Dark brown
COD	15,000–100,000	16,500–190,000	37,000	496,000	118–255
BOD	10,250–43,750	41,300–46,000	6,000	162,000	16.2–44.1
Total solids	11,500–79,000	32,000–300,000	53,000	60,700	-
Volatile solids		-	25,500	419,200	-
Total suspended solids	5,000–54,000	-	1,600	-	-
Total nitrogen	180–1,400	300–1,500	1,400	56,700	2,738–5,035
Ammoniacal nitrogen	4–80	-	-	1,010	1,257–1,973
Phosphorus	180	3,000–11,000	215	2,410–4,400	-
Potassium	2,270	3,000–8,000	9,200	32,800	-
Magnesium	615	600–2,200	-	-	-
Calcium	439	100–800	-	17,300	-
Iron	46.5	-	25	-	-
Boron	7.6	-	-	-	-
Manganese	2.0	-	-	31.8	-
Copper	0.89	-	-	-	-
Zinc	2.3	-	-	-	-
Sulfate	-	-	2,140	21,300	-
Chloride	-	-	8,060	-	-
Organic carbon	-	-	17,220	344,600	-
Bicarbonate	-	-	3,120	-	-
Phenol (g/100 g)	-	2–80,000	-	-	
References	Zahrim et al. (2017); Lee et al. (2019)	Lee et al. (2019); Alavi et al. (2017)	Malik et al. (2019)	Singh et al. (2011); Ji et al. (2014)	Wu et al. (2015)

[a] All parameters are in units of mg/L otherwise stated in parenthesis.

the level of production, and the time between olive picking and extraction (Hassen et al., 2021). According to Kipçak and Akgün (2012), OMW is composed of 83%–96% water, 3.5%–15% organic constituents, and 0.5%–2% mineral compounds.

Distillery spent wash (DSW) or vinasse is usually considered as waste product of distillery processes and can be classified as a dilute organic liquid fertilizer with high potassium content. Alcohol/molasses-based distilleries are among the most polluting industries in the world, producing large volumes of high-concentration wastewater (14–22 L/L of alcohol). DSW contains caramelization products and recalcitrant components having high COD and BOD, along with inorganic salts and low pH (Satyawali and Balakrishanan, 2008; Shinde et al., 2020). The wastewater is dark brown in color due to the presence of melanoidins, an unpleasant odor, and a poor biodegradability index (BI: $BOD_5/COD > 0.2$) (Malik et al., 2018). It also contains nutrients including nitrogen (up to 4.2 g/L), potassium (up to 17.5 g/L), and phosphorus (up to 3.0 g/L) (Hoarau et al., 2018). This characteristic depends on the raw materials used for producing the alcohol.

Swine waste is a type of highly concentrated organic waste that contains a high concentration of carbon, nitrogen, phosphorus, and other elements (Karakashev et al., 2008). According to Fan et al. (2019a), every 10,000 pigs breeding in pig farms produces 190 tons of livestock breeding waste per day, with approximately 40% solid manure and 60% SW/LM with an average MC of 75%, causing major environmental disturbances such as water and soil pollution and pollutants into the atmosphere (Bustamante et al., 2013; Dennehy et al., 2017). Organic contaminants, ammonia nitrogen, and other substances that may be addressed biochemically are common in livestock and poultry breeding wastewaters.

The TL is produced by bio-trickling filters, which is used for the removal of NH_3 in composting site facilities. Following long-term operation, the accumulation of nitrogen-containing compounds and nitrogen species in the TL may have detrimental effects on nitrifying bacteria and reduce bioreactor efficiency. The nitrogen (NH_3–N) in the smells was then preserved in the TL wastewater, which contains a variety of inorganic nitrogen.

As a flavor enhancer, monosodium glutamate (MSG) is extensively used in food products throughout East and Southeast Asia. MSG production in China accounts for about half of the world's total output (Liu and Zhou, 2010). After extraction of MSG from fermentation liquor, the residual dark brown wastewater and effluent have high concentrations of COD, NH_3–N, sulfate, and a strong acidity (Yang et al., 2005; Ji et al., 2014). MSGW has a very low pH and is full of proteins, amino acids, sulfate, and total organic carbon, as well as being free of heavy metal pollution (Bai et al., 2004). It also doesn't have any heavy metal pollution.

NUTRIENT ENHANCER AND MOISTURIZER IN COMPOSTING

POME

In spite of the fact that no chemicals are applied during the oil extraction process, POME is considered non-toxic, yet it has been identified as a major source of aquatic pollution due to the fact that it depletes dissolved oxygen when dumped untreated into water bodies. However, it also contains significant amounts of nitrogen, phosphorus, potassium, magnesium, and calcium (Rupani et al., 2010), all of which are essential nutrients for plant growth. Because of its non-toxic nature and fertilizing characteristics, POME can be utilized as a nutrient enhancer substitute in the sense of providing the mineral requirements of the plant's growing environment and acting as a moisturizer. Moisture loss is attributed to the evaporation of water due to high ambient temperatures and turning during the composting process. Addition of POME can replenish the compost pile and overcome water loss, so that the microbial activity can be sustained (Baharuddin et al., 2009). In POME composting, using a combination of other materials as a bulking agent can further improve the performance of the final compost. Oil palm decanter cake (OPDC), paper, grass clippings, and manures are among the additives that can be used as a source of nitrogen or carbon source and an offset to higher moisture content during composting (Barthod et al., 2018).

A few reports have been using POME for composting in the recent years (Table 6.2). The application of POME helps in accelerating the mineralization of organic matter by microorganisms (Kala et al., 2009). Ahmad et al. (2014) investigated the co-composting of POME and chipped-ground oil palm frond (OPF). The use of POME served as a microbial source and maintained the moisture content at around 55%–65%, which led to a good aeration for bioactivity of the microorganism. The final compost contains 0.1% (P) and 0.9% (K), has a moisture content of 61%, has a pH of 8.1, and has a C/N ratio of 24. The study found that co-composting of OPF and POME produced a compost with an acceptable value of C/N and nutrients. However, the critical nutrients were found to be lower compared to compost made from empty fruit bunches (EFBs) (Baharuddin et al., 2009; Ahmad et al., 2016). This attributed to the low nutrient levels found in the raw OPF (0.85% N, 0.05% P, and 1.73% K) as compared to raw EFB (1.2% N, 0.08%, and 1.73% K).

TABLE 6.2

Industrial Wastewater as Nutrient Enhancer and Moisturizer in Production of Compost in Recent Years by Several Works

Type of Wastewater (%)	Solid Substrate (s) (%)	Composting Conditions	Final Compost Performance	References
		Microbes–Composting		
POME (75)	EFB (25)	Time (50–60 days), turning (once in 3 days), pH (5),	C/N (17), pH (7.8), MC (80%), GI (80%)	Mohammad et al. (2013)
POME (67)	Chipped-ground oil palm frond (33)	Time (60 days), C/N (24), turning (Every 3 days), pH (7.0), T_{max} (56°C)	C/N (17), pH (8.1), MC (61%), N (1.2 %), P (0.1%), K (0.9%)	Ahmad et al. (2014)
Palm oil mill effluent (POME) (71)	EFB (24) Decanter cake (5)	Time (70 days), C/N (24), turning (3 times in 14 days), size (EFB, 150–200 mm), pH (8.0), T_{max} (38°C)	C/N (24), pH (8.4), MC (60%), N (1.57%), P (0.21%), K (0.65%)	Adam et al. (2016)
AnPOME (23)	Paper (31), grass clippings (46)	Time (40 days), C/N (33), pH (7.0), T_{max} (31°C)	C/N (31), pH (7.2), MC (53%), N (1.2%), P (0.1%), K (0.4%), GI (158%)	Zahrim et al. (2016)
POME (6)	EFB (63), recycled compost (31)	Time (43 days), C/N (39), T_{max} (55°C)	C/N (13), MC (59%) N (2.86%)	Alkarimiah and Suja (2020)
POME	EFB	Time (40 days), C/N (25, 35,45), T_{max} (60°C)	C/N (14), pH (8), OM loss (74%)	Hasan et al. (2021)
OMW (58)	Grape marc (14), green waste (14), OMS (14)	Time (98 days), C/N (34), turning (once every 3 days at beginning, then every 7 days, finally once every 15 days), pH (8.5), T_{max} (68°C)	pH (7.0), (MC (30%), EC (1.75 mS/cm), P (0.09%), K (0.28) GI (100%)	Majbar et al. (2017)
OMW (50)	Grape marc (25), Green waste (25)	Time (100 days), C/N (34), turning (once every 3 days at beginning, then every 7 days, finally once every 15 days), pH (6.5), T_{max} (68°C)	C/N (12), pH (8.3), MC (55%), EC (1.92 mS/cm), P (346 mg/kg), K (226 mg/kg), GI (90%)	Majbar et al. (2018)
OMW (50)	Grape marc (50)	Time (90 days), C/N (23), turning (every 14 days), pH (8.6), T_{max} (60°C)	pH (8.6), MC (55%), EC (5.1 mS/cm), N (35.3 g/kg), P (12.1 g/kg), K (64.4 g/kg), GI (156%)	Galliou et al. (2018)
Olive mill wastewater (60)	Municipal green waste (40)	Time (140 days), C/N (21.5), Size (municipal green waste ≤ 30 mm), pH (5.8), Temp (55°C)	C/N (21.5), pH (8.2), MC (50%), N (1.05%), P (0.14%), K (0.79%)	Avidov et al. (2018)
OMW (67)	Household waste (33)	Time (150 days), C/N (35), turning (once in 7 days), pH (6.4), T_{max} (60°C)	C/N (10), pH (7.5), MC (43%), N (3.49%), P (0.32%), GI (94%)	Atif et al. (2020)
OMW (75)	Cotton residue (25)	Time (130 days), C/N (18), turning (64th day), pH (7.3), T_{max} (30°C)	MC (59%), N (3.8 %), GI (90%)	Kefalogianni et al. (2021)

(Continued)

TABLE 6.2 (*Continued*)

Industrial Wastewater as Nutrient Enhancer and Moisturizer in Production of Compost in Recent Years by Several Works

Type of Wastewater (%)	Solid Substrate (s) (%)	Composting Conditions	Final Compost Performance	References
Alcohol/ molasses distillery wastewater (DSW) (80)	Pressmud (20)	Time (40 days), C/N (34), turning (daily), pH (7.8), T_{max} (64°C)	pH (6.8), EC (1.22 dS/m), N (2.0%), P (1.8%), K (3.3%), GI (92%)	Malik et al. (2019)
SW (8.97 L)	Solid pig manure (83.3), Rice husk (16.7)	Time (30 days), turning (every 2 days during thermophilic stage & every 5–10 days during other stages), pH (7.3), T_{max} (64°C)	MC (41%), N (2.11%), P (2.63%), K (1.67%), GI (82%)	Fan et al. (2021)
SW (50)	Solid pig manure (83.3), Rice husk (R) (16.7), Corncob (C) (16.7)	Time (30 days), C/N (23 and turning (every 2 days during thermophilic stage & every 5–10 days during other stages), pH (7.3), T_{max} (64°C and 69°C)	Corncob: pH (8.0), EC (2.71 mS/cm), N (36.5 g/kg), P (29.8 g/kg), K (201.1 g/kg), GI (81%) Rice husk: pH (7.3), EC (3.13 mS/cm), N (24.4 g/kg), P (28.2 g/kg), K (19.8 g/kg), GI (83%)	Fan et al. (2019a)
Composting TL (nd)	Mushroom bran (43), pre-consumer food wastes (43), and post-consumer food wastes (14)	Time (15 days), aeration rate (0.002 m³/min), C/N (36), turning (every 3 days), size (mushroom bran <10 mm, pre-consumer food wastes <10 mm), pH (6.4), T_{max} (63°C)	C/N (15), pH (7.7), MC (38%), N (6.1%), GI (79%)	Wu et al. (2015)
Nitrate-rich STL (67)	Fermentation residue (17), sawdust (10), food waste (3), mushroom brans (3)	Time (30 days), aeration rate (0.100 m³/min), C/N (20), Size (mushroom bran <3 mm, Sawdust <5 mm), pH (6.3), T_{max} (56°C)	pH (8.2), MC (47%), N (2.5%)	Xie et al. (2021a)
Ammonium-rich STL (67)	Fermentation residue (17), sawdust (10), food waste (3), mushroom brans (3)	Time (30 days), aeration rate (0.002 m³/min), C/N (20), turning (every 3 days), size (mushroom bran <3 mm, Sawdust <5 mm), pH (6.2), T_{max} (55°C)	pH (8.1), MC (46%), N (2.4%)	Xie et al. (2021b)
Nitrite-rich STL (67)	Fermentation residue (17), sawdust (10), mushroom (3), food waste (3)	Time (30 days), aeration rate (0.002 m³/min), C/N (20), turning (every 3 days), size (mushroom bran <3 mm, Sawdust <5 mm), pH (7.9), T_{max} (55°C)	pH (8.0), MC (46%), N (2.5%)	Xie et al. (2021c)

(Continued)

TABLE 6.2 (*Continued*)
Industrial Wastewater as Nutrient Enhancer and Moisturizer in Production of Compost in Recent Years by Several Works

Type of Wastewater (%)	Solid Substrate (s) (%)	Composting Conditions	Final Compost Performance	References
MSGW (2)	Cattle manure (60), chicken manure (19), mushroom residue (19)	Time (25 days), turning (daily for the first 10 days), pH (7.8), T_{max} (50°C)	pH (7.7), EC (10.6 mS/ cm), N (28.1 g/kg), GI (85 %)	Liu et al. (2015)
		Integrated microbe–vermicomposting		
Alcohol/ molasses distillery wastewater (DSW) (10,20,40)	Cow manure and bagasse (10, 25, 50) Calcium zeolite (10, 20)	Time (60 days (21 days microbe–composting)), C/N (35–52), pH (6.7–8.2), turning (twice a week), temp (55°C), vermicomposting (15–18 earthworms *Eisenia fetida*)	C/N (9.4–35), pH (8), EC (1.4–1.94 mS/cm), (P (0.06%–0.1%), K (0.06%–0.15%), GI (100%)	Alavi et al. (2017)
		Vermicomposting		
POME (50)	Pressed palm fiber (50)	Time (45 days), C/N (35), T_{max} (26°C), 20 earthworms (*Lumbricus rubellus*)	C/N (17), pH (7.0), N (1.76%), P (0.4%), K (0.8%) GI (75%)	Rupani et al. (2017)
POME	EFB, fishbone, bone meal, bunch ash	Time (52 days), C/N (25,30,35), T_{max} (34°C), 50 earthworms	C/N (22), pH (6), MC (64%)	Lew et al. (2020)
POME (25)	EFB (25), fishbone (50), bone meal (50%), bunch ash (50)	Time (60 days), C/N (25), T_{max} (34°C), 50 earthworms (*Eisenia fetida*)	C/N (16), pH (6.3), MC (59%), N (1.50%), P (1.22%), K (12.01%)	Hau et al. (2020)

Note: C/N, Carbon to nitrogen; T_{max}, Maximum temperature; MC, Moisture content; EC, Electric conductivity; N, Nitrogen; P, Phosphorus; K, Potassium; GI, Germination index.

POME contains high MC but a lower C/N ratio. Mixing POME with EFB could provide better MC and adequate nutrients for microorganism growth and degrade the compost materials. Adam et al. (2016) conducted a study using POME, EFB, and OPDC composting for 10 weeks. The result showed that the optimum compost was achieved with a low C/N ratio (24) and a pH of 8.4. They have found that the addition of the decanter cake can aid in lowering the C/N ratio due to the high nitrogen content of the decanter cake, which enhances the microbial activity and accelerates the rate of compost (Yahya et al., 2010). Furthermore, the NPK values in the final compost have increased significantly (N-1.57%, P-0.21%, K-0.65%).

A similar case of compost with high nutrients was reported by Yahya et al. (2010) upon composting of POME mixed with EFBs and decanter cake slurry.

In another study, Zahrim et al. (2016) investigated the composting of anaerobically treated palm oil mill effluent (AnPOME) mixed with paper and grass clippings at a ratio of 1:1 in composting to balance the ratio of C/N during the 40-day composting process in a bioreactor.

During the composting, the maximum temperature reached was at 31°C, which does not achieve the thermophilic temperature (40°C–65°C). This might be attributed by the lack of green waste, small size of the reactor, insufficient isolation of the composting materials as well as the high humidity. This resulted in easily heat dissipating that slows down the decomposition of organic matter (Lai et al., 2013). The final compost is a C/N of 31 with a pH of 7.2. The results showed a matured compost with the addition of modulates of the nutritional content of compost to an acceptable value when paper and grass clippings are added. The final compost achieved 158% GI values and 18.29% mass reduction with C/N (31) at a pH of 7.2. The NPK values are comparable with the findings of Baharuddin et al. (2009) that used a mixture of partially treated POME and EFB. The compost produced has a C/N of 12 with a considerable amount of NPK and other micro- and macro-nutrients. Apart from that, Mohammad et al. (2013) discovered that composting POME and EFB with multi-enzymatic fungal can accelerate the composting time. The compost reported has the lowest C/N ratio of about 17, stable pH, and MC (80%), and the maximum germination index (GI) obtained was 116% at day 50. The low C/N ratio and the GI of more than 80% show that the compost is matured as well as toxin-free for seed germination and plant growth.

Hasan et al. (2021) have used POME in a periodic addition to maintain the moisture content and enrich the compost product. They investigated the effects of different initial C/N ratios (25, 35, and 45) on the organic matter degradation during active co-composting of EFB and POME in a composter. The two times additions of POME have maintained the moisture level in the range of 70%–80% and are stable during the composting process. Apart from the inclined reactor position, the addition of POME also contributes to water retention in the compost reactor. The study achieved the highest temperature of 60°C with an initial C/N ratio of 35 (reduced to 14) and the best OM degradation (74% OM loss). The composting time has also been shortened to 30 days. The periodic addition of POME assisted in the decomposition of the organic matter through the nutrients needed in metabolism and the growth of microorganisms.

Alkarimiah and Suja (2020) studied EFB, recycled compost, and POME composting using a rotary drum reactor with three different mixture ratios of EFB:POME:recycled compost (10:1:5, 5:0.5:2.5, and 5:2:0.5). They reported that the compost mixture of 10 kg EFB, 1 kg POME sludge, and 5 kg recycled compost with no active aeration has the most favorable composting conditions. The compost mixture recorded a final C/N ratio of 13, with the highest temperature (55°C) and moisture content of 59%. The N value increased by up to 2.86% (from 1.35%) and was also reported as the highest among the mixtures. They concluded that, for temperature rise, the ideal moisture content is around 50% for EFB and the POME composting process.

Another approach to POME composting is through vermicomposting, which uses earthworms. Very limited study can be found on this matter. Lew et al. (2020) and Hau et al. (2020) had adopted vermicomposting in composting the POME and EFB in 7 batches of composts. The study blends the EFB:POME with various organic wastes including fishmeal, bone meal, and bunch ash in the presence of 50 earthworms (*Eisenia fetida*) with different C/N ratios. The fishmeal and bone meal are organic and phosphorus resources while the bunch ash serves as a potassium source. Lew et al. (2020) reported that the EFB-mixed vermicompost has a moisture content in the range of 55%–70% due to constant moisturization by POME and a high mass yield of around 88%–90%. In the extended work, Hau et al. (2020) proposed that EFB:POME mixed with the three organics has an enhanced decomposition rate with increasing compost nutrients as compared to conventional composting. The mixed compost with initial C/N ratio of 25 achieves maturation at day 40 with a pH of 6.3. The final compost has a low C/N ratio (16), and a MC of 59%. The mixed composts also have shown an increase in the nutrients values (N-1.5%, P-1.22%, and K 12.05%) envisages that the organic wastes can be used to enhance the nutrient content of the compost. A similar case was reported by Rupani et al. (2017) where the mixture of POME and pressed palm fiber decomposed using *Lumbricus rubellus* within 45 days, having a low C/N up to 17, raising its improvement in nitrogen, phosphorus, and potassium content.

Among the nutrient values depicted in Table 6.2, EFB mixed together with POME shows higher N values. Apart from the high nutrients in EFB, decanter cake, and the organic wastes, the increase in N value at the end of the composting process is ascribed to the nitrogen-fixing bacteria activity (Jusoh et al., 2013; Zahrim et al., 2015). The deficit in potassium (K) value in compost is said to be due to the high MC of the POME, which might encourage the leaching process to happen. In addition, the K elements are easily leached out, and the usage of POME in liquid form contributes to the high leaching rate of K (Jusoh et al., 2013; Adam et. al., 2016). The deficit in potassium might be attributed to the excessive secretion from maggots and the actions of water washed away during constant moisturizing by POME (Hampton, 2006; Abdullah and Sulaiman, 2013).

OLIVE MILL WASTEWATER (OMW)

The high demand for olive oil and the high volume of OMW waste have led to a variety of ways to reuse OMW, such as composting, which turns the slightly toxic OMW into a safe-to-use product. Because of the high concentration of organic carbon and humic compounds in OMW, as well as the richness of N, P, and K in olive oil mills' wastewater, this material is a suitable soil conditioner and fertilizer with a high fertilizing capacity (Chehab et al., 2019). Aviani et al. (2010) stated that OMW has been used to obtain the initially desired MC of the composting materials and to wet them during the thermophilic phase of the composting process. The majority of the studies reported in this review used solid substrates such as grape marc, green waste, and household wastes.

Majbar et al. (2017, 2018) have investigated the co-composting of OMW with grape marc, olive mill sludge (OMS), and green waste. The composting with the addition of OMS into the mixture has slowed down the decomposition process due to the additional source of organic matter presence and self-insulating capacity of OMS during the degradation process (Hachicha et al., 2009). Both studies have produced compost that is rich in fertilizing and mineral elements and has low levels of heavy metals. Moreover, both studies achieved the GI standard (>80%), and when tested in radish production, the productivity increased by 10%. A study by Atif et al. (2020) describes the pathogen evaluation in composting of OMW, phosphate residue, and household waste where maturity is achieved 5 months later with an increase in total nitrogen. Kefalogianni et al. (2021) stated that diminishing of cotton residue is possible by co-composting with OMW as the addition of OMW gives rise to temperatures, achieving stability and maturity of compost in a shorter time period. With the addition of OMW, the pH was more acidic during the onset of the process. The EC values were higher throughout the process, while the levels of ammonium and nitrate nitrogen, as well as the NH_4^+/NO_3^- ratios, were lower with a final product suitable for improving soil fertility and health.

Avidov et al. (2018) have investigated the composting of OMW with municipal green waste using polyethylene sleeves for a period of 140 days with forced aeration. The excess water in OMW was used to achieve desired moisture content for the absorption of OMW to green waste. The result shows that the compost has a slightly alkaline pH (8.2), a low EC (1.4 dS/m), and low phytotoxicity index. The NPK value was slightly increased in the final compost. The nitrogen loss reported in the study was low (3%) presumably from the recalcitrant nature of the mixture of green waste and OMW. The OMW compost has a phenolic content that was reduced during the composting process (Siles-Castellano et al., 2020), and the final compost having a low total phenol content (0.09 g/L). The phytotoxicity index of the compost is relatively low and remains stable (5%), which is considered non-phytotoxic to the growing media for basil and ornamental plants. It is worth mentioning that the phytotoxicity of OMW is often related to its high phenolic content (Dermeche et al., 2013). However, nonphenolic organic compounds, inorganic salts, and traces of heavy metals are also said to contribute to its overall phytotoxicity (Kistner et al., 2004; Aviani et al., 2010).

A novel approach for OMW application using solar drying followed by composting has been adopted by Galliou et al. (2018). The OMW was gradually added into the manure for 6 months and solar-dried before continuing with the composting process with grape marc for 3 months. Results show that the solar-drying affects the nutrient value of the compost materials, which increases significantly for N (3.5%) and K (6.5%) values due to the repeated addition of OMW into swine manure. In contrast, the phosphorus (1%) value fluctuates with the repetition addition. The compost also has a less phenol content (2.9 g/kg), which was attributed to the addition of grape marc into the mixture. The GI value measured was also greater than 80%, indicating phytotoxic-free compost (Jeong et al., 2017). When the compost was applied to chili pepper cultivation, an improvement in the plant's growth and yield was recorded. The nutrients in the final compost are comparable to those in the commercial chemical NPK fertilizer for plant growth.

ALCOHOL/MOLASSES DISTILLERY WASTEWATER (DSW)

Composting has made use of alcohol/molasses distillery wastewater, also known as vinasse. Limited reports on the direct use of DSW have been found in recent years. Alavi et al. (2017) conducted a study by integrating co-composting and vermicomposting processes using vinasse, cow manure, and bagasse for 60 days. Since the vinasse contains high salinity, natural calcium zeolite was added to improve the quality and decrease the quality on day 21, prior to vermicomposting. The pH was found to be in the alkali range at the end of the investigation, with a declining trend in C/N. The low C/N was expected since an increase in the earthworm population caused more organic compounds to be degraded, resulting in more nitrogen in the compost mixture. The rise in nitrogen in the soil has also been ascribed on the addition of glaze, mucus, growth-stimulating hormones, and enzymes from nitrogen-rich earthworms. Due to the use of zeolite, which entraps the potassium in its pore as well as its ion exchange characteristics (Turan, 2008), the potassium (K) value reveals a deficiency (0.06%–0.15%) The GI was found to be 100%, indicating that the final compost is free of phytotoxins. The weight and number of earthworms have grown after vermicomposting. However, the ratio of vinasse added should be controlled as high vinasse is unfavorable by the earthworms.

In another study, an improved method has been attempted by Malik et al. (2019) on composting the DSW with pressmud. Due to the low biodegradability, recalcitrant nature, and high retention time of the DSW that they used, the pre-treatment was employed to enhance the rate of composting and obtain a good quality of compost. In this case, ozone pretreatment was used to DSW to breakdown the large molecules into short-chain intermediate products that can enter cells and become readily biodegradable (the smaller size enhances the biological oxidation rate). The composting of treated DSW with pressmud was run for 40 days with a different ratio (1:3, 1:4, and 1:5). The results show that the ozone-pretreated DSW indeed increased the percentage of NPK value in the compost. The highest spinach seed GI value of 96% was recorded for a lower ratio of pressmud (1:3) as compared to untreated DSW (21%). Increasing the pressmud ratio gradually increases the EC value, C/N, and pH. The lower ratio also recorded the lowest C/N (13), pH (6.8), and an EC of 1.22 dS/m.

SWINE WASTEWATER

SW or LM can also be used as a moisture conditioner and nutrient enhancer. Both effluent and solid manure are water- and nutrient-rich resources that can be employed in the composting of carbon-rich substrates. According to Vasquez et al. (2013), using swine slurry to water compost that constitutes solid screened manure and poplar wood chips during the thermophilic phase causes some of the water to evaporate and some of the nutrient molecules, particularly nitrogen compounds, to be retained.

Fan et al. (2019a) studied the effects of controlled addition of compost characteristics on the nutrient retention agents and the effects of bulking agent types on water consumption in co-composting of SW and solid manure with two different solid substrates (corncob and rice husk) windrows. Compared to the rice husk, the addition of SW with corncob showed better results in final compost pH (8.0), EC (2.71 mS/cm), and NPK (N-36.5 g/kg, P-29.8 g/kg, and K-20.1 g/kg) values. The dissimilarity between the nutrient value and nutrient retention may be linked to the generation of greater biological heat and maintenance of a longer thermophilic stage, which was more favorable for LM consumption in corncob windrows. The GI value (>80%) implies that phytotoxicity in the compost has been removed, regardless of the two bulking agents. The study revealed that adding LM in a controllable environment is the key to increasing composting efficiency and reducing leachate.

Fan et al. (2021) studied the use of SW in aerobic windrow co-composting of solid pig manure on rice husk. Other than recycling of SW, there is a significant increase in the nutrient amount (N, P, K, Cu, and Zn) of the final compost. They discovered that nutrient enhancement in compost is due to the weight reduction caused by the degradation of organic matter during the aerobic composting process rather than the addition of LM. However, controlled addition of LM may promote the biodegradation of organic matter in the compost, implying that the factors responsible for improved final compost nutrients are not only the nutrients contained in the addition of LM, but also the enhanced biodegradation of organic wastes activated by the controlled composting process, as reported in their study (Fan et al., 2021). The findings reported that with the controlled LM addition, the NPK values were increased by 21% (N), 12.13% (P), and 12.7% (K) as compared to without the LM addition. In addition, the total volume of leachate created during the composting process was less than 0.4% (0.037 L) of the total LM added. The created leachate was reused into the composting windrows, reducing water and nutrient loss during the composting process. Similar findings were also reported in their previous studies (Fan et al., 2019b) where controlled addition of LM to swine manure greatly improved the compost MC, reduced leachate generation, and effectively regulated the thermophilic stage. The final compost is also phytotoxic-free and rich in nutrient retention.

TL Wastewater

The TL has been used in composting techniques by a few researchers. When compared to utilizing tap water as a moisture conditioner for composting, reusing TL with a set nitrogen composition (predominantly NH_4^+-N (1,577 mg/L) and NO_2^+-N (2,271 mg/L)) enhanced the maturity and TN content of the composting product. Wu et al. (2015) explored the application of TL with food waste and mushroom bran composting as a moisture conditioner, where TL compensated for moisture loss during composting. The temperature climbed to almost 55°C and maintained until the composting process was finished. The finished compost has a pH of 7.7, a low C/N ratio of 15%, and a MC content of 38%. According to Wu et al. (2015), adding TL, which contains a significant amount of nitrogen, to TL composting increased aerobic bacterial activity. As a result, nitrogen lost owing to odor emissions can be used to increase the nitrogen content of the compost product during composting in the form of a nitrogen-containing waste liquid (TL). With the addition of TL, the nitrogen mass increased by 3.1 g, contributing to a conservation rate of 47% of the add-in TL. Despite the high carbon-to-nitrogen ratio, their study has the shortest composting time of TL, taking only around 2 weeks (as shown in Table 6.2). The large amount of nitrogen in the TL may have accelerated the aerobic bacteria's activities, resulting in a short composting time.

In another studies, Xie et al. (2021a, 2021b, 2021c) used simulated trickling liquid (STL), which is high in nitrate, ammonium, and nitrite, to simulate the TL for their study as a composting moisturizer agent. Fermentation residue, sawdust, mushrooms, and food waste are examples of solid substrates used for TL waste in composting. The TL wastewater, which contains nitrate-rich, ammonium-rich, and nitrite-rich nutrients, may not only adjust compost moisture content but also improve total nitrogen content without releasing additional greenhouse gases. The use of ozonation to pre-treat complex materials improves biodegradability and reduces color, allowing for better composting (Malik et al., 2019). Furthermore, when used as a moisturizing agent during the thermophilic and cooling periods, the toxicity of the nitrite-rich waste is minimized. According to Xie et al. (2021b), the resulting compost has a nitrate concentration that is 2–10 times higher than when utilizing tap water at high temperatures. Organic decomposition was severely slowed in the mesophilic phase, according to Xie et al. (2021c). Nitrite-rich wastewater increased nitrate nitrogen levels by 10–11 times and TN levels by 2%–9%.

Monosodium Glutamate Wastewater (MSGW)

MSGW contains abundant nutrient substances, and it could be feasible to reuse these organic substances using eco-technological methods. There has only been a little amount of research done on the composting of MSG wastewater from the MSG manufacturing industry. MSG wastewater is a promising conditioner that can compensate for the compost's very high pH as well as the nutrient value loss caused by NH_3 volatilization during the composting process. The effect of MSG dosage and frequency on the amount of N lost due to NH_3 volatilization was investigated by Liu et al. (2015). MSGW is used to increase the acidity of compost and minimize nitrogen loss in cattle due to NH_3 volatilization. MSG is used in the composting process, along with cattle manure, chicken manure, and mushroom waste. With the triple addition of 2% MSG dosage, the study found that MSG reduced the volatilization of NH_3 by up to 71% and boosted the TN. Despite the high total dosage (6%), the GI value was found to be greater than 80%, and no phytotoxicity was noted in the final compost caused by the addition of MSGW.

MSGW may be an excellent conditioner for regulating pH and minimizing nitrogen loss during composting. MSGW nutrients, which provided a readily available carbon source to support microorganism metabolism in compost, are likely to have increased microbial activity in the material. Respiration reduces as it is digested by microbes, but the acidity of the matrix adjusted by MSGW does not rise as quickly when MSGW is depleted, allowing for a longer period of NH_3 volatilization reduction from the compost matrix (Gabhane et al., 2012; Liu et al., 2014).

Due to the abundance of organic ingredients in MSG, Ji et al. (2014) revealed that MSG wastewater can be used as a medium for microbial cultivation in addition to composting. The nutrients in diluted MSG were transferred to the *Chlorella vulgaris* microalgae cultivation in their study, which improved the algae's nutritional quality and stimulated the plant's growth.

SUMMARY OF THE WASTEWATER APPLICATION FOR COMPOSTING

In general, there were significant differences in the final products obtained from each industrial waste stream, which are likely to affect the phytotoxicity and define compost quality. As depicted in Table 6.1, the differences in parameters, including composting time, pH, C/N, temperature, solid substrate addition, aeration, and moisture content, contribute to the final performance of the compost produced. Even though this review paper focuses on the nutrient value and moisture effect of the compost, the parameters chosen in every compost correlate with the stability and maturity of the compost. Co-composting with a range of microorganisms such as bacteria, actinomycetes, fungus, and yeast, as well as earthworms (*Eisenia fetida and Lumbricus rubellus*), expedites the decomposition process. The composting of the wastewater and the addition of the fresh wastewater also influence the composting temperature in obtaining the desired initial moisture content and wetting them during the thermophilic phase. In general, an increasing trend in temperature was recorded, reaching the thermophilic value (> 50°C), before decreasing to a stable temperature for all composting. The temperature evolution trend gives an indication of the efficiency of the composting process (Vasquez et al., 2015). The addition of solid substrate as bulking agents, including animal manures, food waste, grape marc, green waste, and other agricultural waste, increases the compost porosity, which in turn enhances the oxygen availability, which then accelerates the microbial activities.

The final C/N ratio can also be an indicator for determining the maturity of the compost. A satisfactory C/N ratio is less than 20, but the ideal ratio is less than 15 (Huang et al., 2006; Malik et al., 2018). The majority of the final C/N ratios recorded, as reported in Table 6.2, were within an acceptable range. The release of some organic carbon such as carbon dioxide gas, as well as nitrogen mineralization due to microbial decomposition, were both attributed for the decrease in the C/N ratio (Alavi et al. 2017). The final pH was also observed to be within the standard limit for matured compost, which is 5.5 < pH < 8.5.

In comparison to OMW, TL, SE, MSGW, and DSW, POME had the maximum moisture content (80%) and resulted in an average of 63% of final compost as indicated in Table 6.2. Based on the final performance of the wastewater composting, all the compost moisture content is in the range of 40%–80%. Nonetheless, as compared to other types of waste, POME acts as a better humidifier or moisture conditioner for compost. In vinasse wastewater composting, the liquid addition needs to be controlled as it contains high concentrations of soluble salts that can affect the plant growth as well as the reproduction of worms in the integrated composting–vermicomposting process (Owojori et al., 2008; Alavi et al., 2017). The temperature distribution of compost affects its moisture content, with the rise in temperature during composting being the primary cause of moisture content loss (Zakarya et al., 2015).

Aeration is also used to eliminate surplus heat generated by organisms as a result of the metabolic reactions they are carrying out. The optimal oxygen concentration for composting is between 15% and 20% (Bernal et al. 2009). According to Sim et al. (2016), the compost can be provided with oxygen via several methods such as natural convection, physical turning (Raza et al., 2016), or forced aeration (Xie et al., 2021a). As the rate of aeration increases, energy transfer also increases, resulting in temperature reductions. For aerobic decomposition, it is critical that the compost receives steady oxygen at a concentration of at least 5%. Aeration aids in the eradication of anaerobic areas of the pile and ensures that the mature compost is devoid of anaerobic zones (Sim et al., 2016). Composting time is shorter in piles with forced aeration systems than in piles with physical or mechanical turning.

There is a significant increase in the nutrient amount (N, P, and K) of the final compost observed in Table 6.2. The weight loss induced by the decomposition of organic matter during the aerobic composting process, as well as the initial nutrients in the wastewater, contribute to the nutritional enrichment in compost. The nitrogen value was reported to be in the range of 1.05%–6.1%.The highest nitrogen value was recorded in the TL compost (6.1%), contributed by the high nitrogen in the TL content that accelerates the aerobic bacteria's activities, thus enhancing the nitrogen value. Composting wastewater resulted in an increase in phosphorus and potassium concentrations in the ranges of 0.06%–2.8% and 0.06%–12.05%, respectively. This phenomenon can be linked to the creation of acids by microbes during the decomposition of organic substances, which then convert the mixture's insoluble potassium content to soluble form, explaining the increase. The POME composting technique (which uses an integrated co-composting–vermicomposting process) had the highest recorded rate (up to 12.05%). The increased potassium value could be due to earthworms processing waste material into more exchangeable potassium, and the increase in phosphorus could be due to the release of phosphorus from the earthworm gut as well as P-solubilizing microorganisms in the worm cast during organic matter consumption (Hayawin et al., 2012).

Phytotoxicity will reveal the quality or stability of a final compost. As a result, the GI is used to monitor the performance of the composting process and the quality of the compost generated, as well as to identify production issues and enhance waste processing in order to produce high-quality compost (Chowdhury et al., 2013; Siles-Castellano et al., 2020). Compost material with a GI of more than 100% acts as a phytostimulant or a phytonutrient, while compost material with a GI of more than 80% has no phytotoxicity, compost material with a GI of 50% to 80% is moderately phytotoxic, and compost material with a GI of less than 50% is extremely phytotoxic, according to Emino and Warman (2004). The value of GI is determined by electrical conductivity and pH; however, it has no influence on the organic matter content of compost materials (Siles-Castellano et al., 2020). In this review, the majority of compost produced from the wastewater has a GI greater than 80%. The highest GI values recorded were 158% (Zahrim et al., 2016) and 156% (Galliou et al., 2018), which were attained by using POME and OMW in the compost production process. The initial pH of the compost has an effect on the GI tract and the initial C/N ratio (Barral and Paradelo 2011). The average initial pH for all compost is in the range of 5–8, which is favorable for the degradation of organic matter by microbes such as protein (Wang et al. 2012) and speeds up mineralization and humification. As a result, more composted mixed materials are transformed into mature organic fertilizer, directly increasing the GI. Furthermore, composting with a lower C/N ratio helps enhance GI by avoiding NH_4^+ buildup, which prevents seeds from sprouting.

CHALLENGES AND FUTURE PERSPECTIVES

Composting fresh wastewater is still a challenge encountered by researchers that carry out the composting process using wastewater as the main feedstock. The improvement in wastewater consumption in composting could be achieved by the optimization of significant influencing factors, which would be an interesting trend for further research on composting or integrated composting–vermicomposting. Concerns have been expressed concerning the maximum amount of liquids that can be mixed into a given volume of solid waste, as well as the increased risk of leachates (Avidov et al., 2018). The cost of the operation is also a hurdle in wastewater composting, as transportation of the effluent or liquid from the industries to the composting facilities incurs additional costs. Operators must obtain both land and access routes for composting operations. Moving to enclosed facilities instead of open windrows is also not cost-effective because it necessitates higher capital expenditure.

In order to make composting as cheap and efficient as possible, it is recommended to make sure that as much liquid as possible can be absorbed and treated together with the solid substrate. The adsorption of liquid substrates onto solid substrates requires more research. The adsorption of the liquid substrate into the solid substrate is an interesting subject to explore since research into the mechanism and volume of the liquid fresh wastewater used is still limited and requires further investigation. The majority of the reviewed publications in the literature were largely conducted in the laboratory or on a small scale. In order for the large scale to be implemented, many factors need to be considered and smoothed out so that it can be efficiently implemented (Barthod et al., 2018).

Using fresh wastewater in composting can generate unpleasant odors if the plant is not well designed and the processes do not operate properly or efficiently. If not adequately handled, the improper disposal of pollutants and residues removed from the composting process might attract flies and insects and become a breeding place. To kill fly eggs and larvae, temperatures during the aerobic process should be higher than 50°C–55°C. Furthermore, deodorizing the environment is an effective way to reduce the risk of an unpleasant odor. Investing in and maintaining deodorizing systems, on the other hand, will be costly. Even while composting reduces GHG emissions, the aerobic fermentation process produces a certain quantity of CH_4 and nitrous oxide gases. As a result, regular turning and aeration are required during the composting process to maintain aerobic conditions and reduce GHG emissions (Kawai et al., 2020). Further study on the GHG emissions is suggested for wastewater composting.

Apart from the common phytotoxicity test, an assessment of the final compost (humic substances (HSs)) using nondestructive spectroscopic methods such as FTIR, C-NMR, or GC-MS as an index of maturity can be further explored. Only a few reports have used this test to evaluate the compost in this reviewed paper (Malik et al., 2019). The evolution of HS during their various transformations allowed for the identification of a number of groups and functions present during the composting process. The analysis of FTIR spectra in the case of chemical/biochemical reactions can be observed with the disappearance or appearance of certain peaks in the final compost due to the completed degradation of organic compounds and the formation of new compounds as intermediate organic compounds (Raghu et al., 2009; Malik et al., 2019; Aji et al., 2021) or by following the band shifts in the case of reactions that involve structural changes. Many variables, such as the origin of the humic material, the extraction process, and the purification procedures, still influence the FTIR and NMR investigations of humic structure sequences (Fels et al., 2015).

In composting, moisture content, temperature, oxygen supply, pH, C/N ratio, particle size, and addition of solid substrate (bulking agents) should be controlled and maintained within the appropriate range during aerobic composting. Maintaining a suitable moisture level, for example, necessitates a balancing act of competing forces such as biological water production and evaporation

TABLE 6.3

Example of Mathematical Model Used for Composting in Walling et al. (2020) and Walling and Vaneeckhaute (2021)

Kinetic Modeling	Remarks
$R_{\text{degradation; first-order}} = -\dfrac{d[S_i]}{dt} = k_i \cdot [S_i]$	k_i is the (hydrolysis) rate constant (s^{-1}). In both equations, $[S_i]$ is the concentration of the substrate (kg m^{-3}), at is the time (s), μ and $\mu_{\text{max,i}}$ are
$R_{\text{degradation; Monod}} = -\dfrac{d[S_i]}{dt} = \mu_i \dfrac{X_i}{Y_{s_i}} = \dfrac{\mu_{\text{max,i}}[S_i]}{k_{s,i}+[S_i]} \dfrac{X_i}{Y_{s_i}}$	the specific and maximum growth rate of the microorganism (s^{-1}), respectively; $k_{s,i}$ is the half-velocity constant (kg m^{-3}), and Y_{S_i} is the yield coefficient (kg kg^{-1})
Rates of biological oxygen consumption, $Ro_{2 \text{ consumption}} = Yo_2 R_{\text{degradation}}$	Oxygen consumption coefficient (Yo_2)
Rates of water production, $R_{\text{H}_2\text{O consumption}} = Y_{\text{H}_2\text{O}} R_{\text{degradation}}$	
Decay, $R_{\text{decay}} = b_i X_i$	b_i is the microbial death rate (s^{-1})
Correction functions $R_{\text{degradation}} = -\dfrac{d[S_i]}{dt} = k_d f_T f_{\text{MC}} f_{\text{O}_2} [S_i]$	$[S_i]$ is the mass or concentration of biodegradable substrate, t is time, k_d is the (degradation/hydrolysis) rate constant (usually either at 20°C or around 60°C), and f_T, f_{MC}, and f_{O2} are correction functions for temperature, moisture content, and oxygen content.

deduction. Water is required for the degradation of organic substrate to support biological activity, but water evaporation, similar to a cooling system, is also crucial in eliminating excess heat to avoid excessively high temperatures (Klejment and Rosiski, 2008). Adjusting the optimal moisture content is common; however, solutions vary depending on the process and substrate. To optimize the parameters that can pass the quality target specified by the researcher, optimization techniques and modeling might be applied. The number of studies reporting the use of fresh wastewater composting using optimization and kinetics is limited, and more research is needed in this area because the wastewater discussed in this review has a high potential for improving the nutrients in final compost and providing additional moisture during the composting process.

Walling et al. (2020) and Walling and Vaneeckhaute (2021) developed modeling techniques (Table 6.3) that serve as a framework for future composting models, particularly for use in decision-making scenarios where information is limited or where the necessity for quick findings and optimization makes more extensive methods impractical. These kinetic models can also be used as a foundation for more complex models that go beyond only anticipating degradation. The challenge with composting modeling is that the correction functions all vary as functions of time, given the change in these variables throughout the process. According to Wang et al. (2015), it's difficult to say which model is better simply based on simulation behavior or accuracy, but a good way to compare them would be to combine different parameter models for simulating a common composting process, which would also help distinguish the benefits and drawbacks of the various models. Interested readers can further be referred to the works for more details in mathematical models (Walling et al., 2020; Walling and Vaneeckhaute, 2021; Sokač et al., 2022).

Wastewaters generated and composts can be potentially used as nutrient sources for crops, turf grasses, ornamental pants, and landscaping applications. Liquid extracts made from digested organic waste products could be used as supplemental nutrition sources for nursery stock and turf grass species (Michitsch et al., 2007). A few reports have been mentioned of using the wastewater compost as growing media for plants (Majbar et al., 2018; Avidov et al., 2018). Although more research is needed to improve some plant performance (pH fluctuations, nutrient imbalances, etc.), the composts from fresh industrial effluent can still be used as growing media for a variety of agricultural and ornamental plants. In addition, the use of local yeast as an inoculant can be considered as it had a positive effect on the compost temperature and improved the stability and maturity performance of composting (Ugak et al., 2022).

CONCLUSION

Different types of industrial wastewater used as nutrient enhancers and moisturizers showed different performances in the production of compost. To achieve the optimal final compost content, the temperature of the composting process must be maintained between 40°C and 65°C, the initial C/N ratio must be kept below 50, and the pH must be between 6.5 and 8.5. The POME and OMW have been the most frequently used industrial wastes as a nutrient enhancer and moisturizer in the production of compost. The final pH, MC, and EC values were observed in the recommended range, and other physico-chemical parameters were in the acceptable range. The wastewater can compensate for the moisture loss during the composting process. Although the moisture content and GI are lower when using OMW, the average nutrients produced are higher, which may explain why more research is being undertaken with OMW. Gradual and controlled addition of wastewater during the composting process leads to an increment in NPK values in the final compost product. Further research should be conducted by using domestic wastewater and other forms of industrial wastewater as nutrient enhancers and moisturizers in composting.

ACKNOWLEDGMENT

The authors gratefully acknowledge the funding for this project from Universiti Malaysia Sabah under the SDK grant scheme with grant number SDK0164–2020, as well as the postdoctoral awarded to Dr. Junidah Lamaming.

REFERENCES

Abdullah, N., & Sulaiman, F. (2013). The properties of the washed empty fruit bunches of oil palm. *Journal of Physical Science*, 24(2), 117–137.

Adam, S., Ahmad, S. S. N. S., Hamzah, N. M., & Darus, N. A. (2016). Composting of empty fruit bunch treated with palm oil mill effluent and decanter cake. In *Regional Conference on Science, Technology and Social Sciences (RCSTSS 2014)*, Springer, Singapore, pp. 437–445.

Ahmad, A., Buang, A., & Bhat, A.H. (2016). Renewable and sustainable bioenergy production from microalgal co-cultivation with palm oil mill effluent (POME): A review. *Renewable and Sustainable Energy Review*, 65, 214–234.

Ahmad, M. N., Ramlah Ahmad Ali, S., & Ali Hassan, M. (2014). Physico-chemical Changes during co-composting of chipped-ground oil palm frond and palm oil mill effluent. *Oil Palm Bulletin*, 69, 1–4.

Aji, N. A. S., Yaser, A. Z., Lamaming, J., Ugak, M. A. M., Saalah, S., & Rajin, M. (2021). Production of food waste compost and its effect on the growth of Dwarf Crape Jasmine. *Jurnal Kejuruteraan*, 33 (3), 413–424

Alavi, N., Daneshpajou, M., Shirmardi, M., Goudarzi, G., Neisi, A., & Babaei, A. A. (2017). Investigating the efficiency of co-composting and vermicomposting of vinasse with the mixture of cow manure wastes, bagasse, and natural zeolite. *Waste Management*, 69, 117–126.

Alkarimiah, R., & Rahman, R. A. (2014). Co-composting of EFB and POME with the role of nitrogen-fixers bacteria as additives in composting process-A review. *International Journal of Engineering Science and Innovative Technology*, 3(2), 132–145.

Alkarimiah, R, & Suja, F. (2020). Composting of EFB and POME using a step-feeding strategy in a rotary drum reactor: The effect of active aeration and mixing ratio on composting performance. *Polish Journal of Environmental Studies*, 29(4), 2543–2553.

Atif, K., Haouas, A., Aziz, F., Jamali, M. Y., Tallou, A., & Amir, S. (2020). Pathogens evolution during the composting of the household waste mixture enriched with phosphate residues and olive oil mill wastewater. *Waste and Biomass Valorization*, 11(5), 1789–1797.

Aviani, I., Laor, Y., Medina, S., Krassnovsky, A., Raviv, M. (2010). Co-composting of solid and liquid olive mill wastes: Management aspects and the horticultural value of the resulting composts. *Bioresource Technology*, 101, 6699–6706.

Avidov, R., Saadi, I., Krasnovsky, A., Medina, S., Raviv, M., Chen, Y., & Laor, Y. (2018). Using polyethylene sleeves with forced aeration for composting olive mill wastewater pre-absorbed by vegetative waste. *Waste Management*, 78, 969–979.

Ayilara, M. S., Olanrewaju, O. S., Babalola, O. O., & Odeyemi, O. (2020). Waste management through composting: Challenges and potentials. *Sustainability*, 12(11), 4456.

Baharuddin, A. S., Wakisaka, M., Shirai, Y., Abd Aziz, S., Rahman, A. A., Hassan, M. a. (2009). Co-composting of empty fruit bunches and partially treated palm oil mill effluents in pilot scale. *International Journal of Agricultural Research*, 4(2), 69–78.

Bai, Z.H., Zhang, H.X., Qi, H.Y., Peng, X.W., & Li, B.J. (2004). Pectinase production by Aspergillus niger using wastewater in solid state fermentation for eliciting plant disease resistance. *Bioresource Technology*, 95, 49e52.

Barthod, J., Rumpel, C., & Dignac, M-F. (2018). Composting with additives to improve organic amendments. A review. *Agronomy for Sustainable Development*, 38(17), 1–23.

Barral, M.T., Paradelo, R. (2011). A review on the use of phytotoxicity as a compost quality indicator. *Dynamic Soil, Dynamic Plant*, 5, 36e44.

Bernal, M. P., Alburquerque, J. A., & Moral, R. (2009). Composting of animal manures and chemical criteria for compost maturity assessment. A review. *Bioresource Technology*, 100(22), 5444–5453.

Bustamante, M. A., Restrepo, A. P., Alburquerque, J. A., & Perez-Murcia, M. D. (2013). Recycling of anaerobic digestates by composting: Effect of the bulking agent used. *Journal of Cleaner Production*, 47, 61–69.

Chehab, H., Tekaya, M., Ouhibi, M., Gouiaa, M., Zakhama, H., Mahjoub, Z., Laamari, S., Sfina, H., Chihaoui, B., Boujnah, D., & Mechri, B. (2019). Effects of compost, olive mill wastewater and legume cover crops on soil characteristics, tree performance and oil quality of olive trees cv. Chemlali grown under organic farming system. *Scientia Horticulturae*, 253, 163–171.

Chowdhury, A. K. M. M. B., Michailides, M. K., Akratos, C. S., Tekerlekopoulou, A. G., Pavlou, S., & Vayenas, D. V. (2014). Composting of three phase olive mill solid waste using different bulking agents. *International Biodeterioration & Biodegradation*, 91, 66–73.

Dennehy, C., Lawlor, P. G., Jiang, Y., Garddiner, G. E., Xie, S. H., Nghiem, L. D., & Zhan, X. M. (2017). Greenhouse gas emissions from different pig manure management techniques: A critical analysis. *Frontiers in Environmental Science and Engineering*, 11, 1–11.

Dermeche, S., Nadour, M., Larroche, C., Moulti-Mati, F., Michaud, P. (2013). Olive mill wastes: Biochemical characterizations and valorization strategies. *Process Biochemistry*, 48, 1532e1552.

Emino, E.R., & Warman, P.R. (2004). Biological assay for compost quality. *Compost Science and Utilization*, 12, 342–348.

Fan, H.Y., Liao, J., Olusegun, K.A., Liu, L., Huang, X., Wei, L., Xie, W., Yu, H., & Liu, C. (2019a). Effects of bulking material types on water consumption and pollutant degradation in composting process with controlled addition of different liquid manures. *Bioresource Technology*, 288, 121517.

Fan, H.Y., Liao, J., Olusegun, K.A., Liu, L., Huang, X., Wei, L.L., Li, J., Xie, W., & Liu, C.X. (2019b). Effects of compost characteristics on nutrient retention and simultaneous pollutant immobilization and degradation during co-composting process. *Bioresource Technology*, 25, 61–69.

Fan, H., Liao, J., Abass, O. K., Liu, L., Huang, X., Li, J., Tian, S., Liu, X., Xu, K., & Liu, C. (2021). Concomitant management of solid and liquid swine manure via controlled co-composting: Towards nutrients enrichment and wastewater recycling. *Resources, Conservation and Recycling*, 168, 105308.

Fels, L. E., Zamama, M., & Hafidi, M. (2015). Advantages and limitations of using FTIR spectroscopy for assessing the maturity of sewage sludge and olive oil waste co-composts. In: Chamy, R., Rosenkranz, F., & Soler, L. (eds.), *Biodegradation and Bioremediation of Polluted Systems - New Advances and Technologies*. IntechOpen, London, 127–144.

Gabhane, J., William, S.P., Bidyadhar, R., Bhilawe, P., Anand, D., Vaidya, A.N., & Wate, S.R. (2012). Additives aided composting of green waste: Effects on organic matter degradation, compost maturity, and quality of the finished compost. *Bioresource Technology*, 114, 382e388.

Galliou, F., Markakis, N., Fountoulakis, M. S., Nikolaidis, N., & Manios, T. (2018). Production of organic fertilizer from olive mill wastewater by combining solar greenhouse drying and composting. *Waste Management*, 75, 305–311.

Hachicha, S., Sellami, F., Cegarra, J., Hachicha, R., Drira, N., Medhioub, K., & Ammar, E. (2009). Biological activity during co-composting of sludge issued from the OMW evaporation ponds with poultry manure—physico-chemical characterization of the processed organic matter. *Journal of Hazardous Materials*, 162(1), 402–409.

Hampton, M.O. (2006). Soil and nutrient management: Compost and manure. Integrated Pest Management Florida, Florida, pp. 37–40.

Hasan, M. Y., Hassan, M. A., Mokhtar, M. N., Shirai, Y., & Idris, A. (2021). Effect of initial carbon to nitrogen ratio on the degradation of oil palm empty fruit bunch with periodic addition of anaerobic palm oil mill effluent sludge. *Pertanika Journal of Science and Technology*, 29(4), 2435–2449.

Hassen, W., Hassen, B., Werhani, R., Hidri, Y., & Hassen A. (2021). Co-Composting of Various Residual Organic Waste and Olive Mill Wastewater for Organic Soil Amendments. In Makan A (ed) *Humic Subtances*. Intech Open, London, 1–17.

Hau, L. J., Shamsuddin, R., May, A. K. A., Saenong, A., Lazim, A. M., Narasimha, M., & Low, A. (2020). Mixed composting of palm oil empty fruit bunch (EFB) and palm oil mill effluent (POME) with various organics: An analysis on final macronutrient content and physical properties. *Waste and Biomass Valorization*, 11(10), 5539–5548.

Hayawin, Z., Astimar, A.A., Mokhtar, A., Ibrahim, M., Abdul Khalil, H.P.S., Zawawi, I. (2012). Vermicomposting of empty fruit bunch with addition of palm oil mill effluent solid. *Journal of Oil Palm Research*, 24, 1542–1549.

Hoarau, J., Caro, Y., Grondin, I., & Petit, T. (2018). Sugarcane vinasse composting: Toward a status shift from waste to valuable resource. A review. *Journal of Water Process Engineering*, 24, 11–25.

Huang, G. F., Wu, Q. T., Wong, J. W. C., Nagar, B.B. (2006). Transformation of organic matter during co-composting of pig manure with sawdust. *Bioresource Technology*, 97, 1834–1842.

Jeong, K-H., Kim, J-K., Ravindran, B., Lee, D-J., Wong, C.W.C., Selvam, A., Karthikeyan, O.P., & Kwag, J-H. (2017). Evaluation of pilot-scale in-vessel composting for Hanwoo manure management. *Bioresource Technology*, 245, 201–206.

Ji, Y., Hua. W., Li, X., Ma, G., Song, M., Pei, H. (2014). Mixotrophic growth and biochemical analysis of Chlorella vulgaris cultivated with diluted monosodium glutamate wastewater. *Bioresource Technology*, 152, 471–476.

Jusoh, M. L.C., Manaf, L. A., Latiff, P. A. (2013). Composting of rice straw with effective microorganisms (EM) and its influence on compost quality. *Iranian Journal of Environmental Health Science and Engineering*, 10(17), 1–9.

Kala, D.R., Rosenani, A.B., Fauziah, C.I. & Thohirah, L.A. (2009). Composting oil palm wastes and sewage sludge for use in potting media of ornamental plants. *Malaysian Journal of Soil Science*, 13, 77–91.

Karakashev, D., Schmidt, J. E., & Angelidaki, I. (2008). Innovative process scheme for removal of organic matter, phosphorus and nitrogen from pig manure. *Water Research*, 42(15), 4083–4090.

Kawai K., Liu, C., & Gamarallage, P. J. D. (2020). *CCET Guideline Series on Intermediate Municipal Solid Waste Treatment Technologies: Composting*. United Nations Environment Programme, Osaka, 1–38.

Kefalogianni, I., Skiada, V., Tsagou, V., Efthymiou, A., Xexakis, K., & Chatzipavlidis, I. (2021). Co-composting of cotton residues with olive mill wastewater: Process monitoring and evaluation of the diversity of culturable microbial populations. *Environmental Monitoring and Assessment*, 193(10), 1–16.

Kılıç, Z. (2020). The importance of water and conscious use of water. *International Journal of Hydrology*, 4(5), 239–241.

Kipçak, E., & Akgün, M. (2012). Oxidative gasification of olive mill wastewater as a biomass source in supercritical water: Effects on gasification yield and biofuel composition. *Journal of Supercritical Fluids*, 69, 57–63.

Kistner, T., Nitz, G., & Schnitzler, W.H. (2004). Phytotoxic effects of some compounds of olive mill wastewater (OMW). *Fresenius Environmental Bulletin*, 13(11):1360–1361.

Klejment, E., & Rosiński. M. (2008). Testing of thermal properties of compost from municipal waste with a view to using it as a renewable, low temperature heat source. *Bioresource Technology*, 99 (18), 8850–8855.

Krishnan, Y., Bong, C. P. C., Azman, N. F., Zakaria, Z., Abdullah, N., Ho, C. S., et al. (2017). Co-composting of palm empty fruit bunch and palm oil mill effluent: Microbial diversity and potential mitigation of greenhouse gas emission. *Journal of Cleaner Production*, 146, 94–100.

Lai, J. C., Chua, H. B., Saptoro, A., & Ang, H.M. (2013). Effect of isolated mesophilic bacterial consortium on the composting process of pressed-shredded empty oil palm fruit bunch. In: Ravindra, P., Bono, A., Chu, C (eds.), *Developments in Sustainable Chemical and Bioprocess Technology*. Springer, Berlin, 27–33.

Lee, Z. S., Chin, S. Y., Lim, J. W., & Witoon, T. (2019). Treatment technologies of palm oil mill effluent (POME) and olive mill wastewater (OMW): A review. *Environmental Technology and Innovation*, 15, 100377.

Lew, J. H., May, A. K. A., Shamsuddin, M. R, Aqsha, Lazim, A., Narasimba, M. M. (2020). Vermicomposting of palm oil empty fruit bunch (EFB) based fertilizer with various organics additives. *IOP Conference Series: Materials Science and Engineering*, 736, 052014.

Liew, W.L., Kassim, M.A., Muda, K., Loh, S.K., & Affam, A.C. (2015). Conventional methods and emerging wastewater polishing technologies for palm oil mill effluent treatment: A review. *Journal of Environmental Management*, 149, 222–235.

Liu, L., Kong, H., Lu, B., Wang, J., Xie, Y., & Fang, P. (2015). The use of concentrated monosodium glutamate wastewater as a conditioning agent for adjusting acidity and minimizing ammonia volatilization in live-stock manure composting. *Journal of Environmental Management*, 161, 131–136.

Liu, L., Li, T. Y., Wei, X. H., Jiang, B. K., & Fang, P. (2014). Effects of a nutrient additive on the density of func-tional bacteria and the microbial community structure of bioorganic fertilizer. *Bioresource Technology*, 172, 328e334.

Liu, R., & Zhou, Q. (2010). Fluxes and influencing factors of ammonia emission from monosodium glutamate production in Shenyang, China. *Bulletin of Environmental Contamination and Toxicology*, 85, 279–286.

Mahmod, S. S., Arisht, S. N., Jahima, J. M., Takriff, M. S., Tan, J. P., Luthfi, A. A. I., Abdul, P. M. (2021). Enhancement of biohydrogen production from palm oil mill effluent (POME): A review. *International Journal of Hydrogen Energy*, 47(96), 40637–40655.

Majbar, Z., Lahlou, K., ben Abbou, M., Ammar, E., Triki, A., Abid, W., Nawdali, M., Bouka, H., Taleb, M., el Haji, M., & Rais, Z. (2018). Co-composting of olive mill waste and wine-processing waste: An applica-tion of compost as soil amendment. *Journal of Chemistry*, 2018, 7918583.

Majbar, Z., Rais, Z., El Haji, M., Ben Abbou, M., Bouka, H., & Nawdali, M. (2017). Olive mill wastewater and wine by-products valorization by co-composting. *Journal of Materials and Environmental Sciences*, 8(9), 3162–3167.

Malik, S. N., Ghosh, P. C., Vaidya, A.N., & Mudliar, S. N. (2018). Ozone pretreatment of biomethanated distill-ery wastewater in a semi batch reactor: Mapping pretreatment efficiency in terms of COD, color, toxicity and biohydrogen generation. *Biofuels*, 1–9.

Malik, S. N., Ghosh, P. C., Vaidya, A. N., & Mudliar, S. N. (2019). Ozone pre-treatment of molasses-based bio-methanated distillery wastewater for enhanced bio-composting. *Journal of Environmental Management*, 246, 42–50.

Michitsch, R. C., Chong, C., Holbein, B. E., Voroney, R. P., & Liu, H-W. (2007). Use of wastewater and com-post extracts as nutrient sources for growing nursery and Turfgrass species. *Journal of Environmental Quality*, 36, 1031–1041.

Mohammad, N., Alam, M.Z., & Kabbashi, N.A. (2011). Development of compatible fungal mixed culture for composting process. *African Journal of Biotechnology*, 10(81), 18657–18665

Mohammad, N., Alam, M.Z., Kabashi, N.A., & Ahsan, A. (2012). Effective composting of oil palm industrial waste by filamentous fungi: A review. *Resource, Conservation and Recycling*, 58, 69–78.

Mohammad, N., Alam, M.Z., & Kabashi, N.A. (2013). Development of composting process of oil palm indus-trial wastes by multi-enzymatic fungal system. *Journal of Material Cycles and Waste Management*, 15 (3), 348–356.

Owojori, O. J., Reinecke, A. J., & Rozanov, A. B. (2008). Effects of salinity on partitioning uptake and toxicity of zinc in the earthworm Eisenia fetida. *Soil Biology and Biochemistry*, 40, 2385–2393.

Preisner, M., Smol, M., Horttanainen, M., Deviatkin, I., Havukainen, J., Klavins, M., Ozola-Davidane, R., Kruopienė, I., Szatkowska, B., Appels, L., Houtmeyers, S., & Roosalu, K. (2022). Indicators for resource recovery monitoring within the circular economy model implementation in the wastewater sector. *Journal of Environmental Management*, 304, 114261.

Raghu, S., Lee, C. W., Chellamal, S., Palanichamy, S., & Ahmad Basha, C. (2009). Evaluation of electrochemi-cal oxidation techniques for degradation of dye effluents- a comparative approach. *Journal of Hazardous Material*, 171, 748–754.

Raza, S., Raza, S., & Ahmad, J. (2016). Composting process: A review. *International Journal of Biological Research*, 4(2), 102–104.

Research and Markets. (2022). *Water Testing and Analysis - Global Market Trajectory & Analytics*. Retrieved at https://www.researchandmarkets.com/reports/4806422/water-testing-and-analysis-global-market on 12 February 2022.

Rupani, P.F., Embrandiri, A., Ibrahim, M.H., Shahadat, M., Hansen, S. B., Mansor, N. N. A. (2017). Bioremediation of palm industry wastes using vermicomposting technology: Its environmental applica-tion as green fertilizer. *3 Biotech*, 7(3), 155.

Rupani, P. F., Singh, R. P., Ibrahim, M. H., & Esa, N. (2010). Review of current palm oil mill effluent (POME) treatment methods: Vermicomposting as a sustainable practice. *World Applied Sciences Journal*, 11(1), 70–81.

Satyawali, Y., Balakrishanan, M. (2008). Wastewater treatment in molasses based alcohol distilleries for COD and color removal: A review. *Journal of Environmental Management*, 86(3), 481–497.

Shinde, P. A., Ukarde, T. M., Pandey, P. H., & Pawar, H. S. (2020). Distillery spent wash: An emerging chemi-cal pool for next generation sustainable distilleries. *Journal of Water Process Engineering*, 36, 101353.

Siles-Castellano, A. B., López, M. J., López-González, J. A., Suárez-Estrella, F., Jurado, M. M., Estrella-González, & Moreno, J. (2020). Comparative analysis of phytotoxicity and compost quality in industrial composting facilities processing different organic wastes. *Journal of Cleaner Production*, 252, 119820.

Sim, V. J. W., Bing, C. H., Saptoroand, A., & Nandong, J. (2016). Effects of temperature, aeration rate and reaction time on composting of empty fruit bunches of oil-palm. *Iranica Journal of Energy and Environment*, 7(2), 156–162.

Singh, S., Rekha, P. D., Arun, A. B., Huang, Y-M., Shen, F. T., & Young, C.C. (2011). Wastewater from monosodium glutamate industry as low cost fertilizer source for corn (Zea mays L.). *Biomass and Bioenergy*, 35, 4001–4007.

Sokač, T., Valinger, D., Benković, M., Jurina, T., Gajdoš Kljusurić, J., Radojčić Redovniković, I., Jurinjak Tušek, A. (2022). Application of optimization and modeling for the enhancement of composting processes. *Processes*, 10, 229.

Turan, N. G. (2008). The effects of natural zeolite on salinity level of poultry litter compost. *Bioresource Technology*, 99, 2097–2101.

Ugak, M. A. M., Yaser, A. Z., Lamaming, J., Subin, E. K., Rajin, M., Saalah, S., Tze, F. W., H., Abang, S. (2022). Comparative study on passive aerated in-vessel composting of food wastes with the addition of Sabah ragi. *Carbon Resources Conversion*, 5, 200–210.

Vasquez, M. A., de la Varga, D., Plana, R., & Soto, M. (2013). Vertical flow constructed wetland treating high strength wastewater from swine slurry composting. *Ecological Engineering*, 50, 37–43.

Vasquez, M. A., de la Varga, D., Plana, R., & Soto, M. (2015). Integrating, liquid fraction of pig manure in the composting process for nutrient recovery and water re-use. *Journal of Cleaner Production*, 104, 80–89.

Walling, E., Trémier, A., & Vaneeckhaute, C. (2020). A review of mathematical models for composting. *Waste Management*, 113, 379–394.

Walling, E., & Vaneeckhaute, C. (2021). Novel simple approaches to modelling composting kinetics. *Journal of Environmental Chemical Engineering*, 9(3), 105243.

Wang, Y., Ai, P., Cao, H., Liu, Z. (2015). Prediction of moisture variation during composting process: A comparison of mathematical models. *Bioresource Technology*, 193, 200–205

Wang, Y., Huang, G., & Han, L. (2012). Modeling of moisture balance during pig slurry reactor composting. *Transactions of the Chinese Society of Agricultural Machinery*, 43(6), 102–106.

Wang, T., Ni, Z., Kuang, B., Zhou, L., Chen, X., Lin, Z., Guo, B., Zhu, G., & Jia, J. (2022). Two-stage hybrid microalgal electroactive wetland-coupled anaerobic digestion for swine wastewater treatment in South China: Full-scale verification. *Science of The Total Environment*, 820, 153312.

Wu, C., Wang, Q., Shi, S., Xue, N., Zou, D., Pan, S., & Liu, S. (2015). Effective utilisation of trickling liquid discharged from a bio-trickling filter as a moisture conditioning agent for composting. *Biosystems Engineering*, 129, 378–387.

Xie, D., Gao, M., Yang, M., Wu, C., Meng, J., Xu, M., Wang, Q., Liu, S., & Sun, X. (2021a). Nitrate-rich wastewater discharged from a bio-trickling filter can be reused as a moisture conditioning agent for organic waste composting. *Environmental Technology & Innovation*, 24, 101932.

Xie, D., Gao, M., Yang, M., Xu, M., Meng, J., Wu, C., Wang, Q., & Liu, S. (2021b). Re-using ammonium-rich wastewater as a moisture conditioning agent during composting thermophilic period improves composting performance. *Bioresource Technology*, 332, 125084.

Xie, D., Gao, M., Yang, M., Xu, M., Meng, J., Wu, C., Wang, Q., Liu, S., & Sun, X. (2021c). Composting–a solution of eliminating a nitrite-rich wastewater by reusing it as a moisture conditioning agent. *Chemosphere*, 284, 131365.

Yahya, A., Chong, P.S., Ishola, T.A., & Suryanto, H. (2010). Effect of adding palm oil mill decanter cake slurry with regular turning operation on the composting process and quality of compost from oil palm empty fruit bunches. *Bioresource Technology*, 101(22), 8736–8741.

Yang, Q. X., Yang, M., Zhang, J. S., & Lv, W. Z. (2005). Treatment of wastewater from a monosodium glutamate manufacturing plant using successive yeast and activated sludge systems. *Process Biochemistry*, 40, 2483–2488.

Zahrim, A.Y., Asis, T., Hashim, M. A., Al-Mizi, T., & Ravindra, P. (2015). A review on the empty fruit bunch composting: Life cycle analysis and the effect of amendment. In: Ravindra, P (ed.), *Advances in Bioprocess Technology*, Springer, Singapore, 3–15.

Zahrim, A. Y., Leong, P. S., Ayisah, S. R., Janaun, J., Chong, K. P., Cooke, F. M., & Haywood, S. K. (2016). Composting paper and grass clippings with anaerobically treated palm oil mill effluent. *International Journal of Recycling of Organic Waste in Agriculture*, 5(3), 221–230.

Zahrim, A.Y., Yee, I.K.T., Thian, E.S.C., Heng, S.Y., Janaun, J., Chong, K.P., Haywood, S.K., Tan, V., Asis, T., & Al – Mizi, T.M.T.M.A. (2017). Effect of pre-treatment and inoculant during composting of palm oil empty fruit bunches. *ASEAN Journal of Chemical Engineering*, 17(2), 1–16.

Zakarya, I. A., Khalib, S. N. B., Izhar, T. N. T., & Yusuf, S. Y. (2015). Composting of food waste using indigenous microorganisms (IMO) as organic additive. *International Journal of Engineering Research & Technology (IJERT)*, 4(8), 181–184.

7 Biocomposites as Structural Components in Various Applications

Nurjannah Salim
Universiti Malaysia Pahang

Siti Norbaini Sarmin
Universiti Teknologi MARA

CONTENTS

INTRODUCTION ON BIOCOMPOSITE

Biocomposites are materials that are both simple to manufacture and good for the environment. It is a material made up of a matrix and natural fiber or lignocellulose reinforcements such as jute, coir, sisal, pineapple, ramie, bamboo, banana, and bagasse. Biocomposites have been developed as a potentially more environmental-friendly and cost-effective alternative to synthetic fiber-reinforced composite materials. The growing attention and awareness of civilizations toward the environment have significantly increased the demand for long-term sustainable applications, as well as the development of better techniques and procedures for optimum utilization of accessible natural resources (Amiandamhen et al., 2020; Arregi et al., 2020; Rodriguez et al., 2020).

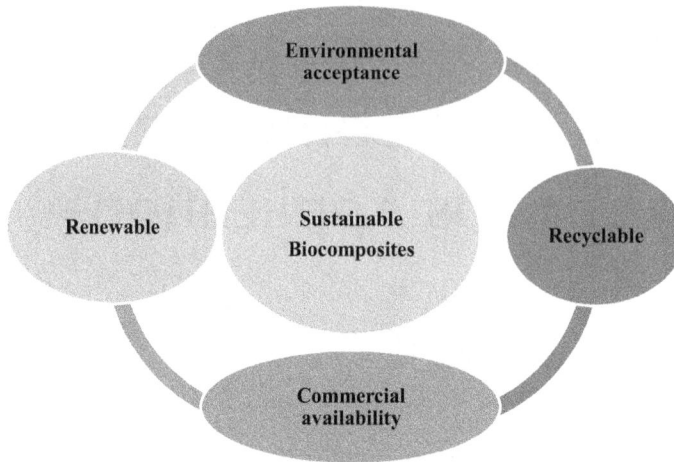

FIGURE 7.1 General attributes associated with the biocomposites.

Globally, the development of sophisticated biocomposite materials is increasing. It will be an alternate method of developing biocomposites that may be used for the everyday needs of ordinary people, such as household furniture, houses, fencing, decking, flooring, low-weight automotive components, and sports equipment (Amiandamhen et al., 2020; Arregi et al., 2020). Many sectors have been driven to adopt natural-resource composites to replace the traditional composites in their designs. The numerous appealing qualities of biocomposites, such as huge availability, low cost, low density, high specific properties, biodegradability, and renewability as in Figure 7.1, sparked this interest (Karimah et al., 2021; Mahmud et al., 2021; Sundararaju et al., 2021). Many research centers across the world are progressively using lignocellulosic materials as fillers or reinforced in polymer composites (Arregi et al., 2020; Hasan et al., 2021; Liu et al., 2019; Zuccarello et al., 2018). This is due to their properties that can be tailored to fulfill specific requirements.

Furthermore, biocomposites have environmental benefits such as lower energy usage, insulation, and sound absorption (Jawaid et al., 2018). With more people becoming aware of the importance of conservation of natural resources, a wood shortage is becoming a major issue in the construction and housing industries, necessitating the development of suitable wood alternatives. Biocomposites would be more widely accepted as a wood alternative, and value-added creative applications of natural fiber composites would secure a worldwide market for cheaper substitutes (Rodriguez et al., 2020).

NATURAL FIBER AS REINFORCED IN A BIOCOMPOSITE

Natural fiber is a renewable source of equivalent qualities to manufactured fiber. Plant, animal, and mineral-based materials are the three types of natural fiber (Figure 7.2). Cellulose, hemicellulose, lignin, pectin, waxes, and water-soluble components are the primary components of plant-based natural fiber. Agricultural wastes (Jawaid et al., 2018; Mahmud et al., 2021), which are currently providing an increasing concern for the global agricultural business, could be used to make plant-based natural fiber. Palm empty fruit bunches, rice straw, pineapple trash, sugarcane bagasse, rice, coffee husks, and coir fiber are only a few examples of agricultural waste sources for natural fiber. Agricultural waste is the most plentiful source of natural fiber, yet only 10% of it is utilized as a substitute raw material such as composites in the automotive and biomedical industries (Hasan et al., 2021).

The characteristics of fibers made from plants and animals typically vary depending on several factors. The matrix, fillers, fiber composition, fiber treatment, and fabrication processes are the main

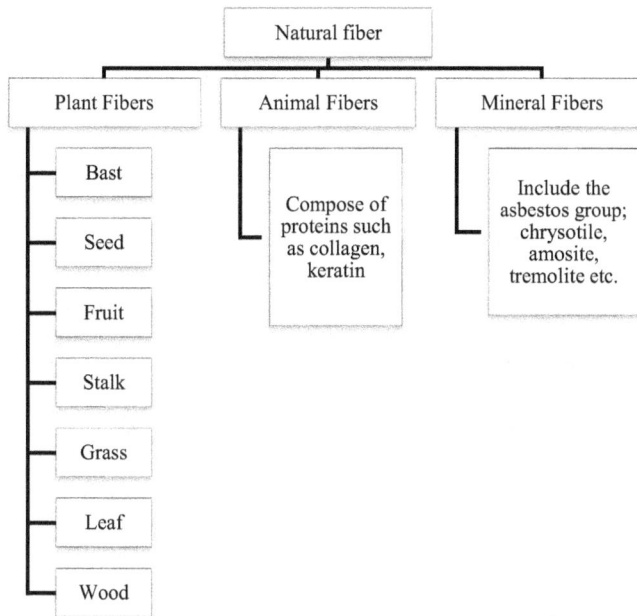

FIGURE 7.2 Classification of natural fiber.

determinants of mechanical qualities. The main component of plant fibers is cellulose, whereas the main component of animal fibers is protein. Due to their scarcity and high cost, both animal- and mineral-based fibers are rarely used as reinforcement media in biocomposite as opposed to plant-based fibers (Jawaid et al., 2018; Mahesh et al., 2021). Wool, silk, feathers, and hair are all examples of animal-based fibers. Conversely, mineral fibers are obtained from minerals, and the most prevalent mineral-based fibers are basal and asbestos.

However, natural fibers are regarded as difficult materials since they have a wide range of mechanical properties based on their source, manner of storage, and nature. Poor water resistance, limited durability, and poor fiber-matrix adhesion (Liu et al., 2019; Mahesh et al., 2021; Sundararaju et al., 2021) are the three main problems encountered when making natural fiber-reinforced polymer composite. Because hemicellulose, pectin, and lignin make up natural fibers, there is weak interfacial adhesion between those fibers and their polymer matrix (Hasan et al., 2021; Liu et al., 2019; Mahesh et al., 2021). When a matrix is hydrophobic, as is the case with these components, the natural fiber filler is easily removed and can crack, fracture, and crate. Additionally, weak bonds between the fibers and matrix are the cause of this occurrence. However, since initiatives to encourage the recyclability of crop wastes into products to replace synthetic fibers were started, the development of biocomposites has kept gaining momentum by modifying the natural fiber through a chemical treatment to improve the fiber-matrix adhesion.

MODIFICATION OF NATURAL FIBER THROUGH CHEMICAL TREATMENT

Natural fibers' main drawbacks in respective composites are their poor compatibility with the matrix and their comparatively high moisture absorption (Arregi et al., 2020; Hasan et al., 2021; Sood & Dwivedi, 2018; Sundararaju et al., 2021). As a result, natural fiber modifications that vary the fiber surface properties to increase their adherence to various matrices are considered. A strong interface with a brittle nature and easy crack propagation across the matrix and fiber could produce exceptional strength and stiffness. A weaker contact could impair the efficiency of stress transfer from the matrix to the fiber (Karthi et al., 2020; Khan et al., 2018; Nurazzi et al., 2021; Wu et al., 2018).

Due to the primary issue with biocomposite materials, known as their hydrophilic nature, Mishra and Chaudhary (2021) found that several chemical treatments can be used to improve the mechanical and surface characteristics between the polymer matrix and natural fiber. Even after hybridization with synthetic fibers, the biocomposite-reinforced natural fiber still have weak mechanical qualities. In comparison to synthetic fibers, natural fibers are weaker and more moisture-absorbent. The importance of the interface and the influence of various forms of surface alterations on the physical and mechanical properties of natural fiber-reinforced composites have been studied extensively and published in the literature (Nurazzi et al., 2021). Chemical modification (alkaline, silane, acetylation, benzoylation, acrylation, and acrylonitrile grafting, maleated coupling, permanganate, peroxide, and isocyanate treatment) appears to be preferred over mechanical treatments (plasma, ultrasound, ultraviolet).

Contrarily, a chemical change to the fibers enhances both their general qualities and the number of their chemical constituents, which enhances water absorption with polymer matrix. Mishra and Chaudhary (2021) and Sood and Dwivedi (2018) stated that undesirable materials including hemi-celluloses, lignin, and pectin that weaken the matrix adhesion are also removed from the surface of the fibers. Chemicals are called compatibilizers to reduce the surface tension of fibers in these predominantly nonpolar and polymer matrices to make them compliant. It is possible to improve the matrix, which is often done by adding binding agents and polymer matrix compatibilizers (Hasan et al., 2021; Mishra & Chaudhary, 2021; Sundararaju et al., 2021). This increases the permeability and excitability of polymers. Coupling agents are primarily responsible for the increased binding between the matrix and reinforcement.

BIODEGRADABLE POLYMER

To correspond with the phrase "green composite," numerous investigations have been undertaken to develop biocomposites that are 100% environmentally. These biocomposites consist of biodegradable polymers and filler substances. According to their origin, biodegradable polymers can be categorized as polymers produced from agro-based monomers such as polylactic acid, agropolymers such as starch, microbial-generated such as polyhydroxyalkanoates, and conventional monomers such as polyester resin (Mittal, 2015). Polylactic acid (PLA) and polyhydroxyalkanoates (PHA) will be further discussed as examples of biodegradable polymers.

Polylactic Acid (PLA)

Polylactic acid (PLA) is a thermoplastic aliphatic polyester that is exceedingly flexible, biodegradable, and biocompatible. PLA is produced from renewable and degradable plant components like glucose and starch (Bhat et al., 2017; Ilyas et al., 2021). L and D-type enantiomers are the two optically active isomers of lactic acid. Lactic acid is most commonly found in fermented milk products, but it can also be produced commercially by a bacterial fermentation process involving a variety of carbohydrate sources. PLA can be created using ring-opening polymerization and direct polycondensation, the two most common synthetic techniques. PLA has superior mechanical, processing, and tensile qualities. PLA has a superior tensile strength to polystyrene PS and polyethylene terephthalate (PET) (Hamad et al., 2018; Ilyas et al., 2021). PET, polypropylene (PP), and high-density polyethylene (HDPE) are inferior to PLA for tensile and flexural modulus characteristics. The ratings for impact strength and elongation at break are lower than PET, PP, and HDPE. PLA has a rapid rate of crystallization due to the rapid growth of its semi-crystalline regions. Crystallinity affects qualities like as tensile strength, melting point, hardness, and rigidity (Hamad et al., 2018; Ilyas et al., 2021; Zwawi, 2021).

Polyhydroxyalkanoates (PHA)

Polyhydroxyalkanoates (PHAs) are a type of biopolyesters that are well-known for their potential application as biodegradable thermoplastics. This characteristic makes them a good choice for replacing conventional polymers and decreasing their environmental effect (Bhat et al., 2017; Cinelli et al., 2019). Various microorganisms produce bacterial polyhydroxyalkanoates as carbon storage and energy sources intra-cellularly. Due to the green- and bio-based character of PHA-derived materials, their potential applications are vast. PHAs can be used as feedstock for the production of thermoplastics and can be combined with other compounds to generate blends and composites, which is another one of its many strengths. Due to its biocompatibility, one of the most important application areas for PHA-derived materials is the medical industry. Due to their antibacterial and antioxidative characteristics, PHAs are particularly popular in tissue engineering (Cinelli et al., 2019; Kaouche et al., 2021). It is possible to customize PHA-based materials to have adaptive mechanical characteristics and enhanced elasticity. Recently, PHA matrices have been employed as implants, drug delivery vehicles, and as tissue for cell growth, as they promote cell proliferation and tissue regeneration without promoting tumor formation. For these medical and pharmaceutical applications, the emphasis is not on inexpensive methods for PHA synthesis but rather on dependable processes and non-hazardous after-treatments. In addition, PHAs are excellent candidates for other applications, such as the production of single-use plastics, with food packaging being one of the primary PHA-market objectives (Cinelli et al., 2019; Kaouche et al., 2021; Zwawi, 2021).

BIOCOMPOSITES AS STRUCTURAL COMPONENTS IN VARIOUS APPLICATIONS

STRUCTURAL COMPONENTS IN AUTOMOTIVE

In the automotive industry, natural fibers such as kenaf, jute, flax, sisal, coir, and hemp are used to manufacture components such as armrests, headrests, bumpers, seat pads, and door panels. In the automotive industry, biocomposites are utilized to produce lightweight automobiles with efficient uses of fuel and also to reduce production costs. Additionally, it is known that damping reduces noise and vibration in biocomposites (Keya et al., 2019; Nagalakshmaiah et al., 2019). There are various automobiles manufactured using biocomposite in their automobiles as below (Zwawi, 2021; Holbery & Houston, 2006):

i. Toyota utilizes kenaf fibers for tire coverings, car seats from soy foams, and components for interior, package trays, and toolbox space are made from PP/PLA-based biocomposites.
ii. Ford employs bio-based seats, soy foam cushions, and hemp fiber composites at the front grilles of multiple automobiles.
iii. Mercedes-Benz utilizes biocomposites derived from jute for interior components, flax fiber biocomposites for storage panels and boot covers, and sisal-based biocomposites for front panel shelves. Compared to synthetic composites, the usage of bio-composites lowered weight by up to 10% and energy consumption by up to 80%.
iv. Volkswagen utilizes biocomposites to produce door frames, flap linings, door inserts, and storage compartments.

STRUCTURAL COMPONENTS IN BUILDING AND CONSTRUCTION

Biocomposites are utilized in the building and construction sector to produce ceilings, doors, roof tiles, windows, window frames, ceilings, and floor matting. However, in load-bearing applications, components such as beams, tanks, floor slabs, and pipes are produced. Besides, biocomposites are used to repair and rehabilitate various structural components. Biocomposites are employed as insulation and dampening materials due to their superior thermal and acoustic qualities. Hemp/lime/concrete biocomposites demonstrate superior sound attenuation compared to conventional binders. Whenever choosing any biocomposite as a building material, a life cycle analysis, durability attributes, and environmental factors must be evaluated (Karimah et al., 2021; Karthi et al., 2020).

For building applications, lighter density and equivalent strength characteristics of synthetic fiber-reinforced composites are essential. Recently, the manufacturers used green materials in designing extremely environmentally solid surfacing materials such as building floors, bathroom products, decking, and paneling (Karimah et al., 2021; Jawaid et al., 2018; Karthi et al., 2020; Saba et al., 2017).

STRUCTURAL COMPONENTS IN FURNITURE AND DECORATIVE PANELS

Rapid technological innovation in the furniture sector enables buyers to select things that suit their preferences. A designer could extend their design experience and apply their creativity more effectively by utilizing more dynamic and superior quality raw materials. Medium density fiberboard (MDF), plywood, and wood plastic composite (WPC) are examples of biocomposites utilized in furniture production (WPC). WPC is a biocomposite composed of wood/lignocellulose waste, recycled plastic, and many other chemical additives. The usage of WPC in the production of furniture is currently on the rise. Because WPC is largely recyclable and eco-friendly; it is appropriate for designing outdoor furniture as it is resistant to stretching, bending, wetness, and extreme temperatures. Examples of biocomposite-made furniture include a coffee table, chair, cabinet, and ornamental panel frame (Saba et al., 2017; Suhaily et al., 2012).

PERFORMANCES OF BIOCOMPOSITE

The physical and chemical composition of the fiber reinforcement heavily influences the qualities of biocomposites. The natural fiber's composition has the biggest impact on mechanical strength. Practically speaking, higher weight percentage fibers are chosen for the construction of composites since they have a stronger influence on tensile strength and modulus (Liu et al., 2019; Sundararaju et al., 2021). The tensile properties of composites are significantly influenced by fiber, which piques researchers' attention. This interest has motivated the researchers to examine the tensile characteristics of the biocomposite by adjusting the fiber's length, composition, and surface. When compared to a composite with 0% fiber content, the tensile modulus of the 50% fiber composite increased by almost two-and-a-half times. Further, a decrease is seen when the fiber content is raised to 70%. This finding indicated that the biocomposite provided the best tensile behavior when reinforced with 50% fiber (Khan et al., 2018; Liu et al., 2019; Zuccarello et al., 2018).

The composite's flexural strength will not be the same as its tensile strength, which means it will not follow the same pattern. In this case, the fiber characteristics have a greater impact on the variables. Tensile strength is more than flexural strength when the fiber reinforcement is more robust; however, if the fiber reinforcement is weak and quickly breaks, flexural strength is greater (Hasan et al., 2021; Mahesh et al., 2021; Mahmud et al., 2021). This phenomenon depends on the composite's homogeneous and heterogeneous qualities. For developing constructions like beams, shafts, and other components, the composite's flexural strength is particularly crucial. According to Jawaid et al. (2018), the most important factor determining the flexural strength of fiber composites is the number of fibers present. In addition to void formation and fiber orientation, the interaction between the fibers decreased the flexural modulus at a high fiber volume proportion.

The energy that a material absorbs under a sudden load over a unit area determines its impact strength. Due to their brittle nature, pure polymers typically have very low impact strengths. The addition of fiber to polymers makes composites more resilient, which is beneficial for boosting impact strength. According to Nurazzi et al. (2021), one aspect that affects fiber-matrix bonding is the impact strength of the fiber composites. Due to the favorable adhesion between the fiber and matrix, it was discovered in this study that the impact strength increased as the fiber loading increased. In their investigation of several natural fiber-reinforced epoxy composites, Siakeng et al. (2019) found that the hydrogen bonding between the fibers had a significant impact on the composites' ability to absorb impact energy. For greater impact strength, the fiber type and its binding characteristics were essential.

One significant reason for the biocomposite's poor moisture-sorption behavior is the fiber's hydrophilic character. The presence of moisture in the fiber may cause the matrix to crack due to insufficient interfacial bonding, which could shorten the lifespan of the composites by reducing the strength of the composite (Amiandamhen et al., 2020; Rodriguez et al., 2020; Sundararaju et al., 2021). The process of absorbing moisture by polymer composites principally employs three diffusion techniques. As a result of the microscopic gaps or vacancies in the polymer chain, the first mechanism involves the diffusion of water molecules. Second is the capillary transport across the voids at the fiber-matrix contact. The third approach is water diffusion through the microscopic gap in the composite surface.

The properties of the natural fiber have a significant impact on the thermal properties of the fiber-reinforced polymer composites. Differential scanning calorimetry (DSC), derivative thermogravimetry (DTG), and thermogravimetric analysis (TGA) are a few significant testing techniques to assess the thermal behavior of the biocomposite. The weight loss of the composite, the breakdown of the fibers' components including hemicellulose, cellulose, and lignin, and depolymerization are the five main parameters used to characterize the thermal degradation of biocomposite material (Karimah et al., 2021; Liu et al., 2019; Zuccarello et al., 2018). The extrusion temperature for thermoplastic polymers and the curing temperature for thermoset polymers, which are both factors in the manufacturing of fiber-reinforced polymer composites, have a big effect on the thermal deterioration of the fiber and matrix.

MANUFACTURING OF BIOCOMPOSITE

Generally, biocomposite can be manufactured using various techniques depending on the application of the end product. The manufacturing method includes compression molding, extrusion process, injection molding, long fiber thermoplastic-direct (LFT-D) method, resin transfer molding (RTM), and pultrusion process.

COMPRESSING MOLDING

The manufacturing process is the most influential factor in determining the performance and attributes of the composite produced. Compression molding is one of the primary processes for fabricating natural fiber composites, and it is very compatible with the production of big and complex objects (Jaafar et al., 2019). Compression molding offers numerous benefits, including low cost and minimal material waste. The compression molding process has a comparatively high production rate because the mold cycle time requires only a few minutes (Gopanna et al., 2019; Jaafar et al., 2019). Figure 7.3 depicts a compression molding diagram.

In the compression molding process, the preparation prior to compression molding and process parameters has a considerable impact on the composites' mechanical properties. In preparation for compression molding, the removal of moisture and mixing operations is also vitally important. The optimal temperature for the drying process and the precise selection of drying time are necessary for removing all moisture from the fiber (Khalid et al., 2021; Sormunen & Kärki, 2019). It is necessary to select the optimal mixing time and temperature in order to make a homogeneous composite. All the parameters such as the pressure of the compression, duration of compression, and temperature are proven to be interdependent, crucial elements affecting the performance of composites. The selection of the molding temperature must be around the melting temperature of the polymer and the temperature of fiber breakdown. It is very important to find suitable compressive pressure applied during making composite to ensure strong inter-linkage between the polymer and fiber. This is also to prevent possible damage to the structure of the fiber so that good performance of biocomposites can be produced (Jaafar et al., 2019; Khalid et al., 2021; Sormunen & Kärki, 2019).

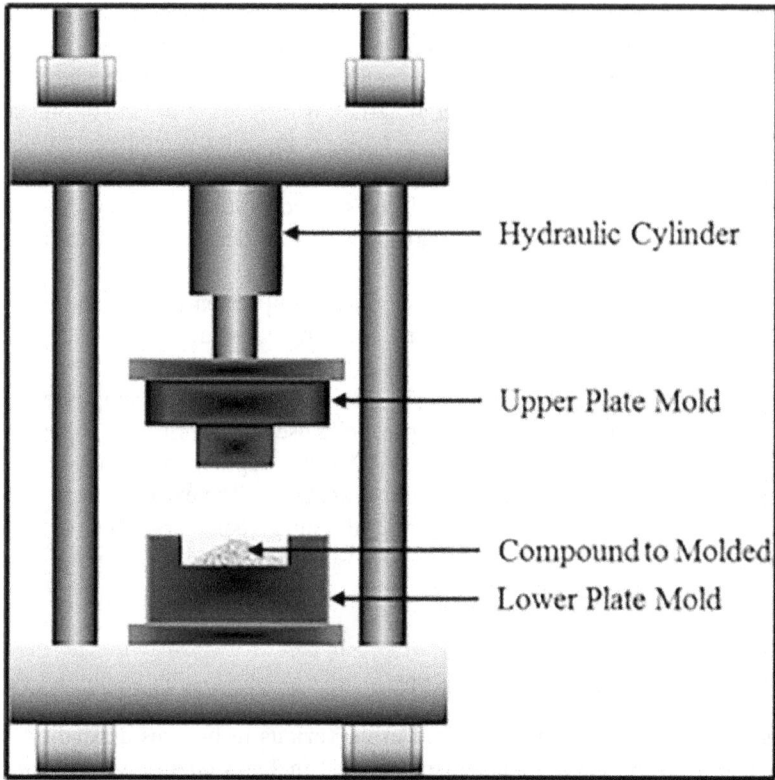

FIGURE 7.3 Compression molding. Reproduced with permission from Jaafar et al. (2019). Copyright, Springer Nature License.

EXTRUSION

Extrusion is utilized in the polymer and composite sector for the manufacturing of pellets and semi-finished materials that are created with different shapes depending on the mold shape through the die. There are two types of extruders, which are single-screw extruders and twin-screw extruders. Single-screw extruders will produce a homogeneous mixture however using a single screw; the mixing impact is not necessary to be very strong. In twin-screw extruders, materials can be evenly disseminated and melted due to the screw's excellent mixing (Cao et al., 2018; Dhaval et al., 2020).

The procedure for extrusion begins with feeding the sample into the heated barrel of the extruder. After each sample run was complete, the procedure was discarded and the extruded filaments were cut into pellets. Some of the extruders required a water bath to return to room temperature; therefore, samples were chopped using a pelletizer and thoroughly dried. Extrusion requires temperatures over the polymer's melting point, so deformation and flow will occur. The role of the screw speed is to stabilize and mix molten samples so that they can enter the extruder barrel and die (Cao et al., 2018; Dhaval et al., 2020).

INJECTION MOLDING

In plastic injection molding, biocomposite pellets are melted and injected under pressure into a mold cavity. Through this method, the specified types of samples, such as dumbbell-shaped samples, were created. In most cases, the extruder-pelletized composite was subsequently injection-molded into tensile coupons (Holbery & Houston, 2006; Singh & Chaitanya, 2015). According to Holbery and

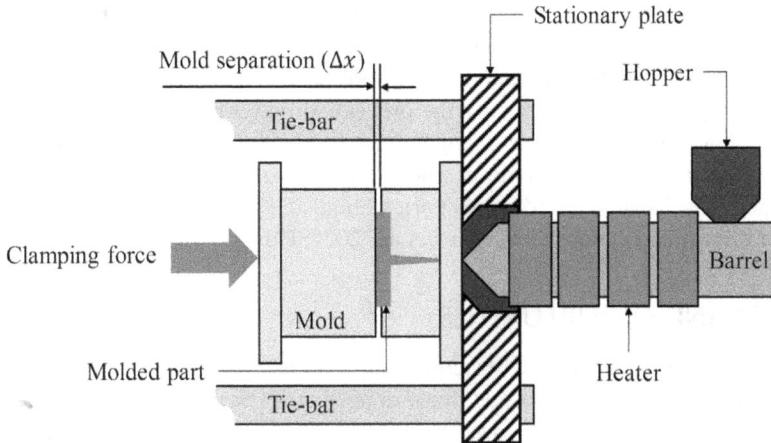

FIGURE 7.4 Injection molding process. Reproduced with permission from Chen et al. (2020). Copyright, Elsevier.

Houston (2006), injection molding is a preferable technique in fabricating composite-based goods, particularly when complex shapes and high-volume production are required. This technique's benefits include high dimensional resistance, shorter cycle durations, and minimum post-processing operations (Holbery & Houston, 2006). Figure 7.4 illustrates the three primary components of an injection molding machine: the injection unit, the mold, the ejection unit, and the clamping unit (Chen et al., 2020; Singh & Chaitanya, 2015).

Long Fiber Thermoplastic-Direct (LFT-D) Method

This approach is a combination of extrusion and pressing in which natural fiber mats are directly pressed with polymer melt in the compression equipment. By using an adjustable extruder, a film of melted polymer is poured into the compression equipment, and then the fiber mat is put into the molten mass. In the next step, the layers are compressed together (Faruk et al., 2012; Knezevic et al., 2022).

Resin Transfer Molding (RTM)

This method is by pumping a resin-catalyst mixture into a mold that is containing a fiber pack using a low-pressure molding. Once the resin is hardened, the mold will be opened and the samples will be taken out from the mold. Using a pressure gradient within a closed vessel, dry semi-finished fiber pieces are infused with reaction resin during the RTM operation. Transfer molding employs higher pressures than injection molding to fill the mold cavity uniformly. This permits thicker matrices of reinforcing fibers to be more thoroughly soaked with resin. In addition, in comparison with the injection molding technique, resin-transfer molding can start with a solid material. Thus, it can decrease the expenditures on tool and time dependence. The fill rate of the resin transfer molding method may be a bit slower than a similar injection molding technique (Ageyeva et al., 2019; Faruk et al., 2012).

Pultrusion

Pultrusion is a low-cost and continuous method for manufacturing composites with close-dimensional cross-sections. This method is suitable for producing solid or hollow profiles such as flat bars, channels, pipelines, tubes, and rods (Faruk et al., 2012). Continuous, pre-heated fiber strands are drawn into the saturated device to wet out the fibers in thermoplastic pultrusion. The melt-impregnated fibers are then driven through a cooling die that controls the final product's form, size, and finish. A puller is used to control the processing power and produce pulling force on the items. Afterward, a pelletizing process is applied to trim the finished product. The pultrusion technique is suited for components that require, among other things, precise dimensional stability, large fiber volume portions, flawless reinforcement orientation, and precise control of resin and fiber, and elimination or minimization rates (Esfandiari et al., 2022; Faruk et al., 2012).

CHALLENGES AND OPPORTUNITIES

Biocomposites will be used more in structural applications in the future (Mahesh et al., 2021; Mishra & Chaudhary, 2021). Various additional applications rely on their continued development. However, several issues must be addressed before these composites can be considered fully competitive with synthetic fiber composites. Biocomposite is environmental-friendly and potentially totally recyclable, although it may be more expensive if it is fully bio-based and biodegradable, and it is particularly sensitive to moisture and temperature. Biocomposites maybe 100% biodegradable with the right matrix, but their biodegradation might be challenging to control. Biocomposite materials have good particular qualities; however, they are highly variable. Their flaws can and will be overcome as more advanced processing of natural fibers and composites is developed.

For mass production of these biocomposites, it is required to solve challenges including cost reduction and performance dependability, and increase the mechanical properties. Biocomposites have a substantial amount of application potential despite these constraints. To successfully commercialize biocomposites, additional research and development are required. Even though the study has produced promising outcomes, extra research and development are required. The goal is to obtain properties comparable to those of synthetic composites. Biocomposites have great advantages such as being produced from renewable material, biodegradable, and natural biocomposites with less impact on the environment and substantially can decrease the pollution of carbon emissions. Increasing environmental awareness among the populace and the passage of new environmental protection rules will result in major biocomposite advancements. Additionally, advancements in agricultural sciences will ease the gathering of fibers with enhanced properties for these biocomposites (Faruk et al., 2012; Karimah et al., 2021). In the future, biocomposites can eliminate the requirement for synthetic materials.

Biocomposite production takes substantially less energy than synthetic fiber production. The manufacturing process in making synthetic composite is energy-intensive, but biocomposites are energy-efficient (Faruk et al., 2012). To tackle trash and environmental problems, several governments are urging businesses to employ biodegradable materials. Variability in the mechanical properties of plant fibers is one of the biocomposites' primary constraints. Changes in geography, habitat, and even the same type of fiber from a different planet are likely to affect the properties of the composite produced. Various processing and chemical treatments are utilized to compensate for these inadequacies. The automotive, packaging industries, building and construction, and textile industries are the most common biocomposite users (Karthi et al., 2020; Keya et al., 2019).

As it governs the ultimate properties of composites, fiber bonding between the lignocellulose fibers and matrix will continue to be the most important aspect in contributing to the properties of the biocomposite. Factors such as the significance of the interface, the types of surface treatment, the types of polymer matrix used, the processing methods, and the performance of composites need to be analyzed, reviewed, and emphasized (Faruk et al., 2012; Halim & Salim, 2021). Outdoor applications necessitate extra study to reduce the weakness of biocomposites such as water absorption, low hardness, and decreased long-term stability. Temperature, humidity, and ultraviolet light, among other environmental factors, affect the product's service life. Property deterioration, staining, and distortion are the most harmful impacts of exposure to hydrothermal and UV radiation.

Globally, significant research is currently being performed to address and overcome the aforementioned challenges. This effort is currently underway to manufacture biocomposites with enhanced quality for better usability (Faruk et al., 2012; Lotfi et al., 2021; Mishra & Chaudhary, 2021).

CONCLUSION

Unquestionably, developed biomaterial has a stronger impact on the global economy by enabling the creation of energy-saving goods that enhance living quality. Given the large number of problems that remain unsolved, study in this area is extremely valuable. Without a doubt, proper design and selection of the production method will enable biocomposites to become one of the dominant structural materials in engineering industries in the future. Despite the fact that a variety of factors influence the usage of natural fibers as reinforcement in polymer-based composites, their renewability and cost competitiveness continue to entice all business sectors to look for chances to substitute natural fibers for conventional materials. Many environmental problems can be resolved by combining natural fibers with polymers made from renewable resources. Industry leaders and senior government officials must collaborate to foster global growth in this cutting-edge class of materials for favorable societal, environmental, and economic benefits. These leaders' leadership is required in addition to the economic and functional advantages of composites derived from renewable and sustainable materials.

REFERENCES

Ageyeva, T., Sibikin, I., & Kovács, J. G. (2019). A review of thermoplastic resin transfer molding: Process modeling and simulation. *Polymers*, *11*(10), 1555.

Amiandamhen, S., Meincken, M., & Tyhoda, L. (2020). Natural fibre modification and its influence on fibre-matrix interfacial properties in biocomposite materials. *Fibers and Polymers*, *21*(4), 677–689.

Arregi, B., Garay-Martinez, R., Astudillo, J., García, M., & Ramos, J. C. (2020). Experimental and numerical thermal performance assessment of a multi-layer building envelope component made of biocomposite materials. *Energy and Buildings*, *214*, 109846.

Bhat, A., Dasan, Y., Khan, I., & Jawaid, M. (2017). Cellulosic biocomposites: Potential materials for future. In Jawaid, M., Salit, M.S., & Alothman, O.Y (eds) *Green Biocomposites* (pp. 69–100). Springer, Switzerland.

Cao, Q., Howard, J. L., Crawford, D. E., James, S. L., & Browne, D. L. (2018). Translating solid state organic synthesis from a mixer mill to a continuous twin screw extruder. *Green Chemistry*, *20*(19), 4443–4447.

Chen, J.-Y., Yang, K.-J., & Huang, M.-S. (2020). Optimization of clamping force for low-viscosity polymer injection molding. *Polymer Testing*, *90*, 106700.

Cinelli, P., Mallegni, N., Gigante, V., Montanari, A., Seggiani, M., Coltelli, M. B., ... Lazzeri, A. (2019). Biocomposites based on polyhydroxyalkanoates and natural fibres from renewable byproducts. *Applied Food Biotechnology*, *6*(1), 35–43.

Dhaval, M., Sharma, S., Dudhat, K., & Chavda, J. (2020). Twin-screw extruder in pharmaceutical industry: History, working principle, applications, and marketed products: An in-depth review. *Journal of Pharmaceutical Innovation*, 17, 294–318.

Esfandiari, P., Silva, J. F., Novo, P. J., Nunes, J. P., & Marques, A. T. (2022). Production and processing of pre-impregnated thermoplastic tapes by pultrusion and compression moulding. *Journal of Composite Materials*, *56*(11), 1667–1676.

Faruk, O., Bledzki, A. K., Fink, H.-P., & Sain, M. (2012). Biocomposites reinforced with natural fibers: 2000–2010. *Progress in Polymer Science*, *37*(11), 1552–1596.

Gopanna, A., Rajan, K. P., Thomas, S. P., & Chavali, M. (2019). Chapter 6- Polyethylene and polypropylene matrix composites for biomedical applications. In Grumezescu, V. & Grumezescu, A. M. (eds.), *Materials for Biomedical Engineering* (pp. 175–216). Elsevier, Amsterdam.

Halim, N., & Salim, N. (2021). Study on effect of different fiber loadings on properties of seaweed/polypropylene blend composite. *Paper Presented at the AIP Conference Proceedings*, 2339, 020184.

Hamad, K., Kaseem, M., Ayyoob, M., Joo, J., & Deri, F. (2018). Polylactic acid blends: The future of green, light and tough. *Progress in Polymer Science*, *85*, 83–127.

Hasan, K. F., Horváth, P. G., Bak, M., & Alpár, T. (2021). A state-of-the-art review on coir fiber-reinforced biocomposites. *RSC Advances*, *11*(18), 10548–10571.

Holbery, J., & Houston, D. (2006). Natural-fiber-reinforced polymer composites in automotive applications. *JOM*, *58*(11), 80–86.

Ilyas, R., Sapuan, S., Harussani, M., Hakimi, M., Haziq, M., Atikah, M., ... Nurazzi, N. (2021). Polylactic acid (PLA) biocomposite: Processing, additive manufacturing and advanced applications. *Polymers*, *13*(8), 1326.

Jaafar, J., Siregar, J. P., Tezara, C., Hamdan, M. H. M., & Rihayat, T. (2019). A review of important considerations in the compression molding process of short natural fiber composites. *The International Journal of Advanced Manufacturing Technology*, *105*(7), 3437–3450.

Jawaid, M., Thariq, M., & Saba, N. (2018). *Durability and Life Prediction in Biocomposites, Fibre-Reinforced Composites and Hybrid Composites*. Woodhead Publishing, Sawston.

Kaouche, N., Mebrek, M., Mokaddem, A., Doumi, B., Belkheir, M., & Boutaous, A. (2021). Theoretical study of the effect of the plant and synthetic fibers on the fiber-matrix interface damage of biocomposite materials based on PHAs (polyhydroxyalkanoates) biodegradable matrix. *Polymer Bulletin*, 79, 7281–7301

Karimah, A., Ridho, M. R., Munawar, S. S., Adi, D. S., Damayanti, R., Subiyanto, B.,... Fudholi, A. (2021). A review on natural fibers for development of eco-friendly bio-composite: Characteristics, and utilizations. *Journal of Materials Research and Technology*, *13*, 2442–2458.

Karthi, N., Kumaresan, K., Sathish, S., Gokulkumar, S., Prabhu, L., & Vigneshkumar, N. (2020). An overview: Natural fiber reinforced hybrid composites, chemical treatments and application areas. *Materials Today: Proceedings*, *27*, 2828–2834.

Keya, K. N., Kona, N. A., Koly, F. A., Maraz, K. M., Islam, M. N., & Khan, R. A. (2019). Natural fiber reinforced polymer composites: History, types, advantages and applications. *Materials Engineering Research*, *1*(2), 69–85.

Khalid, M. Y., Arif, Z. U., Sheikh, M. F., & Nasir, M. A. (2021). Mechanical characterization of glass and jute fiber-based hybrid composites fabricated through compression molding technique. *International Journal of Material Forming*, *14*(5), 1085–1095.

Khan, M. Z., Srivastava, S. K., & Gupta, M. (2018). Tensile and flexural properties of natural fiber reinforced polymer composites: A review. *Journal of Reinforced Plastics and Composites*, *37*(24), 1435–1455.

Knezevic, D., Tutunea-Fatan, O. R., Gergely, R., Okonski, D. A., Ivanov, S., & Dörr, D. (2022). Thermographic analysis of a long fiber–reinforced thermoplastic compression molding process. *The International Journal of Advanced Manufacturing Technology*, *119*(9), 6119–6133.

Liu, W., Chen, T., Fei, M.-e., Qiu, R., Yu, D., Fu, T., & Qiu, J. (2019). Properties of natural fiber-reinforced biobased thermoset biocomposites: Effects of fiber type and resin composition. *Composites Part B: Engineering*, *171*, 87–95.

Lotfi, A., Li, H., Dao, D. V., & Prusty, G. (2021). Natural fiber–reinforced composites: A review on material, manufacturing, and machinability. *Journal of Thermoplastic Composite Materials*, *34*(2), 238–284.

Mahesh, V., Joladarashi, S., & Kulkarni, S. M. (2021). Damage mechanics and energy absorption capabilities of natural fiber reinforced elastomeric based bio composite for sacrificial structural applications. *Defence Technology*, *17*(1), 161–176.

Mahmud, S., Hasan, K., Jahid, M., Mohiuddin, K., Zhang, R., & Zhu, J. (2021). Comprehensive review on plant fiber-reinforced polymeric biocomposites. *Journal of Materials Science*, *56*(12), 7231–7264.

Mishra, S., & Chaudhary, V. (2021). Chemical treatment of reinforced fibers used for bio composite: A review. In: Sharma, B.P., Rao, G.S., Gupta, S., Gupta, P., & Prasad, A. (eds) *Advances in Engineering Materials Lecture Notes in Mechanical Engineering*. Springer, Singapore, 137–147.

Mittal, K. L. (2015). *Progress in Adhesion and Adhesives*.485p, John Wiley & Sons. New Jersey.

Nagalakshmaiah, M., Afrin, S., Malladi, R. P., Elkoun, S., Robert, M., Ansari, M. A., ... Karim, Z. (2019). Chapter 9- Biocomposites: Present trends and challenges for the future. In Koronis, G. & Silva, A. (eds.), *Green Composites for Automotive Applications* (pp. 197–215). Woodhead Publishing, Sawston.

Nurazzi, N., Asyraf, M., Khalina, A., Abdullah, N., Aisyah, H., Rafiqah, S. A., ... Ilyas, R. (2021). A review on natural fiber reinforced polymer composite for bullet proof and ballistic applications. *Polymers*, *13*(4), 646.

Rodriguez, L. J., Peças, P., Carvalho, H., & Orrego, C. E. (2020). A literature review on life cycle tools fostering holistic sustainability assessment: An application in biocomposite materials. *Journal of Environmental Management*, *262*, 110308.

Saba, N., Jawaid, M., Sultan, M., & Alothman, O. Y. (2017). Green biocomposites for structural applications. In Jawaid, M., Salit, M.S., & Alothman, O.Y. (eds) *Green Biocomposites* (pp. 1–27). Springer, Switzerland.

Siakeng, R., Jawaid, M., Ariffin, H., Sapuan, S., Asim, M., & Saba, N. (2019). Natural fiber reinforced polylactic acid composites: A review. *Polymer Composites*, *40*(2), 446–463.

Singh, I., & Chaitanya, S. (2015). Injection molding of natural fiber reinforced composites. In Thakur, V. K., & Kessler, M. R. (eds) *Green Biorenewable Biocomposites From Knowledge to Industrial Applications*. (pp. 273–288). Taylor & Francis Group, Florida.

Sood, M., & Dwivedi, G. (2018). Effect of fiber treatment on flexural properties of natural fiber reinforced composites: A review. *Egyptian Journal of Petroleum*, *27*(4), 775–783.

Sormunen, P., & Kärki, T. (2019). Compression molded thermoplastic composites entirely made of recycled materials. *Sustainability*, *11*(3), 631.

Suhaily, S., Jawaid, M., Khalil, H. A., Mohamed, A. R., & Ibrahim, F. (2012). A review of oil palm biocomposites for furniture design and applications: Potential and challenges. *BioResources*, *7*(3), 4400–4423.

Sundararaju, P. A., Karuppusamy, M., & Ramar, K. (2021). Mechanical and water transport characterization of Indian almond–banana fibers reinforced hybrid composites for structural applications. *Journal of Natural Fibers* 19(13), 7049–7059.

Wu, Y., Xia, C., Cai, L., Garcia, A. C., & Shi, S. Q. (2018). Development of natural fiber-reinforced composite with comparable mechanical properties and reduced energy consumption and environmental impacts for replacing automotive glass-fiber sheet molding compound. *Journal of Cleaner Production*, *184*, 92–100.

Zuccarello, B., Marannano, G., & Mancino, A. (2018). Optimal manufacturing and mechanical characterization of high performance biocomposites reinforced by sisal fibers. *Composite Structures*, *194*, 575–583.

Zwawi, M. (2021). A review on natural fiber bio-composites, surface modifications and applications. *Molecules*, *26*(2), 404.

8 Natural Fiber Composites for Automotive Applications

Noor Afeefah Nordin
Universiti Tenaga Nasional

CONTENTS

INTRODUCTION

Over the last few decades, the concern for green products and environment has become a main global issue to reduce dependency on petroleum-based materials and simultaneously provides environmental sustainability. One of the sectors that requires sophisticated advanced materials with excellent properties is the automotive industry, as to reduce the dependency on the petroleum and non-biodegradable materials. Dates back to the past century, vehicles were built using steels attributing to its reliable strength and stiffness. As technology advances, the development of vehicle parts using fiberglass and polymers came into the picture to reduce the utilization of conventional materials. However, this did not solve the problem on environmental sustainability as it has become one of the main concerns since the past five decades. Therefore, manufacturers have opted for new alternative materials derived from natural resources and developed natural fiber composites from various raw materials.

Conventionally, glass fibers were used as reinforcements in thermoplastic resin ascribing to its high strength-to-weight ratio, corrosion-free and ease of handling, which makes it desirable for such applications (Venkateshwaran et al., 2011). However, there are some drawbacks of the glass fibers such as its non-renewability and requirement of a lot of energy consumption during the production. Besides that, glass fibers are abrasive to the processing equipment and impart potential health risk

DOI: 10.1201/9781003358084-8

toward production workers (Zhang et al., 2015). Therefore, research works were conducted to find an alternative material to replace glass fiber without compromising its properties, and natural fibers came into the sight as a potential material. A review by Joshi et al. was done on the life cycle assessment of natural fiber and glass fiber composite which concluded that natural fibers are environmentally superior to glass fiber (Joshi et al., 2004).

Natural fiber is obtained from the natural resources such as plants, animals and minerals but the common fiber used in the composite manufacture is derived from plants. Since it is a renewable source, this would lead to reduce the dependency on foreign and domestic petroleum oil. Two major issues in using synthetic/petroleum-based fiber are their high carbon dioxide emission (CO_2) and non-biodegradability (Elseify et al., 2021; Simon et al., 1996). Therefore, since the nature of plant is to absorb CO_2, this would counter the problem of CO_2 emission and subsequently replacing the conventional composite with a biodegradable one for various applications, particularly in automotive sector. There are a few factors affecting the properties of natural fibers, namely, origin, age, soil and chemical compositions. However, the structure, microfibrillar angle, fiber defects, fiber length and cell dimensions would give a significant effect on the overall properties of fibers and their composites (Bhattacharyya et al., 2015).

CLASSIFICATIONS OF NATURAL FIBER

Plant fibers are classified into wood and non-wood based on its structure. Wood has been extensively used in the past decades, but since the issue on deforestation raised concern toward global warming, non-wood fibers have been given attention to be exploited and utilized as a substitution to hardwood and softwood. Figure 8.1 shows the classifications of natural fibers that are commonly available and used in various natural fiber-based products.

Cellulose, hemicellulose and lignin are the main constituents of plant fibers. These components are found in the plant cell walls with the ratio of 4:3:3, respectively, but it varies depending on the wood types either hardwood, softwood or non-wood categories (Chen et al., 2011). For instance, hardwood has a greater amount of cellulose while leaves' and wheat straw's major constituent is hemicellulose. Cellulose structure is composed of glucose residues, with cellobiose as the basic coupling unit that is arranged regularly and gathered into bundles with a degree of polymerization of ~10,000. Technically, the basic chemical structures are similar for any types of plants except for their degrees of polymerization. In the case of cell geometry, it varies for different cellulose types depending on the fibers, and this compact structure determines the framework of the cell wall

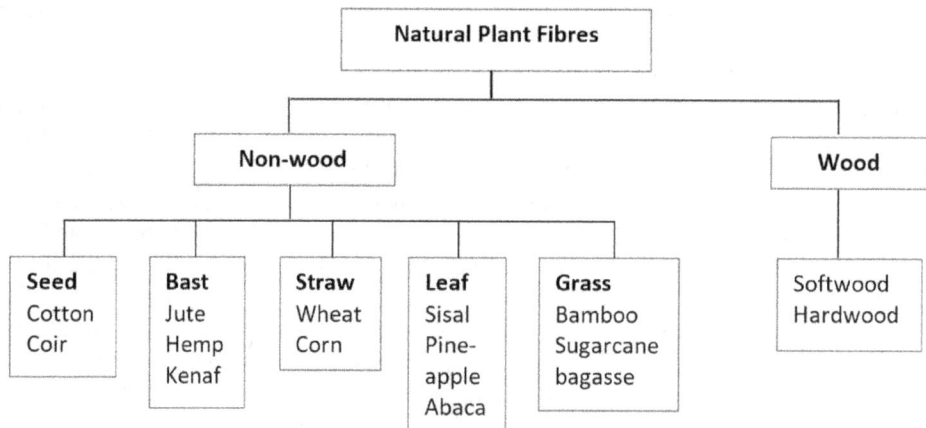

FIGURE 8.1 Classifications of natural fiber.

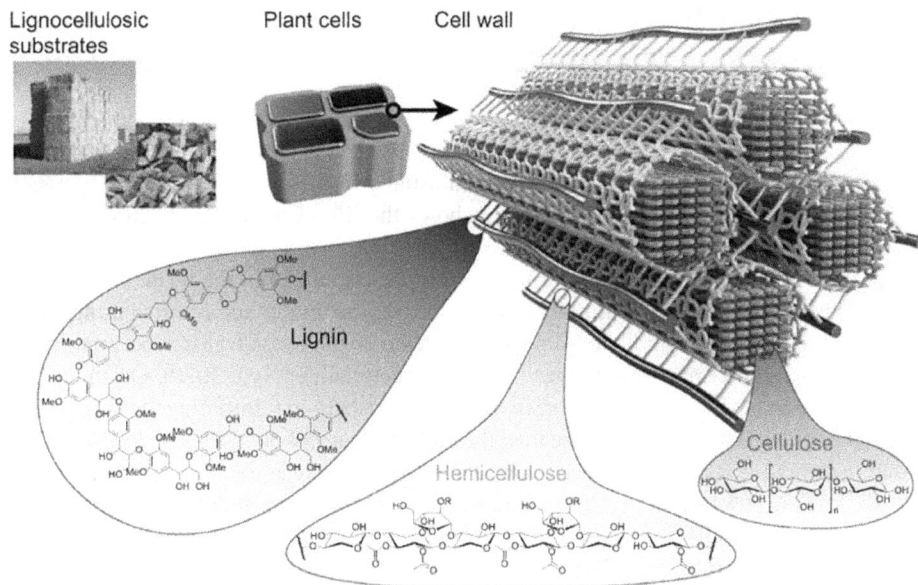

FIGURE 8.2 Components in plant cell wall. Reproduced with permission from Brethauer et al. (2020). Copyright, CC BY License.

(Jarvis, 2018). Hence, cellulose is responsible to provide basic strength of the fiber, which is the most important component when a composite is manufactured.

Cellulose is generally divided into two major types, namely, cellulose type I and type II. The former is a native cellulose which is originated from plants; meanwhile, the latter is derived from cellulose type I, after undergoing chemical treatment. Since cellulose contains crystalline and amorphous regions within the microfibrils, it is categorized as a semi-crystalline polymer. The intrinsic degree of crystallinity for cellulose typically ranges from 40% to 70%, depending on the chemical, mechanical or biological treatments during the cellulose isolation as well as the source of cellulose (French et al., 2011). Figure 8.2 illustrates the components in plant cell wall, with a major constituent of cellulose, hemicellulose and lignin, and their chemical structures.

Other than cellulose, hemicellulose is categorized as a heterogeneous polysaccharide formed through biosynthetic routes, with a degree of polymerization of approximately 200 and functions in providing support to the cell wall (Komuraiah et al., 2014). Hemicellulose is mainly composed of hexoses (mannose, glucose and galactose), pentoses (xylose, arabinose) and acetylated sugars (Danish & Ahmad, 2018). Besides providing support to the cell wall, hemicellulose possesses great industrial potential since it has good physical, mechanical and biological properties. However, its complex physical and chemical structure makes it quite difficult to isolate and requires multiple treatments (Seghini et al., 2020).

Lignin with its complex structure is one of the components in the plant cell wall which contributes to 15%–35% of the lignocellulosic material and functions to provide physical barrier for the plant cell wall (Erfani Jazi et al., 2019). Lignin has an amorphous structure due to its highly complex branched configuration (Xu et al., 2014; Zhou et al., 2016). Unlike cellulose and hemicellulose, the molecular architecture of lignin makes it insoluble in water and highly recalcitrant toward chemical and biological degradation (Jarvis et al., 2018). In the sense of composite fabrication, the presence of lignin somehow causes interference in the uniformity of mixture and subsequently would negatively affect the mechanical and physical properties of the end product (Yang et al., 2019). Therefore, the fibers were usually pre-treated to remove the presence of lignin in order to create better interfacial adhesion between the matrix and filler.

NATURAL FIBER COMPOSITES

REINFORCED POLYMER COMPOSITES

Composite is defined as two or more materials (with different physical and chemical properties) combined together to produce a stronger/superior product. It is usually the combination of matrices and fillers that are compatible with each other and usually results in a composite with outstanding physical and mechanical properties. Figure 8.3 shows the life cycle of composites from its initial development until composting.

Addition of fillers which is aimed to reinforce composites can be in the form of particles, flakes or fibers. However, fibers are commonly used in composite manufacture attributing to its high aspect ratio. This is due to the fibers' ability to provide shear-stress transfer between the matrix fibers, which eventually improved its mechanical characteristics (Ngo, 2020). On the other hand, short fibers are found to be less effective compared to long fibers in reinforcing the composite. Since long fibers have the ability to transmit the load through the matrix better, they are commonly chosen to be used in critical-/high-performance applications, particularly at elevated temperatures (Capela et al., 2017; Venkateshwaran et al., 2011). It is worth noting that the quality of natural fiber is significantly influenced by the species, growth environment, quality/type of soil, harvesting method and humidity of the surrounding. Therefore, the fibers need to be carefully chosen and analyzed before it is further utilized as a reinforcing material in the development of composites (Ngo, 2020). Generally, the composites made with fibers aligned in the direction of tension force would result in the highest stiffness and strength attributing to the effective stress transfer between the matrix and filler (Zhang et al., 2021). Thus, the reinforced composites can be tailored to impart specific properties for various applications.

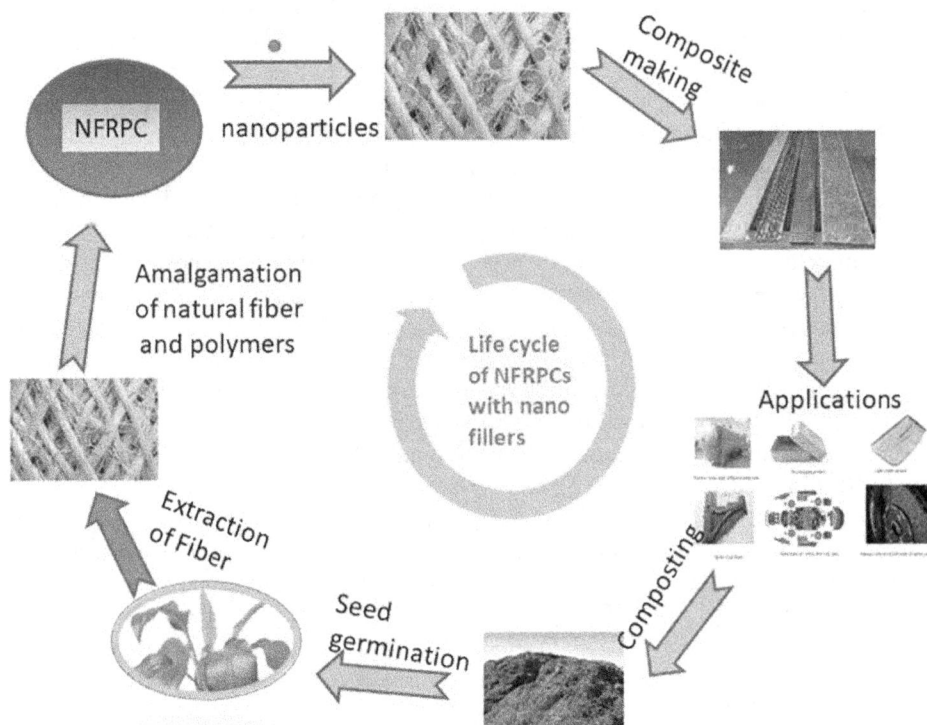

FIGURE 8.3 Life cycle of natural fiber-reinforced polymer composite with nano-fillers. Reproduced with permission from Mishra et al. (2022). Copyright, CC BY License.

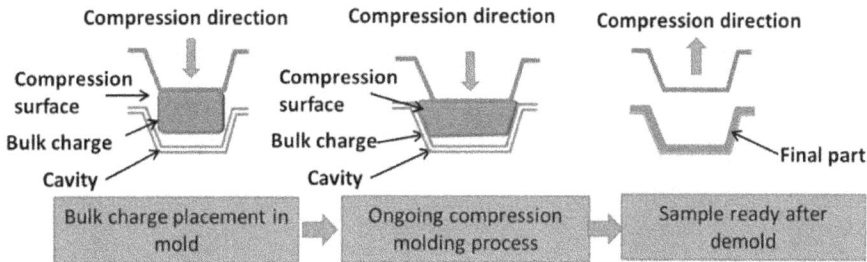

FIGURE 8.4 Schematic diagram of compression molding process. Reproduced with permission from Song et al. (2018). Copyright, CC BY License.

Manufacturing Process

Compression Molding

In compression molding, matrix materials either thermoplastic or thermoset are melted in the mold cavity under heat and pressure, followed by cooling and part removal after it is cured as illustrated in Figure 8.4 (Salit et al., 2015a). The amount of filler content can be limited to 60%. There are a few factors that need to be monitored while using compression molding which are raw material, shape (mold), temperature, pressure applied and curing time. In the case of thermoset resins, the hydraulic would seal the mold once heated and set the material to a shape that may not be changed. Whereas for thermoplastic resin, it will harden as a result of being heated to liquid state and subsequently cooled. The process can be repeated as much as it is required (Ismail et al., 2015; Kumar et al., 2018). The right amount of heat, time and pressure required to produce a composite by compression molding may vary depending on the materials and complexity of the mold. Therefore, it is essential to have a good technical development in order to attain uniformity in die matching.

Resin-Transfer Molding (RTM)

Resin-transfer molding is commonly used to manufacture complex three-dimensional parts in large scale with good surface finish. The process begins with reinforcing the pre-forms into the thermo-setting resin and placed in the mold cavity in the closed mold. The mold will then be filled with resin by applying pressure/vacuum, and the fibers will be wetted. The curing will take place in the mold, and the curing time required for the composites depends on the types of resin used. The type of resin that is commonly used in RTM is a low-viscosity resin. The reason is to ensure that the resin will permeate through the preform rapidly and thoroughly before it gels and cured, particularly when it involves thick composite parts (Ngo, 2020). The schematic diagram illustrating RTM process is shown in Figure 8.5.

The RTM process is also said to be cost-effective for the fabrication of composites because it reduces the manufacturing cycle times in comparison with other methods and found to be feasible for the mass production of composites (Salit et al., 2015b; Verma et al., 2016). In addition, the utilization of natural fibers as reinforcing material to substitute glass fibers was reported to have a positive contribution to the stiffness and strength of composites (Dai & Fan, 2013).

Extrusion

The extrusion process begins when the polymer/resin is forced through a die to produce composites as can be seen in Figure 8.6. The matrix material or resin is softened and mixed with a bundle of fibers which are continuously transported into a single- or twin-screw extruder. It is crucial to ensure a good fiber dispersion during the extrusion process in order to produce a product with superior performance (Lotfi et al., 2021). Different types of extruders are utilized to compress and remove the air/bubble entrapped between the polymers. Extrusion is a continuous process and proven to have

1. Film-bagging
or preform

2. Sealing and
evacuation of preform

3. Loading of film-sealed
preform package

Film

Preform

4. Injection phase

Resin injection

Injection gap

Tool

5. Compression phase

X

6. Curing and film removal

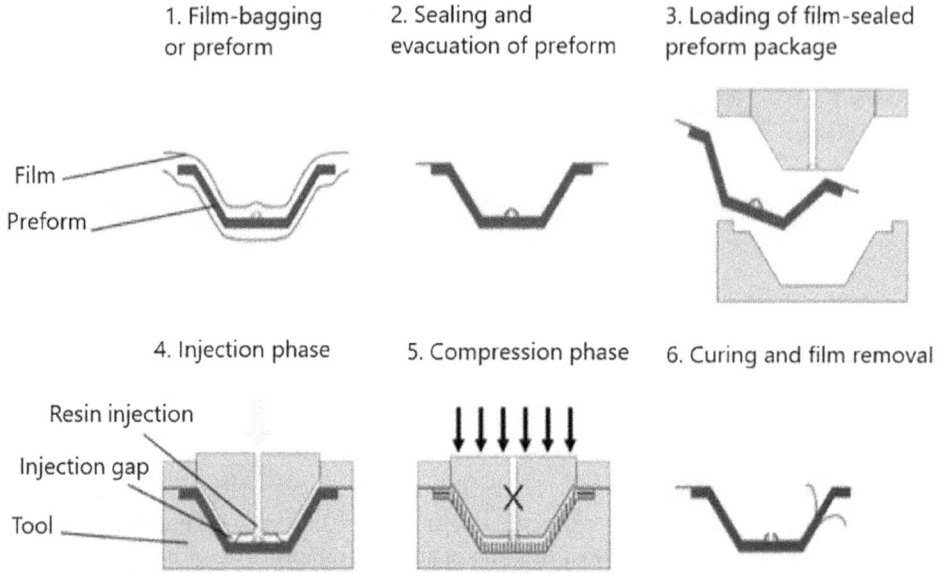

FIGURE 8.5 Schematic diagram of resin transfer molding (RTM) process. Reproduced with permission from Vollmer et al. (2021). Copyright, CC BY License.

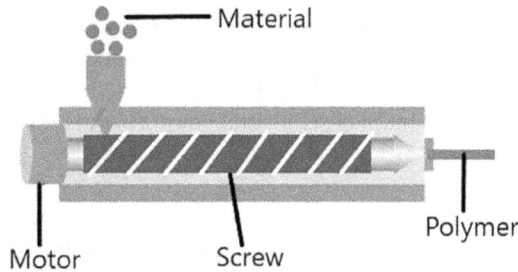

Material

Polymer

Motor   Screw

FIGURE 8.6 Schematic diagram of extrusion process.

good economic benefits. The current technology has now executed extrusion process fully controlled by computer with several functions and parameters that can be manipulated according to the suitability of polymer such as pressure, drying temperatures and delivery rate (Gallos et al., 2017).

Injection Molding

Initially, injection molding is used to make components that are using thermoplastics polymer, but with technological advancements and modifications, it can also be used to develop components by using thermoset polymers. The process is technically related to pressure die casting technique, in which it is usually used when the precision and accuracy of the parts are required (Elseify et al., 2021). The process is initiated when a neat or reinforced polymer, usually used in the form of granules/powder, is passed from a feed hopper into the barrel where it will be heated and gradually softened. The mixture will then be forced to pass through a nozzle into a cold mold that is tightly closed. Upon solidification, the mold will be opened and the component will be removed as illustrated in Figure 8.7.

The cycle continues, and the foremost benefit of injection molding is its short cycle time, in the range of 20–60s per cycle, making it feasible for high-volume production of composites. Besides,

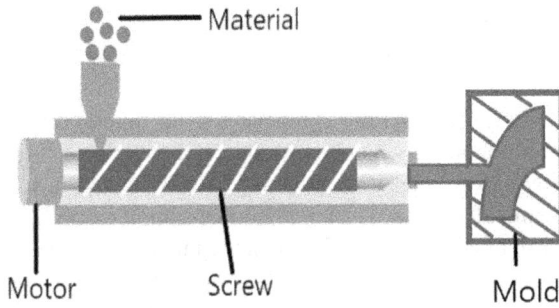

FIGURE 8.7 Schematic diagram of injection molding process.

components with complex shapes are feasible to be done by this technique with superior dimensional tolerances and fine finishing. In addition, the process has a very minimal scrap loss which makes it economically viable for various industries (Jordan & Chester, 2017; Rabbi et al., 2021).

MATRIX MATERIAL

THERMOPLASTIC MATRIX

Thermoplastic polymers can be softened or melted by heating and are able to be processed in either softened or liquid state. When sufficient heat is subjected to the polymer above its melting point, the plastic will melt, whereas when the heating stops and the temperature decreases below the plastic's melting point, it will solidify and becomes solid. This process is reversible, and thus, the polymer can be recycled. However, it should be worth bearing in mind that repeating the process may cause deterioration toward some of the polymer properties. Hence, there is guidance on practical limits for the plastic to be reprocessed. In addition, thermoplastic polymer requires lesser processing time in comparison with thermosets because it does not involve cross-linking chemical reaction in the mold as needed by thermoset polymers. Therefore, it is more favorable to be used in most applications owing to its fast-processing time and imparts good mechanical properties (Legrand et al., 2016; Uddin & Mousa, 2013). In the case of natural fiber-reinforced composite, the main attribute is ascribed to the ability of this thermoplastic resin to transfer the external stress from one fiber to another. Consequently, it would result in a mechanically strong composite that could withstand fracture and absorb energy from the subjected load.

THERMOSETS MATRIX

Dissimilar to thermoplastic, thermoset polymers undergo chemical reaction/cross-linking or curing and would transform from liquid to solid state, and the process is not reversible. Thermoset material is initially in a single form called as monomer. Upon addition of materials such as cross-linker, catalyst, curing agent and the presence of heat, it would induce the chemical reaction/curing, form cross-link network and the material will eventually solidify (Ngo, 2018). Since the process is irreversible, further exposure to heat after solidification would degrade the material. Polyester and vinyl ester resins have been widely used in the automotive industries but polyesters have been of great demand nowadays due to its low viscosity and rapid curing time. This will in turn reduce its cost due to easy processing which favors the manufacturers in the automotive industry (Ha et al., 2019). Another high-potential thermoset resin available in the market is epoxy, which possesses better mechanical properties compared to polyesters and vinyl esters. However, it is not often used for automotive parts since it is more expensive and the curing time is much longer, up to several hours. Therefore, it is more commonly used for the production of composites in aerospace applications.

APPLICATIONS IN AUTOMOTIVE INDUSTRY

HISTORY

The history of the utilization of natural fiber composites in the automotive industry began in 1930s, where Henry Ford initiated the move when he introduced the world's first car made by incorporating flax and straw into soy-based panels (Mann et al., 2020; Witayakran et al., 2017). However, the production was not economical and they started manufacturing the cars from metal back. In 1990s, Daimler Chrysler made use of coir fibers and latex in the manufacturing of car's interior parts such as backseats, bunk cushions, head rest and sun visors. In 1994, the R&D team of Mercedes Benz made an attempt to use jute-reinforced polymers for door panels in their E-class models which contributed to 20% weight reduction of the total vehicle's weight.

Weight reduction is the main important factor to look upon as it would result in lesser assembly/maintenance cost imposed. Moving on from that, Benz started utilizing natural fibers for the manufacturing of some components for Evobus, Setra Top-class and Travego Busses. Among the commonly used natural fibers in automotive parts, kenaf fiber was widely used for car interiors as reported by Ford Montero in 1996 (Bajwa & Bhattacharjee, 2016). Table 8.1 shows the list of natural fibers that were used to reinforce the composites utilized for automobile components.

The history shows that the idea of using natural fiber-reinforced composites dated back about 92 years ago, but the implementation in the current automotive industry is still underutilized, especially for exterior parts.

ADVANTAGES OF NATURAL FIBER COMPOSITES IN AUTOMOTIVE INDUSTRY

Light-weight material and satisfactory specific strength of natural fibers are the key factors that recommend natural fiber composites to be used in the automotive industry. For the record, German has been using flax as the most relevant natural fiber for manufacturing automotive parts while hemp is the second choice of natural fiber for such applications (Karus et al., 2003). Even though natural fiber in its initial form has lower strength compared to synthetic fiber, it could be improved by means of chemical treatments and proper mixture with resins/polymers and would eventually result in comparable strength properties to the conventional materials. Some of the distinct advantages of natural fiber composites for automotive applications are listed below:

TABLE 8.1

Types of Natural Fibers Used in Automotive Industry (Ilyas & Nurazzi, 2022; Ramdhonee & Jeetah, 2017; Sreenivas et al., 2020)

Fibers	Automobile Components
Flax	Door panel cover, instrument panels, arm rest, seat back panels
Hemp	Dashboard, door panels, seat back panels
Jute	Door panels, dashboard
Sisal/Flax	Interior linings, door panels
Cotton	Insulation materials, trunk panel, sound proofing
Abaca	Body panel, floor panel
Coir	Mats, seat mattresses, seat covers
Wood	Seat back panels, covered inserts and components, foamed instrument panels, fiber in the seatback cushions
Kenaf	Dashboards, door interiors, under-floor components
Ramie	Insulation (sound proofing)

- Weight reduction: lightweight material possessed by natural fiber-reinforced composite is the greatest advantage for automobiles. The density of natural fiber is also half of the conventional glass fiber that is commonly used for automotive components. Lower weight vehicle only requires less fuel to propel itself forward and is said to be fuel-efficient (Ngo, 2020). Hence, this leads to cost-effectiveness in terms of fuel consumption.
- Abundant availability: The natural fibers are abundantly available in any parts of the world throughout the year. This makes it a sustainable resource since it can be obtained at any time of the year.
- Comparable strength values to glass fiber composites: This is achieved by the natural fiber composites without reducing its performance, either mechanical or acoustic properties. Besides, the composites also have great thermal-insulation property (Venkateshwaran et al., 2011).
- Safer: Natural fiber composite is said to have safer crash behavior. This is attributed to the effective stress transfer between the fibers and matrix once crash occurs (Ahmad et al., 2015; Jawaid et al., 2017). Therefore, it will give less impact to the driver/passengers compared to the steel.
- Sustainable: Besides its availability, natural fiber is also responsible to absorb CO_2 which contributes to carbon dioxide neutrality. In contrary, the use of the conventional glass fiber and petroleum to produce vehicle parts would result in greater carbon emission and greenhouse gasses into the atmosphere. Therefore, this substitution is an excellent move toward achieving the SDG as mandated by the united nation.

MARKET DEMAND AND FUTURE PERSPECTIVES

The growth of natural fiber composite market has shown tremendous improvement, attributing to its increasing utilizations in aerospace, defense and transportation applications. Statistics have shown that the global composite market is expected to reach \$40.2 billion by 2024, with a forecast to grow at CAGR of 3.3% from 2019 to 2024 (Ngo, 2020). This is based on the increasing market demand in various industries, including automotive sector. For instance, Ford used soybean-based cushions in all of its North American vehicles and projected a saving of about 5 million pounds of petroleum annually. This is a great move toward substitution of petroleum-based material with natural resources to cater the need or reduce carbon footprints. Apart from that, they have also found that using natural fiber composites significantly reduced the processing time up to 40%. This means that it contributes to cost reduction on processing/manufacturing of the vehicle components.

Despite the proven excellent properties of natural fiber composites, there is still some room for improvements especially on reducing the hydrophilicity of fibers by means of chemical or mechanical treatments that should be given fair attention. The moisture absorption is also another factor that should be addressed when developing the composites especially for exterior applications. With that, proper treatment and choice or polymers/resins in the mixture play an important role to enhance its physical properties in order to withstand chemical, mechanical and climatic and fire degradation when exposed to outdoor environment.

The natural fiber composite's application in automobile sector has substantially increased ascribing to their ability to meet divergent prerequisites like significant weight reductions, accountable specific strength and stiffness as well as its contribution toward environmental benefits. However, there are still some doubts among the consumers especially in developing countries on the utilization of this material to replace conventional metals/steel, where their resources are more abundant than the developed countries. Hence, more exposure should be given on its potential with proper guidelines and knowledge to the consumers in order to convince them about the reliability of this composite which in turn will bring our world into a greener place to live in, especially for future generations.

CONCLUSION

Natural fiber composites have proven their significant benefits and established infrastructure over synthetic fibers particularly in the sense of lower weight material, cost-effectiveness, good surface finish and, most importantly, biodegradable. Since there are a wide range of materials that can be used for composite manufacturing, this allows for design flexibility and they can also be customized to tailor specific properties aimed for specific applications. These criterions have opened the doors for its utilization in automotive industry since 1930s to date. Benz, Ford, Volvo and Audi have taken the initiatives to move a step forward toward focusing on renewable and sustainable materials for their automobiles without compromising its strength properties. With the main aim to design a lightweight and powerful heavy-duty composite, it greatly affects the fuel consumption and subsequently reduces the CO_2 emissions in the environment. The increasing number of research conducted and published in this field has also indicated that the importance of replacing vehicle parts with natural fiber composites makes a significant contribution to the automotive industry. However, great effort must be placed in order to convince the consumers about the reliability of this natural fiber-reinforced composite in automotive sectors since many are still having doubts about its performance in comparisons to conventional materials. With all factors being taken into account, the breakthrough of excellent potential of these natural fiber composites for automotive application would finally contribute to improvements in societal well-being and eventually meet the requirement of sustainable development goal.

REFERENCES

Ahmad, F., Choi, H. S., & Park, M. K. (2015). A review: Natural fiber composites selection in view of mechanical, light weight, and economic properties. *Macromolecular Materials and Engineering, 300*(1), 10–24.
Bajwa, D. S., & Bhattacharjee, S. (2016). Current progress, trends and challenges in the application of biofiber composites by automotive industry. *Journal of Natural Fibers, 13*(6), 660–669.
Bhattacharyya, D., Subasinghe, A., & Kim, N. K. (2015). Natural fibers: Their composites and flammability characterizations. In Friedrich, K. & Breuer, U. (eds) *Multifunctionality of Polymer Composites: Challenges and New Solutions*. Elsevier Inc. Amsterdam,102–143.
Brethauer, S., Shahab, R. L., & Studer, M. H. (2020). Impacts of biofilms on the conversion of cellulose. *Applied Microbiology and Biotechnology, 104*(12), 5201–5212.
Capela, C., Oliveira, S. E., Pestana, J., & Ferreira, J. A. M. (2017). Effect of fiber length on the mechanical properties of high dosage carbon reinforced. *Procedia Structural Integrity, 5*, 539–546.
Chen, W., Yu, H., Liu, Y., Hai, Y., Zhang, M., & Chen, P. (2011). Isolation and characterization of cellulose nanofibers from four plant cellulose fibers using a chemical-ultrasonic process. *Cellulose, 18*(26), 433–442.
Dai, D., & Fan, M. (2013). Wood fibres as reinforcements in natural fibre composites: Structure, properties, processing and applications. In Hodzic, A. & Shanks, R. (eds) *Natural Fibre Composites: Materials, Processes and Applications*. Woodhead Publishing Limited, Sawston, 3–65.
Danish, M., & Ahmad, T. (2018). A review on utilization of wood biomass as a sustainable precursor for activated carbon production and application. *Renewable and Sustainable Energy Reviews, 87*(2017), 1–21.
Elseify, L. A., Midani, M., El-Badawy, M., & Jawaid, M. (2021). *Manufacturing Automotive Components from Sustainable Natural Fiber Composites (SpringerBriefs in Materials)*, Springer, Singapore, 83p.
Erfani Jazi, M., Narayanan, G., Aghabozorgi, F., Farajidizaji, B., Aghaei, A., Kamyabi, M. A., Navarathna, C. M., & Mlsna, T. E. (2019). Structure, chemistry and physicochemistry of lignin for material functionalization. *SN Applied Sciences, 1*(9), 1–19.
French, A. D., Rajasekaran, K., & Condon, B. (2011). Book review: "Cellulose science and technology. *Cellulose, 18*(3), 851–852.
Gallos, A., Paës, G., Allais, F., & Beaugrand, J. (2017). Lignocellulosic fibers: A critical review of the extrusion process for enhancement of the properties of natural fiber composites. *RSC Advances, 7*(55), 34638–34654.
Ha, G. X., Bernaschek, A., & Zehn, M. W. (2019). Experimentally examining the mechanical behaviour of nap-core sandwich material: A novel type of structural composite. *Journal of Reinforced Plastics and Composites, 38*(8), 369–378.

Ilyas, R. A., & Nurazzi, N. M. (2022). Fiber-reinforced polymer nanocomposites. *Nanomaterials*, 12(17),3045, 10–13.

Ismail, N. F., Sulong, A. B., Muhamad, N., Tholibon, D., MdRadzi, M. K. F., & WanIbrahim, W. A. S. (2015). Review of the compression moulding of natural fiber-reinforced thermoset composites: Material processing and characterisations. *Pertanika Journal of Tropical Agricultural Science*, 38(4), 533–547.

Jarvis, M. C. (2018). Structure of native cellulose microfibrils, the starting point for nanocellulose manufacture. *Philosophical Transactions of the Royal Society A: Mathematical, Physical and Engineering Sciences*, 376(2112), 20170045.

Jawaid, M., Alothman, O. Y., & Salit, M. S. (2017). Preface. In Jawaid, M., Salit, M. S., & Alothman, O. Y. (eds) *Green Biocomposites*. ,. Springer, Switzerland, vii–viii.

Jordan, W., & Chester, P. (2017). Improving the properties of banana fiber reinforced polymeric composites by treating the fibers. *Procedia Engineering*, 200, 283–289.

Joshi, S. V., Drzal, L. T., Mohanty, A. K., & Arora, S. (2004). Are natural fiber composites environmentally superior to glass fiber reinforced composites? *Composites Part A: Applied Science and Manufacturing*, 35(3), 371–376.

Karus, M., Kaup, M., & Ortmann, S. (2003). Use of natural fibres in composites in the German and Austrian automotive industry–market survey 2002: Status, analysis and trends. *Journal of Industrial Hemp*, 8(2), 73–78.

Komuraiah, A., Kumar, N. S., & Prasad, B. D. (2014). Chemical composition of natural fibers and its influence on their mechanical properties. *Mechanics of Composite Materials*, 50(3), 359–376.

Kumar, T. S. K., Ponnala, R. P., & Ravi, S. (2018). Behavior of natural fiber-polymer composites using compression moulding process, *International Journal of Engineering Research and Technology*, 7(12), 175–177.

Legrand, X., Cochrane, C., & Koncar, V. (2016). A complex shaped-reinforced thermoplastic composite part made of commingled yarns with an integrated sensor. In Koncar, V. (ed) *Smart Textiles and Their Applications*. Elsevier Ltd, Amsterdam, 353–374.

Lotfi, A., Li, H., Dao, D. V., & Prusty, G. (2021). Natural fiber–reinforced composites: A review on material, manufacturing, and machinability. *Journal of Thermoplastic Composite Materials*, 34(2), 238–284.

Mann, G. S., Singh, L. P., Kumar, P., & Singh, S. (2020). Green composites: A review of processing technologies and recent applications. *Journal of Thermoplastic Composite Materials*, 33(8), 1145–1171.

Mishra, T., Mandal, P., Rout, A. K., & Sahoo, D. (2022). A state-of-the-art review on potential applications of natural fiber-reinforced polymer composite filled with inorganic nanoparticle. *Composites Part C: Open Access*, 9(July), 100298.

Ngo, T.-D. (2018). Natural fibers for sustainable bio-composites. In Gunay, E. (ed) *Natural and Artificial Fiber-Reinforced Composites as Renewable Sources*. IntechOpen, London, 107–126.

Ngo, T. (2020). Composite and nanocomposite materials: From knowledge to industrial applications. In Ngo, T. (ed) *Composite and Nanocomposite Materials - From Knowledge to Industrial Applications*. Intech Open, London, 240p.

Rabbi, M. S., Islam, T., & Islam, G. M. S. (2021). Injection-molded natural fiber-reinforced polymer composites: A review. *International Journal of Mechanical and Materials Engineering*, 16(1), 1–21.

Ramdhonee, A., & Jeetah, P. (2017). Production of wrapping paper from banana fibres. *Journal of Environmental Chemical Engineering*, 5(5), 4298–4306.

Salit, M. S., Jawaid, M., Yusoff, N. B., & Hoque, M. E. (2015a). Manufacturing of natural fibre reinforced polymer composites. *Manufacturing of Natural Fibre Reinforced Polymer Composites*, 2015, 1–383.

Salit, M. S., Jawaid, M., Yusoff, N. B., & Hoque, M. E. (2015b). Manufacturing of natural fibre reinforced polymer composites. *Manufacturing of Natural Fibre Reinforced Polymer Composites*, 2016, 1–383.

Seghini, M. C., Tirillò, J., Bracciale, M. P., Touchard, F., Chocinski-Arnault, L., Zuorro, A., Lavecchia, R., & Sarasini, F. (2020). Surface Modification of flax yarns by enzymatic treatment and their interfacial adhesion with thermoset matrices. *Applied Sciences (Switzerland)*, 10(8), 1–17.

Simon, F., Loussert-Ajaka, I., Damond, F., Saragosti, S., Barin, F., & Brun-Vézinet, F. (1996). HIV type 1 diversity in Northern Paris, France. *AIDS Research and Human Retroviruses*, 12(15), 1427–1433.

Song, Y., Gandhi, U., Sekito, T., Vaidya, U. K., Hsu, J., Yang, A., & Osswald, T. (2018). A novel cae method for compression molding simulation of carbon fiber-reinforced thermoplastic composite sheet materials. *Journal of Composites Science*, 2(2), 33.

Sreenivas, H. T., Krishnamurthy, N., & Arpitha, G. R. (2020). A comprehensive review on light weight kenaf fiber for automobiles. *International Journal of Lightweight Materials and Manufacture*, 3(4), 328–337.

Uddin, N., & Mousa, M. A. (2013). Innovative fiber-reinforced polymer (FRP) composites for disaster-resistant buildings. In Uddin, N. (ed) *Developments in Fiber-Reinforced Polymer (FRP) Composites for Civil Engineering*. Woodhead Publishing, Sawston, 272–301, 302e.

Venkateshwaran, N., Elayaperumal, A., & Jagatheeshwaran, M. S. (2011). Effect of fiber length and fiber content on mechanical properties of banana fiber/epoxy composite. *Journal of Reinforced Plastics and Composites*, *30*(19), 1621–1627.

Verma, D., Gope, P. C., Zhang, X., Jain, S., & Dabral, R. (2016). Green composites and their properties: A brief introduction. *Green Approaches to Biocomposite Materials Science and Engineering*, *2016*, 148–164.

Vollmer, M., Zaremba, S., Mertiny, P., & Drechsler, K. (2021). Edge race-tracking during film-sealed compression resin transfer molding. *Journal of Composites Science*, *5*(8), 195.

Witayakran, S., Smitthipong, W., Wangpradid, R., Chollakup, R., & Clouston, P. L. (2017). Natural fiber composites: Review of recent automotive trends. *Encyclopedia of Renewable and Sustainable Materials*, 2, 166–174.

Xu, C., Arancon, R. A. D., Labidi, J., & Luque, R. (2014). Lignin depolymerisation strategies: Towards valuable chemicals and fuels. *Chemical Society Reviews*, *43*(22), 7485–7500.

Yang, J., Ching, Y. C., & Chuah, C. H. (2019). Applications of lignocellulosic fibers and lignin in bioplastics: A review. *Polymers*, *11*(5), 1–26.

Zhang, H., Zhu, L., Zhang, F., & Yang, M. (2021). Effect of fiber content and alignment on the mechanical properties of 3d printing cementitious composites. *Materials*, *14*(9), 1–16.

Zhang, S., Cao, J., Shang, Y., Wang, L., He, X., Li, J., Zhao, P., & Wang, Y. (2015). Nanocomposite polymer membrane derived from nano TiO2-PMMA and glass fiber nonwoven: High thermal endurance and cycle stability in lithium ion battery applications. *Journal of Materials Chemistry A*, *3*(34), 17697–17703.

Zhou, S., Xue, Y., Sharma, A., & Bai, X. (2016). Lignin valorization through thermochemical conversion: Comparison of hardwood, softwood and herbaceous lignin. *ACS Sustainable Chemistry and Engineering*, *4*(12), 6608–6617.

9 Cellulose Nanocrystals from Agro Residues as Reinforcing Agent in Nanocomposites

Santhana Krishnan
Prince of Songkla University

Dianah Mazlan
Universiti Sains Malaysia

Mohd Fadhil MD Din
Universiti Teknologi Malaysia

Mohd Nasrullah
Universiti Malaysia Pahang (UMP)

Sumate Chaiprapat
Prince of Songkla University

CONTENTS

INTRODUCTION

With its quality, concrete is one of the most commonly used materials in the construction industry and can be engineered with the requisite strength and durability. In the middle of the nineteenth century when Portland cement was discovered, reinforced concrete structures have been built and used in the construction (Haga, et al., 2021; Crow, 2008). The hydration products of the Portland cement react with water will help on binding the cement particles together to form a hardened

DOI: 10.1201/9781003358084-9

cement paste. When cement, water, and sand were mixed, a product called mortar formed. If the mixture also contains coarse aggregate, the concrete is produced. Nowadays, cement-based materials have become a well-known material in the world that consists of 70% of total usage around the world. Human beings had been using concrete in their pioneering architectural feats for millennia. Therefore, much research has been done by researchers throughout the years to improve the cement-based materials' (mortar and concrete) weakness at the same time to be suitable to the present condition and environment. The properties of concrete which are known as strong in compression and 8% to 10% relatively weak in tension, and if subjected to bending or in tension will easily crack and cause a structure failure (Min et al., 2014; Rahuman & Yeshika, 2015). The failure may vary depending on the concrete strength and materials used.

A wide range of poor construction practices can cause concrete structures to crack. The most common method to improve workability is by adding water to concrete (Aliabdo et al., 2016). The addition of water reduces strength, increases settling, and increases drying shrinkage. Supported by a higher cement content to help alleviate the loss of strength, an increase in water content would also raise the temperature difference between the inner and outside portions of the structure, resulting in increased thermal stress and the possibility of cracking. The degree of cracking within a concrete building might grow after curing. Early cure termination might result in significant shrinkage during a period of low concrete strength (José et al., 2015; Shen et al., 2016). The shortage of hydration of the cement due to drying would result not only in reduced long-term strength but also in reduced structural durability. Furthermore, a higher water-cement ratio during the concrete mix can increase the air void in hardened concrete later after the curing process. This air void can affect the concrete structure as shown in Figure 9.1. The fine cracks also known as micro-cracks may occur due to shrinkage deformations of the cement paste (Khushefati & Demirbog, 2015). Micro-cracks are unavoidable in ordinary concrete, where if the micro-cracks get bigger, it may contribute to the high penetration and permeability of the concrete. With high permeability of concrete, it may allow penetration of aggressive elements such as chloride ions, sulfate, and water. Thus, it may reduce the strength of concrete structures and durability as well.

New types of concrete such as high-strength concrete, nowadays, are being constantly developed to meet the increasing demand for improved mechanical properties and durability (Venkatesan & Tamizhazhagan, 2016). The higher water content during mixing will affect the concrete or mortar strength, and this can be solved by using additives called plasticizers

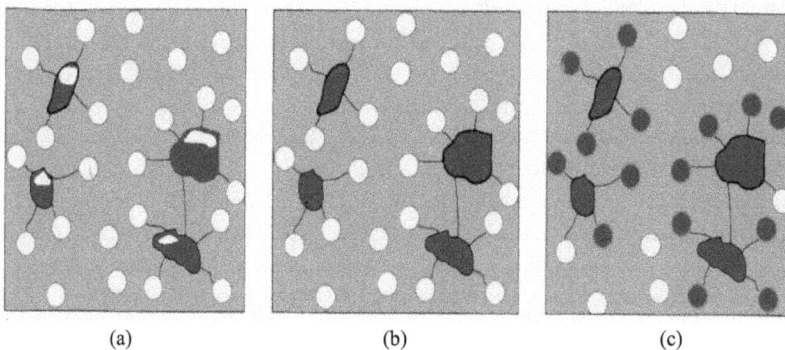

FIGURE 9.1 Small air voids cause a connecting fissure in the cement matrix. (a) Temperature drop causes water to fill the void. (b) In the winter, the water freezes and expands. (c) Under pressure, water moves to surrounding air pores and causes the break in the concrete matrix. Modified with permission from Goguen (2012).

(Aliabdo et al., 2016). The invention of renewable and sustainable infrastructure materials with new property combinations that profoundly alter standard engineering paradigms is one of the newest frontiers of engineering. Nano-reinforced materials, which include graphene and carbon nanofiber, are a potential family of materials that can improve attributes such as elastic modulus, compressive strength, tensile strength, flexural strength, impact resistance, and fracture energy. Due to the nanomaterials having a high surface area per volume unit compared to macro-size materials, the nanomaterials combined well with other materials compared to macro-size materials. Hence, this results in novel properties, which are capable of meeting or exceeding design expectations (Ling et al., 2020). Nano-reinforced materials, on the other hand, offer exceptional possibilities for tailoring mechanical, chemical, and electrical properties (Cao et al., 2015). On the other hand, substantial effort in the usage of nano-reinforcements has been questioned because of potential environmental, financial, health, and safety problems. Over 15 billion tons of aggregates and 4.2 billion tons of cement are consumed annually as a result of the high yearly output of concrete, resulting in total annual consumption of approximately 30 billion tons (Tosic et al., 2017).

'Greener' materials, especially those made from renewable and sustainable resources, are gaining popularity (Khalil et al., 2006). In addition, the objective is to lower the carbon footprint of infrastructure materials to increase the demand for biodegradable, non-petroleum-based, and low-impact environmental materials (Cao et al., 2015). By enhancing the efficiency of infrastructure materials, it is possible to drastically reduce the amount of these materials used, hence reducing the demand for raw materials. Thus, various new engineered materials have been discovered and proposed as potential materials that can be improved or overcome the concrete weakness. Secondary materials (such as pulverized fuel ash (PFA), a by-product of coal-fired power generation) are gaining in popularity to ensure success on both counts, as they have a proven track record in construction and good environmental credentials by diverting waste from landfills and conserving our quarried resources. Many types of waste can potentially be used to increase the performance of cement materials as fiber-reinforcing agents, such as palm oil waste, rice husk, banana tree waste, pineapple fiber, and many more. However, in this new era, these types of wastes can be reused as the best source of cellulose to extract cellulose nanofiber and cellulose nanocrystals that can help in improving the cement matrix structure. (Givi et al., 2010; Dungani et al., 2016; Khalid et al., 2016). Therefore, the objective of this chapter is to identify important performance criteria and parameters of research that had been done and compared with current research. This chapter then discusses the science and different approaches to the utilization of CNCs and their ability to enhance the properties of materials. The differing performance of CNCs as admixture evaluation methods is discussed by looking at the different admixtures that each researcher reported.

SUSTAINABILITY FACTORS

Since sustainability being an increasingly important part of every building project it is becoming standard practice to use recycled materials in concrete. Increased legislation and customer demands drive sustainability in every construction process for the construction industry as a whole and concrete have an important part to play as one of the most widely used building materials. It's not enough to be sustainable, though. Every building material must perform and follow all industry requirements to give clients the confidence that it does not adversely affect the structure's long-term performance. Secondary materials (such as pulverized fuel ash (PFA), the by-product of coal-fired power generation), are seeing greater uptake to ensure success on both counts, as they have both an established track record in construction as well as good environmental credentials by diverting waste from landfill and protecting our quarried resources.

However, many types of waste can be potentially been used to increase the performance of cement materials as fiber reinforcing agents, such as palm oil waste, rice husk, banana tree waste, pineapple fiber and many more. However, in this new era, these types of agro-wastes such as empty palm fruit bunches (EFB), can be reused as the best source of cellulose to extract nanofibers such as cellulose nanofiber and cellulose nanocrystals that can help in improving the cement matrix structure.

CELLULOSE MATERIALS

Overdependence on petroleum products has steadily increased during the last decade. Biodegradable goods from renewable resources are becoming ever more desirable due to higher prices and crude oil shortages.

Natural cellulosic fibers, particles, fibrils (micro- and nano-scale), crystals, and whiskers are appropriate replacements for man-made fibers (e.g., glass and aramid fibers) used as reinforcements and fillers in ecologically friendly products. These cellulosic materials have numerous benefits, such as being renewable, inexpensive, low density, low energy consumption, high specific strength and modulus, excellent sound attenuation, abrasion-resistant, and having a relatively reactive surface (Cheng et al., 2006). Cellulose is a linear homopolysaccharide composed of -D-glucopyranose units held together by -1–4 bonds (Cheng et al., 2017). There are three hydroxyl groups on each monomer. These hydroxyl groups and their ability to establish hydrogen bonds play a significant role in controlling the crystalline packing and determining the physical properties of cellulose (Siqueira et al., 2010a).

During biosynthesis, cellulose microfibril is generated; it is the fundamental structural component of cellulose. As shown in Figure 9.2, the poly-(14)-D-glucosyl residue chains assemble to

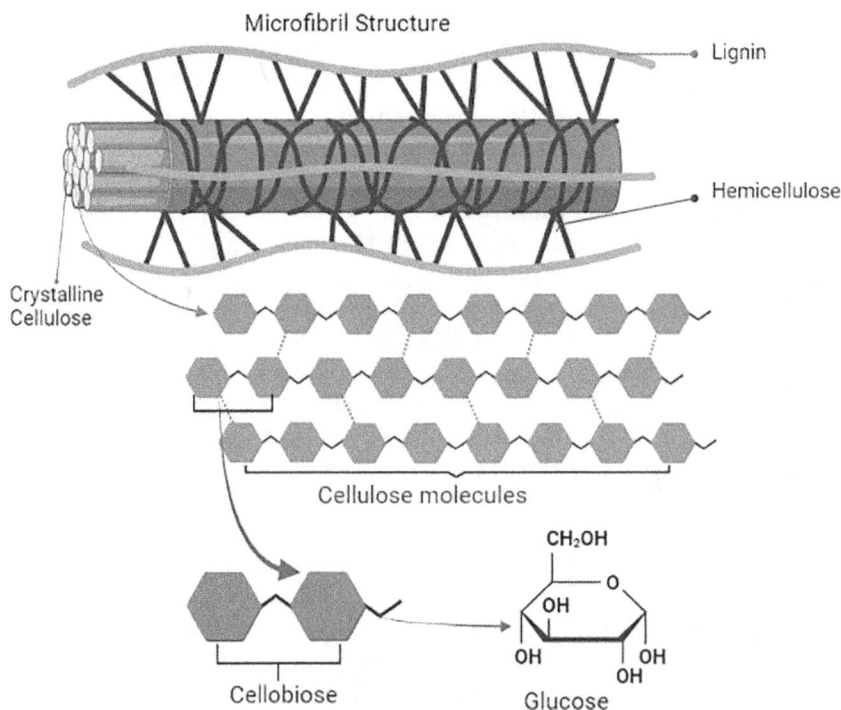

FIGURE 9.2 Chemical structure of the natural fiber. Reproduced with permission from Siqueira et al. (2010a). Copyright, CC BY license.

form a fibril, which is a long thread-like bundle of molecules held together by inter-molecular hydrogen bonds (Cheng et al., 2006; Siqueira et al., 2010b). Individual cellulose microfibrils have sizes ranging from 2 to 20 nm. Each microfibril can be thought of as a string of cellulose crystals connected by disordered amorphous domains along the microfibril axis, such as kinks and twists (Abdul Khalil et al., 2012). Cellulose comprises different crystalline structures. There are a variety of crystalline structures for cellulose. Every object has its unique diffraction pattern. These cellulose polymorphs indicated as cellulose I, II, IIII, IIIII, IVI, and IVII are inter-convertible based on the chemical treatment and source. Nanofiller cellulose is composed of native cellulose (Cellulose I), which is obtained from traditional lignocellulosic fiber-bleaching processes. Cellulose I is responsible for mechanical qualities because of its high modulus and crystallinity. Two types of nano-scale cellulose particles have been discovered as possible reinforcing elements. The first consists of cellulose nanocrystals, whereas the second is microfibrillated cellulose (MFC).

Cellulose also has bundles of amorphous regions, aside from the crystalline region. Amorphous cellulose is also present between the microfibrils of cellulose. In the amorphous regions, the cellulose chains are arbitrarily oriented and entangled in a spaghetti-like structure, resulting in lower density in these areas. This leaves the amorphous regions vulnerable to acid attack. Wide-angle X-ray scattering was used to investigate possible amorphous cellulose structures. Studies of diffraction reveal light and dark areas in a microfibril of cellulose due to crystalline and amorphous cellulose, respectively. Cotton cellulose has a high volume of crystalline cellulose and a low amount of amorphous cellulose. On the other hand, there is a comparatively higher amount of amorphous cellulose found in regenerated cellulose. The high amorphous fraction in cellulose means that chemicals are rather visible to the cellulose structure. The cellulose chains are not so tightly packed in the amorphous or less ordered regions and are thus more accessible for hydrogen bonding to other molecules such as water.

CELLULOSE NANOCRYSTALS (CNCs)

In the 1950s, Ranby showed for the first time that controlled sulfuric acid-catalyzed breakdown of cellulose fibers can yield colloidal suspensions of cellulose nanocrystals (Ranby, 1951; Mao et al., 2017). The origin of cellulose affects the crystal size and degree of crystallinity of elementary fibrils from various cellulose samples. Algal and tunicate cellulose microfibrils, for instance, generate nanocrystals that are several micrometers in length, but wood microfibrils generate nanocrystals that are only a few hundred nanometers in length (Janardhnan & Sain, 2006; Gu et al., 2017). Wood whiskers are 3–15 nm in width and 100–200 nm in length, while those of the sea plant Valonia are 20 nm in width and 1,000–2,000 nm in length, according to reports. Similarly, the usual dimensions of cotton crystallites are 4–10 nm in diameter and 100–300 nm in length (De Souza Lima and Borsali, 2004), whereas tunicate produces particles that are 10–20 nm in width and 500–2,000 nm in length (Shen et al., 2022; Peng et al., 2011). The length and width of cellulose whiskers obtained from various sources are listed in Table 9.1 (Saxena, 2013; Kargarzadeh et al., 2017).

Several researchers have investigated the morphology of cellulose nanocrystals. Figure 9.3 depicts a few typical images from published literature. Table 9.2 presents the moduli and specific moduli (modulus/density) of several extensively used engineering materials, demonstrating that crystalline cellulose has a higher specific modulus than steel, concrete, glass, and aluminum. The low-modulus fiber can increase the tensile strength only to some extent, whereas the high-modulus fiber can greatly increase the tensile strength (Suksawang et al., 2018).

TABLE 9.1

Physical Characteristics of Cellulose Nanocrystal (CNCs) from Different Sources (Saxena, 2013 and Mazlan et al., 2020)

Source	L (nm)	D (nm)	Reaction Conditions
Bacterial	100–1,000	10–50	65% sulfuric acid; 75°C
	100–1,000	5–10×30–50	
Cotton	100–150	5–10	65% sulfuric acid, 60 min, 45°C
	70–170	~7	64% sulfuric acid, 120 min, 60°C
	200–300	8	64% sulfuric acid, 120 min, 45°C
	255	15	
	150–210	5–11	
Cotton liner	100–200	10–20	
	25–320	6–70	
	300–500	15–30	
Microcrystalline cellulose	35–265	3–48	63.5% sulfuric acid, 2 h
	250–270	23	
	~500	10	
Ramie	150–250	6–8	
	50–150	5–10	
Sisal	100–500	3–5	65% sulfuric acid, 15 min, 60°C
	100-several microns	10–20	55% sulfuric acid, 20 min, 60°C
Turnicate	500–1–2 microns	15	60%, 30 min, 65°C
	1–2 microns	8–15 nm	65%, overnight, room temperature
EFB (MCC)	100–150 nm	3–5 nm	70%, sulphuric acid, 45°C, 60 min

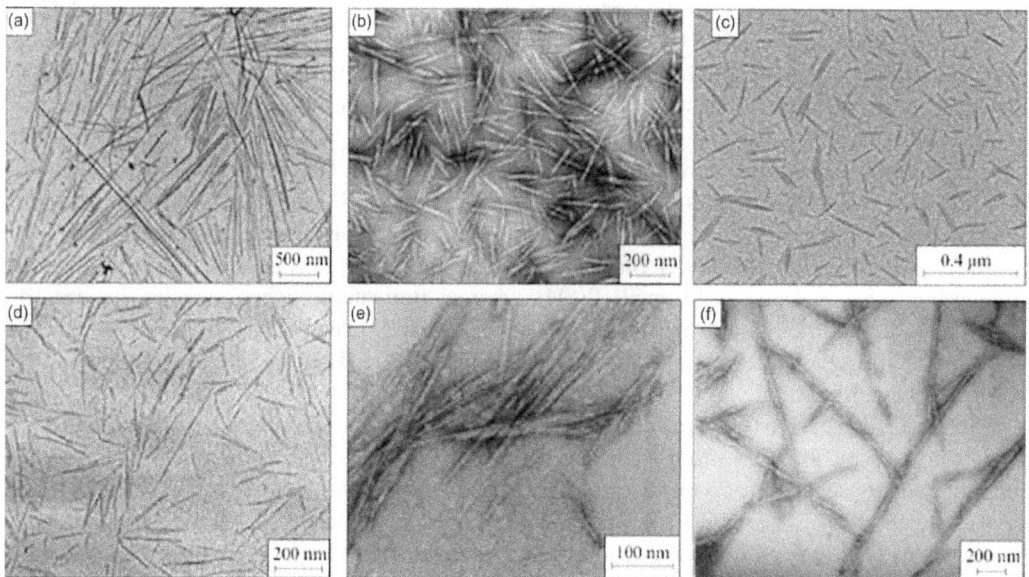

FIGURE 9.3 NCC particles produced by acid hydrolysis of (a) tunicin, (b) China grass (ramie), (c) cotton, (d) sugar beet, (e) microcrystalline cellulose, and (f) bacterial cellulose are depicted by transmission electron microscopy. Reproduced with permission from Siqueira et al. (2010a).Copyright, CC BY License.

TABLE 9.2

Moduli of Engineering Materials Compared to Cellulose Nanocrystals (Siqueira et al., 2010a)

Material	Modulus (GPa)	Density (mg/m³)	Specific Modulus (GPa/mg⁻¹/m³)
Aluminum	69	2.7	26
Steel	200	7.8	26
Glass	69	2.5	28
Cellulose nanocrystals	138	1.5	92

CELLULOSE NANOCRYSTALS (CNCs) EXTRACTION

The fundamental notion underlying all three extraction methods is that CNCs are often created by a two-step process, beginning with early hydrolysis to eliminate amorphous portions of cellulosic materials and subsequently breaking the nanocrystals' aggregation (Saxena, 2013).

OXIDATION

Mascheroni et al. (2016) and Goh et al. (2016) carried out the oxidation of high oxidants, including ammonium persulfate. Persulfate oxidation has resulted in the generation of CNCs by dissolving lignin, hemicellulose, and other contaminants in a single-pot technique, thereby reducing the number of labor-intensive procedures required to produce biomass CNCs (Chen & Lee, 2018). From the oxidation, free radicals and hydrogen peroxide permeated the amorphous areas and severed the -1,4 connection to generate CNCs. Goh et al. (2016) reported that persulfate oxidation defibrillated cellulose and successfully removed amorphous regions, resulting in CNCs with a narrow size distribution. 2,2,6,6-tetramethylpiperidine-1-oxyl (TEMPO), which selectively oxidizes primary alcohol, is a second oxidative method for creating cellulosic nanomaterials. TEMPO-mediated oxidation generates carboxylated CNCs in an aqueous suspension that is stable and uniformly disseminated (Cheng et al., 2017). The degree of cellulose crystallinity determines the degree of TEMPO oxidation. While TEMPO can efficiently defibrillate cellulose without an acid treatment, it cannot entirely degrade amorphous areas (Isogai & Zhou, 2019).

ENZYMATIC

Enzymatic hydrolysis of cellulose is a heterogeneous multi-step reaction in which cellulose is degraded by a complex of enzymes: endoglucanase, cellobiohydrolase, and cellobiase that work synergistically (Satyamurthy et al., 2011). Fungi, specifically *Trichoderma, Aspergillus*, and *Penicillium* species, contain commercial cellulases (Gao et al., 2017; Reiniati, 2017). *Trichodermareesei* is one of the most efficient producers of extracellular cellulase and the cellulase of choice for the creation of CNCs (Satyamurthy et al., 2011; Bischof et al., 2016). The enzymatic reaction is affected by the surface area of the cellulose substrate, the temperature of the reaction, the enzyme concentration, and the duration of enzyme activity. Due to the anticipated increase in enzyme adsorption to cellulose fibrils, enzyme concentrates with a greater rate of hydrolysis. CNCs produced from *Trichodermareesei* cellulase enzymatic hydrolysis were shorter (120 nm) than those originating from sulfuric acid hydrolysis (287 nm) (Bischof et al., 2016). The thermal and mechanical properties of CNCs treated with enzymes were superior to those of CNCs treated with sulfuric acid, even with the addition of low quantities of CNCs (Kruyeniski et al., 2019). It was previously believed that the degree of cellulose crystallinity influenced the rate of enzymatic hydrolysis; however, a recent study concluded otherwise.

Acid Hydrolysis

Leong et al. (2022) described the findings of creating stable cellulose nanocrystal suspensions with sulfuric acid or hydrochloric acid. Sulfuric acid produces nanocrystals with more stable aqueous solutions than hydrochloric acid (Tian et al., 2016). Utilizing sulfuric acid results in the esterification of the surface hydroxyl groups of cellulose nanocrystals, which results in the formation of charged sulfate groups (Dole et al., 2004; Beck-Candanedo et al., 2005; Tian et al., 2016) According to Dufresne (2012) and Song et al. (2018), the application of a negative charge on the surface of sulfuric acid-prepared nanocrystals results in a more stable dispersion than hydrochloric acid-prepared cellulose nanocrystals with a lower surface charge. The acid hydrolysis of cellulose fibers is a heterogeneous acid diffusion process in which hydronium ions (H_3O+) can enter the less-ordered amorphous domains of cellulose chains and enhance the hydrolytic cleavage of the glycosidic linkages. The reaction continues until all glycosidic linkages that are accessible are hydrolyzed (Mazlan et al., 2020; Chen et al., 2016; Julie et al., 2016). The settings of 45°C and 60 min are optimal for achieving complete hydrolysis of the amorphous portions of cellulose, resulting in particle lengths in the order of 200 nm (Brown, 2006; Negar & Milad, 2017). In recent years, cellulose nanowhiskers were also synthesized by sulfuric acid hydrolysis of cotton fiber (Sun et al., 2016), microcrystalline cellulose (MCC) (Bai et al., 2009), and sisal fiber (Siqueira et al., 2010b) using the same general technique as stated previously. Table 9.3 provides an exhaustive summary of preparative settings involving sulfuric acid hydrolysis and the average diameters of cellulose nanowhiskers derived from various sources.

TABLE 9.3
Dimension of Cellulose Nanocrystals Prepared Under Different Sulfuric Acid Hydrolysis Conditions (Saxena, 2013 and Mazlan et al., 2020)

Cellulose Source	H_2SO_4 Conc. (% w/w)	Time (min)	T (°C)	Acid/Cellulose (mL/g)	Dimension, Length (nm)×Width (nm)
Pulp from bleached softwood kraft	64	10	70	8.78	~200×5
	60	20	<70	8.75	~200×5
	65	10	70	10	185±75×3.5
	65	60	45	8.75	185±75×3.5
	64	45	45	17.5	100–250×5.15
Pulp from bleached softwood kraft	64	25	45	8.75	147±7×3.5
	64	25	45	8.75	141±6×3±0.3
	64	45	45	17.3	120±5×4.9±0.3
	64	45	45	8.75	105±4×4.5±0.3
Cotton	64	120	45	8.75	~200×5
	64	60	45	8.75	115±10×~7
	64	45	45	17.5	176±21×13±3
	64	120	60	8.33	70–130×10–20
	65	60	45	8.75	100–150×5–10
Sisal	65	15	60	16.2	~250×4
Flax	64	240	45	8.33	327±108×21±7
Wheat straw	65	60	25	34.3	150–300×~5
MCC	63.5	130.3	44	10	200–400×<10
	64	300	45	8.75	41–320×<100
	64	180	45	17.5	60–120×8–10
	64	120	45	8.75	100,225×10–15

CELLULOSE NANOCRYSTALS AS NANOCOMPOSITES

The inherent renewability of cellulose nanocrystals is one of their greatest selling points. CNCs are the ultimate environmental-friendly material since they are composed of infinitesimally minute bits of cellulose, the most prevalent biomaterial on Earth. Sources include bamboo, trees, and algae. Tree-based cellulose can also be extracted from lumber and paper scraps. Due to their ten-fold invasiveness and rapid development, algae and bamboo are suitable cellulose sources. Cao et al. (2016) investigated how the incorporation of CNCs altered the performance of cement paste. With only 0.2% of the volume of CNCs compared to cement, mechanical tests reveal a gain in flexural strength of around 30%. Due to the strong anisotropies in cellulose nanocrystals' density, quantum mechanics demonstrates that their stiffness is comparable to that of steel, based on experimental young's moduli values. Because cellulose nanocrystals have a high value for Young's modulus and a potentially high ratio of crystals per unit of density, they have the potential to aid in preventing material deformation in response to an applied force. Peters et al. (2010) conducted a study in which reactive powder concrete, also known as ultra-high-performance concrete, was reinforced with nano-cellulose and micro-cellulose fibers to boost the tensile strength of an otherwise fragile material. These fibers could offer the same advantages as other micro- and nano-fiber-reinforcing methods at a fraction of the price. To evaluate fracture energy under conditions of steady crack propagation, notched-beam tests were conducted with crack-mouth opening displacement control. Preliminary results indicate that the addition of up to 3% micro- and nano-fibers in combination enhanced the fracture energy by more than 50% in comparison to the unreinforced material, with few processing modifications. Physical property improvement for cement composite.

The majority of earlier work on fiber-reinforced cement composites, regardless of fiber size, attributed the improvement in mechanical performance to the fiber-bridging mechanism (Pickering et al., 2016). The majority of claims are predicated on the notion that fibers can postpone or even stop crack propagation. Figure 9.4 illustrates how the bridging process in steel fiber-reinforced cement composites behaves according to Geus (2014).

As this is well known, traditional long fibers (1–5 cm) are excellent for enhancing the toughness or energy absorption of cement composites because they may bridge macro-level cracks and limit their growth. The short fibers (1–5 mm) lack the length to bridge large-width cracks and are therefore not intended to improve the composites' durability (Madsen and Gamstedt, 2013). The short fibers, however, can bridge the cracks at the micro-level with widths less than the fiber length, effectively preventing the fissures from expanding and merging into a macro-level crack when the material fails and the ultimate strength is obtained. Therefore, the short

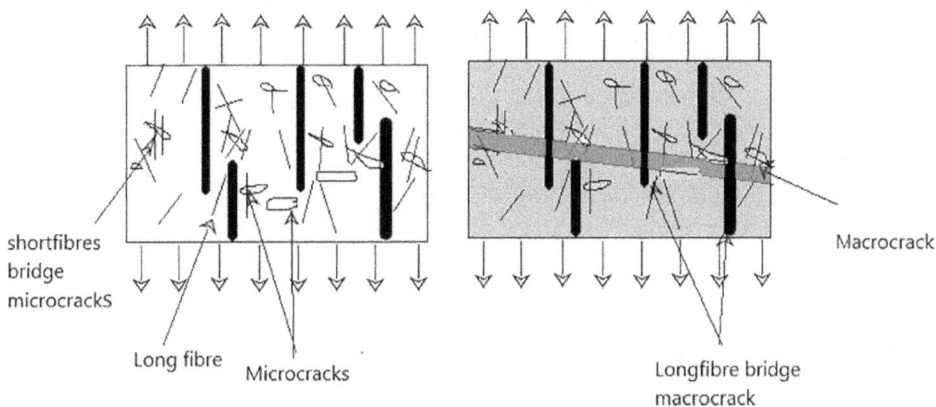

FIGURE 9.4   Bridging effect of fiber in concrete.

fibers are meant to increase the material's strength while having a negligible impact on its toughness. Extremely short fibers, such as CNCs, which are shorter than 0.5 m, constitute the final category of fibers in terms of length. At this scale, it is nearly impossible to notice fissures in cement composites.

THERMAL BEHAVIOR OF CELLULOSE NANOCRYSTALS

The thermal stability of CNCs is a critical aspect in the field of polymer composites, as the processing temperatures of popular polymers are close to or surpass the degradation initiation temperature of cellulose (Leszczyńska et al., 2018). It was discovered that the thermal stability of cellulose nanoparticles is dependent on several chemical compositions and structural characteristics of the nanomaterial. The most important are hemicellulose and lignin residues, as well as the kind and quantity of surface groups formed during the production and modification processes. For instance, sulfate esters formed during sulfuric acid-catalyzed hydrolysis or carboxyl groups in TEMPO-oxidized cellulose, crystal polymorph, crystallinity content, degree of polymerization, and large surface area of nanocellulose (Eyley & Thielemans, 2014; Leszczyńska et al., 2018). The high crystallinity index and mechanical properties may affect the thermal properties, such as thermal conductivity and thermal expansion, and may improve the use of nanocellulose in applications such as electrical devices and packaging. Low thermal stability can limit the usage of nanocellulose and the creation of nanocomposites at high temperatures. When fiber components lack heat stability, degradation accelerates (Sheltami et al., 2017). Extractives, hemicellulose, and lignin begin to break down at lower temperatures than cellulose, and the mechanism of degradation is also distinct among individual components. A thermal degradation is a form of molecular degeneration caused by over-heating. At high temperatures, polymers undergo molecular scission, which increases their molecular weight and diminishes their desirable physical properties (e.g., mechanical and optical properties). Thus, thermal degradation determines the maximum service temperature at which polymer composites can be manufactured and utilized. Consequently, the heat stability of the fillers used in composite manufacture and applications is crucial.

CONCLUSIONS

The usage of nanomaterials in cement composite materials has become a new interest in construction industries. CNCs are one of the nanomaterials which should not be left behind. Since the CNCs carry good properties, they have the potential to become compatible with the cement composites structure as a strengthening agent. However, few researchers have conducted studies on the effect of CNCs in cement composites in terms of strength (compressive, flexural and tensile). However, many reasons are still uncovered for how the CNCs react and improve the structure of the cement matrix. Nowadays, sustainable building materials research has been quite in the spotlight. And more demand for materials that carry a 'green' factor such as good thermal performance has been in demand especially in a country with a tropical climate. As far as today, the study on CNCs on their thermal behavior shows good potential for the CNCs to become the cooling agent in cement composites. CNCs as the cooling agent in the cement matrix will become one of the important research gaps still attempted to be uncovered.

ACKNOWLEDGMENTS

This research is financially supported by FRGS R.J130000.7801.5F221, PDRU Grant – Vot No. Q.J130000.21A2.04E53, Hitachi Global Foundation 2019, and MRUN R.J130000.7805.4L886, which are gratefully acknowledged. This research acknowledges the support by Postdoctoral Fellowship from Prince of Songkla University, Thailand.

REFERENCES

Abdul Khalil, H. P. S., Bhat, A.H., & Ireana Yusra, A. F. (2012). Green composites from sustainable cellulose nanofibrils: A review. *Carbohydrate Polymers*, 87(2), 963–979.

Aliabdo, A. A., Elmoaty, A., Elmoaty, M. A., & Salem, H. A. (2016). Effect of water addition, plasticizer and alkaline solution constitution on fly ash based geopolymer concrete performance. *Construction and Building Materials*, 121, 694–703.

Bai, W., Holbery, J., & Li, K. (2009). A technique for production of nanocrystalline cellulose with a narrow size distribution. *Cellulose*, 16, 455–465.

Beck-Candanedo, S., Roman, M., & Gray, D. G. (2005). Effect of reaction conditions on the properties and behavior of wood cellulose nanocrystal suspensions. *Biomacromolecules*, 6(2), 1048–54.

Bischof, R. H., Ramoni, J., & Seiboth, B. (2016). Cellulases and beyond: The first 70 years of the enzyme producer Trichoderma reesei. *Microbial Cell Factories*, 15(106), 1–13.

Brown, R. M. (2006). The biosynthesis of cellulose. *Journal of Macromolecular Science (Part A)*, 33 (10), 1345–1373.

Cao, Y., Zavaterri, P., Youngblood, J., Moon, R., & Weiss, J. (2015). The influence of cellulose nanocrystal additions on the performance of cement paste. *Cement and Concrete Composites*, 56, 73–83.

Cao, Y., Zavattieri, P., Youngblood, J., Moon, R., & Weiss, J. (2016). The relationship between cellulose nanocrystal dispersion and strength. *Construction and Building Materials*, 119, 71–79.

Chen, Y. W., & Lee, H. V. (2018). Revalorization of selected municipal solid wastes as new precursors of "Green" nanocellulose via a novel one-pot isolation system: A source perspective. *International Journal of Biological Macromolecules*, 107, 78–92.

Chen, Y. W., Lee, H. V., Juan, J. C., & Phang, S.-M. (2016). Production of new cellulose nanomaterial from red algae marine biomass Gelidium elegans. *Carbohydrate Polymers*, 151, 1210–1219.

Cheng, F., Liu, C., Wei, X., Yan, T., Li, H., He, J., & Huang, Y. (2017). Preparation and characterization of 2,2,6,6-tetramethylpiperidine-1-oxyl (TEMPO)-Oxidized cellulose nanocrystal/alginate biodegradable composite dressing for hemostasis applications'. *Sustainable Chemical Engineering*, 5, 3819–3828.

Cheng, Q., Devallance, D., Wang, J., & Wang, S. (2006). Advanced cellulosic nanocomposite materials. In Attaf, B. (ed) *Advances in Composite Materials for Medicine and Nanotechnology*. London: IntechOpen, pp. 547–564.

Crow, J. M. (2008). The concrete conundrum. *Chemistry World*, (March), 62–66.

De Souza Lima, M. M., & Borsali, R. (2004). Rodlike cellulose microcrystals: Structure, properties, and applications. *Macromolecular Rapid Communications*, 25(7), 771–787.

Dole, P., Joly, C., Espuche, E., Alric, I., & Gontard, N. (2004). Gas transport properties of starch based films. *Carbohydrate Polymers*, 58 (3), 335–343.

Dufresne, A. (2012). Processing of nanocellulose-based materials. In Dufresne, A. (ed) *Nanocellulose: From Nature to High Performance Tailored Materials*, Berlin, Boston: De Gruyter, pp. 351–418.

Dungani, R., Karina, M., Subyakto, Sulaeman, A., Hermawan, D., & Hadiyane, A. (2016) Agricultural waste fibers towards sustainability and advanced utilization: A review. *Asian Journal of Plant Sciences*, 15 (1–2), 42–55.

Eyley, S., & Thielemans, W. (2014). Surface modification of cellulose nanocrystals. *Nanoscale*, 6(14), 7764–7779.

Gao, G., Yue, R. M., Jing, X., Liu, Z. Y., & Li, G. (2017). Efficient yeast cell-surface display of an endoglucanase of Aspergillus flavus and functional characterization of the whole- cell enzyme. *World Journal of Microbiology and Biotechnology*, 33(6), 1–10.

Geus, J. V. De (2014). *Ultra High Performance Fibre Reinforced Concrete Applied in Railway Bridges*. Master Thesis, Delft University of Technology, Netherlands.

Givi, A. N., Rashid, S. A., Aziz, F. N. A., & Salleh, M. A. M. (2010). Assessment of the effects of rice husk ash particle size on strength, water permeability and workability of binary blended concrete. *Construction and Building Materials*, 24(11), 2145–2150.

Goguen, C. (2012). *Air Entrainment versus Air Entrapment*. National Precast Concrete Association, Indiana.

Goh, K. Y., Ching, Y. C., Chuah, C. H., Abdullah, L. C., & Liou, N.-S. (2016). Individualization of microfibrillated celluloses from oil palm empty fruit bunch: Comparative studies between acid hydrolysis and ammonium persulfate oxidation. *Cellulose*, 23(1), 379–390.

Gu, J., Hu, C., Zhong, R., Tu, D., Yun, H., Zhang, W., & Leu, S. (2017). Isolation of cellulose nanocrystals from medium density fiberboards. *Carbohydrate Polymers*, 167, 70–78

Haga, S., Fawwaaz, M., & Bucks, G. (2021). Crosly's concrete conundrum. *Undergraduate Scholarly Showcase*, 3(1), 2021.

Isogai, A., & Zhou, Y. (2019). Current opinion in solid state & materials science diverse nanocelluloses prepared from TEMPO-oxidized wood cellulose fibers: Nanonetworks, nano fibers, and nanocrystals. *Current Opinion in Solid State & Materials Science*, 23(2), 101–106.

Janardhnan, S., & Sain, M. M. (2006). Isolation of cellulose microfibrils – An enzymatic approach. *BioResources*, 1(2), 176–188.

José, M., Bettencourt, A., & Garrido, F. (2015). Curing effect in the shrinkage of a lower strength self-compacting concrete. *Construction and Building Materials*, 93, 1206–1215.

Julie, J. C. S., George, N., & Narayanankutty, S. K. (2016). Isolation and characterization of cellulose nanofibrils from arecanut husk fibre. *Carbohydrate Polymers*, 142, 158–166

Kargarzadeh, H., Ioelovich, M. & Ahmad, I. (2017). Methods for extraction of nanocellulose from various sources. In *Handbook of Nanocellulose and Cellulose Nanocomposites*, eds. H. Kargarzadeh, I. Ahmad, S. Thomas and A. Dufresne. John Wiley & Sons, Weinheim, pp. 1–49.

Khalid, N. H. A., Hussin, M. W., Mirza, J., Ariffin, N. F., Ismail, M. A., Lee, H. S., Mohamed, A., & Jaya, R. P. (2016). Palm oil fuel ash as potential green micro-filler in polymer concrete. *Construction and Building Materials*, 102, 950–960.

Khalil, H. P. S. A., Alwani, M. S., & Omar, A. K. M. (2006). Distribution, and cell wall structure of malaysian plant waste fibers. *Bioresources*, 1, 220–232.

Khushefati, W. H., & Demirbog, R. (2015). Effects of nano and micro size of CaO and MgO, nano-clay and expanded perlite aggregate on the autogenous shrinkage of mortar. *Construction and Building Materials*, 81, 268–275.

Kruyeniski, J., Ferreira, P. J. T., Videira, G., & Carvalho, S. (2019). Industrial crops & products physical and chemical characteristics of pretreated slash pine sawdust influence its enzymatic hydrolysis. *Industrial Crops & Products*, 130, 528–536.

Leong, S. L., Tiong, S. I. X., Siva, S. P., Ahamed, F., Chan, C. H., Lee, C. L., ... & Ho, Y. K. (2022). Morphological control of cellulose nanocrystals via sulfuric acid hydrolysis based on sustainability considerations: An overview of the governing factors and potential challenges. *Journal of Environmental Chemical Engineering*, 10(4), 108145.

Leszczyńska, A., Radzik, P., Haraźna, K., & Pielichowski, K. (2018). Thermal stability of cellulose nanocrystals prepared by succinic anhydride assisted hydrolysis. *Thermochimica Acta*, 663, 145–156.

Ling, Y.-F., Zhang, P., Wang, J., & Shi, Y. (2020). Effect of sand size on mechanical performance of cement-based composite containing pva fibers and Nano-SiO2. *Materials*, 13(2), 1–14.

Madsen, B., & Gamstedt, E. K. (2013). Wood versus plant fibers: Similarities and differences in composite applications. *Advances in Materials Science and Engineering*, 2013, 564346.

Mao, J., Abushammala, H., Brown, N., & Laborie, M. (2017). Comparative assessment of methods for producing cellulose I nanocrystals from cellulosic sources. In Agarwal, U. P., Atalla, R. H., & Isogai, A. (eds) *Nanocelluloses: Their Preparation, Properties, and Applications*, American Chemical Society, pp. 19–53.

Mascheroni, E., Rampazzo, R., Aldo, M., Piva, G., Bonetti, & S.Piergiovanni, L. (2016). Comparison of cellulose nanocrystals obtained by sulfuric acid hydrolysis and ammonium persulfate, to be used as coating on flexible food-packaging materials. *Cellulose*, 23(1), 779–793.

Mazlan, D., Krishnan, S., Md Din, M. F., Tokoro, C., Abd Khalid, N. H., Ibrahim, I. S., Takahashi, H., & Komori, D. (2020). Effect of cellulose nanocrystals extracted from oil palm empty fruit bunch as green admixture for mortar. *Scientific Reports*, 10(1), 1–11.

Mazlan, D., Krishnan, S., Md Din, M. F., et al. (2020). Effect of cellulose nanocrystals extracted from oil palm empty fruit bunch as green admixture for mortar. *Scientific Reports*, 10, 6412.

Min, F., Yao, Z., & Jiang, T. (2014). Experimental and numerical study on tensile strength of concrete under different strain rates. *The Scientific World Journal*, 2014, 173531.

Negar, K., & Milad, F. (2017). Production of cellulose nanocrystals from pistachio shells and their application for stabilizing Pickering emulsions. *International Journal of Biological Macromolecules*, 106, 1023–1031

Peng, B. L., Dhar, N., Liu, H. L., & Tam, K. C. (2011). Chemistry and applications of nanocrystalline cellulose and its derivatives: A nanotechnology perspective. *The Canadian Journal of Chemical Engineering*, 89 (5), 1191–1206.

Peters, S. J., Rushing, T. S., Landis, E. N., & Cummins, T. K. (2010). Nanocellulose and microcellulose fibers for concrete. *Transportation Research Record: Journal of the Transportation Research Board*, 2142(1), 25–28.

Pickering, K. L., Efendy, M. G. A., & Le, T. M. (2016). A review of recent developments in natural fibre composites and their mechanical performance. *Composites Part A: Applied Science and Manufacturing*, 83, 98–112.

Rahuman, A., & Yeshika, S. (2015). Study on properties of sisal fiber reinforced concrete with different mix proportions and different percentage of fiber addition. *International Journal of Research in Engineering and Technology*, 4(3), 474–477.

Ranby, B. G. (1951). The colloidal properties of cellulose micelles. Fibrous macromolecular systems. *Discussions of Faraday Society*, 11, 158–164

Reiniati, I. (2017). Bacterial cellulose nanocrystals: Production and application. *Electronic Thesis and Dissertation Repository*, 4826, The University of Western Ontario, Canada, pp. 1–187.

Satyamurthy, P., Jain, P., Balasubramanya, R. H., & Vigneshwaran, N. (2011). Preparation and characterization of cellulose nanowhiskers from cotton fibres by controlled microbial hydrolysis. *Carbohydrate Polymers*, 83(1), 122–129.

Saxena, A. (2013). *Nanocomposites Based on Nanocellulose Whiskers*. PhD Thesis. Georgia Institute of Technology, USA.

Sheltami, R., Kargarzadeh, H., Abdullah, I., & Ahmad, I. (2017). Thermal properties of cellulose nanocomposites. *Journal of Thermal Analysis*, 35(7), 2235–2242.

Shen, D., Jiang, J., Shen, J., Yao, P., & Jiang, G. (2016). Influence of curing temperature on autogenous shrinkage and cracking resistance of high-performance concrete at an early age. *Construction and Building Materials*, 103, 67–76.

Shen, P., Tang, Q., Chen, X., & Li, Z. (2022). Nanocrystalline cellulose extracted from bast fibers: Preparation, characterization, and application. *Carbohydrate Polymers*, 290, 119462.

Siqueira, G., Bras, J., & Dufresne, A. (2010a). Cellulosic bionanocomposites: A review of preparation, properties and applications. *Polymers*, 2(4), 728–765.

Siqueira, G., Bras, J. & Dufresne, A. (2010b). New process of chemical grafting of cellulose nanoparticles with a long chain isocyanate. *Langmuir*, 26, 1, 402–411.

Song, X., Zhou, L., Ding, B., Cui, X., Duan, Y. & Zhang, J. (2018). Simultaneous improvement of thermal stability and redispersibility of cellulose nanocrystals by using ionic liquids. *Carbohydrate Polymers*, 186, 252–259.

Suksawang, N., Wtaife, S. and Alsabbagh, A. (2018). Evaluation of elastic modulus of fiber-reinforced concrete. *ACI Materials Journal*, (March), 115(2), 239–249.

Sun, B., Zhang, M., Hou, Q., Liu, R., Wu, T., & Si, C. (2016). Further characterization of cellulose nanocrystal (CNC) preparation from sulfuric acid hydrolysis of cotton fibers. *Cellulose*, 23(1), 439–450.

Tian, C., Yi, J., Wu, Y., Wu, Q., Qing, Y., & Wang, L. (2016). Preparation of highly charged cellulose nanofibrils using high-pressure homogenization coupled with strong acid hydrolysis pretreatments. *Carbohydrate Polymers*, 136, 485–492.

Tosic, N., Marinkovic, S., & Stojanovic, A. (2017). Sustainability of the concrete industry: Current trends and future outlook. *Tehnika*, 72(1), 38–44.

Venkatesan, G. & Tamizhazhagan. T. (2016). Ultra high strength concrete. *International Journal of Innovative Research in Science*, 5(3), 4412–4418.

10 All-Cellulose Composites
Processing, Properties, and Applications

Supachok Tanpichai
King Mongkut's University of Technology Thonburi

CONTENTS

Introduction 145
Processing Routes 146
   Impregnation 147
   Partial Surface Dissolution 150
Applications 153
   Composites 153
   Packaging 153
   Fibers and Filaments 154
   Bioengineering and Biomedical Materials 154
   Dye and Heavy Metal Ion Adsorbents 156
Recycling, Sustainability, and Biodegradability 156
Conclusions and Prospects 158
References 158

INTRODUCTION

Cellulose, a linear polymer comprising numerous repeating units of two anhydroglucose rings bonded with a β-1,4-glycosidic linkage, is the most abundant biopolymer on earth, with an annual production rate of 1.5×10^{12} tons (Acharya et al., 2021; Gindl and Keckes, 2005; Huber et al., 2012b). With many hydroxyl groups in its molecules, cellulose possesses superior mechanical properties along with biocompatibility, biodegradability, renewability, and chemical stability (Foroughi et al., 2021; Tanpichai, 2022; Tanpichai et al., 2022c). The elastic modulus and strength of the cellulose I crystallite have been measured to be ~140 GPa and 10 GPa, respectively, which are higher than those of glass fiber and aluminum (Azizi Samir et al., 2004; Huber et al., 2012b; Nishino et al., 1995; Rusli and Eichhorn, 2008; Sakurada et al., 1962). Therefore, several attempts have been made to use cellulose in various forms, such as fibers and particles, to enhance the properties of polymers (Abe et al., 2009; Akter et al., 2020; Bulota et al., 2012; Tanpichai et al., 2012; Tanpichai and Wimolmala, 2022d; Yuan et al., 2008; Yuwawech et al., 2015). Unfortunately, improving the mechanical performance of such composites using cellulose is challenging as poor interfacial bonding between the polymer matrix and cellulose reinforcement phase causes inefficient stress transfer from the matrix to cellulose (Pullawan et al., 2010; Rusli et al., 2010; Tanpichai et al., 2012, 2014). However, this concern is largely mitigated when both the reinforcement and polymer matrix are derived from the same material (Nishino et al., 2004).

The concept of using cellulose as both the reinforcement and matrix was originally proposed by Nishino et al. (2004) to prepare all-cellulose composites (ACCs). Since then, ACCs have been

DOI: 10.1201/9781003358084-10

prepared from various cellulose sources, such as wood, cotton, eucalyptus, and agricultural waste (e.g., pineapple leaves, corn stalk, rice straw, grass, and water hyacinth) (Huber et al., 2012b; Senthil Muthu Kumar et al., 2018; Tang et al., 2021; Tanpichai, 2018; Tanpichai and Witayakran, 2016). Although the regenerated cellulose (also called the matrix) and reinforcing cellulose are derived from the same cellulose material, these two components in ACCs have different structures and mechanical properties. Notably, it is impossible to melt cellulose and mold it into a desirable shape owing to extensive H-bonding between the hydroxyl groups of cellulose molecules, only thermal degradation occurs when cellulose is heated (Acharya et al., 2021; Nishino et al., 2004; Tanpichai et al., 2022a). Therefore, dissolution of cellulose with a specific solvent, such as lithium chloride/ N,N-dimethyl acetamide (LiCl/DMAc) (Nishino et al., 2004; Tanpichai et al., 2022a; Tanpichai and Witayakran, 2017), NaOH/urea (Hildebrandt et al., 2017; Tanpichai et al., 2022c), NaOH (Labidi et al., 2019), N-methylmorpholine-N-oxide monohydrate (Labidi et al., 2019), and ionic liquids (Spörl et al., 2018; Swatloski et al., 2002; Yousefi et al., 2011b; Zhao et al., 2009), is used to prepare ACCs. Each solvent requires different conditions, such as temperature, dissolution time, and pre-treatment, to dissolve cellulose (Nishino et al., 2004; Pullawan et al., 2014; Tanpichai et al., 2022a, 2022c). For example, during cellulose dissolution using NaOH/urea, the disruption of inter- and intramolecular H-bonding between cellulose molecules is caused by OH anions and intercalation of Na cations into the hydroxyl groups of cellulose, which prevents cellulose chains from interacting with adjacent chains. This leads to the dissolution of cellulose, and urea molecules accumulate around the dissolved cellulose chains, which transforms cellulose I into cellulose II (Piltonen et al., 2016). Alternatively, the intermolecular bonding between cellulose chains (in-plane and out-of-plane) is fully disrupted by LiCl/DMAC or ionic liquids, and the disruption time is longer in this system than in the NaOH/urea system (Duchemin et al., 2016).

PROCESSING ROUTES

As shown in Figure 10.1, there are two main processing routes to fabricate ACCs: impregnation and partial surface dissolution. The impregnation approach primarily requires a cellulose solution made from a cellulose material, and another cellulose material as the reinforcement is impregnated into the cellulose solution (matrix) to form an ACC. By contrast, an ACC prepared via partial surface

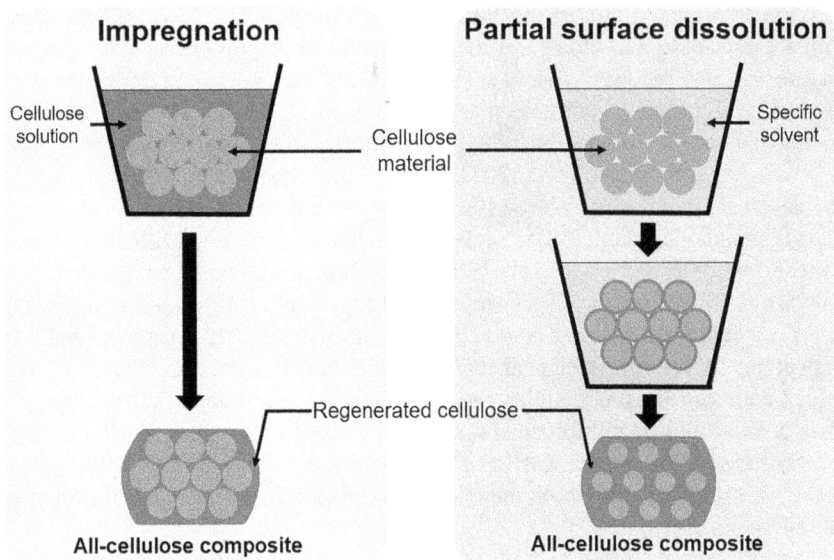

FIGURE 10.1 Fabrication of ACCs: impregnation and partial surface dissolution. Figure by Author.

dissolution is formed by transforming an external surface of a cellulose material into the cellulose matrix covered by the undissolved part of the cellulose material, which acts as the reinforcement (Nishino et al., 2004; Pullawan et al., 2010, 2014).

IMPREGNATION

Impregnation is the conventional processing method to prepare polymer composites with cellulose fibers (Pullawan et al., 2014). The development of ACCs via the impregnation approach requires two processing steps. A cellulose solution is initially prepared by dissolving purified cellulose fibers or particles in a specific solvent, as discussed previously, and a cellulose material, such as fibers, textiles, and particles, is subsequently impregnated into the cellulose solution to act as a reinforcing phase. Then, the ACC is formed via the conversion of the cellulose solution into the regenerated cellulose matrix reinforced by the cellulose reinforcement (Nishino et al., 2004; Pullawan et al., 2014; Tanpichai and Witayakran, 2017).

Nishino et al. (2004) originally developed ACCs via this impregnation method. Prior to preparing the cellulose solution, wood pulp was pretreated via immersion in various solvents for 24 h each (distilled water, acetone, and DMAc). The pretreated pulp was subsequently dissolved in a mixed solvent of LiCl and DMAc to yield the cellulose solution. Then, aligned ramie fibers were directly impregnated in the cellulose solution. The cellulose solution with aligned ramie fibers was exposed to ambient atmosphere for gelation, and the solidified gel incorporated with fibers was submerged in methanol to remove LiCl and DMAc. The sample was hot-air-dried, and the ACC was obtained. The high crystallinity of the ramie fibers was maintained in the composite, while the cellulose solution was completely converted into a noncrystalline material. This indicated that the incorporation of the ramie fibers in the cellulose solution did not damage the structure of the fibers. Scanning electron microscopy (SEM) revealed that in the composite, the ramie fibers were completely surrounded by the cellulose matrix regenerated from the pulp dissolved in LiCl and DMAc, and breakage of the fiber and matrix was observed for the fractured surface of the ACC after tensile deformation. This implied strong interfacial interaction between the fibers and cellulose matrix. Consequently, the ACCs showed superior mechanical performance with an average tensile strength of 480 GPa and dynamic storage modulus of 45 GPa at 25°C. These great mechanical properties of the ACCs were attributed to the promising mechanical properties of ramie fibers (tensile strength of 730±190 MPa and Young's modulus of 42±9 GPa), which were aligned in a parallel direction during mechanical testing and imparted strong interfacial adhesion between the cellulose matrix and reinforcing fibers. Notably, the interface between the fibers and matrix still existed in the composites owing to the occurrence of interfacial cracks between the reinforcement and matrix.

Surface activation of the cellulose-reinforcing phase with a series of solvents is advised before impregnation into a cellulose solution to obtain greater interfacial interactions between the matrix and reinforcement (interface-less) in ACCs (Nishino et al., 2004; Wei et al., 2016). Surface activation would partially swell or dissolve the fiber surface, and interdiffusion of cellulose molecules along the fiber surface would ensure that cellulose molecules from the cellulose solution interact with the fiber surface (Nishino et al., 2004; Wei et al., 2016). Wei et al. (2016) investigated the efficiency of activation of straw fibers as a reinforcing agent in the regenerated cellulose matrix. Straw fibers were activated via immersion in distilled water, ethanal, and DMAc for 3 h each. The tensile strength of the composites with activated straw fibers was 650 MPa, which was considerably higher than that of the composite with inactivated straw fibers (143 MPa). Moreover, the introduction of raw fibers in the cellulose matrix could cause poor interfacial interactions between the cellulose matrix and raw fibers owing to the introduction of impurities and noncellulosic materials on the fiber surface (Wei et al., 2016).

The stiffness of the cellulose reinforcement is considered important as cellulose reinforcement with greater stiffness generally improves the mechanical properties of composites. Moreover, greater stress transfer from the soft matrix to stiff cellulose can be achieved when cellulose reinforcement with a higher aspect ratio is introduced. A 57% improvement in tensile strength of composites was reported with the introduction of high-aspect-ratio cellulose particles compared with the 12% improvement in tensile strength achieved with the introduction of low-aspect-ratio cellulose particles (Tanpichai, 2018). Additionally, the reinforcing abilities of two different types of cellulose nanocrystals (CNCs) extracted from tunicates (T-CNCs) and cotton linters (C-CNCs) as the reinforcing phase in ACCs were compared (Pullawan et al., 2014). T-CNCs and C-CNCs had aspect ratios of 72.8±40.8 and 17.2±6.2, and the estimated Young's moduli were ~140 and 50–105 GPa, respectively (Rusli and Eichhorn, 2008; Sturcova et al., 2005). The ACCs with T-CNCs presented better mechanical properties than the composites with C-CNCs. The Young's moduli of the composites with T-CNCs and C-CNCs at 15 v/v% were higher than those of the cellulose matrix by 293% and 183%, respectively. The greater enhancement in mechanical properties of the composites with T-CNCs was thought to be due to the higher aspect ratio and Young's modulus of T-CNCs compared with those of C-CNCs (Sturcova et al., 2005).

The concentration of the reinforcing phase also plays a key role in determining the mechanical properties of ACCs (Labidi et al., 2019; Qi et al., 2009; Yang et al., 2015). As expected, better mechanical properties of ACCs have been reported by increasing the content of the reinforcing phase (Pullawan et al., 2010, 2014; Qi et al., 2009; Yang et al., 2015). The increase in the mechanical properties got more pronounced with increasing volume fraction of CNCs (Pullawan et al., 2014). The tensile strength and Young's modulus of ACCs with 5 v/v% T-CNCs were 134.0±4.2 MPa and 6.3±0.3 GPa, respectively; in comparison, the cellulose matrix had a strength of 97.7±3.9 MPa and Young's modulus of 3.0±0.1 GPa. When the T-CNC content increased to 15 v/v%, the strength and Young's modulus of the ACCs were considerably enhanced by 68% and 293% to 165.4±3.9 MPa and 11.8±0.2 GPa, respectively. These increments were attributed to efficient stress transfer between the cellulose matrix and reinforcement owing to the high stiffness of T-CNCs (140 GPa). However, a decrease in tensile strain was found with increasing concentration of the stiff reinforcement. This was due to embrittlement of the composites. A similar increase in mechanical properties of ACCs was reported with increasing concentrations of cellulose nanofibers (CNFs) when they were used as a filler (Zhao et al., 2014). With the introduction of 5 wt% CNFs, the tensile stress and Young's modulus of the composites increased from 61.6 MPa and 0.8 GPa to 84.2 MPa and 2.2 GPa, respectively, and those of the composites with 10 wt% CNFs were 96.3 MPa and 2.9 GPa, respectively. Furthermore, the thermal properties of ACCs could be modified by modifying the content of the cellulose reinforcement (Yang et al., 2010). With increasing content of the cellulose reinforcement, the maximum thermal decomposition temperature shifted to a higher temperature owing to the greater thermal stability of the cellulose reinforcement (Yang et al., 2010; Zhao et al., 2014).

Although the incorporation of high concentrations of the cellulose reinforcement could enhance the mechanical and thermal properties of the composites, when the quantity of the reinforcement in the composites is too high, a reduction in mechanical properties of the ACCs is observed. This is due to the presence of voids generated between the aggregate of the cellulose reinforcement (Labidi et al., 2019). Additionally, by increasing a concentration of the cellulose reinforcing phase, the optical transmittance of the composites decreased (Yang et al., 2010). The transmittance of the composites at 800 nm drastically reduced from 86.9% to 16.2% with the addition of 25 wt% cellulose fibers. Zhao et al. (2014) reported a similar decrease in transmittance of ACCs reinforced with a high content of CNFs; the cellulose matrix had a transmittance of 96.4% at 800 nm, which decreased to 76.0% when 20 wt% CNFs were added.

Fiber surfaces can govern stress transfer and failure deformations in ACCs. Tanpichai and Witayakran (2017) prepared ACCs incorporated with a cellulose network via the impregnation route. Pineapple leaf fibers were subjected to alkaline treatment or steam explosion. The steam explosion process reduced the average fiber width from 36 to 3.4 µm, while the width of alkaline-treated fibers was similar to that of untreated fibers with small amounts of microfibers loosened. Then, these two types of cellulose networks (alkaline-treated and steam-exploded fibers) were impregnated into a cellulose solution prepared by immersing microcrystalline cellulose (MCC) in a mixed solvent of

LiCl and DMAc, and an ACC incorporated with a cellulose mat was fabricated. Mechanical properties (tensile strength of 24.8 MPa and Young's modulus of 1.3 GPa) of the ACCs with a network of steam-exploded fibers were considerably better than those of the ACCs with a network of alkaline-treated fibers (strength of 13.9 MPa and Young's modulus of 0.9 GPa). The better mechanical performances of the ACCs were attributed to the interaction between a large quantity of finer fibers and the cellulose matrix, allowing stress to efficiently transfer from the cellulose matrix to the stiff fibers. Owing to better interaction between the fibers and matrix, failure deformation mechanisms such as fiber breakage and pull-out were observed for the ACCs incorporated with steam-exploded fibers, whereas only the fiber pull-out failure mechanism was observed for the ACCs with alkaline-treated fibers. During the deformation, the debonding between the matrix and reinforcing phases occurs. When the interfacial adhesion between the matrix and reinforcing phases is low, fibers are extracted from the matrix by the applied stress after debonding. This failure mechanism is called fiber pull-out. On the other hand, with a strong interaction between matrix and reinforcing phases in ACCs, fiber breakage occurs owing to the crack propagation through the matrix phase to the fibers (Huber et al., 2012c; Hull and Clyne, 1996).

The alignment of the cellulose reinforcing phase affects the mechanical properties of ACCs (Pullawan et al., 2012). A magnetic field was applied to align CNCs in a dissolved cellulose matrix, and the mechanical properties of the ACCs prepared within the magnetic field were considerably better than those of the composites prepared without a magnetic field (Pullawan et al., 2012). The tensile stress and Young's modulus of the ACCs prepared within the magnetic field deformed perpendicular to the magnetic field direction were 160.6 ± 4.6 MPa and 8.3 ± 0.3 GPa, respectively, whereas the stress and Young's modulus of the composites prepared without a magnetic field were 130.7 ± 3.1 MPa and 6.4 ± 0.3 GPa, respectively. The better mechanical performance was attributed to the orientation of CNCs in the composite under a magnetic field.

Raman spectroscopy was used to monitor the orientation of the reinforcing and matrix phases after the magnetic field was applied. The intensity of the Raman band at ~1095 cm^{-1}, corresponding to the C–O stretching mode along the cellulose backbone, was recorded to monitor the orientation of CNCs in the composites with various rotation angles, while the peak at ~895 cm^{-1} was used to determine the orientation of the cellulose matrix phase as this peak was not observed for CNCs. The intensities of the peaks at ~895 and 1,095 cm^{-1} were similar with respect to rotation angles for the composites in the absence of a magnetic field; however, in the presence of a magnetic field, a considerable change in the intensity of the Raman band at ~1,095 cm^{-1} was observed, but the intensity of the band at ~895 cm^{-1} remained unchanged. This suggested that the use of a magnetic field aligned CNCs but did not align cellulose molecules in the cellulose matrix.

Furthermore, the orientation of the ACCs in a wet state is possibly controlled by mechanical drawing (Fujisawa et al., 2016; Pullawan et al., 2013). With increasing draw ratio, the tensile strength and Young's modulus of the composites increased, whereas a dramatic loss in tensile strain was observed. CNFs were found to well align along the drawing direction; however, the random orientation of CNFs was observed for the undrawn composites (Fujisawa et al., 2016). Moreover, ACCs with 15 v/v% CNFs were wet-stretched by 1% and 3% strain, and the tensile strength and Young's modulus of the composites wet-stretched by 1% strain increased from 148.9 MPa and 9.8 GPa (unstretched sample) to 155.7 MPa and 11.2 GPa (wet-stretched sample), respectively (Pullawan et al., 2013). When stretching was increased to 3% strain, the composites showed a considerable increase in the strength and Young's modulus, which increased to 170.2 MPa and 13.6 GPa, respectively. The orientation of the CNCs and cellulose matrix was monitored using Raman spectroscopy. Slight reorientation of CNCs occurred in the composites when the wet composites were stretched with a small strain of 1%, whereas considerable realignment of CNCs was observed when the samples were stretched with 3% strain. However, considerable reorientation of cellulose chains in the cellulose matrix was observed for both low (1% strain) and high (3% strain) stretching. Therefore, the improved mechanical properties of the stretched ACCs were mainly attributed to the combined reorientation of the reinforcement and cellulose molecule chains in the cellulose matrix. Moreover, the degree of polymerization (DP) of the cellulose reinforcement affects the mechanical properties of ACCs (Tang et al., 2021).

PARTIAL SURFACE DISSOLUTION

Partial surface dissolution is another way to prepare ACCs. When cellulose materials, such as fibers or particles, are immersed in a specific solvent, the outer surface of the cellulose material is partially dissolved and transformed into the matrix, while the undissolved core part of the cellulose material containing its original structure acts as a reinforcing phase in the composite. Partial surface dissolution only requires a single cellulose material to prepare ACCs, whereas cellulose solution and cellulose reinforcement are the main components required to develop composites via impregnation.

The original process of preparing ACCs via partial surface dissolution was introduced by Gindl and Keckes (2005). Typically, 2, 3, and 4g of MCC as raw materials were initially activated in ethanol, acetone, and DMAc for 4h each and partially dissolved in LiCl/DMAc for 5min to afford cellulose solutions. After leaving these solutions at room temperature for 12h, transparent gels were formed, and the gels were dried to obtain ACCs. The crystalline structure of MCC is the same as that of cellulose I and was partially converted into the structure of cellulose II after dissolution. The ratio of the cellulose I to cellulose II structure in the composites depended on the content of original cellulose. Composites prepared from higher MCC content presented a higher ratio, indicating a higher proportion of cellulose I over cellulose II. This indicated that low amounts of MCC were dissolved and regenerated (less matrix), whereas the proportion of undissolved MCC acting as a reinforcing material was high. ACCs with a higher ratio of cellulose I to cellulose II exhibited higher Young's modulus and tensile strength. Later, many researchers used this concept to develop ACCs from various cellulose sources, from wood to agricultural waste residues (Nishino and Arimoto, 2007; Soykeabkaew et al., 2009a; Tanpichai and Witayakran, 2015, 2016; Yousefi et al., 2011a).

The time of cellulose dissolution in a solvent plays a vital role in controlling the mechanical properties of ACCs as larger fractions of fibers are transformed into the matrix with increasing immersion time (Arévalo et al., 2010; Nishino and Arimoto, 2007; Tanpichai and Witayakran, 2018a), which forms composites with different volume fractions of fibers. Nishino and Arimoto (2007) took the number of layers in natural fibers into consideration and applied partial dissolution processing to convert the outer less-ordered layer of natural fibers into the matrix, which was reinforced by the undissolved aligned layer. Effects of dissolution time on the crystallinity and mechanical properties of the ACCs were investigated. With increasing immersion time in LiCl/DMAc, higher dissolution was observed. Figure 10.2 presents the stress–strain curves, fractured surface after tensile deformation, and optical photographs of the filter paper (immersion time of 0h) and ACCs made using partial surface dissolution for 6 and 12h. When the filter paper was immersed in solvent for 6 and 12h, the crystallinity of the ACCs decreased from 86% (undissolved paper) to 19% and 14%, respectively; the proportions of the cellulose matrix formed were 77vol% and 84vol%, respectively, and the diameter of the fibers decreased, as observed from the fractured surface of the composites (Figure 10.2). Although the fibers within the filter paper were several micrometers in diameter, dissolution could generate a good interface between the matrix and undissolved fibers, and filling cavities within the cell wall could provide a material with higher transparency when the immersion time was longer.

The tensile strength and Young's modulus of the ACCs were higher than those of the filter paper when the immersion time was >6h, whereas the ACCs prepared with lower immersion times (2 and 4h) exhibited lower or similar strength and Young's modulus compared to those of the filter paper. This was attributed to a decrease in crystallinity and the availability of less quantity of the matrix to bind the undissolved part of the fibers. The tensile strength and Young's modulus of the ACCs prepared with 12h immersion time were 211MPa and 8.2GPa, respectively, which are considerably higher than those of conventional glass-fiber-reinforced composites. This outstanding mechanical performance was due to the better interface between the reinforcement and matrix. A

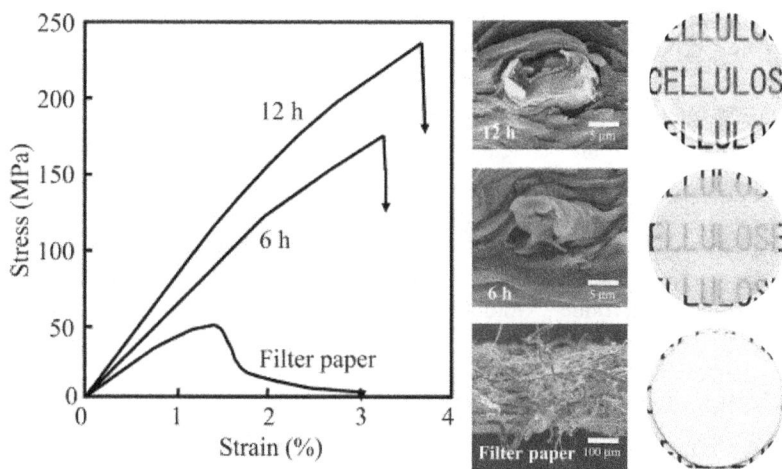

FIGURE 10.2 Stress–strain curves, SEM images of the fractured surface after tensile deformation, and optical photographs of the filter paper and ACCs prepared via partial dissolution for 6 and 12 h. Reproduced with permission from Nishino and Arimoto (2007). Copyright (2007) American Chemical Society.

similar increase in mechanical properties of ACCs prepared from pineapple leaf and cotton mats was observed with increasing dissolution time (Arévalo et al., 2010; Tanpichai and Witayakran, 2018a). Similarly, the higher conversion (from 91% to 58%) of cellulose I to cellulose II in wood pulp fibers was observed for the first 30 s of dissolution with NaOH and urea (Piltonen et al., 2016). After that, the conversion rate reduced. A considerable increase in the tensile strength and Young's modulus of ACCs was found when the dissolution time was increased to 30 s. The tensile strength and Young's modulus of the ACCs prepared with 2 s dissolution time were 9.4 MPa and 1.6 GPa, whereas those with 30 s of dissolution time increased to 48.8 MPa and 5.7 GPa, respectively.

Although longer dissolution time can develop all-cellulose MCC-derived composites with better mechanical properties (Abbott and Bismarck, 2010), inferior thermal stability of the composites was observed. The onset and maximum degradation temperatures shifted toward lower temperature with increasing dissolution treatment time. Moreover, the higher char content (23.5%) and faster degradation rate (1.43% min^{-1}) were observed for all-cellulose MCC-derived composites with a dissolution of 12 h than those of undissolved MCC, which has a char yield of 6.7% and degradation rate of 2.28% min^{-1}. This was attributed to the lower degree of crystallinity of the composites. The moisture content is also affected by dissolution time. Water molecules can diffuse faster into the amorphous phase of cellulose with longer dissolution time, affording higher moisture content in composites prepared with longer dissolution time.

Furthermore, natural fibers with different widths require different dissolution times to obtain ACCs with optimum mechanical performance (Soykeabkaew et al., 2009b; Yousefi et al., 2011a). Yousefi et al. (2011a) compared ACCs made from cellulose microfibers with widths of 26 μm and nanofibers with widths of 32 nm as a function of dissolution time up to 120 min. As the immersion time increased, gradual loss in crystallinity was observed for the composites prepared from both micro- and nanofibers. The tensile strength of the nanofiber-based composites with 10 min dissolution time was 164 MPa, whereas with 120 min dissolution time, the microfiber-based composites had strength of 59 MPa. CNFs required an immersion time of 10–20 min to develop ACCs with optimum mechanical performance owing to the high surface area of the nanofibers that rapidly interacted with the solvent, whereas the microfibers required longer dissolution time to form a matrix (Soykeabkaew et al., 2009b; Yousefi et al., 2011a). Similarly, 10 min dissolution time was suitable to prepare ACCs from networks of bacterial cellulose nanofibers and nanofibrillated cellulose (NFC) with excellent mechanical properties (Soykeabkaew et al., 2009b; Yousefi et al., 2010). However, a dissolution time of >40 min had no impact on the mechanical properties of ACCs made from micro-fibrillated cellulose (MFC) using an ionic liquid (Duchemin et al., 2009a). This could be due to the compactness of the sheet with highly defibrillated MFC, which decelerated the penetration of the solvent. The tensile strength and Young's modulus were optimized at a dissolution time of 80 min. After this time, a decrease in mechanical properties was observed owing to cellulose degradation. Although CNFs as a starting material could yield ACCs with optimum mechanical properties in a shorter time than those made from microfibers, the processing methods required an additional step to prepare CNFs (Yousefi et al., 2011b, 2015). The structure and mechanical properties of the composites prepared from microfibers and nanofibers were comparable. Therefore, the preparation of ACCs from cellulose microfibers would be beneficial owing to their simpler, faster, and cheaper preparation compared to that of CNFs.

The fiber orientation direction has a huge effect on the mechanical properties of ACCs. Soykeabkaew et al. (2008) investigated the effects of dissolution time on the physical and mechanical properties of ACCs prepared from aligned ramie fibers. The composites prepared with immersion time of 2 h in LiCl/DMAc presented the highest mechanical properties in the longitudinal direction owing to a suitable amount of regenerated matrix to adhere the undissolved fibers, improving stress transfer in the composites and reducing cavities within the composites. Crack and voids in the composites were observed with 1 h immersion time, and inferior tensile strength and Young's modulus were observed for composites with an immersion time >2 h due to fewer remaining reinforcement fibers. The tensile strength and Young's modulus of composites prepared with 2 h immersion time in the longitudinal direction were 460 MPa and 28 GPa, respectively, which are considerably higher than those of composites prepared from randomly oriented filter paper with 12 h immersion time (strength of 211 MPa and Young's modulus of 8.2 GPa) (Nishino and, 2007). Moreover, the transverse mechanical properties of the composites could be enhanced by increasing immersion time. The composites prepared with 3–6 h immersion times showed a similar transverse strength of 29–33 MPa, while the transverse strength increased to 40 MPa when the immersion time was 12 h. Interestingly, the longitudinal tensile strength of the composites prepared by 2 h immersion (460 MPa) was similar to that of ACCs prepared by impregnating aligned ramie fibers into the cellulose matrix (480 MPa). This suggested that the mechanical properties of composites prepared by dissolution predominantly originate from the highly oriented core fibers, which exhibit a highly efficient reinforcing effect in the composites.

In contrast, when crystalline cellulose fibers highly oriented on the skin and less oriented in the core were used to develop ACCs via dissolution, a rapid decrease in their mechanical properties was reported (Soykeabkaew et al., 2009a). ACCs prepared from lyocell fibers (regenerated cellulose) with a highly oriented skin and less-oriented core structure showed inferior mechanical properties with longer immersion times (Soykeabkaew et al., 2009a). This suggested that fibers with the skin–core structure are not a suitable cellulose material for developing ACCs (Soykeabkaew et al., 2008, 2009a). During composite preparation via partial dissolution, the oriented cellulose in the skin was dissolved to form a matrix, while the less-oriented core served as a reinforcing agent, leading to a tremendous decrease in the mechanical properties of the composites with increasing immersion time. This is distinctly different from composites prepared with ramie fibers, where the less-oriented skin was dissolved to form a matrix.

Moreover, the mechanical properties, thermal stability, and crystallinity of ACCs prepared from various cellulose sources via partial dissolution could be altered due to the difference in the DP of cellulose (De Silva and Byrne, 2017). De Silva and Byrne (2017) investigated the relationship between the DP and mechanical properties of cellulose fibers made from waste cotton lint dissolved in an ionic liquid. A linear relation between the tensile strength and DP was observed until the DP reached 1,150, and when the DP was >1,150, no considerable change in tensile strength was observed. This increase resulted from increased chain entanglement induced by the increased molecular weight of cellulose. However, the Young's modulus of the fibers increased steadily with increasing DP. Additionally, the cellulose concentration in the solvent, purification of raw materials, postprocessing such as drawing, processing conditions, solvents used, and precipitation rate affect the final properties of ACCs (Duchemin et al., 2009b, 2016; Gindl and Keckes, 2007; Tervahartiala et al., 2018).

APPLICATIONS

Various processing approaches such as calendaring, infusion, electrospinning, and three-dimensional (3D) printing have been used to prepare ACCs for various applications from general to advanced usages (Dormanns et al., 2016; Sirviö et al., 2017; Tenhunen et al., 2018; Yousefi et al., 2015). Potential applications of ACCs are discussed below.

COMPOSITES

ACCs exhibit extremely high mechanical performance in comparison with petroleum-based polymers, such as high-density polyethylene (tensile strength of 15–38 MPa and Young's modulus of 0.4–1.5 GPa), polypropylene (tensile strength of 26–42 MPa and Young's modulus of 0.95–1.77 GPa), and poly(vinyl chloride) (tensile strength of 10–25 MPa and Young's modulus of 1.12 GPa), bio-based polymers, including poly(lactic acid) (PLA) (tensile strength of 32.8 MPa and Young's modulus of 2.4 GPa) and banana starch (tensile strength of 5.1 MPa and Young's modulus of 0.05 GPa), and polypropylene reinforced with natural fibers, such as fax and sisal (Arévalo et al., 2010; Chawalitsakunchai et al., 2021; Khouaja et al., 2021; Reddy et al., 2014; Senthil Muthu Kumar et al., 2018; Tanpichai et al., 2022a; Tanpichai and Wootthikanokkhan, 2018b; Torres et al., 2011). Additionally, ACCs prepared from aligned ramie fibers exhibit better mechanical properties than conventional glass-fiber-reinforced composites (Nishino et al., 2004; Soykeabkaew et al., 2008), and ACCs made from nonwoven mats of flax stable fibers via partial surface dissolution present mechanical performances comparable to those of epoxy composites with flax mats (Gindl-Altmutter et al., 2012). Moreover, ACC laminates with five layers of Cordenka textile exhibited an impact strength of 41.5 ± 4.4 kJ/m^{-2}, greater than that of jute-fiber-reinforced polyester composites (27–37.9 kJ/m^{-2}) (Huber et al., 2012a). Notably, the mechanical properties of ACCs were reported to depend on temperature (Nishino et al., 2004). The mechanical properties of ACCs decreased with increasing temperature. Although the storage modulus of the ACCs at 300°C dropped considerably to 20 GPa, it is considerably higher than that of petroleum-based polymer composites (Nishino et al., 2004). Furthermore, ACCs exhibited superior thermal dimensional stability. The coefficient of thermal expansion (CTE) of the cellulose matrix was 18.5 ppm/K^{-1}, and with the introduction of 1 wt% CNFs, the CTE of ACCs plummeted to 6.8 ppm/K^{-1} (Yang et al., 2015), which is dramatically lower than that of petroleum-based plastics, such as polycarbonate and poly(ethylene terephthalate) (>150 ppm/K^{-1}). This manifests the improved thermal dimensional stability of ACCs (Biswas et al., 2019; Nishino et al., 2004; Tanpichai et al., 2020; Yang et al., 2015). Therefore, owing to their advantages of biodegradability, sustainability, and superior mechanical properties, ACCs can substitute petroleum-based composites in many composite applications over a wide temperature range.

PACKAGING

ACC films possess high oxygen-barrier properties (Yang et al., 2015). The oxygen permeability of ACCs was found to be lower than that of high-density polyethylene, poly(ethylene terephthalate), and poly(vinyl alcohol) films, whereas the water vapor permeability of ACCs (5.2×10^{-11} g m^{-1}/ s^{-1}/Pa^{-1}) was slightly higher than that of low-density polyethylene (0.036×10^{-11} g m^{-1}/s^{-1}/Pa^{-1}) and high-density polyethylene (0.02×10^{-11} g m^{-1}/s^{-1}/Pa^{-1}) but lower than that of protein and carbohydrate films, such as corn starch, methyl cellulose, agar, and casein (Ghaderi et al., 2014). Although ACCs cannot compete with high-density and low-density polyethylene in terms of water vapor permeability, ACCs can replace protein and carbohydrate films. Their promising thermal dimensional stability and high oxygen permeability suggest that ACCs can be used not only for composite applications but also for efficient packaging materials (Yang et al., 2015).

Moreover, porous-structured ACCs, known as aerocellulose, have been prepared for protective packaging applications (Duchemin et al., 2010). After dissolving MCC in a mixed LiCl/DMAc solution, the cellulose gel was precipitated and freeze-dried. The densities of aerocelluloses depend on cellulose concentration, which are in the range of 120–350 kg/m^{-3}. Aerocelluloses exhibit specific flexural strength and Young's modulus similar to those of bio- and synthetic polymeric foams, such as starch and polypropylene foams.

FIBERS AND FILAMENTS

Fibers or filaments with high strength and toughness along with biodegradability and biocompatibility can be efficiently prepared via wet-spinning or electrospinning processes (Qiu et al., 2018). Qiu et al. (2018) prepared superstrong ACC filaments by incorporating NFC (Figure 10.3a–b). Cotton pulp was dissolved in a LiOH and urea solvent to obtain a transparent cellulose solution, and the suspension of NFC was mixed with the cellulose solution. This dope passed through a spinneret cylinder and was immersed into an acid coagulation bath. Filaments with various NFC contents were formed, as shown in Figure 10.3c. The tensile strength and elongation-at-break of the ACC filaments with 3 wt% NFC (CF-3) were 3.92 cN/dtex and 14.6%, respectively, which were considerably higher than those of filaments without NFC (CF-0). This was attributed to strong interfacial interaction between the cellulose matrix and NFC and high alignment of NFC in the matrix caused by the drawing process during spinning. However, with the introduction of >3 wt% NFC, a decrease in tensile strength and Young's modulus of the composite filaments was observed (Figure 10.3b) due to NFC aggregation. SEM images of the fractured surface of the ACC filaments revealed that NFC was pulled out from the matrix, as shown in Figure 10.3d-e. This fiber pull-out failure mechanism increased the elongation-at-break of CF-3 compared to that of CF-0. Additionally, yarns and final garments were prepared from the dissolved cellulose solution using an ionic liquid (Ma et al., 2016).

BIOENGINEERING AND BIOMEDICAL MATERIALS

Hydrogels, a 3D material containing a large volume of liquid, have been extensively developed for bioengineering and biomedical applications (Tanpichai et al., 2022c; Wang and Chen, 2011). Wang and Chen (2011) developed ACC hydrogels reinforced with CNCs via a phase separation and regeneration process. Gels with and without CNCs before and after the regeneration process are shown in Figure 10. 3f–g. The compressive stress and Young's modulus of the hydrogels increased linearly with increasing CNC content owing to the formation of additional H-bonding in the gels because CNCs acted as a cross-linker. The equilibrium swelling ratio decreased from 9.88 for the gels without CNCs to 6.50 for the gels with 60 wt% CNCs. The increased cross-linking density induced by CNCs restricted the movement of cellulose molecules in the matrix. The drug-release properties of the cellulose composite hydrogels were further investigated. Gels without CNCs released ~36% of the loaded drug after immersion in simulated body fluid at 37°C for 60 min, and their diffusion coefficient was $0.42 \times 10^{-4} mm^2/s^{-1}$. This rapid drug release could be delayed using CNCs. Hydrogels with 20 wt% CNCs showed gradual drug release with a diffusion coefficient of $0.21 \times 10^{-4}/mm^2/s^{-1}$. Rapid drug release of the gels without CNCs was attributed to their large cavities, whereas a homogeneous porous structure was observed in the hydrogels with 20 wt% CNCs.

Scaffolds for tissue engineering were developed from electrospun cellulose and CNC nanofibers (He et al., 2014). Cotton was dissolved in a LiCl/DMAc solvent, and CNCs were mixed with the cellulose solution. Scaffolds of aligned electrospun cellulose and CNC nanofibers were fabricated via electrospinning. The tensile strength and Young's modulus of the scaffolds increased with increasing CNC content. 3-(4,5-Dimethylthiazol-2-yl)-2,5-diphenyl tetrazolium bromide (MTT) assay results confirmed that the scaffolds of aligned electrospun cellulose and CNC nanofibers can

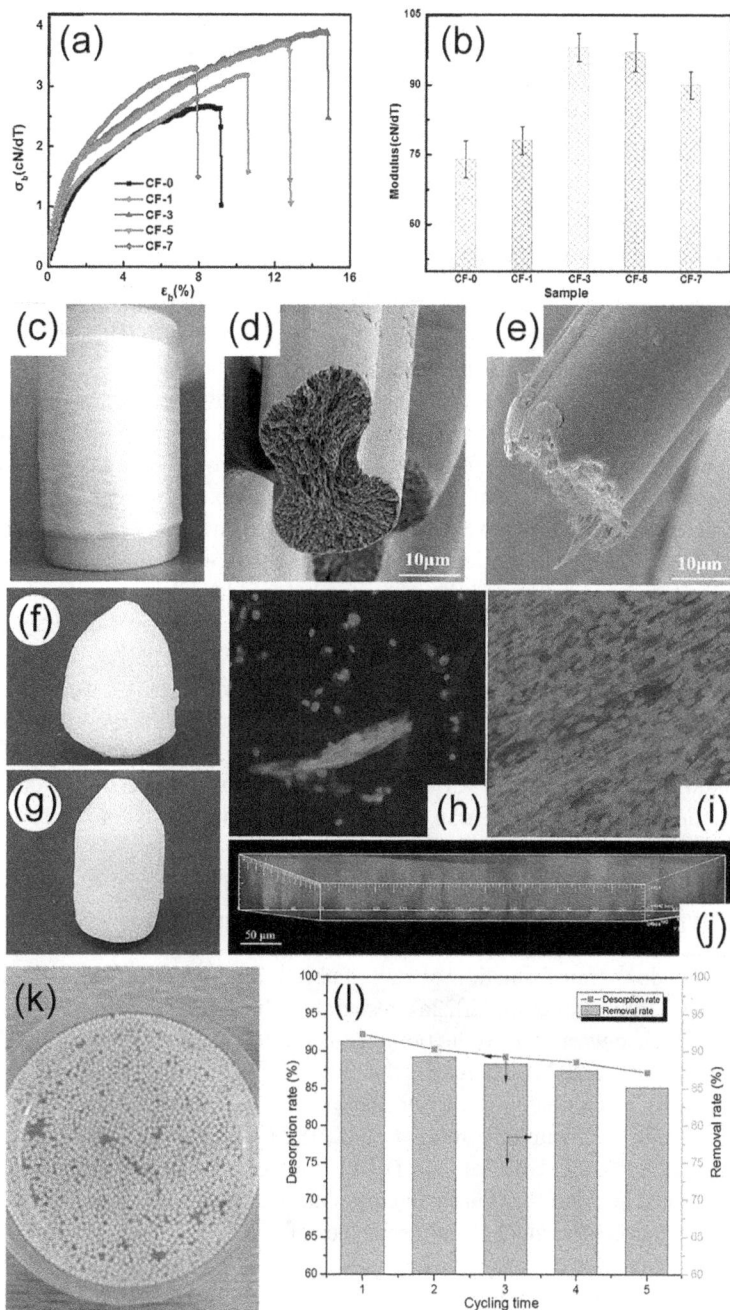

FIGURE 10.3 (a) Stress–strain curves and (b) tensile modulus of the ACC filaments. (c) Photograph of the ACC filaments and SEM images of the fractured surface of the ACC filaments (d) without (CF-0) and (e) with 3 wt% NFC (CF-3) after tensile deformation. Reproduced with permission from Qiu et al. (2018). Copyright (2018) American Chemical Society. Photographs of ACC gels with (f) 0 and (g) 50 wt% CNCs. Reproduced with modification from Wang and Chen (2011). Copyright (2011), with permission from Elsevier. Cell proliferation inside the scaffolds of aligned electrospun cellulose and CNC nanofibers captured using CLSM after (h) 3 and (i) 7 days of culture, and (j) 3D view of the scaffold with cells. Reproduced with permission from He et al. (2014). Copyright (2014) American Chemical Society. (k) ACC spheres for dye adsorption and (l) their adsorption capacity for methylene blue as a function of reuse cycles. Adapted with permission from Li et al. (2018). Copyright (2018) American Chemical Society.

support cell proliferation because no toxic signs were observed on the fibroblast cells. Figure 10. 3h–i presents the growth of primary human dental follicle cells on the scaffold on day 3 and highly aligned cells grown along the aligned fibers on day 7. Confocal laser scanning microscopy (CLSM) showed cells grown on the surface of and inside the scaffold. This scaffold of aligned electrospun cellulose and CNC nanofibers could be applied for artificial tissues and organs, such as blood vessels, nerves, and tendons.

DYE AND HEAVY METAL ION ADSORBENTS

Wastewater released from industrial and technological activities containing harmful chemicals, such as dyes and heavy metal ions, causes water pollution, which threatens the environment and living organisms, including human beings (Li et al., 2018, 2019; Wang et al., 2017). Cellulose composites as adsorbents for removing dyes and heavy metal ions have been widely prepared (Li et al., 2014, 2018, 2019). Li et al. (2018) developed porous cellulose spheres by dissolving cellulose fibers in a mixed solvent of NaOH and urea, followed by the addition of MFC and nano-sized $CaCO_3$ as a pore-forming agent. The porous cellulose spheres were precipitated under acidic conditions and modified using maleic acid, as shown in Figure 10.3k. The maximum adsorption capacity for methylene blue was 303 mg/g^{-1}, which is considerably higher than that of other bioadsorbents. Figure 10.3l presents the reusability of the cellulose spheres for methylene blue adsorption as a function of reuse cycles. Although the desorption and removal rates of the cellulose spheres decreased slightly with increased recycling, those at the fifth cycle were as high as 87.1% and 85.1%, respectively. Later, Li et al. (2019) modified the porous cellulose beads by dissolving MCC in an ionic liquid solution with glutaric anhydride and grafted with aminoguanidine hydrochloride. The modified porous cellulose beads had very high adsorption capacities for Hg(II) and Cu(II) of 625 and 98.5 mg/g, respectively. This was attributed to the interaction of various functional groups on the cellulose chains with metal ions and suggested that cellulose beads can be used as green adsorbents for removing heavy metal ions and dyes from contaminated water reservoirs.

RECYCLING, SUSTAINABILITY, AND BIODEGRADABILITY

Although the tremendous transformation of raw materials into products has attracted increasing attention worldwide (Tanpichai et al., 2022a), researchers have majorly focused on circular economy to manage and reuse waste for production due to limited resources (Costa and Broega, 2022; Khan et al., 2023; Wei et al., 2023). Therefore, the recyclability and reprocessability of ACCs are in accordance with this environmental issue (Spörl et al., 2018). Properties of ACCs as a function of recycling number were investigated by Spörl et al. (2018). Recycling had little influence on the DP and molecular weight of ACCs. A decrease in DP and molecular weight was observed owing to the shorter length of cellulose chains caused by the dissolution and regeneration of cellulose. However, this reduction did not affect the mechanical properties of the ACCs. Recycling the ACC up to two times did not alter its mechanical properties (tensile strength, Young's modulus, flexural modulus, and impact strength). However, after the third recycling, a slight increase in the Young's and flexural moduli was observed owing to a decrease in tensile strain.

Although wood is a sustainable and renewable material, trees take time to grow (Tang et al., 2021). Therefore, the development of ACCs from agricultural waste is a promising alternative to efficiently use waste that would otherwise go directly to a landfill or be used as a fertilizer or animal food (Tanpichai et al., 2019a, 2019b, 2022b; Wei et al., 2016). Using agricultural waste (corn stalk) from corn harvesting as a raw material yielded ACCs with mechanical properties similar to those of composites made from eucalyptus (Tang et al., 2021). However, ACCs from corn stalk showed slightly lower mechanical properties (tensile strength of 9.6 MPa and Young's modulus of 0.2 GPa) than those from pine wood (tensile strength of 16.9 MPa and Young's modulus of 0.32 GPa), which was attributed to the higher DP and longer fibers of pine wood pulp compared with those from corn stalk pulp. The DPs of wood pulp and agricultural residue pulp were 382 and 270, and the fiber lengths of wood pulp and rice straw pulp were 1.145 and 0.499 mm, respectively. A similar comparison of raw materials between wood pulp and alfa grass to prepare ACCs was made, and mechanical results showed that ACCs with cellulose matrixes were made from wood pulp and alfa glass pulp had similar mechanical properties as those with the same content of reinforcing cellulose fibers (Labidi et al., 2019). Therefore, agricultural waste has great potential as an alternative cellulose source to prepare ACCs.

Cellulose waste can be an alternative source to prepare ACCs (De Silva and Byrne, 2017; Hai et al., 2017; Kale and Gorade, 2019). Deinked copy/printing paper (DIP) as office waste was dissolved to prepare the cellulose matrix, and CNFs made from these DIP fibers (DIP-CNFs) or wood pulp (W-CNFs) were incorporated in the matrix to develop green ACCs (Hai et al., 2017). The introduction of W-CNFs improved the mechanical properties of the composites compared to those with DIP-CNFs. The tensile strength and Young's modulus of the composites with 3 wt% W-CNFs were 41.5 MPa and 2.47 GPa, respectively, while the composites with DIP-CNFs at the same concentration showed a slight lower tensile strength of 32.8 MPa and Young's modulus of 2.15 GPa. Therefore, W-CNFs showed better reinforcing abilities than DIP-CNFs owing to the higher crystallinity and aspect ratio of W-CNFs. Kale and Gorade (2019) used waste medical-grade cotton collected from a local hospital as a reinforcing phase in ACCs. Incorporating 10 wt% of this cotton waste yielded composites with a tensile strength of 245 MPa and Young's modulus of 5.1 GPa compared to a strength of ~25 MPa and Young's modulus of >0.25 GPa for the regenerated cellulose matrix. Similar results were reported using cotton lint waste as a feedstock to prepare cellulose fibers (De Silva and Byrne, 2017). The mechanical properties of the cellulose fibers from cotton lint waste were comparable to those of fibers prepared from wood pulp. Therefore, recycling cellulose waste is a potential way to prepare composites with high mechanical performance and reduce waste (De Silva and Byrne, 2017).

Additionally, the decomposition of cellulose is performed using either enzymes or water molecules (Kalka et al., 2014). Figure 10.4 presents the biodegradation of ACCs after 70 days of soil burial testing at soil temperatures of 20.83°C (T1) and 33.45°C (T2), respectively, as well as the change in mass of these composites after incubation for 70 days. The matrix phase of the ACCs showed more biodegradation than the reinforcement fibers owing to its less-ordered structure. More

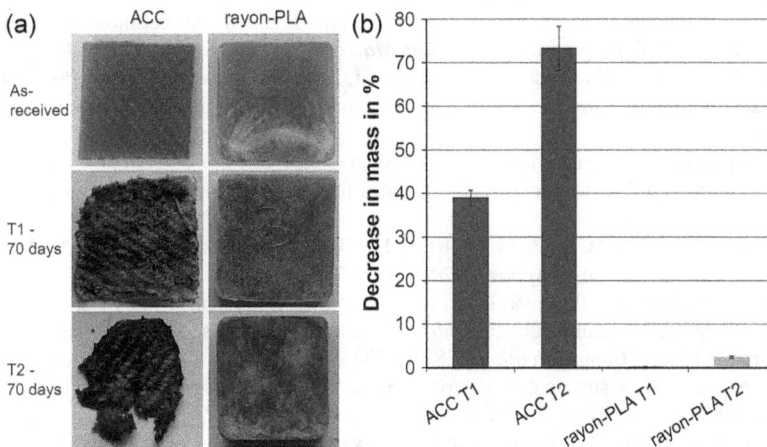

FIGURE 10.4 (a) Photographs of the ACCs and PLA composites with 20 wt% rayon fibers (rayon–PLA) before and after soil burial biodegradation testing at soil temperatures of 20.83°C (T1) and 33.45°C (T2) for 70 days. (b) Mass change (%) of the ACCs and rayon–PLA composites after 70 days of biodegradation testing as a function of soil temperature. Reprinted with modification from Kalka et al. (2014), Copyright (2014), with permission from Elsevier.

severe signs of degradation, including mass loss and change in color, were observed for the composites incubated at T2 than those at T1 as microorganisms are more active at higher temperatures (Pérez et al., 2002). At T1, ~39% of the composite mass degraded after soil burial, whereas the mass of the composites decreased by ~73% after soil burial at T2. However, no sign of degradation was observed for PLA composites with 20 wt% rayon fibers (rayon–PLA) at both T1 and T2. Although PLA is a biodegradable material, its efficient degradation requires high humidity and a temperature of 40°C–60°C. Additionally, fragmentation of the ACC films into small pieces with fungal mycelia on the film surface was observed after 10 days of soil burial owing to the attack of microorganisms (Yang et al., 2010).

CONCLUSIONS AND PROSPECTS

The two main processing approaches (impregnation and partial surface dissolution) to fabricate ACCs as well as the effects of vital factors on their mechanical and physical properties, including transparency and crystallinity, have been discussed. ACCs are completely composted by disposal, with no industrial compositing required. Their degradation rate is faster than that of biodegradable materials, and after the end of their life, these composites can be recycled to fabricate composites with high mechanical performance. The promising properties of ACCs, such as high mechanical properties, transparency, and thermal expansion, as well as the upscaled production of ACCs should generate more opportunities for the use of ACC materials in several advanced applications, such as composites, packaging, textiles, hydrogels, membranes, electronic devices, and biomaterials.

REFERENCES

Abbott, A. & Bismarck, A. (2010). Self-reinforced cellulose nanocomposites. *Cellulose*, 17(4), 779–791.

Abe, K., Nakatsubo, F. & Yano, H. (2009). High-strength nanocomposite based on fibrillated chemi-thermomechanical pulp. *Composites Science and Technology*, 69(14), 2434–2437.

Acharya, S., Hu, Y. & Abidi, N. (2021). Cellulose dissolution in ionic liquid under mild conditions: Effect of hydrolysis and temperature. *Fibers*, 9(1), 1–14.

Akter, T., Nayeem, J., Quadery, A. H., Abdur Razzaq, M., Tushar Uddin, M., Shahriar Bashar, M. & Sarwar Jahan, M. (2020). Microcrystalline cellulose reinforced chitosan coating on kraft paper. *Cellulose Chemistry and Technology*, 54(1–2), 95–102.

Arévalo, R., Picot, O., Wilson, R. M., Soykeabkaew, N. & Peijs, T. (2010). All-cellulose composites by partial dissolution of cotton fibres. *Journal of Biobased Materials and Bioenergy*, 4(2), 129–138.

Azizi Samir, M. A. S., Alloin, F., Sanchez, J.-Y., El Kissi, N. & Dufresne, A. (2004). Preparation of cellulose whiskers reinforced nanocomposites from an organic medium suspension. *Macromolecules*, 37(4), 1386–1393.

Biswas, S. K., Tanpichai, S., Witayakran, S., Yang, X., Shams, M. I. & Yano, H. (2019). Thermally superstable cellulosic-nanorod-reinforced transparent substrates featuring microscale surface patterns. *ACS Nano*, 13(2), 2015–2023.

Bulota, M., Tanpichai, S., Hughes, M. & Eichhorn, S. J. (2012). Micromechanics of TEMPO-oxidized fibrillated cellulose composites. *ACS Applied Materials & Interfaces*, 4(1), 331–337.

Chawalitsakunchai, W., Dittanet, P., Loykulnant, S., Sae-oui, P., Tanpichai, S., Seubsai, A. & Prapainainar, P. (2021). Properties of natural rubber reinforced with nano cellulose from pineapple leaf agricultural waste. *Materials Today Communications*, 28, 102594.

Costa, J. & Broega, A. C., New sustainable materials for the fashion industry: The button in the circular economy. *Springer Series in Design and Innovation*, 2022, 342–356.

De Silva, R. & Byrne, N. (2017). Utilization of cotton waste for regenerated cellulose fibres: Influence of degree of polymerization on mechanical properties. *Carbohydrate Polymers*, 174, 89–94.

Dormanns, J. W., Weiler, F., Schuermann, J., Müssig, J., Duchemin, B. J. C. & Staiger, M. P. (2016). Positive size and scale effects of all-cellulose composite laminates. *Composites Part A: Applied Science and Manufacturing*, 85, 65–75.

Duchemin, B., Le Corre, D., Leray, N., Dufresne, A. & Staiger, M. P. (2016). All-cellulose composites based on microfibrillated cellulose and filter paper via a NaOH-urea solvent system. *Cellulose*, 23(1), 593–609.

Duchemin, B. J. C., Mathew, A. P. & Oksman, K. (2009a). All-cellulose composites by partial dissolution in the ionic liquid 1-butyl-3-methylimidazolium chloride. *Composites Part A: Applied Science and Manufacturing*, 40(12), 2031–2037.

Duchemin, B. J. C., Newman, R. H. & Staiger, M. P. (2009b). Structure–property relationship of all-cellulose composites. *Composites Science and Technology*, 69(7), 1225–1230.

Duchemin, B. J. C., Staiger, M. P., Tucker, N. & Newman, R. H. (2010). Aerocellulose based on all-cellulose composites. *Journal of Applied Polymer Science*, 115(1), 216–221.

Foroughi, F., Rezvani Ghomi, E., Morshedi Dehaghi, F., Borayek, R. & Ramakrishna, S. (2021). A review on the life cycle assessment of cellulose: From properties to the potential of making it a low carbon material. *Materials*, 14(4), 714.

Fujisawa, S., Togawa, E. & Hayashi, N. (2016). Orientation control of cellulose nanofibrils in all-cellulose composites and mechanical properties of the films. *Journal of Wood Science*, 62(2), 174–180.

Ghaderi, M., Mousavi, M., Yousefi, H. & Labbafi, M. (2014). All-cellulose nanocomposite film made from bagasse cellulose nanofibers for food packaging application. *Carbohydrate Polymers*, 104(1), 59–65.

Gindl, W. & Keckes, J. (2005). All-cellulose nanocomposite. *Polymer*, 46(23), 10221–10225.

Gindl, W. & Keckes, J. (2007). Drawing of self-reinforced cellulose films. *Journal of Applied Polymer Science*, 103(4), 2703–2708.

Gindl-Altmutter, W., Keckes, J., Plackner, J., Liebner, F., Englund, K. & Laborie, M. P. (2012). All-cellulose composites prepared from flax and lyocell fibres compared to epoxy-matrix composites. *Composites Science and Technology*, 72(11), 1304–1309.

Hai, L. V., Kim, H. C., Kafy, A., Zhai, L., Kim, J. W. & Kim, J. (2017). Green all-cellulose nanocomposites made with cellulose nanofibers reinforced in dissolved cellulose matrix without heat treatment. *Cellulose*, 24(8), 3301–3311.

He, X., Xiao, Q., Lu, C., Wang, Y., Zhang, X., Zhao, J., Zhang, W., Zhang, X. & Deng, Y. (2014). Uniaxially aligned electrospun all-cellulose nanocomposite nanofibers reinforced with cellulose nanocrystals: Scaffold for tissue engineering. *Biomacromolecules*, 15(2), 618–627.

Hildebrandt, N. C., Piltonen, P., Valkama, J. & Illikainen, M. (2017). Self-reinforcing composites from commercial chemical pulps via partial dissolution with NaOH/urea. *Industrial Crops and Products*, 109, 79–84.

Huber, T., Bickerton, S., Müssig, J., Pang, S. & Staiger, M. P. (2012a). Solvent infusion processing of all-cellulose composite materials. *Carbohydrate Polymers*, 90(1), 730–733.

Huber, T., Müssig, J., Curnow, O., Pang, S., Bickerton, S. & Staiger, M. P. (2012b). A critical review of all-cellulose composites. *Journal of Materials Science*, 47(3), 1171–1186.

Huber, T., Pang, S. & Staiger, M. P. (2012c). All-cellulose composite laminates. *Composites Part A: Applied Science and Manufacturing*, 43(10), 1738–1745.

Hull, D. & Clyne, T. W. (1996). *An Introduction to Composite Materials*. Cambridge: Cambridge University Press.

Kale, R. D. & Gorade, V. G. (2019). Potential application of medical cotton waste for self-reinforced composite. *International Journal of Biological Macromolecules*, 124, 25–33.

Kalka, S., Huber, T., Steinberg, J., Baronian, K., Müssig, J. & Staiger, M. P. (2014). Biodegradability of all-cellulose composite laminates. *Composites Part A: Applied Science and Manufacturing*, 59, 37–44.

Khan, M. A. A., Cárdenas-Barrón, L. E., Treviño-Garza, G. & Céspedes-Mota, A. (2023). Optimal circular economy index policy in a production system with carbon emissions. *Expert Systems with Applications*, 212, 118684.

Khouaja, A., Koubaa, A. & Ben Daly, H. (2021). Dielectric properties and thermal stability of cellulose high-density polyethylene bio-based composites. *Industrial Crops and Products*, 171, 113928.

Labidi, K., Korhonen, O., Zrida, M., Hamzaoui, A. H. & Budtova, T. (2019). All-cellulose composites from alfa and wood fibers. *Industrial Crops and Products*, 127, 135–141.

Li, B., Pan, Y., Zhang, Q., Huang, Z., Liu, J. & Xiao, H. (2019). Porous cellulose beads reconstituted from ionic liquid for adsorption of heavy metal ions from aqueous solutions. *Cellulose*, 26(17), 9163–9178.

Li, Y., Xiao, H., Chen, M., Song, Z. & Zhao, Y. (2014). Absorbents based on maleic anhydride-modified cellulose fibers/diatomite for dye removal. *Journal of Materials Science*, 49(19), 6696–6704.

Li, Y., Xiao, H., Pan, Y. & Wang, L. (2018). Novel composite adsorbent consisting of dissolved cellulose fiber/microfibrillated cellulose for dye removal from aqueous solution. *ACS Sustainable Chemistry & Engineering*, 6(5), 6994–7002.

Ma, Y., Hummel, M., Määttänen, M., Särkilahti, A., Harlin, A. & Sixta, H. (2016). Upcycling of waste paper and cardboard to textiles. *Green Chemistry*, 18(3), 858–866.

Nishino, T. & Arimoto, N. (2007). All-cellulose composite prepared by selective dissolving of fiber surface. *Biomacromolecules*, 8(9), 2712–2716.

Nishino, T., Matsuda, I. & Hirao, K. (2004). All-cellulose composite. *Macromolecules*, 37(20), 7683–7687.

Nishino, T., Takano, K. & Nakamae, K. (1995). Elastic modulus of the crystalline regions of cellulose polymorphs. *Journal of Polymer Science Part B: Polymer Physics*, 33(11), 1647–1651.

Pérez, J., Muñoz-Dorado, J., De La Rubia, T. & Martínez, J. (2002). Biodegradation and biological treatments of cellulose, hemicellulose and lignin: An overview. *International Microbiology*, 5(2), 53–63.

Piltonen, P., Hildebrandt, N. C., Westerlind, B., Valkama, J. P., Tervahartiala, T. & Illikainen, M. (2016). Green and efficient method for preparing all-cellulose composites with NaOH/urea solvent. *Composites Science and Technology*, 135, 153–158.

Pullawan, T., Wilkinson, A. N. & Eichhorn, S. J. (2010). Discrimination of matrix-fibre interactions in all-cellulose nanocomposites. *Composites Science and Technology*, 70(16), 2325–2330.

Pullawan, T., Wilkinson, A. N. & Eichhorn, S. J. (2012). Influence of magnetic field alignment of cellulose whiskers on the mechanics of all-cellulose nanocomposites. *Biomacromolecules*, 13(8), 2528–2536.

Pullawan, T., Wilkinson, A. N. & Eichhorn, S. J. (2013). Orientation and deformation of wet-stretched all-cellulose nanocomposites. *Journal of Materials Science*, 48(22), 7847–7855.

Pullawan, T., Wilkinson, A. N., Zhang, L. N. & Eichhorn, S. J. (2014). Deformation micromechanics of all-cellulose nanocomposites: Comparing matrix and reinforcing components. *Carbohydrate Polymers*, 400, 31–39.

Qi, H., Cai, J., Zhang, L. & Kuga, S. (2009). Properties of films composed of cellulose nanowhiskers and a cellulose matrix regenerated from alkali/urea solution. *Biomacromolecules*, 10(6), 1597–1602.

Qiu, C., Zhu, K., Yang, W., Wang, Y., Zhang, L., Chen, F. & Fu, Q. (2018). Super strong all-cellulose composite filaments by combination of inducing nanofiber formation and adding nanofibrillated cellulose. *Biomacromolecules*, 19(11), 4386–4395.

Reddy, K. O., Zhang, J., Zhang, J. & Rajulu, A. V. (2014). Preparation and properties of self-reinforced cellulose composite films from Agave microfibrils using an ionic liquid. *Carbohydrate Polymers*, 114, 537–545.

Rusli, R. & Eichhorn, S. J. (2008). Determination of the stiffness of cellulose nanowhiskers and the fiber-matrix interface in a nanocomposite using Raman spectroscopy. *Applied Physics Letters*, 93(3), 033111.

Rusli, R., Shanmuganathan, K., Rowan, S. J., Weder, C. & Eichhorn, S. J. (2010). Stress-transfer in anisotropic and environmentally adaptive cellulose whisker nanocomposites. *Biomacromolecules*, 11(3), 762–768.

Sakurada, I., Nukushina, Y. & Ito, T. (1962). Experimental determination of elastic modulus of crystalline regions in oriented polymers. *Journal of Polymer Science*, 57(165), 651–660.

Senthil Muthu Kumar, T., Rajini, N., Obi Reddy, K., Varada Rajulu, A., Siengchin, S. & Ayrilmis, N. (2018). All-cellulose composite films with cellulose matrix and Napier grass cellulose fibril fillers. *International Journal of Biological Macromolecules*, 112, 1310–1315.

Sirviö, J. A., Visanko, M. & Hildebrandt, N. C. (2017). Rapid preparation of all-cellulose composites by solvent welding based on the use of aqueous solvent. *European Polymer Journal*, 97, 292–298.

Soykeabkaew, N., Arimoto, N., Nishino, T. & Peijs, T. (2008). All-cellulose composites by surface selective dissolution of aligned ligno-cellulosic fibres. *Composites Science and Technology*, 68(10–11), 2201–2207.

Soykeabkaew, N., Nishino, T. & Peijs, T. (2009a). All-cellulose composites of regenerated cellulose fibres by surface selective dissolution. *Composites Part A: Applied Science and Manufacturing*, 40(4), 321–328.

Soykeabkaew, N., Sian, C., Gea, S., Nishino, T. & Peijs, T. (2009b). All-cellulose nanocomposites by surface selective dissolution of bacterial cellulose. *Cellulose*, 16(3), 435–444.

Spörl, J. M., Batti, F., Vocht, M. P., Raab, R., Müller, A., Hermanutz, F. & Buchmeiser, M. R. (2018). Ionic liquid approach toward manufacture and full recycling of all-cellulose composites. *Macromolecular Materials and Engineering*, 303(1), 1700335.

Sturcova, A., Davies, G. R. & Eichhorn, S. J. (2005). Elastic modulus and stress-transfer properties of tunicate cellulose whiskers. *Biomacromolecules*, 6(2), 1055–1061.

Swatloski, R. P., Spear, S. K., Holbrey, J. D. & Rogers, R. D. (2002). Dissolution of cellose with ionic liquids. *Journal of the American Chemical Society*, 124(18), 4974–4975.

Tang, X., Liu, G., Zhang, H., Gao, X., Li, M. & Zhang, S. (2021). Facile preparation of all-cellulose composites from softwood, hardwood, and agricultural straw cellulose by a simple route of partial dissolution. *Carbohydrate Polymers*, 256, 117591.

Tanpichai, S. (2018). A comparative study of nanofibrillated cellulose and microcrystalline cellulose as reinforcements in all-cellulose composites. *Journal of Metals, Materials and Minerals*, 28(1), 10–15.

Tanpichai, S. (2022). Recent development of plant-derived nanocellulose in polymer nanocomposite foams and multifunctional applications: A mini-review. *Express Polymer Letters*, 16(1), 52–74.

Tanpichai, S., Biswas, S. K., Witayakran, S. & Yano, H. (2019a). Water Hyacinth: A sustainable lignin-poor cellulose source for the production of cellulose nanofibers. *ACS Sustainable Chemistry & Engineering*, 7(23), 18884–18893.

Tanpichai, S., Biswas, S. K., Witayakran, S. & Yano, H. (2020). Optically transparent tough nanocomposites with a hierarchical structure of cellulose nanofiber networks prepared by the Pickering emulsion method. *Composites Part A: Applied Science and Manufacturing*, 132, 105811.

Tanpichai, S., Boonmahitthisud, A., Soykeabkaew, N. & Ongthip, L. (2022a). Review of the recent developments in all-cellulose nanocomposites: Properties and applications. *Carbohydrate Polymers*, 286, 119192.

Tanpichai, S., Mekcham, S., Kongwittaya, C., Kiwijaroun, W., Thongdonsun, K., Thongdeelerd, C. & Boonmahitthisud, A. (2022b). Extraction of nanofibrillated cellulose from water hyacinth using a high speed homogenizer. *Journal of Natural Fibers*, 19(13), 5676–5696.

Tanpichai, S., Phoothong, F. & Boonmahitthisud, A. (2022c). Superabsorbent cellulose-based hydrogels crossliked with borax. *Scientific Reports*, 12(1), 8920.

Tanpichai, S., Sampson, W. W. & Eichhorn, S. J. (2012). Stress-transfer in microfibrillated cellulose reinforced poly(lactic acid) composites using Raman spectroscopy. *Composites Part A: Applied Science and Manufacturing*, 43(7), 1145–1152.

Tanpichai, S., Sampson, W. W. & Eichhorn, J. S. (2014). Stress transfer in microfibrillated cellulose reinforced poly(vinyl alcohol) composites. *Composites Part A: Applied Science and Manufacturing*, 65, 186–191.

Tanpichai, S. & Wimolmala, E. (2022d). Facile single-step preparation of cellulose nanofibers by TEMPO-mediated oxidation and their nanocomposites. *Journal of Natural Fibers*, 19(15), 10094–10110.

Tanpichai, S. & Witayakran, S. (2015). Mechanical properties of all-cellulose composites made from pineapple leaf microfibers. *Key Engineering Materials*, 659, 453–457.

Tanpichai, S. & Witayakran, S. (2016). All-cellulose composites from pineapple leaf microfibers: Structural, thermal, and mechanical properties. *Polymer Composites*, 39(3), 895–903.

Tanpichai, S. & Witayakran, S. (2017). All-cellulose composite laminates prepared from pineapple leaf fibers treated with steam explosion and alkaline treatment. *Journal of Reinforced Plastics and Composites*, 36(16), 1146–1155.

Tanpichai, S. & Witayakran, S. (2018a). All-cellulose composites from pineapple leaf microfibers: Structural, thermal, and mechanical properties. *Polymer Composites*, 39(3), 895–903.

Tanpichai, S., Witayakran, S. & Boonmahitthisud, A. (2019b). Study on structural and thermal properties of cellulose microfibers isolated from pineapple leaves using steam explosion. *Journal of Environmental Chemical Engineering*, 7(1), 102836.

Tanpichai, S. & Wootthikanokkhan, J. (2018b). Reinforcing abilities of microfibers and nanofibrillated cellulose in poly(lactic acid) composites. *Science and Engineering of Composite Materials*, 25(2), 395–401.

Tenhunen, T., Moslemian, O., Kammiovirta, K., Harlin, A., Kääriäinen, P., Österberg, M., Tammelin, T. & Orelma, H. (2018). Surface tailoring and design-driven prototyping of fabrics with 3D-printing: An all-cellulose approach. *Materials & Design*, 140, 409–419.

Tervahartiala, T., Hildebrandt, N. C., Piltonen, P., Schabel, S. & Valkama, J.-P. (2018). Potential of all-cellulose composites in corrugated board applications: Comparison of chemical pulp raw materials. *Packaging Technology and Science*, 31(4), 173–183.

Torres, F. G., Troncoso, O. P., Torres, C., Díaz, D. A. & Amaya, E. (2011). Biodegradability and mechanical properties of starch films from Andean crops. *International Journal of Biological Macromolecules*, 48 (4), 603–606.

Wang, F., Pan, Y., Cai, P., Guo, T. & Xiao, H. (2017). Single and binary adsorption of heavy metal ions from aqueous solutions using sugarcane cellulose-based adsorbent. *Bioresource Technology*, 241, 482–490.

Wang, Y. & Chen, L. (2011). Impacts of nanowhisker on formation kinetics and properties of all-cellulose composite gels. *Carbohydrate Polymers*, 83(4), 1937–1946.

Wei, D. R., Irshad, M., Noman, S. M., Murthy, A., Hu, B., Khayrillo, N. & Olawale, O. A., Sustainable designing of reusable waste sources from the transport sector, smart innovation. *Systems and Technologies*, 2023, 329–338.

Wei, X., Wei, W., Cui, Y. H., Lu, T. J., Jiang, M., Zhou, Z. W. & Wang, Y. (2016). All-cellulose composites with ultra-high mechanical properties prepared through using straw cellulose fiber. *RSC Advances*, 6(96), 93428–93435.

Yang, Q., Lue, A. & Zhang, L. (2010). Reinforcement of ramie fibers on regenerated cellulose films. *Composites Science and Technology*, 70(16), 2319–2324.

Yang, Q., Saito, T., Berglund, L. A. & Isogai, A. (2015). Cellulose nanofibrils improve the properties of all-cellulose composites by the nano-reinforcement mechanism and nanofibril-induced crystallization. *Nanoscale*, 7(42), 17957–17963.

Yousefi, H., Faezipour, M., Nishino, T., Shakeri, A. & Ebrahimi, G. (2011a). All-cellulose composite and nanocomposite made from partially dissolved micro- and nanofibers of canola straw. *Polymer Journal*, 43(6), 559–564.

Yousefi, H., Mashkour, M. & Yousefi, R. (2015). Direct solvent nanowelding of cellulose fibers to make all-cellulose nanocomposite. *Cellulose*, 22(2), 1189–1200.

Yousefi, H., Nishino, T., Faezipour, M., Ebrahimi, G. & Shakeri, A. (2011b). Direct fabrication of all-cellulose nanocomposite from cellulose microfibers using ionic liquid-based nanowelding. *Biomacromolecules*, 12(11), 4080–4085.

Yousefi, H., Nishino, T., Faezipour, M., Ebrahimi, G., Shakeri, A. & Morimune, S. (2010). All-cellulose nanocomposite made from nanofibrillated cellulose. *Advanced Composites Letters*, 19(6), 190–195.

Yuan, Q., Wu, D. Y., Gotama, J. & Bateman, S. (2008). Wood fiber reinforced polyethylene and polypropylene composites with high modulus and impact strength. *Journal of Thermoplastic Composite Materials*, 21(3), 195–208.

Yuwawech, K., Wootthikanokkhan, J. & Tanpichai, S. (2015). Effects of two different cellulose nanofiber types on properties of poly(vinyl alcohol) composite films. *Journal of Nanomaterials*, 2015, 908689.

Zhao, J., He, X., Wang, Y., Zhang, W., Zhang, X., Zhang, X., Deng, Y. & Lu, C. (2014). Reinforcement of all-cellulose nanocomposite films using native cellulose nanofibrils. *Carbohydrate Polymers*, 104(1), 143–150.

Zhao, Q., Yam, R. C. M., Zhang, B., Yang, Y., Cheng, X. & Li, R. K. Y. (2009). Novel all-cellulose ecocomposites prepared in ionic liquids. *Cellulose*, 16(2), 217–226.

11 Cellulose-Based Bioadhesive for Wood-Based Composite Applications

Ahsan Rajib Promie
Université de Technologie de Troyes

Afroza Akter Liza
Nanjing Forestry University

Md Nazrul Islam
Khulna University

Atanu Kumar Das
Swedish University of Agricultural Sciences

Md Omar Faruk, Sumaya Haq Mim, and Kallol Sarker
Khulna University

CONTENTS

INTRODUCTION

Wood adhesive is an integral part of wood products industry. The use of wood adhesives in wood industry for making composites or panel products is increasing day by day (Wu et al., 2020; Lei et al., 2020). It is estimated that the market for wood adhesives is anticipated to reach $21.9 billion by 2028 (Hussin et al., 2022). Currently, the majority of the commercially available adhesives are

DOI: 10.1201/9781003358084-11

petroleum-based and have good physical, mechanical, and adhesion qualities. (Singh et al., 2017). However, most of these petroleum-based adhesives are produced from nonrenewable resources (McDevitt and Grigsby, 2014). On the other side, some synthetic adhesives, such as formaldehyde-based adhesives, emit low molecular toxic gases that are carcinogenic for humans, and recycling of these synthetic adhesives is quite difficult (Rashid et al., 2020; Nazrul et al., 2019). Therefore, scientists are now developing green adhesive instead of petroleum-based adhesives to reduce these challenges by using renewable, biodegradable, environmentally friendly, and biocompatible resources.

In this modern era, bio-based adhesives have been developed using plant oils, soybean meal, cellulose, lignin, starch, etc. Some of them have even found commercial success (Gouveia et al., 2020; Karagiannidis et al., 2020; Zhu et al., 2016). Due to imperative for advancements in the fabrication of products from biodegradable polymers, the storage of natural materials derived from fossil sources, and the mitigation of the level of carbon dioxide emissions, bioadhesives are the subject of extensive research. One of the motives green composites have gained so much scientific curiosity is the use of these adhesives in the manufacture of green products (Yue et al., 2020).

Besides this, reactivity bonding strength, molecular weight, and degree of penetration must take into consideration before commercialization of these bioadhesives (Ferdosian et al., 2017). Another important factor is the bonding strength because it influences the physical and mechanical characteristics of the wood-based products (Yin et al., 2020). Therefore, keeping these factors in mind, researchers are trying to formulate different bio-based adhesives that can meet all these issues, and currently, cellulose-based adhesives have received much attention which can lead the modern era of wood-based industries.

Cellulose-based green adhesive gets more attention due to its renewability, versatility, and non-toxic nature. It is the most abundant renewable resources mostly found in plants in larger quantity (Dufresne, 2013). Moreover, it is also biodegradable and biocompatible with outstanding mechanical performance (Kumar et al., 2021). Since cellulose is made of glucose units, which contain hydroxyl groups, it can be used as an adhesive for attaching particular materials (Wang et al., 2016; Zhang et al., 2015). However, celluloses are extensively endowed with hydroxyl groups that make them hydrophilic, which reduces their adhesive power in a moist or wet environment. Since cellulose contains hydroxyl groups, it exhibits great adherence to polar compounds but weak adhesion to non-polar compounds (Wang et al., 2016). Now, as cellulose has this reactive hydroxyl groups, it can interact with the suitable chemicals to strengthen its adhesive properties and increase the range of applications for it (Heise et al., 2021).

Various studies are going on for formulating new wood adhesives modifying cellulose most importantly using different nanocellulose (Tang et al., 2021; Kumar et al., 2021). Chemically modifying the crystalline cellulose into dialdehyde cellulose (DAC) is the prominent and successful bio adhesive formulated so far from cellulose (Madivoli et al., 2019). Along with that, different nanocellulose like cellulose nanofiber, cellulose nanocrystal, cellulose nanowhisker, etc. are being used incorporated with different wood adhesives as fillers or hardeners to provide excellent mechanical and physical properties. Thus, this chapter covers the present use of cellulose-based bioadhesive for the fabrication of wood-based composites. The modification of cellulose for the implication as bioadhesive in wood-based composites has been discussed. The formulation process of cellulose-based bioadhesives and their performance have also been pointed out.

CELLULOSE

Cellulose is the most crucial organic polymer mainly produces by living organisms. The lignocellulosic material that abounds in forests, of which wood is thought to be the most significant source of cellulose (Khandelwal and Windle, 2013). D-glucose units condense through β (1→4) glycosidic linkages to form cellulose. In order to hold the chains firmly together and create microfibrils with great tensile strength, the numerous hydroxyl groups in one chain of glucose form hydrogen bonds with oxygen atoms on either the same chain as the one that is attached to or on a neighboring chain.

Cellulose contains crystalline and amorphous region where crystalline region is very strong due to inter- and intramolecular bonding and amorphous region is weak due to weaker bond (Wang et al., 2016). Cellulose is the main chemical constituent of wood where the monomeric units are glucose. In both softwood and hardwood, cellulose accounts of 45%–50%. However, in industrial commercialization, cellulose is thought to be the most prevalent polymer that has been used for ages in various applications (Simon et al., 1998).

Cellulose was first identified by Anselm Payen in 1838 (Purves, 1946). In cotton fiber, it occurs in almost pure form at 98% in combination with lignin and hemicellulose. The function of cellulose microfibrils is to impart strength to the cell wall. Research into cellulose's physico-chemical properties has been extensive. However, cellulose may be observed in plant cell walls, and its arrangement in vascular bundles is thought to act as a foundation to withstand any applied external pressure in the formation of wood (Yu et al., 2005).

There are several different forms and arrangements of cellulose, but most of the physical and chemical attributes of cellulosic compounds rely on two main factors: production of cellulose microfibrils, which varies depending on the precursor, and the separation processes, which includes pretreatment, breakdown, and/or deconstruction (Aravamudhan et al., 2014). Depending on the extraction method utilized, there might be significant differences in the degree of polymerization, morphologies, surface areas, porosities, crystallinities, thermal stabilities, and mechanical properties. Cellulose is a great source of fermentable sugar since it is a homopolymer of a glucose derivative. It is grown as biomass feedstocks so that ethanol, ethers, acetic acid, etc., can be produced. In addition to energy requirements, wood pulp and cotton crops meet the industrial demands for cellulose (Aunina et al., 2010).

Due to the significant commercial benefit of organic cellulose, its biosynthesis is currently receiving increased attention from scientists because it is still hotly debated (Li et al., 2014). Most recent research on the molecular process of cellulose production in higher plants was conducted on model herbaceous plants and fiber crops, and these findings have recently been analyzed. All these elements pique a researcher's interest, leading them to desire to learn more and uncover other cellulose attributes and possibilities.

PROPERTIES OF CELLULOSE

Since hydrogen bonding between the network of hydroxy groups in cellulose define its structure, the topic has always required much investigation (Gardner and Blackwell, 1974). Structure analysis techniques including X-ray diffraction, NMR spectroscopy, and electron microscopy have undergone extensive advancement for more than a century. For the steps of synthetic reactions and the production of man-made goods based on cellulose that have several applications, a thorough, in-depth analysis is necessary. The hydroxyl groups of -1,4-glucan cellulose are located at C_2, C_3, and C_6 in the structure of cellulose as seen in Figure 11.1 (Shanks, 2014). In addition to a shear relationship with O_5-C_5 bonds, the CH_2OH group is oriented in relation to the C_4 and C_5 bonds. The frequency of the crystalline (high order) and amorphous forms of the solid form being depicted is similar (low order). Particularly, the monoclinic unit cell with a two-fold screw axis and two parallel cellulose strands is used in X-ray diffraction to establish the crystal structure (Klemm et al., 2005).

The existence of triclinic and monoclinic unit cells in the cellulose crystal structures has been thoroughly confirmed by research using combined X-ray, neutron, and electron microbeam diffraction. However, recent research on the I crystal structure have already shown many H-bonds and neighboring chain conformations. Recent studies on the I crystal structure, however, have found several H-bonds and nearby chain conformations. The thermodynamically stable cellulose II phase, which also occurs in other kinds of crystal forms, is the most stable form of cellulose. Aqueous sodium hydroxide can be used to convert cellulose I into cellulose II.

Besides this, cellulose possesses other different properties like physical, chemical, thermal, and electrical properties. Properties of cellulose are mentioned below.

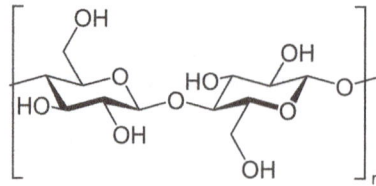

FIGURE 11.1 Molecular structure of cellulose. Reproduced with permission from Shanks (2014). Copyright, Elsevier.

PHYSICAL AND MECHANICAL PROPERTIES OF CELLULOSE

Cellulose is a solid homo-biopolymer that is non-toxic and biodegradable. Depending on the hydroxyl behavior of cellulose, it can have two distinct regions. One is crystalline cellulose and the other is amorphous region cellulose. Crystalline cellulose has evenly ordered hydroxyl molecules, so it does not dissolve in most solvents and is less reactive and more stable (Wang et al., 2016). However, amorphous regions of cellulose have a large number of hydrogen bonds that are unevenly distributed. Therefore, it is more reactive and less stable (Zhang et al., 2015). Fully hydrated cellulose is very flexible but dry cellulose is inflexible and brittle. Cellulose shows high tensile and compressive strength (Lin et al., 2022).

CHEMICAL PROPERTIES OF CELLULOSE

Cellulose is a long polymeric chain of glucose monomer. This polymeric structure can be measured by degree of polymerization, which gives the mechanical strength of the cellulose chain (Samir et al., 2005). Degree of polymerization of cellulose is within the range of 8,000–10,000. Generally, cellulose is soluble in organic solvents but insoluble in water. Though cellulose is insoluble in water, it can absorb water at a rate of 8%–14% at 20°C and 60% relative humidity. Cellulose gives anhydrous reaction with concentrated acids (Simon et al., 1998).

THERMAL AND ELECTRICAL PROPERTIES OF CELLULOSE

Cellulose is relatively stable in thermal conduction, but not a thermoplastic polymer. It shows thermal softening at 231°C–153°C. Thermal decomposition of cellulose occurs at >180°C. It is one of the oldest electrically insulating material. The electrical property of cellulose shows high resistivity and low conductivity. Water content of cellulose decreases the resistivity. Electric resistivity of cellulose is also affected by relative humidity (Yarar et al., 2017).

CELLULOSE MODIFICATION FOR ADHESIVES PRODUCTION

Cellulose can be an excellent source of bio-based wood adhesive. But its chemical and physical structure shows some difficulties. To be able to compete with synthetic adhesive or other bio-based adhesive, cellulose must possess some properties adequate for the formulation of such kinds of adhesives (Yao et al., 2017). The modification of cellulose improves the chemical profile of the cellulose for adhesive production and improves physical or mechanical properties of adhesive. Chemical modification is the core modification method for adhesive production because it modifies the chemical structure of cellulose in favor of its use as adhesive. Different chemical modification methods can be implemented for adhesive production, whereas the process of acetylation involves replacing the hydroxyl groups in cellulose with less hydrophilic acetyl groups (Eslah et al., 2018).

Chemical modification of cellulose involves the change in their chemical structure mostly reacting with hydroxyl groups. The most widely utilized chemical to alter cellulose is

3,4-dihydroxyphenylalanine (DOPA), which is a type of catechol-containing compound (Karabulut et al., 2012; Moulay, 2014; Yao et al., 2017). Catechol groups are introduced into cellulose through chemical modification. Modified cellulose, as opposed to unmodified cellulose, can firmly attach to diverse surfaces under varied environmental circumstances, even in a water environment (Tang et al. 2021).

The grafting DOPA reaction cannot be performed on all cellulosic OH groups due to the complex and time-consuming nature of chemical modification (Madison and Carnali, 2013; Nakajima and Ikada 1995; Staros et al. 1986). Since most DOPA-modified cellulose cannot achieve a very high grafting rate, it has low water resistance and poor underwater adhesion ability (Madison and Carnali, 2013).

Apart from chemical modification, surface modification of cellulose is becoming an important method for formulating cellulose-based adhesives. In surface modification, the mechanical and physical strength of the cellulose are enhanced, and the modified cellulose is used as filler for adhesive production in most of the cases. Cellulose in the form of nanocellulose has a huge application in adhesive industry due to its enhanced mechanical and thermal properties. Nanocellulose (NC) is nanostructured cellulose that has a minimum of one nanometer-sized dimension (Moon et al., 2011). It can be in nanofibrillated cellulose (NFC) or cellulose nanocrystals (NCC or CNC) or bacterial cellulose forms. CNC contains a lot of OH groups on its surface, where chemical reactions can take place and polymerized on the CNC surface (Salas et al., 2014). The OH group at the sixth carbon is mainly responsible for this kind of modification (Hajlane, 2014). Now, using epoxy resin or different coupling agents like different silane agents, surface modification of the nanocellulose can be done (Eyley and Thielemans, 2014).

CELLULOSE-BASED ADHESIVES FOR WOOD-BASED COMPOSITES

DIALDEHYDE CELLULOSE (DAC)

Dialdehyde cellulose (DAC) is a cellulose-based, environmentally friendly bioadhesive that can be extensively used in the wood industry. It can be produced by periodate or selective oxidation on C_2 and C_3 carbon of anhydroglucose units (AGUs) (Kim et al., 2000; Liu et al., 2017). Periodate oxidation is the easiest and most effective way of producing DAC because of its nature in aqueous solution (Kim et al., 2000). The C-C bond between C_2 and C_3 carbon of the anhydroglucose units breaks down and transformed the OH groups into aldehyde group (Figure 11.2) (Zhang et al., 2019).

The conversion of OH group into aldehyde group initially decreases the crystallinity of cellulose and produces highly oxidized DAC (Kim et al., 2004), and the aldehyde group formed in DAC makes an intra and intermolecular cross-linking, which gives DAC the polymer chain structure (Potthast et al., 2009). Depending on the type of oxidation, temperature, and reaction time, the DAC formation varies. Besides this, the degree of oxidation (DO) plays a great role for the formulation and characterization of DAC. Table 11.1 represents the reaction temperature and time for DAC formation using different DOs.

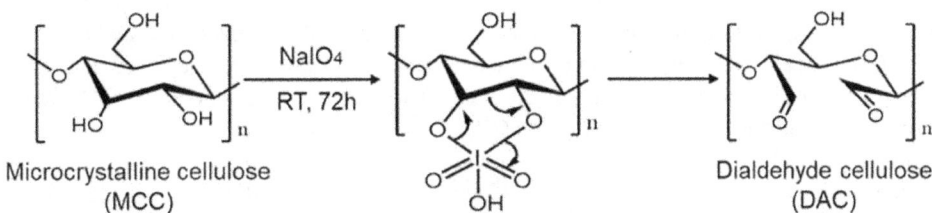

Microcrystalline cellulose (MCC)   NaIO₄ RT, 72h   Dialdehyde cellulose (DAC)

FIGURE 11.2 Chemical formation of DAC. Reproduced with permission from (Zhang et al., 2019). Copyright, Elsevier.

TABLE 11.1

Summary of Degrees of Oxidation and Reaction Conditions for Obtained DAC reaction

DAC	DO	Reagents	Temp (°C)	Time (h)
DAC-1.88	1.88±0.002	MCC (1g), NaIO$_4$ (1.65g), and NaCl (3.87g)	25	72
DAC-1.75	1.75±0.002	MCC (1g) and NaIO$_4$ (1.65g)	25	72
DAC-1.25	1.25±0.003	MCC (1g) and NaIO$_4$ (1.65g)	55	4
DAC-0.78	0.78±0.008	MCC (1g) and NaIO$_4$ (1.65g)	45	4

DAC-x, Dialdehyde cellulose (DAC) with certain number of degree of oxidation (DO) that was determined by a titration method.

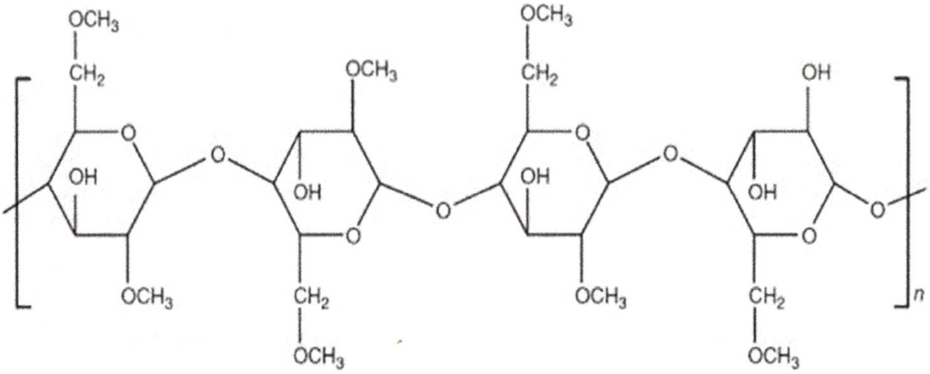

FIGURE 11.3 Chemical structure of methyl cellulose.

In wood composite industry, DAC can be an excellent alternative to the synthetic toxic adhesives. Apart from these, DAC along with amino groups can give more strong cross-linking and excellent adhesion property. Li et al. (2022) investigated the Schiff base linkage between the amino groups in polyurea (PEI-U) and the aldehyde groups in DAC and found that it could form strong cross-linking networks and could good adhesion property.

METHYL CELLULOSE

Methyl cellulose (MC), the methyl ether of cellulose (see in Figure 11.3), is synthesized when alkali cellulose is reacted with methyl chloride (Brady et al., 2017). It contains 27.58%–31.5% of methoxy groups. It is a perfect adhesive for use in food packaging because it is fully harmless, flavorless, and odorless. It can produce high-viscosity solutions at extremely low concentrations, making it a potential thickening agent for water-soluble adhesives and a component of waterborne composites (Brady et al., 2017). Besides this, it has a greater use in cosmetics, pharmaceutics, and the chemical industry (Ding et al., 2014). It consists of numerous linked glucose molecules used as a stabilizer and as an emulsifying and suspending agent. The average number of substituted OH groups per glucose is known as the degree of substitution (DS). Pure MC is white powdery hydrophilic material and generally dissolves in cold water but insoluble in hot water and forms viscous solution, which can be used as adhesive. Different types of MC are commercially available depending on different viscosity grades ranging from 5 to 75,000 cP (Brady et al., 2017).

MC has unique and remarkable ability to set in mild conditions and melt in cooler conditions. MC has a significant use in construction materials as a performance additive. Tile adhesives are well-known examples of dry mixed mortars that make use of MC. But MC can also be used as filler or cross-linking copolymer in wood adhesives and can be used in composites. Like carboxymethyl cellulose (CMC), it is used along with starch-based wood adhesives at a concentration of 0.375%, and it improves the initial viscosity and solids content of pure starch. It lowers the curing temperature and enhances the interfacial characteristics and compatibility between the starch adhesive and the PAPI pre-polymer in terms of mechanical property (Qiao et al., 2014). Moreover, CMC can also be used with sodium silicate (Na_2SiO_3)-based wood adhesive. It improves the bonding strength of Na_2SiO_3-based wood adhesive with 7.5%. The stronger molecular structure increased thermal stability, and desirable changes in the flexibility of Na_2SiO_3 wood adhesive all contributed to the improved performance of CMC/ Na_2SiO_3-based wood adhesive (Qiao et al., 2014). The influence of CMC on the structure and characteristics of the adhesive suggests that CMC can be employed to create a high-performing, ecologically friendly Na_2SiO_3 wood adhesive (Zhang et al., 2012).

CELLULOSE NANOFIBRILS

Solid wood adhesion is a significant problem in the production of many wood products. In this context, one viable way to boost strength is by mixing fiber fillers into the liquid adhesive. Polyvinyl acetate (PVAc) is a water-soluble, thermoplastic, and biodegradable polymer often used as an adhesive. But it has poor performance in humid condition and higher temperature. Furthermore, to improve its performance, different studies have been conducted. Jiang et al. (2018) have conducted a study where they have shown the enhanced bonding strength by the incorporation of cellulose nanofibrils (CNF) with conventional adhesive like starch-based adhesive or polyvinyl acetate (PVAc). The incorporation of PVAc and CNF also enhanced the adhesion property with greater effects (Kaboorani et al., 2012). Lap-joint strength of the incorporation has been increased by 74.5% (Vineeth et al., 2019). Another study has found that the incorporation of nanocrystalline cellulose (NCC) with PVAc also improves the performance. Vineeth et al. (2019) have shown through block shear test that this incorporation can enhance the bonding strength of the adhesive in all conditions. On the other hand, mechanical properties of wood-based products can be enhanced by NCC incorporation. Along with PVAc, NCC or CNF can also be incorporated with polyvinyl alcohol, and it increases the reinforcing effects. CNF can also be used in particleboard incorporating with tannin-based adhesive. It enhances the high performance with the higher mechanical properties of the particleboard (Zhang et al., 2011). Another type of nanocellulose named cellulose nanowhiskers (CNW) also has an ability to enhance the performance of soya bean-based adhesive (Gao et al., 2012). This incorporation also improves the water resistance of plywood.

PERFORMANCE OF CELLULOSE-BASED BIOADHESIVE FOR WOOD-BASED COMPOSITES

Nanocellulose being cheaper and sustainable adhesive can compete with synthetic adhesives without compromising its permeance in the wood composite industry. It can act like binder incorporating with other adhesives and also enhanced the structural reinforcement. It also reduces the harmful emissions like leaching of toxic chemical from the composites in its working condition. Different bio-based adhesives provide different types of mechanical properties and enhance their working performance as adhesives. Among them the most promising one is dialdehyde cellulose (DAC).

DAC enhances the composite's water resistance, enabling usage as waterborne composites. In the glue line between two substates, DAC creates a hyperbranched cross-linking structure and dense cross-sections to cure the adhesive (Li et al., 2022; Zhang et al., 2019). In indoor and outdoor wood-based panel products, cellulose-based wood adhesives can be utilized as an alternative to synthetic adhesives due to their improved wet bonding strength (Li et al. 2022). Moreover, Zhang et al. (2019) have found that the best bonding performance is demonstrated by DAC with DO of 1.75 and a concentration of 40 wt%, with bonding strengths of 9.53 MPa for beech and 5.75 MPa for spruce wood, respectively. In this case, DAC adhesives form a stronger bond in the wood composites, and it is strong enough than the wood itself. The shear strength of DAC was found ranged from 1.34 to 1.51 MPa, while the bond strength was found to be 2.71 MPa (Zhang et al., 2019).

Apart from DAC, different wood adhesive incorporated with nanocellulose provides greater mechanical properties, physical properties of panels or composites. It was observed that cellulose nanofiber (CNF) when incorporated with cottonseed protein-based adhesive provides 22% improvement in dry adhesive strength compared with only cottonseed protein-based adhesive (Gadhave et al., 2021). Another study showed that adding CNF in urea-formaldehyde (UF) or melamine formaldehyde (MF) improves the reinforcing nature of CNF (Veigel et al., 2012). Moreover, this reinforcement of CNF improves fracture energy and toughness of the adhesive which improves the mechanical property of particleboard panel (Amini et al., 2017). When CNF is incorporated with polyvinyl acetate adhesive, it increases the viscosity and shear strength. It also makes the wood joints water-resistant and improves the mechanical properties (Chaabouni and Boufi, 2017).

CHALLENGES AND OPPORTUNITIES

Due to the presence of hydroxyl groups, cellulose has a significant adhesion property to polar compounds, but it has a weak adhesion strength to non-polar materials. The transformation of cellulose adhesives is constrained by this aspect. Many scientists are working to increase adhesion behavior of cellulose-based adhesives in waterborne conditions (Bai et al., 2022; Tang et al., 2021). It takes a lot of effort and time to successfully chemically modify cellulose to add catechol groups. Due of the weak water resistance and hydrographic adhesive capabilities of cellulose, all of its hydroxyl groups are unable to undergo the grafting DOPA reaction (Madison and Carnali, 2013; Nakajima and Ikada, 1995; Staros et al., 1986).

Again, the characteristics of cellulose fibers are influenced by a number of variables, including type, climate, harvest, maturity, decortications, fiber modification, disintegration, and technological processes (Velde and Kiekens, 2001). It is extremely difficult to create cellulose-based adhesive because of moisture absorption, poor heat stability, quality fluctuations, and poor compatibility with the hydrophobic matrix of cellulose fibers (Georgopoulos et al., 2005; Saheb and Jog, 1999). In addition, a lot of time is needed for mixing cellulose with tannic acid (TA) to make TA firmly incorporate with cellulose (Chen et al., 2021; Shao et al., 2019). However, more research is still needed to fully understand the enhancement of DAC adhesives' water resistance. It can be said that DAC has the potential to be a promising and new bio-based adhesive for wood bonding especially in indoor use. But there are no experiments on the nature of the water resistance ability of DAC in service life. Research is going on how to improve it like using amino groups but need more studies. Besides, there is scope to improve the cross-linking of nanocellulose by undertaking suitable chemistry using different coupling agents and nanocellulose grafting. These findings will motivate other researchers in the field to think outside the box and to reject the idea that cellulose-based bioadhesives can only be described in terms of H-bonding. We think that this would be extremely advantageous for the cellulose community as a whole.

CONCLUSION

The wood products like panel or composites traditionally use formaldehyde-based adhesives, which is produced from petroleum-based origin and very hazardous for the environment. Different countries are taking actions against these synthetic adhesives, and many are concerning about their end use. As a result, scientists are working to develop a variety of bio-based wood adhesives, including those based on starch, lignin, and tannin. However, cellulose is the most abundant and sustainable material. It has a lot of uses depending on the demand. But so far, the wood industry has totally relied on toxic, synthetic adhesives like UF and MF. The formulation of wood adhesive from cellulose has still not been established yet except for DAC, though it has huge potential to be used as adhesive in the wood industry. Among all the cellulose-based adhesives, the use of DAC through chemical modification as wood adhesive is the best cellulose-based wood adhesive discovered so far. It provides better strength, good penetration, and low cost but has a problem with water resistance ability. New research is going on to improve water resistibility, like cross-linking with amino groups, but more research is needed. Apart from DAC, which is a sustainable source, nanocellulose usage in wood adhesive can reduce the formaldehyde emission from synthetic adhesive and can provide mechanical and physical strength to the adhesion by creating cross-linking. The application of nanocellulose as wood adhesive in wood composites has a very bright future, but not well established yet. More research is needed for improving and optimizing the use of nanocellulose-based bioadhesives by increasing the bonding with other base polymeric matrix. Moreover, the water resistance of nanocellulose-based bioadhesives has to be improved through further study.

REFERENCES

Amini, E., Tajvidi, M., Gardner, D.J., & Bousfield, D.W. (2017). Utilization of cellulose nanofibrils as a binder for particleboard manufacture. *BioResources*, 12(2), 4093–4110.

Aravamudhan, A., Ramos, D.M., Nada, A., & Kumbar, S. (2014). Natural polymers: Polysaccharides and their derivatives for biomedical applications. In Kumbar, S.G., Laurencin, C.T., & Deng, M (eds) *Natural and Synthetic Biomedical Polymers*, Elsevier, Amsterdam, pp. 67–89.

Aunina, Z., Bazbauers, G., & Valters, K. (2010). Feasibility of bioethanol production from Lignocellulosic biomass. *Scientific Journal of Riga Technical University Environmental and Climate Technologies*, 4, 11–15.

Bai, H., Yu, C., Zhu, H., Zhang, S., Ma, P., & Dong, W. (2022). Mussel-inspired cellulose-based adhesive with underwater adhesion ability. *Cellulose*, 29, 893–906.

Brady, J., Dürig, T., Lee, P.I., & Li, J.X. (2017). Polymer properties and characterization. In Qiu, Y., Zhang, G.G.Z., Mantri, R. V., Chen, Y., & Yu, L (eds) *Developing Solid Oral Dosage Forms*, Elsevier, Amsterdam, pp. 181–223.

Chaabouni, O., & Boufi, S. (2017). Cellulose nanofibrils/polyvinyl acetate nanocomposite adhesives with improved mechanical properties. *Carbohydrate Polymers*, 156, 64–70.

Chen, Y., Zhang, Y., Mensaha, A., Li, D., Wang, Q., & Wei, Q. (2021). A plant-inspired long-lasting adhesive bilayer nanocomposite hydrogel based on redox-active Ag/Tannic acid-Cellulose nanofibers. *Carbohydrate Polymers*, 255, 117508.

Ding, C., Zhang, M., & Li, G. (2014). Rheological properties of collagen/hydroxypropyl methylcellulose (COL/HPMC) blended solutions. *Journal of Applied Polymer Science*, 131, 40042.

Dufresne, A. (2013). Nanocellulose: A new ageless bionanomaterial. *Materials Today*, 16(6), 220–227.

Eslah, F., Jonoobi, M., Faezipour, M. and Ashori, A. (2018). Chemical modification of soybean flour-based adhesives using acetylated cellulose nanocrystals. *Polymer Composites*, 39(10), 3618–3625.

Eyley, S., & Thielemans, W. (2014). Surface modification of cellulose nanocrystals. *Nanoscale*, 6(14), 7764–7779.

Ferdosian, F., Pan, Z., Gao, G., & Zhao, B. (2017). Bio-based adhesives and evaluation for wood composites application. *Polymers*, 9(2), 70.

Gadhave, R.V., Dhawale, P.V., & Sorate, C.S. (2021). Surface modification of cellulose with silanes for adhesive application. *Open Journal of Polymer Chemistry*, 11(2), 11–30.

Gao, Q., Li, J., Shi, S.Q., Liang, K., & Zhang, X. (2012). Soybean meal-based adhesive reinforced with cellulose nano-whiskers. *BioResources*, 7(4), 5622–5633.

Gardner, K.H., & Blackwell, J. (1974). The structure of native cellulose. *Biopolymers*, 13, 1975–2001.

Georgopoulos, S.T., Tarantili, P.A., Avgerinos, E., Andreopoulos, A.G., & Koukios, E.G. (2005). Thermoplastic polymers reinforced with fibrous agricultural residues. *Polymer Degradation and Stability*, 90(2), 303–312.

Gouveia, J.R., Garcia, G.E., Antonino, L.D., Tavares, L.B., & Dos Santos, D.J. (2020). Epoxidation of kraft lignin as a tool for improving the mechanical properties of epoxy adhesive. *Molecules*, 25(11), 2513.

Hajlane, A. (2014). Development of hierarchical cellulosic reinforcement for polymer composites. Doctoral dissertation, Luleå tekniska universitet.

Heise, K., Kontturi, E., Allahverdiyeva, Y., Tammelin, T., Linder, M.B., & Ikkala, O. (2021). Nanocellulose: Recent fundamental advances and emerging biological and biomimicking applications. *Advanced Materials*, 33(3), 2004349.

Hussin, M.H., Abd Latif, N.H., Hamidon, T.S., Idris, N.N., Hashim, R., Appaturi, J.N., Brosse, N., Ziegler-Devin, I., Chrusiel, L., Fatriasari, W., & Syamani, F.A. (2022). Latest advancements in high-performance bio-based wood adhesives: A critical review. *Journal of Materials Research and Technology*, 2, 3909–3946.

Jiang, W., Tomppo, L., Pakarinen, T., Sirviö, J.A., Liimatainen, H., & Haapala, A.T. (2018). Effect of cellulose nanofibrils on the bond strength of polyvinyl acetate and starch adhesives for wood. *BioResources*, 13(2), 2283–2292.

Kaboorani, A., Riedl, B., Blanchet, P., Fellin, M., Hosseinaei, O., & Wang, S. (2012). Nanocrystalline cellulose (NCC): A renewable nano-material for polyvinyl acetate (PVA) adhesive. *European Polymer Journal*, 48(11), 1829–1837.

Karabulut, E., Pettersson, T., Ankerfors, M., & Wagberg, L. (2012). Adhesive layer-by-layer films of carboxy-methylated cellulose nanofibril–dopamine covalent bioconjugates inspired by marine mussel threads. *ACS Nano*, 6(6), 4731–4739.

Karagiannidis, E., Markessini, C., & Athanassiadou, E. (2020). Micro-fibrillated cellulose in adhesive systems for the production of wood-based panels. *Molecules*, 25(20), 4846.

Khandelwal, M., & Windle, A. (2013). Hierarchical organisation in the most abundant biopolymer–Cellulose. *MRS Online Proceedings Library*, 1504, 10.1557.

Kim, U.J., Kuga, S., Wada, M., Okano, T., & Kondo, T. (2000). Periodate oxidation of crystalline cellulose. *Biomacromolecules*, 1(3), 488–492.

Kim, U.J., Wada, M., & Kuga, S. (2004). Solubilization of dialdehyde cellulose by hot water. *Carbohydrate Polymers*, 56(1), 7–10.

Klemm, D., Fink, B.H.H-P., & Bohn, A. (2005). Cellulose: Fascinating biopolymer and sustainable raw material. *Angewandte Chemie International Edition*, 44(22), 3358–3393.

Kumar, R., Rai, B., Gahlyan, S., & Kumar, G. (2021). A comprehensive review on production, surface modification and characterization of nanocellulose derived from biomass and its commercial applications. *Express Polymer Letters*, 15(2), 104–120.

Lei, Y.F., Wang, X.L., Liu, B.W., Ding, X.M., Chen, L., & Wang, Y.Z. (2020). Fully bio-based pressure-sensitive adhesives with high adhesivity derived from epoxidized soybean oil and rosin acid. *ACS Sustainable Chemistry & Engineering*, 8(35), 13261–13270.

Li, S., Bashline, L., Lei, L., & Gu, Y. (2014). Cellulose synthesis and its regulation. *The Arabidopsis Book/American Society of Plant Biologists*, 12, e0169.

Li, Z., Du, G., Yang, H., Liu, T., Yuan, J., Liu, C., Li, J., Ran, X., Gao, W., & Yang, L. (2022). Construction of a cellulose-based high-performance adhesive with a crosslinking structure bridged by Schiff base and ureido groups. *International Journal of Biological Macromolecules*, 223, 971–979.

Lin, Q., Jiang, P., Ren, S., Liu, S., Ji, Y., Huang, Y., Yu, W., Fontaine, G. and Bourbigot, S. (2022). Advanced functional materials based on bamboo cellulose fibers with different crystal structures. *Composites Part A: Applied Science and Manufacturing*, 154, 106758.

Liu, P., Mai, C., & Zhang, K. (2017). Formation of uniform multi-stimuli-responsive and multiblock hydrogels from dialdehyde cellulose. *ACS Sustainable Chemistry & Engineering*, 5(6), 5313–5319.

Madison, S.A., & Carnali, J.O. (2013). pH optimization of amidation via carbodiimides. *Industrial & Engineering Chemistry Research*, 52(38), 13547–13555.

Madivoli, E.S., Kareru, P.G., Gachanja, A.N., Mugo, S.M., & Makhanu, D.S. (2019). Synthesis and characterization of dialdehyde cellulose nanofibers from O. sativa husks. *SN Applied Science*, 1, 723.

McDevitt, J.E., & Grigsby, W.J. (2014). Life cycle assessment of bio-and petro-chemical adhesives used in fiberboard production. *Journal of Polymers and the Environment*, 22(4), 537–544.

Moon, R.J., Martini, A., Nairn, J., Simonsen, J., & Youngblood, J. (2011). Cellulose nanomaterials review: Structure, properties and nanocomposites. *Chemical Society Reviews*, 40(7), 3941–3994.

Moulay, S. (2014). Dopa/catechol-tethered polymers: Bioadhesives and biomimetic adhesive materials. *Polymer Reviews*, 54(3), 436–513.

Nakajima, N., & Ikada, Y. (1995). Mechanism of amide formation by carbodiimide for bioconjugation in aqueous media. *Bioconjugate Chemistry*, 6(1), 123–130.

Potthast, A., Schiehser, S., Rosenau, T., & Kostic, M. (2009). *Oxidative Modifications of Cellulose in the Periodate System–Reduction and Beta-Elimination Reactions 2nd ICC 2007*, Tokyo, Japan, October 25–29, 2007.

Purves, C.B. (1946). Chemical Nature of Cellulose and its Derivatives. New York: InterScience.

Qiao, Z., Gu, J., Zuo, Y., Tan, H., & Zhang, Y. (2014). The effect of carboxymethyl cellulose addition on the properties of starch-based wood adhesive. *BioResources*, 9(4), 6117–6129.

Rashid, K., Mohammadi, K., & Powell, K. (2020). Dynamic simulation and techno-economic analysis of a concentrated solar power (CSP) plant hybridized with both thermal energy storage and natural gas. *Journal of Cleaner Production*, 248, 119193.

Saheb, D.N., & Jog, J.P. (1999). Natural fiber polymer composites: A review. *Advances in Polymer Technology: Journal of the Polymer Processing Institute*, 18(4), 351–363.

Salas, C., Nypelö, T., Rodriguez-Abreu, C., Carrillo, C., & Rojas, O.J. (2014). Nanocellulose properties and applications in colloids and interfaces. *Current Opinion in Colloid & Interface Science*, 19(5), 383–396.

Samir, M.A.S.A., Alloin, F., & Dufresne, A. (2005). Review of recent research into cellulosic whiskers, their properties and their application in nanocomposite field. *Biomacromolecules*, 6(2), 612–626.

Simon, J., Muller, H.P., Koch, R., & Muller, V. (1998). Thermoplastics and biodegradable polymers of cellulose. *Polymer Degradation and Stability*, 59, 107–115.

Singh, A.A., Afrin, S., & Karim, Z. (2017). Green composites: Versatile material for future. In: Jawaid, M., Salit, M., Alothman, O. (eds) *Green Biocomposites. Green Energy and Technology*. Springer, Cham, pp. 29–44.

Shanks, R.A. (2014). Chemistry and structure of cellulosic fibres as reinforcements in natural fibre composites. In Hodzic, A., & Shanks, R. (eds) *Natural Fibre Composites*, Elsevier, Amstedam, pp. 66–83.

Shao, C., Meng, L., Wang, M., Cui, C., Wang, B., Han, C.R., Xu, F., & Yang, J. (2019). Mimicking dynamic adhesiveness and strain-stiffening behavior of biological tissues in tough and self-healable cellulose nanocomposite hydrogels. *ACS Applied Materials & Interfaces*, 11(6), 5885–5895.

Staros, J.V., Wright, R.W., & Swingle, D.M. (1986). Enhancement by N-hydroxysulfosuccinimide of water-soluble carbodiimide-mediated coupling reactions. *Analytical Biochemistry*, 156(1), 220–222.

Tang, Z., Bian, S., Lin, Z., Xiao, H., Zhang, M., Liu, K., Li, X., Du, B., Huang, L., Chen, L., & Ni, Y. (2021). Biocompatible catechol-functionalized cellulose-based adhesives with strong water resistance. *Macromolecular Materials and Engineering*, 306(9), 2100232.

Van de Velde, K., & Kiekens, P. (2001). Thermoplastic pultrusion of natural fibre reinforced composites. *Composite Structures*, 54(2–3), 355–360.

Veigel, S., Rathke, J., Weigl, M., & Gindl-Altmutter, W. (2012). Particle board and oriented strand board prepared with nanocellulose-reinforced adhesive. *Journal of Nanomaterials*, 2012(1), 158503.

Vineeth, S.K., Gadhave, R.V., & Gadekar, P.T. (2019). Nanocellulose applications in wood adhesives. *Open Journal of Polymer Chemistry*, 9(4), 63–75.

Wang, C., Xiong, Y., Fan, B., Yao, Q., Wang, H., Jin, C., & Sun, Q. (2016). Cellulose as an adhesion agent for the synthesis of lignin aerogel with strong mechanical performance, Sound-absorption and thermal Insulation. *Scientific Reports*, 6(1), 1–9.

Wu, Q., Shao, W., Xia, N., Wang, P., & Kong, F. (2020). A separable paper adhesive based on the starch——lignin composite. *Carbohydrate Polymers*, 229, 115488.

Yao, K., Huang, S., Tang, H., Xu, Y., Buntkowsky, G., Berglund, L.A., & Zhou, Q. (2017). Bioinspired interface engineering for moisture resistance in nacre-mimetic cellulose nanofibrils/clay nanocomposites. *ACS Applied Materials & Interfaces*, 9(23), 20169–20178.

Yarar Kaplan, B., Işıkel Şanlı, L., & Alkan Gürsel, S. (2017). Flexible carbon–cellulose fiber-based composite gas diffusion layer for polymer electrolyte membrane fuel cells. *Journal of Materials Science*, 52(9), 4968–4976.

Yin, H., Zheng, P., Zhang, E., Rao, J., Lin, Q., Fan, M., Zhu, Z., Zeng, Q., & Chen, N. (2020). Improved wet shear strength in eco-friendly starch-cellulosic adhesives for woody composites. *Carbohydrate Polymers*, 250, 116884.

Yue, L., Shi, R., Yi, Z., Shi, S.Q., Gao, Q., & Li, J. (2020). A high-performance soybean meal-based plywood adhesive prepared via an ultrasonic process and using significantly lower amounts of chemical additives. *Journal of Cleaner Production*, 274, 123017.

Zhang, H., Liu, P., Musa, S.M., Mai, C., & Zhang, K. (2019). Dialdehyde cellulose as a bio-based robust adhesive for wood bonding. *ACS Sustainable Chemistry & Engineering*, 7(12), 10452–10459.

Zhang, H., Zhang, J., Song, S., Wu, G., & Pu, J. (2011). Modified nanocrystalline cellulose from two kinds of modifiers used for improving formaldehyde emission and bonding strength of urea-formaldehyde resin adhesive. *BioResources*, 6(4), 4430–4438.

Zhang, J., Choi, Y.S., Yoo, C.G., Kim, T.H., Brown, R.C., & Shanks, B.H. (2015). Cellulose–hemicellulose and cellulose–lignin interactions during fast pyrolysis. *ACS Sustainable Chemistry & Engineering*, 3(2), 293–301.

Zhang, X., Liu, X., Yang, S., Long, K., & Wu, Y. (2012). Effect of carboxyl methyl cellulose on the adhesion properties of sodium silicate wood adhesive. In *Proceedings of 2012 International Conference on Biobase Material Science and Engineering*, 230–233. IEEE, Changsha, China.

Zhu, Y., Romain, C., & Williams, C.K. (2016). Sustainable polymers from renewable resources. *Nature*, 540 (7633), 354–362.

12 Deriving Renewable Feedstock from Palm Oil Mill Effluent for Polyhydroxyalkanoate (PHA) Production

Gobi Kanadasan
Universiti Tunku Abdul Rahman

Vel Murugan Vadivelu and Muaz Mohd Zaini Makhtar
Universiti Sains Malaysia

CONTENTS

INTRODUCTION

Palm oil (*Elaeis guineensis*) is a significant commodity for producers such as Malaysia, Indonesia, Nigeria, and Thailand. In recent years, the stature of palm oil industry has been ever growing in the global market due to its versatility. In 2021, Malaysia produced about 18 million tonne of crude palm oil (CPO) (Kadir, 2022). Active trading of CPO has been on going with various countries such as China, India, Pakistan, Netherlands, Japan, and South Africa (MPOB, 2022).

Apart from economic gain, the socioeconomic status of the residents has been elevated as well. It provided ample job opportunities for the CPO-producing countries. Job opportunities are available in abundance in the palm oil industry due to the palm oil tree plantation in large scale. The ever-expanding applications of the palm oil in daily life and the lucrative CPO price have led to the mass plantation of it. Currently, the palm oil is being widely used as raw material to produce cooking oil, soap, margarine, and biodiesel. As the natural resources like petroleum and natural gas are on the verge of depleting in the coming years, the biodiesel production gained the interest among the researchers to test it as an alternative fuel. Researchers have proven that the palm oil can be one of the excellent raw materials to produce biodiesel (Maleki et al., 2022; Pirouzfar et al., 2022; Rawindran et al., 2022).

DOI: 10.1201/9781003358084-12

As a result, the demand for the palm oil has been steadily increasing for the past few years in the international market. Mass production of palm oil needs large plantation of palm oil trees. These trees are sensitive toward the climate and rainfall of the plantation area. They need 2,000 mm of rain per year on average (Oettli et al., 2018). The ripe fruit from the tree is called fresh fruit branches (FFB). From this FFB, the crude palm oil will be extracted. In the process of extracting the crude palm oil from the FFB, several significant processes are involved. Among the processes that the FFB needs to undergo are the sterilization, clarification, and centrifugation. During these processes, water is being used extensively to extract the oil. It has been reported that the production of 1 ton of crude palm oil needs 5–7.5 tons of water. From this amount of water, more than 50% of it will end up as palm oil mill effluent (POME) (Mohammad et al., 2021). Based on 2021 palm oil production, approximately 108 million tons of POME were released in Malaysia alone (MPOB, 2022). The main sources of POME from a factory processing FFB are the clarification (60%), and it comprised of 96% of water, 0.6%–0.7% of oil, and 2%–4% of suspended solids (Mohammad et al., 2021). The characteristics of POME are given in Table 12.1.

From Table 12.1, the chemical oxygen demand (COD) value around 60,000 mg/L indicates that POME is highly environment polluting waste. Hence, POME could not be directly channeled into receiving body. It must be adequately treated before being released. If it is directly channeled into a river or any other water source, problems such as eutrophication, disruption in the food chain in the water and clean water scarcity might occur. The chain effects of these problems could affect the end users such as human beings and animals. Therefore, POME must undergo treatment for the compliance of standards set by the Department of Environment. This treatment is required to be done within the boundaries of the factories before being discharged into the receiving body. For the palm oil factories nearby water catchment area, stringent rules are applied. The Department of Environment has set the Standard A for the area nearby water catchment, while the Standard B for the inland waters. Standard A and Standard B are represented in Table 12.2.

TABLE 12.1
Characteristics of POME From Various Works

pH	Temperature (°C)	Biochemical Oxygen Demand (mg/L)	Chemical Oxygen Demand (mg/L)	Oil and Grease (mg/L)	Suspended Solids (mg/L)	Nitrogen Content (mg/L)	Reference
4.67	86	20,838	42,482	1,927	14,721	340	Kadier et al. (2022)
3.60	-	15,600	25,000	2,000	4,500	800	Nasrullah et al. (2017)
4.20	85	25,000	51,000	6,000	18,000	750	Bello and Abdul Raman (2017)
4.50	-	15,700	26,000	2,500	13,100	800	Mohamad et al. (2022)
4.10–5.20	80–90	25,000	57,500	9,000	30,000	750	Mahmod et al. (2021)
4.63	-	34,771	89,591	37,883	36,560	-	Rosa et al. (2020)
4.50	-	30,100	70,000	10,540	28,900	980	Lok et al. (2020)
4.35	63	-	28,260	6,020	70,320	1,400	Abdullah et al. (2020)

TABLE 12.2
Environmental Quality (Sewage and Industrial Effluents)
Regulations, 1979. Maximum Effluent Parameter Limits
Standards A and B (Sekitar, 2010)

Parameters	Standard A	Standard B
pH	6.0–9.0	5.5–9.0
Temperature	40°C	40°C
Chemical Oxygen Demand (COD)	80 mg/L	200 mg/L
Biological Oxygen Demand (BOD)	20 mg/L	40 mg/L
Oil and grease	1.0 mg/L	10 mg/L
Suspended solids	50 mg/L	100 mg/L

In order to meet the regulations set by the Department of Environment Malaysia (2010), the palm oil producing factory has to design a POME treating plant. This treatment method utilized in the treatment plant varies from one company to another. This variation in treatment has resulted in different percentage of efficiency in treating the POME. Generally, there would be only a minimum investment on the wastewater treatment plant. No revenues and lack of environmental concerns have been the reasons for the lack of investment in wastewater treatment plant.

Thus, there is a need to strategize a way to conserve the environment and at the same time to drive the palm oil millers to invest in wastewater treatment plant. Value adding to the waste produced from the palm oil mill has huge weightage to coerce investment in the wastewater treatment plant. As seen in Table 12.1, the COD value in the POME is in between 16,000 and 50,000. High COD value indicates that the waste contains large amount of organic content in it. These abundant amounts of organic content shall be utilized as sustainable and renewable resources for PHA production.

As per the recent statistics, the amount of plastic waste produced yearly stands at 381 million tons and it sets to be doubled by 2034 (Ferries, 2022). On average, a garbage truck of plastics is dumped into the ocean every minute. As of now, 25 trillion macro plastics and 51 trillion microplastics have polluted the ocean. This huge amount of plastic waste has adverse effects on the aquatic livings. Lately, enormous reports have been found on the discovery of microplastics in the aquatic livings. It was found that over 1 million seabirds and 100,000 sea mammals were killed by the plastic pollution (Ferries, 2022). Marine turtles, fishes, whales, seals, and seabirds are the most affected by the plastics pollution (Ferries, 2022). It is also projected that there will be 850–950 million tons of plastic waste in ocean by 2050. Unfortunately, it outweighs the projected volume of fishes, which is expected to be around 812–899 million tons only (Gatehouse, 2019). Substantive act must be taken to curtail the plastic waste in ocean and simultaneously secure the food supply for the future generations. In fact, the microplastics have ended up in human beings without our knowledge due to the food cycle chain. One third of the fishes that human beings consume has been reported to contain microplastics. It is a serious concern, and in the future, the protein resources for humans could be hugely depleted as 700 species of marine animals are in danger of extinction due the plastic waste.

In addition to the marine pollution, the plastic waste also pollutes the land. Conventionally, the plastic wastes are dumped into landfills. It may take years for it to degrade, and it may require more landfills to dump the plastics in future. As the world population is projected to increase to 2 billion in 2050, the lands should be available for the food crops. Polluted soil could hamper the cultivation of food crops and threatens the food security for the future generation. To make it worse, the landfill dumping creates a secondary problem which is the microplastic pollution. Microplastics are the plastics with sizes below 5 mm (Lim, 2021). It infiltrates into the water resources, and it was also found in livestock. Eventually, the microplastics would end up in human bodies, and the adverse effects of their ingestion are not fully understood to this day. Thus, it warrants the usage of biodegradable plastics in our daily activities to reduce the adverse effect of conventional plastics to0 the land and sea.

BIODEGRADABLE POLYMER

Biodegradable polymer utilization in daily life has been on the upward trend for the past 5 years. The importance of conserving nature from plastic pollution has increased the utilization of biodegradable polymers. Figure 12.1 shows the global market share of the biodegradable plastic (Bioplastic, 2014). It is expected to dominate about 40% of the plastic in the year 2030. The increasing trend of biodegradable plastic usage has demanded a method to produce it sustainably. Currently, the biodegradable plastic has been produced using either the sugarcane or corn. These crops are rich with carbohydrate, which provides the basis to produce the biodegradable plastic. However, it creates a food versus fuel debate to produce the biodegradable plastic as they are edible crops. Thus, there is a need to find a resource which is non-edible and sustainable in nature. Researchers figured out that the biodegradable plastic could be produced using microorganisms and organic rich waste streams. In line with that, POME fits the criteria to be the feedstock for the biodegradable plastic production. For the past decade, numerous works have been published on the production of biodegradable plastic from POME using microorganisms. The biodegradable polymer of particular interest is polyhydroxyalkanoate (PHA).

Global Bioplastic Market

■ Bioplastic   Non-bioplastic

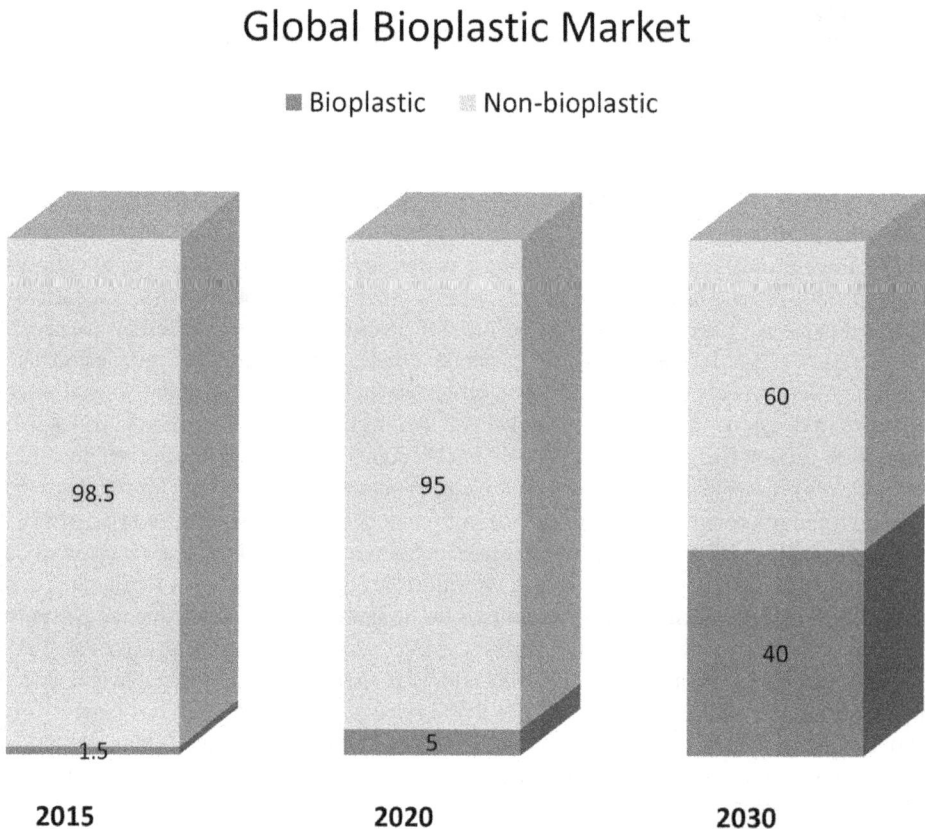

FIGURE 12.1 Global bioplastic market.

TABLE 12.3

Chemical Structure of Some of the Common Types of PHA

PHA Type	Chemical Structure	Type of Polymer
PHB		Homopolymer
PHV		
P3(HB-*co*-HV)		Co-polymer
P3(HB-*co*-HHx)		

Polyhydroxyalkanoate (PHA)

Polyhydroxyalkanoate is a biodegradable polymer that has been used for various purposes. The chemical structure of the PHA is shown in Table 12.3. PHA is a general term used for the accumulation of total monomers such as hydroxybutyrate (HB), hydroxyvalerate (HV), hydroxyhexanoate (HHx), and co-polymers such as hydroxybutyrate-hydroxyvalerate (HB-HV). The type of PHA synthesized is influenced by the chain length of the carbon source (Muthuraj et al., 2021).

Conventionally, waste streams that contain a high content of organic compounds could be converted into VFA. Thereafter, it can be used as the feedstock to produce PHA. POME, which is rich with organic compounds, perfectly suits the criteria for being the feed to produce VFA.

VFA GENERATION

As aforementioned, POME can be anaerobically digested to produce VFA. Generally, a complete anaerobic digestion of POME will result in methane gas and CO_2 gas production. A complete anaerobic digestion consists of four theoretical stages, namely, hydrolysis, acidogenesis, acetogenesis, and methanogenesis. In general, POME is rich with carbohydrates, lipids, proteins, and cellulose (Bala et al., 2018). During the hydrolysis process, these complex compounds will be degraded into simpler compounds such as sugar, alcohol, amino acids, fatty acids, and peptides. In the hydrolysis process, the hydrolytic microorganisms secrete enzymes such as lipase and cellulase that cleave the complex compounds into simpler compounds (Shafwah et al., 2019). In the next stage, the acidogenesis microorganisms convert the produced simple compounds into intermediary products such as VFA, ethanol, and lactate. Then, the intermediary products will be converted into biohydrogen, acetate, and carbon dioxide. This conversion takes place during the acetogenesis phase. The final stage of the anaerobic digestion process is the methanogenesis. In this methanogenesis phase, the methanogenic microorganisms convert the acetate via acetoclastic methanogens, while the CO_2-reducing methanogens convert the CO_2 produced earlier (Mobilian & Craft, 2022). The end-products of the methanogenesis stage are methane and CO_2 gas.

With the intention of producing VFAs, the anaerobic process must be ceased at the acidogenesis stage. In other words, the acetogenesis process needs to be suppressed. The common strategy adopted to cease the anaerobic digestion at the acidogenesis stage is the pH control. It will be maintained at 5.5 or below to ensure the acetogenesis process is suppressed (Atasoy & Cetecioglu, 2022). The unutilized VFA (due to acetogenesis phase suppression) will be converted into PHA by specific microorganisms. The types of VFA produced from the waste streams are dependent on the organic compounds present in the wastewater. In POME, acetic acid, propionic acid, and butyric acid were found to be dominant in the produced VFA (Chinwetkitvanich & Jaikawna, 2015).

PHA PRODUCTION PATHWAY

In general, VFAs consist of acetic acid, propionic acid, butyric acid, isobutyric, valeric acid, isovaleric, and hexanoic acid. At the end of acidogenesis stage of POME, the most common VFAs produced were propionic acid, butyric acid, and acetic acid. VFA could be transformed into PHA via several pathways (Chen, 2010). In each pathway, VFA will be converted into intermediates. These intermediates will transform into precursors. Precursors, such as 3-hydroxybutyryl-CoA and R-3-hydroxyacyl-CoA, are commonly found in PHA production. In each pathway, enzymes play a significant role. The intermediates formed in each pathway are catalyzed by the enzymes, and thereafter, the enzymes play a role in catalyzing the formation of PHA from the precursor. For the formation of PHA, the key enzyme is the PHA synthase (Sudesh et al., 2000). The produced PHA will be stored within the cell structure of the microorganisms. It is termed as PHA accumulation. PHA accumulated inside the aerobic granules cells is shown in Figure 12.2. In Figure 12.2, the bright and refractive granules observed under the scanning transmission electron microscopy (STEM) are the images of the PHA.

PHA ACCUMULATING MICROORGANISMS

The conversion of VFA into PHA by the microorganisms is catalyzed by the enzymes secreted by the microorganisms. Till date, pure culture microorganisms have been favored for the conversion

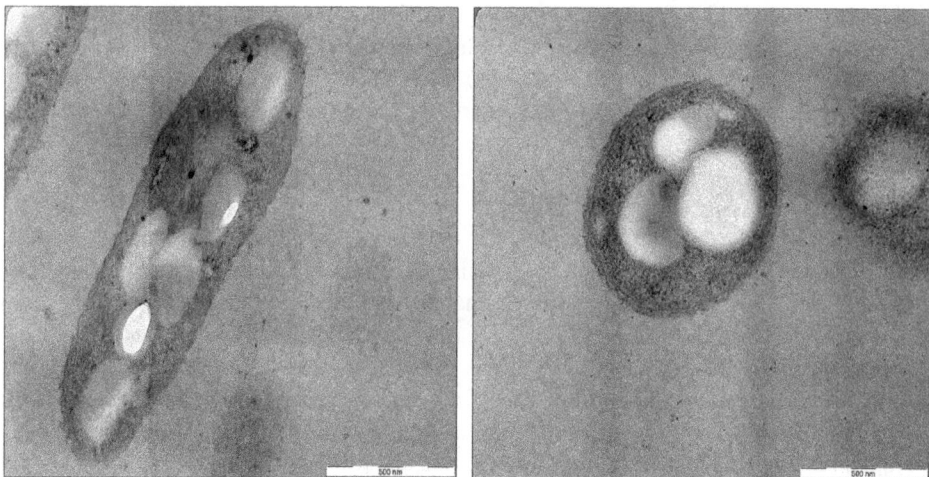

FIGURE 12.2   STEM image of the PHA inside the aerobic granules cells. Reproduced with permission from Gobi (2015).

of VFA into PHA. Hong et al. (2019) have isolated the Vibrio proteolyticus strain from the Korean marine to accumulate PHA. The accumulation of PHA using this isolated strain was 54.8% of the cell dry weight. Fructose and yeast extract were used as the medium for the Vibrio proteolyticus strain to accumulate PHA. Meanwhile, a novel strain of Bacillus sp. YHY22 was used by Lee et al. (2022) to accumulate PHB using lactate as the feedstock. Up to 64.7% of the cell dry weight was occupied by the PHB. On the other hand, soybean oil and corn starch were used as the feedstock to produce PHB and addition of propionate produced the P(HB-co-HV). Photobacterium sp. TLY01, a novel strain of Photobacterium genus that was isolated from an oil field, was used for this purpose. 16.28 g/L of the PHBV was produced using the Photobacterium sp. TLY01 strain (Tian et al., 2022). In addition to that, P(HB-co-HV) was found to be produced from starchy water using *Cupriavidus Necator* DSM 545. About 0.95 g/L of PHA was produced with this strain (Brojanigo et al., 2022).

Though the pure culture method is favored, researchers have explored the idea of producing PHA using mixed culture (Amanat et al., 2022; Li et al., 2022; Silva et al., 2022). Mixed culture is a co-existence of various microorganisms (PHA and non-PHA producing microorganism) in a reactor. Mixed culture has the edge over pure culture due to several reasons. Relative to pure culture, mixed culture is more robust and less labor-intensive (Silva et al., 2022). Pure culture requires sterile environment, and it results in relatively high operating cost (Silva et al., 2022). On the other hand, the mixed culture does not require such facilities. However, the common drawback of using the mixed culture is the lower yield of PHA in relative to the pure culture. Thus, the mixed culture has been subjected to selection process whereby the PHA accumulating microorganisms will be enriched in the reactor (Inoue et al., 2021). As such, the yield of PHA could be improved to match up with the pure culture strains.

PHA APPLICATIONS

PHA is biocompatible, non-carcinogenic, biodegradable, and has good mechanical properties. Applications of PHA have been gaining diversification over the past 10 years. One of the primary usages of the PHA has been in the medical field (Lingle, 2018). Majorly, they are used as sutures, artificial heart valves, blood vessels, tissue engineering scaffolds, and bone (Dhania et al., 2022; Gregory et al., 2022; Guo et al., 2022). Insertion of the PHA into the human bodies supports the healing process, and it will slowly degrade along the healing process. The nature of the PHA does not trigger immune response upon implantation. PHA has promoted the cell proliferation and supported the tissue regrowth. It will naturally degrade and secreted out as carbon dioxide. In some circumstances, the PHA will be used as the energy substrate also by the human cells. It is also safe to be used as there is no evidence has been reported on the cancer cell growth from the cell proliferation due to the PHA insertion. Conventionally, a support system would be inserted into the body for the healing process, and later, it will be removed. With the usage of PHA, this process could be circumvented.

PHA has also been used in the packaging field. Food packaging requires the material to be high heat withstanding. Monomers such as PHBV have high operating temperature, stiffness, and toughness, which makes them suitable for the food container. PHB can be used as the food container for food which undergoes sterilization process. The properties of the PHB remain intact upon the gamma radiation (sterilization process). P(HB-*co*-HV) has also been reported to be used for the packaging of milk, vegetable oil, and sour cream.

The feedstock for the 3D printing has been extensively based on PHA. PHBV found to be flexible and stiff at the same time. It was found to be suitable to be used for the 3D printing as it could be shaped as needed.

Meanwhile, medium chain length (mcl-PHA) such as P(HHx) and P(HO) have exhibited low melting temperatures and low crystallinity. As such, the mcl-PHAs are useful in applications with no requirement of high mechanical strengths such as coatings, adhesives, biocarriers, and medical devices (Muthuraj et al., 2021). At the same time, the mcl-PHA also biodegrades quicker compared with the scl-PHA due to the low crystallinity value.

On the other hand, the co-polymers consist of scl-PHA and mcl-PHA exhibited better properties in terms of flexibility, crystallinity, and processability compared with PHB. Co-polymers such as P3(HB-co-HHx) have been extensively used in food packaging, tissue engineering, and consumables. P3(HB-co-HHx) has shown biocompatibility and also quick biodegradability, which increases its applications in medical field in recent years (Muthuraj et al., 2021).

LATEST TRENDS IN PRODUCING PHA FROM POME

In recent years, various developments have been noticed on the PHA production using POME. Aerobic granules, which are a mixed consortia of microorganisms, have been used to produce the PHA. Up to 89% of the dry weight of the aerobic granules was occupied by the PHA (Gobi & Vadivelu, 2015). The biggest advantage of using aerobic granules is the minimal footprint occupation for the PHA production. Moreover, the amount of PHA produced is comparable with the pure culture strains. Naturally, during the formation of aerobic granules, it has undergone the selection process (Gobi & Vadivelu, 2014). During this process, the PHA accumulating microorganisms will naturally proliferate and dominate in the reactor. As such, the aerobic granules were able to accumulate PHA within its cell structure.

Activated sludge also has been used to produce PHA from POME. Bio-PORec® system found to produce 80% yield of PHA using the activated sludge (Din et al., 2012). The high accumulation of PHA was facilitated by the short feeding rate, limited substrate, and nutrient deficient. Pure culture strain also being used to produce PHA from POME. Isolates from the soil were used to accumulate PHA, and gram-positive heterotrophs (part of the isolates) were found to produce the largest amount of PHA using POME (Paulraj et al., 2019).

CONCLUSIONS

In a nutshell, the production of PHA from POME will create circular economy in which zero waste would be generated. The enormous amount of POME generated shall be benefited by converting it into biodegradable plastic. The demand for the biodegradable plastic to be produced from economical resources has been catalyzed by the diverse application of PHA in recent years. Malaysia, as one of the largest producers of palm oil, has the potential to be the leader in PHA production from waste as well. Overall, production of PHA from POME would simultaneously resolve plastic pollution and water pollution due to POME.

ACKNOWLEDGMENT

UTAR is gratefully acknowledged for funding this research.

REFERENCES

Abdullah, M.F., Md Jahim, J., Abdul, P.M., Mahmod, S.S. (2020). Effect of carbon/nitrogen ratio and ferric ion on the production of biohydrogen from palm oil mill effluent (POME). *Biocatalysis and Agricultural Biotechnology*, 23, 101445.

Amanat, N., Matturro, B., Villano, M., Lorini, L., Rossi, M.M., Zeppilli, M., Rossetti, S., Petrangeli Papini, M. (2022). Enhancing the biological reductive dechlorination of trichloroethylene with PHA from mixed microbial cultures (MMC). *Journal of Environmental Chemical Engineering*, 10(2), 107047.

Atasoy, M., Cetecioglu, Z. (2022). The effects of pH on the production of volatile fatty acids and microbial dynamics in long-term reactor operation. *Journal of Environmental Management*, 319, 115700.

Bala, J.D., Lalung, J., Al-Gheethi, A.A.S., Hossain, K., Ismail, N. (2018). Microbiota of palm oil mill wastewater in Malaysia. *Tropical Life Sciences Research*, 29(2), 131–163.

Bello, M.M., Abdul Raman, A.A. (2017). Trend and current practices of palm oil mill effluent polishing: Application of advanced oxidation processes and their future perspectives. *Journal of Environmental Management*, 198, 170–182.

Bioplastic, G.D. (2014). What growth in bioplastics industry means for investors and the economy. https://www.greendotbioplastics.com/growth-bioplastics-industry-means-investors-economy/

Brojanigo, S., Alvarado-Morales, M., Basaglia, M., Casella, S., Favaro, L., Angelidaki, I. (2022). Innovative co-production of polyhydroxyalkanoates and methane from broken rice. *Science of the Total Environment*, 825, 153931.

Chen, G. (2010). *Plastics Completely Synthesized by Bacteria: Polyhydroxyalkanoates*, Berlin, Heidelberg: Springer-Verlag.

Chinwetkitvanich, S., Jaikawna, H. (2015). Volatile Fatty Acids (VFAs) Production from Palm Oil Mill Effluent (POME) Fermentation. In David, C. (ed) *Environmental Science and Information Application Technology*, CRC Press, London, pp. 151–155.

Department of Environment. (2010). *Environmental Requirements: A Guide for Investors*. Department of Environment, Ministry of Natural Resources and Environment, Putrajaya, pp. 1–82.

Dhania, S., Bernela, M., Rani, R., Parsad, M., Grewal, S., Kumari, S., Thakur, R. (2022). Scaffolds the backbone of tissue engineering: Advancements in use of polyhydroxyalkanoates (PHA). *International Journal of Biological Macromolecules*, 208, 243–259.

Din, M.M.F., Mohanadoss, P., Ujang, Z., van Loosdrecht, M., Yunus, S.M., Chelliapan, S., Zambare, V., Olsson, G. (2012). Development of Bio-PORec® system for polyhydroxyalkanoates (PHA) production and its storage in mixed cultures of palm oil mill effluent (POME). *Bioresource Technology*, 124(0), 208–216.

Ferries, C. (2022). *Plastic in the Ocean Statistics 2020–2021*. https://www.condorferries.co.uk/plastic-in-the-ocean-statistics. Accessed on April 20, 2022

Gatehouse, J. (2019). *Will there be more Plastic than Fish in the Ocean by 2050?* CBC News. https://www.cbc.ca/news/politics/ocean-plastic-liberals-fact-check-1.5212632.

Gobi, K. (2015). Accumulation and extraction of polyhydroxyalkanoate from aerobic granules treating palm oil mill effluent. Doctoral Thesis, Universiti Sains Malaysia, Penang Malaysia.

Gobi, K., Vadivelu, V.M. (2014). Aerobic dynamic feeding as a strategy for in situ accumulation of polyhydroxyalkanoate in aerobic granules. *Bioresource Technology*, 161, 441–445.

Gobi, K., Vadivelu, V.M. (2015). Dynamics of polyhydroxyalkanoate accumulation in aerobic granules during the growth–disintegration cycle. *Bioresource Technology*, 196, 731–735.

Gregory, D.A., Taylor, C.S., Fricker, A.T.R., Asare, E., Tetali, S.S.V., Haycock, J.W., Roy, I. (2022). Polyhydroxyalkanoates and their advances for biomedical applications. *Trends in Molecular Medicine*, 28(4), 331–342.

Guo, W., Yang, K., Qin, X., Luo, R., Wang, H., Huang, R. (2022). Polyhydroxyalkanoates in tissue repair and regeneration. *Engineered Regeneration*, 3(1), 24–40.

Hong, J.-W., Song, H.-S., Moon, Y.-M., Hong, Y.-G., Bhatia, S.K., Jung, H.-R., Choi, T.-R., Yang, S.-y., Park, H.-Y., Choi, Y.-K., Yang, Y.-H. (2019). Polyhydroxybutyrate production in halophilic marine bacteria Vibrio proteolyticus isolated from the Korean peninsula. *Bioprocess and Biosystems Engineering*, 42(4), 603–610.

Inoue, D., Fukuyama, A., Ren, Y., Ike, M. (2021). Optimization of aerobic dynamic discharge process for very rapid enrichment of polyhydroxyalkanoates-accumulating bacteria from activated sludge. *Bioresource Technology*, 336, 125314.

Kadier, A., Wang, J., Chandrasekhar, K., Abdeshahian, P., Islam, M.A., Ghanbari, F., Bajpai, M., Katoch, S.S., Bhagawati, P.B., Li, H., Kalil, M.S., Hamid, A.A., Abu Hasan, H., Ma, P.-C. (2022). Performance optimization of microbial electrolysis cell (MEC) for palm oil mill effluent (POME) wastewater treatment and sustainable Bio-H2 production using response surface methodology (RSM). *International Journal of Hydrogen Energy*, 47(34), 15464–15479.

Kadir, A.P.H.G. (2022). *Overview of the Malaysian Oil Palm Industry 2021*. Malaysia Palm Oil Board (MPOB), Selangor.

Lee, H.-J., Kim, S.-G., Cho, D.-H., Bhatia, S.K., Gurav, R., Yang, S.-Y., Yang, J., Jeon, J.-M., Yoon, J.-J., Choi, K.-Y., Yang, Y.-H. (2022). Finding of novel lactate utilizing Bacillus sp. YHY22 and its evaluation for polyhydroxybutyrate (PHB) production. *International Journal of Biological Macromolecules*, 201, 653–661.

Li, D., Yan, X., Li, Y., Ma, X., Li, J. (2022). Achieving polyhydroxyalkanoate production from rubber wood waste using mixed microbial cultures and anaerobic–aerobic feeding regime. *International Journal of Biological Macromolecules*, 199, 162–171.

Lim, X. (2021). Microplastics are everywhere — but are they harmful? *Nature*, 593, 22–25.

Lingle, R. (2018). *PHA Bioplastics a 'Tunable' Solution for Convenience Food Packaging*. Plastics Today. https://www.plasticstoday.com/industry-trends/us-plastics-processing-equipment-shipments-post-slight-year-over-year-increase. Accessed on May 13, 2022.

Lok, X., Chan, Y.J., Foo, D.C.Y. (2020). Simulation and optimisation of full-scale palm oil mill effluent (POME) treatment plant with biogas production. *Journal of Water Process Engineering*, 38, 101558.

Mahmod, S.S., Arisht, S.N., Jahim, J.M., Takriff, M.S., Tan, J.P., Luthfi, A.A.I., Abdul, P.M. (2021). Enhancement of biohydrogen production from palm oil mill effluent (POME): A review. *International Journal of Hydrogen Energy*, 47(960), 40637–40655.

Maleki, F., Torkaman, R., Torab-Mostaedi, M., Asadollahzadeh, M. (2022). Optimization of grafted fibrous polymer preparation procedure as a new solid basic catalyst for biodiesel fuel production from palm oil. *Fuel*, 329, 125015.

Mobilian, C., Craft, C.B. (2022). Wetland soils: Physical and chemical properties and biogeochemical processes. In: *Encyclopedia of Inland Waters* (Second Edition), (Eds.) Mehner, T. & Tockner, K. Oxford: Elsevier, pp. 157–168.

Mohamad, Z., Razak, A.A., Krishnan, S., Singh, L., Zularisam, A.W., Nasrullah, M. (2022). Treatment of palm oil mill effluent using electrocoagulation powered by direct photovoltaic solar system. *Chemical Engineering Research and Design*, 177, 578–582.

Mohammad, S., Baidurah, S., Kobayashi, T., Ismail, N., Leh, C.P. (2021). Palm oil mill effluent treatment processes—A review. *Processes*, 9(5), 739.

MPOB. (2022). *Export of Palm Oil by Destination: 2022 (Tonnes)*. Malaysian Palm Oil Board (MPOB), Selangor.

Muthuraj, R., Valerio, O., Mekonnen, T.H. (2021). Recent developments in short- and medium-chain-length Polyhydroxyalkanoates: Production, properties, and applications. *International Journal of Biological Macromolecules*, 187, 422–440.

Nasrullah, M., Singh, L., Mohamad, Z., Norsita, S., Krishnan, S., Wahida, N., Zularisam, A.W. (2017). Treatment of palm oil mill effluent by electrocoagulation with presence of hydrogen peroxide as oxidizing agent and polialuminum chloride as coagulant-aid. *Water Resources and Industry*, 17, 7–10.

Oettli, P., Behera, S.K., Yamagata, T. (2018). Climate based predictability of oil palm tree yield in Malaysia. *Scientific Reports*, 8(1), 2271.

Paulraj, P., Shukri, H.A., Nagiah, V., Suryadevara, N., Ganapathy, B. (2019). Use of bacteriostatic antibiotics for the optimization of new media in view of supplementing POME as an alternative carbon source for PHA production: A statistical approach. *Materials Today: Proceedings*, 16, 1692–1701.

Pirouzfar, V., Sakhaeinia, H., Su, C.-H. (2022). Power generation using produced biodiesel from palm oil with GTG, STG and combined cycles; process simulation with economic consideration. *Fuel*, 314, 123084.

Rawindran, H., Leong, W.H., Suparmaniam, U., Liew, C.S., Raksasat, R., Kiatkittipong, W., Mohamad, M., Ghani, N.A., Abdelfattah, E.A., Lam, M.K., Lim, J.W. (2022). Residual palm kernel expeller as the support material and alimentation provider in enhancing attached microalgal growth for quality biodiesel production. *Journal of Environmental Management*, 316, 115225.

Rosa, D., Medeiros, A.B.P., Martinez-Burgos, W.J., do Nascimento, J.R., de Carvalho, J.C., Sydney, E.B., Soccol, C.R. (2020). Biological hydrogen production from palm oil mill effluent (POME) by anaerobic consortia and Clostridium beijerinckii. *Journal of Biotechnology*, 323, 17–23.

Shafwah, O.M., Suhendar, D., Hudiyono, S. (2019). Pretreatment of palm oil mill effluent (POME) using Lipase and Xylanase to improve biogas production. *Advances in Biological Sciences Research* (*10th International Seminar and 12th Congress of Indonesian Society for Microbiology (ISISM 2019)*), 15, 86–190.

Silva, F., Matos, M., Pereira, B., Ralo, C., Pequito, D., Marques, N., Carvalho, G., Reis, M.A.M. (2022). An integrated process for mixed culture production of 3-hydroxyhexanoate-rich polyhydroxyalkanoates from fruit waste. *Chemical Engineering Journal*, 427, 131908.

Sudesh, K., Abe, H., Doi, Y. (2000). Synthesis, structure and properties of polyhydroxyalkanoates: Biological polyesters. *Progress in Polymer Science*, 25(10), 1503–1555.

Tian, L., Li, H., Song, X., Ma, L., Li, Z.-J. (2022). Production of polyhydroxyalkanoates by a novel strain of Photobacterium using soybean oil and corn starch. *Journal of Environmental Chemical Engineering*, 10(5), 108342.

13 Recent Developments in Pre-Treatment and Nanocellulose Production From Lignocellulosic Materials

Asniza Mustapha and Mohd Fahmi Awalludin
Forest Research Institute Malaysia

Wan Noor Aidawati Wan Nadhari
Universiti Kuala Lumpur

Nur Izzaati Saharudin
Universiti Sains Malaysia

CONTENTS

DOI: 10.1201/9781003358084-13

INTRODUCTION

Environmental awareness has risen due to environmental pollution, the depletion of natural resources, fluctuating fossil fuel prices and a growing population. It leads to great interest in searching sustainable and renewable materials as an alternative. This issue can be solved by promoting biomass materials, such as agricultural sources, as substitute sources for various products and applications. Agricultural biomass, also known as lignocellulosic, is the plant residue left in the plantation field after harvesting. This renewable plant waste is abundant, biodegradable, low cost and known as the largest cellulose resource in the world. Annually, 181.5 billion tons of lignocellulosic biomass are generated. 8.2 billion tons are used, with 42% and 43% of the biomass coming from grasslands, dedicated forests and agricultural land, respectively, while the rest is agricultural waste (Dahmen et al. 2019).

The uses of these materials are not only limited to composite, pulp and paper industry but are also expending to other applications such as medical, nanotechnology, electronic, biofuel and pharmaceutical (Sharma & Kuila 2017). For example, oil palm empty fruit bunch and sugarcane bagasse have been the most researched waste-based sources for nanocellulose production. Nanocellulose from abundant sources has received attention due to its nature in nanoscale dimension, unique morphology, large and flexible surface area, high mechanical strength and low density (Kargarzadeh et al. 2017).

In nanotechnology applications, nanofibrillated cellulose (NFC) and nanocrystalline cellulose (NCC) are two general types of nanocelluloses with a diameter of 1–100 nm. NFC and NCC can be distinguished by the structure and isolation methods (Abitbol et al 2016; Klemm et al 2018). Numerous methods were established for nanocellulose extraction, and each method produced different types of nanocellulose. Basically, the isolation process is done mechanically, chemically and a combination of the mechanical and chemical methods, which is known as chemo-mechanical process. Researchers have combined mechanical processes such as refining and cryocrushing with pre-treatment of cellulosic fiber by means of physical, chemical or enzymatic hydrolysis.

The main challenges are related to efficient separation of nanocellulose from natural resources. In addition to increasing the fibrillation rate and yield, researchers are also progressing immensely to improve the incompatible nature of nanocellulose with most of the polymers that restrict its application. Hydrophilic nature of nanocellulose limits good dispersion of these materials in hydrophobic polymers and therefore affects the desired properties. Appropriate isolation techniques will lead to promoting the excellent properties of nanocellulose. This chapter will concentrate on a general overview of nanocellulose from various lignocellulosic sources, recent developments in pretreatment and nanocellulose production where researchers can gain insight into the various methods available for nanocellulose isolation.

LIGNOCELLULOSIC MATERIAL

Lignocellulosic material is the most plentiful material on earth and is known as a renewable energy source. Sources of lignocellulosic materials such as agricultural residues (e.g. wheat straw, sugarcane bagasse, corn stover), forest products (e.g. softwood, hardwood), forestry residues (e.g. sawdust, mill wastes), dedicated crops (salix, switchgrass), municipal solid waste portions (e.g. used paper) and various industrial wastes. These raw materials have great likely to be used in the industrial processes since they are sufficiently available and generate very low greenhouse gas emissions.

Lignocellulose is the main building block of plant cell walls. The plant is composed primarily of cellulose, hemicellulose, lignin and minor portion of pectin, proteins, extractives (e.g. soluble non-structural materials, i.e. non-structural sugars, chlorophyll, nitrogenous material, waxes) and ash. Cellulose linear chains in plants are responsible for tensile strength, while hydrophobic amorphous lignin contributes in chemical resistance and protection against water. Hemicellulose on the other hand provides bonding between cellulose and lignin.

The percentage of these elements varies from one plant species to another. For example, hardwood has a higher amount of cellulose, while leaves and wheat straw contain more hemicelluloses.

In addition, the composition of these elements also varies by plant, plant age and growth stage (Kumar et al. 2009). This chemical component affects the hydrolysis of cellulose. Basically, there are two main obstacles that prevent hydrolysis of cellulose in lignocellulosic materials. Known as crystalline cellulose recalcitrance, they arise from the linear structure of cellulose chains tightly bound into microfibrils and the highly protective lignin surrounding the microfibrils that acts as a physical barrier to hydrolysis.

CLASSIFICATION OF LIGNOCELLULOSIC RAW MATERIALS

Lignocellulosic, also known as agricultural biomass, is produced in billions of tons every year. Various types of agricultural biomass have the potential to be used as raw materials in various applications, especially for the development of environmental renewable materials. For instance, oil palm empty fruit bunch, oil palm trunks, bamboo, bagasse, coconut coir, kenaf and pineapple leaves. This biomass is mostly in the form of plant stem residues, roots, leaves, seeds, seed skins, etc. They are classified into large groups according to the part of the plant from which they are extracted. For example, bast (stem), leaves, fruit (seeds) and husks, as illustrated in Figure 13.1. Its properties vary according to the type of plant and the part of the plant itself. The researchers concluded that the general properties of agricultural biomass vary by location, geography, plant age, climate and soil conditions (Hakeem et al. 2015).

NANOCELLULOSE

Nanocellulose is a term that refers to cellulosic materials that possess at least one dimension in the nanometer range or nanometer-size cellulose. It can be produced by several methods from various sources of lignocellulosic including plants, agricultural biomass, crops and forest residues, algae, tunicates and some bacteria. In recent years, great interest has been directed to nanocellulose. They show low thermal expansion, high aspect ratio, high reinforcement effect and flexibility, good mechanical and optical properties that make nanocellulose used in many applications such as nanocomposites, pulp and paper industry, food packaging, coating additives and gas barriers (Bacakova et al. 2019; Lee 2018; Wei et al. 2014).

CLASSIFICATION OF NANOCELLULOSE

Nanocellulose is grouped based on preparation/isolation methods, sources and dimensions/average size. The various nomenclature of nanocellulose is summarized in Table 13.1.

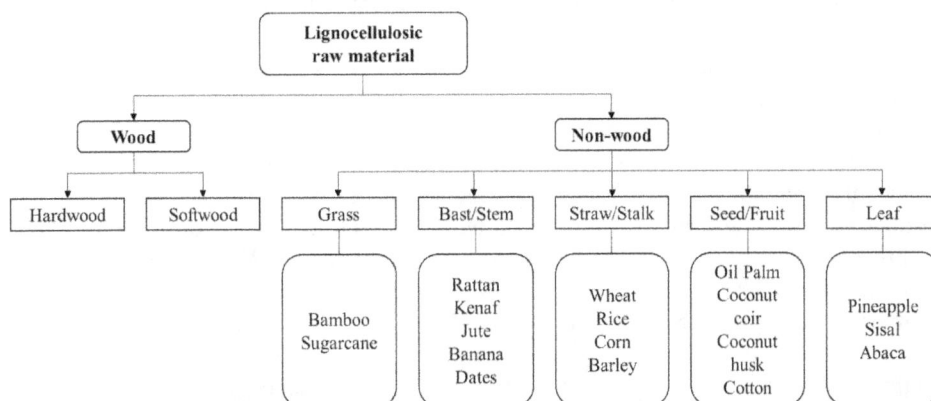

FIGURE 13.1 Classification of lignocellulosic raw materials.

TABLE 13.1

Classification and Types of Nanocellulose

Nanocellulose/Synonyms	Typical Sources	Isolation Process	Average Size
Nanofibrillated cellulose (NFC), cellulose nanofibers (CNF), nanofibrils, microfibrils, microfibrillated cellulose (MFC)	Wood, sugar beet, potato tuber, hemp, flax, etc.	Mechanical shearing, homogenization, enzymatic hydrolysis	Diameter: 5–60 nm Length: several μm
Cellulose nanocrystal (CNC), nanocrystalline cellulose (NCC), nanowhiskers, rod-like cellulose, microcrystals	Wood, cotton, hemp, flax, wheat straw, rice straw, mulberry bark, ramie, avicel, tunicin, algae, bacteria, etc.	Strong acid hydrolysis, sonication	Diameter: 5–70 nm Length: 100–250 nm (plant); 100 nm - several μm (tunicates, algae, bacteria)

NANOCRYSTALLINE CELLULOSE (NCC)

NCC is called cellulose nanocrystals or microcrystals depending on the size and dimensions. In the literature, NCCs are also known as nanowhiskers, rod-shaped cellulose crystals and nanorods. NCC is generally isolated from cellulose fibers by an acid hydrolysis process. They have relatively low aspect ratio and typically 2–20 nm in diameter with lengths ranging from 100 nm to several micrometers. NCC is 100% cellulose and highly crystalline with 54%–88% crystalline zones. The level of crystallinity, morphology and dimensionality are affected by the source of the cellulosic material, the preparation conditions and the technique used.

NCC does not have gel-like properties like NFC due to its rod shape, lower aspect ratio and super crystalline structure. In addition to having an elongated rod-like crystal shape, NCC also exhibits very limited flexibility compared to NFC due to the absence of amorphous regions. The less elongated form of NCC facilitates its use as a reinforcing agent in polymer composite materials. Crystals can also form a chiral nematic liquid crystalline phase in concentrated solution. Common applications for NCC cast films are ink pigments and optically variable films for cosmetics, pharmaceutical capsules, security papers and food additives, including stabilizers, fat replacers and texturing agents.

NANOFIBRILLATED CELLULOSE (NFC)

NFC is also recognized as nanofibrillar cellulose, nanofibrillated cellulose, cellulose nanofibers and cellulose nanofibrils. NFC consists of a set of stretched cellulose chain molecules. They are long, flexible interwoven cellulose nanofibers with a diameter of about 1–100 nm in size with a high aspect ratio. Unlike NCC, NFC consists of alternating crystalline and amorphous domains. It is usually produced by delamination of wood pulp through mechanical process before and/or after enzymatic/chemical treatment.

Conventionally, fibrils aggregate with a diameter of 30–100 nm and a length of several micrometers (μm) are called microfibrillated cellulose (MFC). Therefore, MFC can also be considered NFC, since the definition of NFC is any aggregate of fibrils with a size less than 100 nm in one dimension. However, the term microfibril is often misleading because it does not represent the actual dimensions of nanometer-sized fibrils. At present, the terms CNF or NFC are widely used to replace classical microfibrils or MFCs, because the nanoscale properties of individual fibrils are well defined. Furthermore, conventional NFC and MFC are not only distinguished by size. It also involves a pre-treatment process, where MFC is referred to as NFC when pre-treatment is used during the fibrillation process, because it produces homogenous nanoscale individual fibrils. NFC leads to intensive and promising research by expanding the area of potential applications, including nanocomposite materials, additives for paper and board, biomedical applications and as adsorbents.

ISOLATION OF NANOCELLULOSE

Nanocellulose can be isolated from cellulosic materials using several methods, including chemical treatment, mechanical treatment, enzymatic treatment and chemo-mechanical treatment. Different methods lead to different properties and dimensions of nanocellulose, depending on the cellulosic raw material and pre-treatment used. Cellulose pre-treatment or combination of two or more methods is usually used to obtain desired properties of nanocellulose, efficient fibrillation, high yield and minimal power consumption. Figure 13.2 illustrates an example of the isolation method of NCC from bamboo processing residue via a chemo-mechanical process.

Pre-Treatment of Lignocellulosic Raw Materials

The aim of the pre-treatment stage is to break down and remove non-cellulosic materials, which are lignin, hemicellulose and waxy materials from the lignocellulosic feedstock (Figure 13.3). Cellulose crystallinity, hemicellulose coating, lignin protection and accessible surface area all contribute to cellulose's resistance to hydrolysis. In this case, lignocellulosic needs to undergo appropriate treatment to modify its structure and make cellulose more accessible. Pre-treatment changes the physical and chemical structure of the lignocellulosic and increases the hydrolysis rates.

Pre-Treatment of Cellulosic Material

Pre-treatment methods can be classified as physical treatment (milling, grinding), physico-chemical treatment (steam/auto-hydrolysis, wet oxidation, hydrothermolysis), chemical treatment (liquefied acids, alkalis, organic solvents, oxidizing agents), ionic liquids, biological treatment (enzymatic hydrolysis), electrical treatment or a combination of any of these (Table 13.2). Pre-treatment processes that are effective for certain lignocellulosic may not be suitable for other materials. The

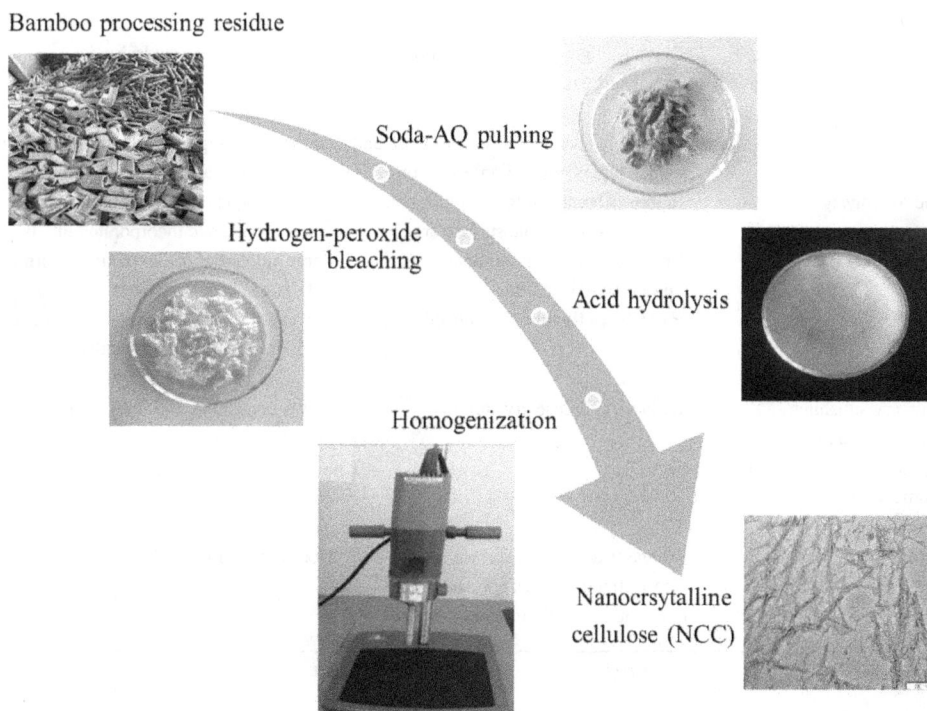

FIGURE 13.2   Isolation of NCC from bamboo processing residue.

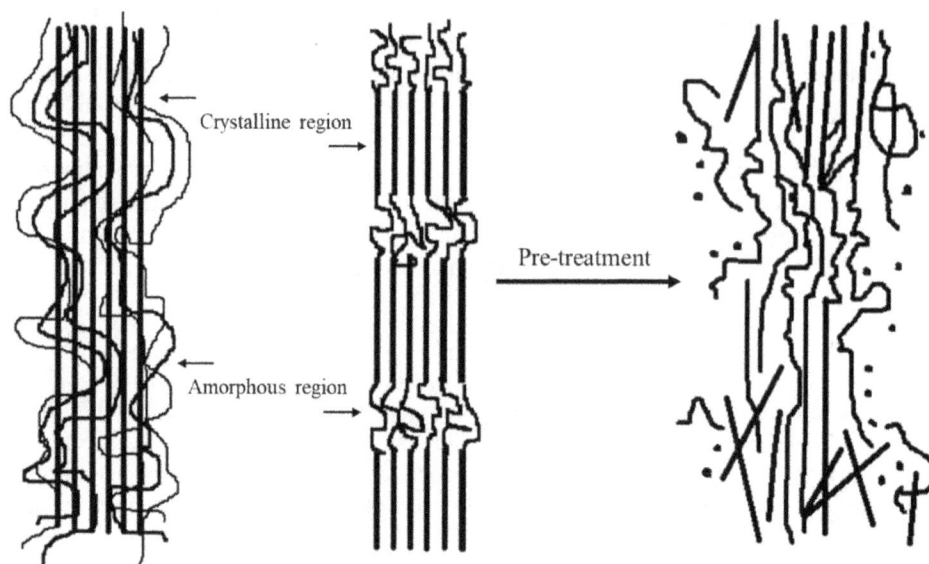

FIGURE 13.3 Pre-treatment of cellulosic material.

TABLE 13.2
Various Pre-Treatment Process of Lignocellulosic Raw Material

Pre-Treatment	Advantages	Limitations
Biological		
Enzyme-assisted hydrolysis	Degrades lignin and hemicelluloses; low energy requirements	Very low in rate of hydrolysis
Chemical		
Acid hydrolysis	Hydrolyzes hemicellulose to xylose and other sugars; alters lignin structure.	High cost; equipment corrosion; formation of toxic substances
Alkaline hydrolysis	Removes hemicelluloses and lignin; increases accessible surface area.	Long residence time required; irrecoverable salts formed and incorporated into biomass.
Ozonolysis	Reduces lignin content; does not produce toxic residues	Large amount of ozone required, formation of toxic substances.
Organosolv	Hydrolyzes lignin and hemicelluloses.	Solvents need to be drained from the reactor, evaporated, condensed, high cost.
Mechanical		
Grinding, cryocrushing, microfluidization, high-pressure homogenization, high-intensity ultrasonication	Reduces cellulose crystallinity	Power consumption usually higher than inherent biomass energy
Physicochemical		
CO_2 explosion	Increases accessible surface area; cost-effective; does not cause formation of inhibitory compounds	Does not modify lignin or hemicelluloses

choice of pre-treatment used depends on the composition, structure of the substrate, hydrolysis agent and by-products produced from the pre-treatment process. These features are important due to the conditions used in the chosen pre-treatment method influence the characteristics of the substrate and control the susceptibility of the substrate to hydrolysis. Hence, each pre-treatment process must be properly selected and critically justified.

CHEMICAL PRE-TREATMENT

Chemical pre-treatment of lignocellulosic has traditionally been used in the pulp and paper industry for pulp delignification. The main advantages of this pre-treatment method are high efficiency and minimal inhibitor formation. On the other hand, the need for extensive post-washing, the need for specialized corrosion-resistant equipment and the disposal of chemical waste are the main disadvantages of this process.

Alkaline Pre-Treatment

Alkaline pre-treatment is used to dissolve lignin, hemicellulose and pectin of cellulose fibers. This pre-treatment method is performed at ambient conditions, at a lower temperature and pressure than other pre-treatment. This process takes place over several hours or days. This process includes pulping and bleaching stages, which promote fiber to swelling and facilitate the defibrillation process. Less energy is required to break the macrofibrils into microfibrils because the swelling of the fibers increases the surface area and reduces the hydrogen bonds in the fibers. Basically, the goal of pulping is to remove lignin from cellulosic fibers. Also known as delignification, where lignin is broken down in solution and washed away with water.

Sodium, potassium, calcium and ammonium hydroxides are suitable for use as alkaline pre-treatment agents. However, of these four, sodium hydroxide (NaOH) has been commonly used. Pre-treatment with liquid NaOH results in the rearrangement of microfibrils due to inhomogeneous swelling, which leads to an increase in internal surface area, a decrease in the degree of polymerization, an increase in the crystallinity index, the separation of structural bonds between lignin and carbohydrates and disruption of lignin structure (Kumar et al. 2009). The degree of crystallinity gradually increases with delignification due to the removal of amorphous components such as lignin and hemicellulose.

However, the low yield percentage is the biggest challenge in this process. It is due to the amount of carbohydrates that will be lost in the pre-treatment process, because a complete and efficient removal of non-cellulosic materials especially lignin is possible without damaging the carbohydrates. Soda pulping is widely used as a pulping method, which uses NaOH as a cooking chemical (Wan Rosli et al. 2003). Bleaching, on the other hand, is applied to remove excess impurities while increasing the yellowness and brightness of cellulosic pulp fibers, depending on the end use of the pulp. There are various stages of bleaching process, with different selection of chemicals such as hydrogen peroxide, chlorine dioxide, ozone or peracetic acid (Ek et al. 2009).

Furthermore, alkaline pre-treatment has been shown to play an important role in exposing cellulose to enzymatic hydrolysis. The rate of enzymatic hydrolysis has been reported to be significantly affected by enzyme adsorption and the efficiency of adsorbed enzymes. Removal of lignin increases enzyme efficiency by eliminating waxy and non-cellulosic areas along with increasing enzyme accessibility to cellulose and hemicellulose. Chang and Holtzapple (2000) reported correlations between lignin content, crystallinity and acetyl content of lignocellulosic materials towards enzymatic digestibility. Authors claimed that (1) extensive delignification is adequate to obtain high digestibility of enzymatic hydrolysis regardless of crystallinity and acetyl content; (2) parallel barriers to enzymatic hydrolysis are removed through delignification and deacetylation; and (3) the crystallinity index significantly affects the initial hydrolysis rates. It also stated that all the acetyl groups must be removed and the lignin content reduced to around 10% in the pre-treated lignocellulosic to achieve effective pre-treatment. Authors also mentioned that the acetyl group of hemicellulose was removed by alkali, which reduced the steric hindrance of the enzyme and significantly increased carbohydrate digestibility.

BIOLOGICAL PRE-TREATMENT

Biological treatment involves the use of enzymes obtained from various types of rot fungi, such as brown-, white- and soft-rot fungi, to degrade lignin and hemicellulose in lignocellulosic materials (Galbe & Zacchi 2007). The rate of hydrolysis in most biological pre-treatment is very low. However, it is a safe, environmentally friendly method, with low energy requirements and moderate environmental conditions for large-scale removal of lignin from lignocellulosic (Okano et al. 2005). Brown-rot fungi primarily attack cellulose, whereas white- and soft-rot fungi attack both lignin and cellulose. White-rot fungi are the most effective and broadly used for biological pre-treatment of lignocellulosic materials. Degradation of lignin by white-rot fungi occurs through the action of lignin-decomposing enzymes namely laccase and peroxidases (Lee et al. 2007).

Enzyme-Assisted Hydrolysis

In enzyme-assisted pre-treatment, enzymes are used to degrade and/or modify hemicellulose and lignin while maintaining cellulose content. Enzymes help in limited hydrolysis of some elements or selective hydrolysis of certain components in cellulosic fibers (Janardhnan & Sain 2006). Normally, one enzyme cannot break down the fibers because cellulosic fibers contain diverse organic compounds. Enzyme pre-treatment therefore often involves a range of cellulase enzymes as in the Table 13.3.

Many studies have been performed on the NFC production via enzymatic pre-treatment. For example, Paakko et al. (2007) isolated NFC from sulphite-bleached softwood pulp using a combination of mild enzymatic hydrolysis, mechanical shearing and homogenization. Selective and mild hydrolysis using the mono-component endoglucanase enzyme was found to be less aggressive and allow a higher aspect ratio than acid hydrolysis. Combining these processes to crush the fibers results in long, highly bonded, high-aspect-ratio networks of nanoscale cellulose I elements.

In another study, enzyme pre-treatment was reported to have a minimal effect on nanofiber size, but increased the level of solids to pass the high-pressure homogenizer without clogging when enzymes A and B were used at different concentrations. The optimal enzyme concentration for the best fibril size reduction and smooth flow passage through the homogenizer is 1%, producing nanofibers with a diameter of 38–42 nm after three passes (Siddiqui et al. 2011). Janardhnan and Sain (2011) investigated a combination of bio-treatment with fungi isolated from Dutch elm, followed by high shear filtration in bleached kraft pulp for NFC fibrillation. The results show that more than 90% of bio-treated NFCs have a diameter of less than 50 nm, a higher aspect ratio and are more distinct compared to untreated fibers. The internal defibrillation of the fiber is enhanced when the bio-treatment increases the structural disruption in the crystalline region.

PHYSICOCHEMICAL PRE-TREATMENT

Physicochemical pre-treatment is a method of pre-treatment of lignocellulosic materials that involves a combination of physical and chemical processes. The process includes steam or autohydrolysis,

TABLE 13.3
Set of Cellulose Enzymes Commonly Used in Enzymatic Hydrolysis

Type of Cellulase Enzyme	Details
Cellobiohydrolases	Attack greatly the crystalline cellulose
Endoglucanases	Need some disordered structure in cellulose to attack

wet oxidation and hydrothermolysis. For example, the following section will briefly discuss the hydrothermolysis technique.

Hydrothermolysis

Liquid water is also sometimes used as a pre-treatment. This type of pre-treatment is called hydrothermolysis, aqueous or steam cracking, aquasolv and uncatalyzed solvolysis (Kumar et al. 2009). Water pre-treatment is done at high temperature with pressure to keep the water in a liquid state (Mosier et al. 2005). The residence time for this method is approximately ±15 min at 200°C–230°C. About 40%–60% of the total cellulosic material is dissolved throughout the process, i.e. 4%–22% cellulose, 35%–60% lignin and all hemicelluloses are removed. Before hot water pre-treatment, size reduction of the cellulosic material is not necessary because the particles will break during the treatment process.

Effect of liquid hot water parameters on wheat straw pre-treatment, including temperature (170°C and 200°C), solid concentration [5% and 10% (w/v)], residence time (0 and 40 min) and overpressure in the reactor (30 bar), was evaluated by Pérez et al. (2007). The effectiveness of pre-treatment was studied based on the solid composition and liquid fraction obtained after the filtration process and the susceptibility of the solid fraction to enzymatic hydrolysis using commercial cellulase enzymes. The authors reported that temperature and time affected the pre-treatment, where the yield of enzymatic hydrolysis increased with increasing temperature and time.

Various treatment systems have been tried and developed including appropriate treatment time, pressure and temperature to optimize the process. It is important to choose an appropriate treatment system to control the decomposition pathway of the material, because the properties of hydrothermolysis can be controlled by temperature, pressure and treatment time.

MECHANICAL TREATMENT

Common mechanical approaches used in the defibrillation of cellulose fibers in NFC include filtration, exfoliation, homogenization, microfluidization, milling, freeze-grinding, high-intensity ultrasound, steam explosion, electrospinning or a combination of any of these methods. All these mechanical isolation methods involve high energy consumption which causes a drastic decrease in fiber length and yield. The characteristics of the cellulosic fibers and the final particle size determine the power and energy consumption for mechanical grinding of the material. The researchers measured that the energy input for shredding was less than the 30 kWh per ton of lignocellulosic required for particle size reduction in the 3–6 mm range. In addition to inefficient fiber fragmentation, this technique also tends to damage the microfibril structure by reducing both the degree of crystallinity and polymerization of cellulose.

Along with this, the cellulose fiber treatment approach through chemical or enzymatic pre-treatment before the mechanical insulation process was introduced to overcome this problem (Chauhan & Chakrabarti 2012). The combination of chemical and mechanical treatment is called chemo-mechanical treatment process. The pre-treatment process is reported to help reduce energy consumption to a total of 1,000 kWh/ton from 20,000 to 30,000 kWh/ton of cellulose fibers (Siró & Plackett 2010). On the other hand, it is worth noting that the combination of two or more mechanical isolation methods also results in promising NFC properties. Qing et al. (2013) in their study stated that the combination of refining and microfluidization process led to the production of uniform NFC with high fibrillation. Another study by Wang and Cheng (2009) also reported that the combination of high-intensity ultrasonication and high-pressure homogenization produced efficient fibrillation and uniform NFC. Furthermore, cryogrinding and refining are generally used in combination with homogenization to increase the efficiency of the production process.

Homogenization

Homogenization is defined as a process of achieving homogeneity throughout a product by alteration of particle size. The term homogenous is described as owing the same composition, structure or character. Homogenizing is also known as blending, mixing, dispersing, disrupting, stirring, emulsifying, etc. There are three major categories: pressure homogenization, ultrasonic pressure homogenization and mechanical homogenizer (Dhankhar 2014).

Mechanical Homogenization

Mechanical homogenizer can be classified into rotor-stator homogenizer and blade type homogenizer. Fiber disruption by rotor-stator homogenizer involves hydraulic, mechanical shear and cavitation. It is also known as high shear homogenization (HSH) or high shear mechanical dispersion (HSMD), which involves high energy sonic and ultrasonic pressure gradients. Cavitation is generated as a solid object is moving through a flowing liquid that circulate at a high rate of speed. Fibers are drawn up by a rapidly rotating rotor (blade) in a static head or tube (stator), which contain slots or holes (Figure 13.4). Fibers are centrifugally thrown outward in a pump to exit through the slots or holes.

Combination of extreme turbulence, scissor-like mechanical shearing and cavitation that occur within the narrow gap between stator and rotor rapidly reduce the fiber size as the rotor turns at a very high speed. In other words, HSMD is referred as a mechanical homogenization technique involving two forms of mechanical actions. The first is a shearing process, which promotes fiber shortening, and the second is a shear force mechanism, which releases the microfibrils. The maximum efficiency of homogenization process depending on design and size of rotor-stator, rotor tip speed, initial size of sample, viscosity of medium, processing time or flow rate, volume of medium and concentration of sample (Dhankhar 2014; Zhao et al. 2013).

In this process, the cellulosic fibers are subjected to a crushing treatment that tends to form fibers of smaller dimensions and release microfibrils. HSMD can produce NFC in both small amounts (gram-scale) and large amounts (kilogram-scale) depending on the equipment, conditions and methods used. A commonly used HSMD for the gram-scale is the Ultra Turrax high-speed dispersion unit designed for batch operation referring to small amounts of input volume. In contrast, kg-scale equipment is designed for industrial scale. The Ultra Turrax high-speed diffusion unit consists of a diffusion chamber that has a ring generator and a chamber volume. Pulp of very thin consistency should be used to ensure good mixing as viscosity tends to increase during the crushing process.

Zhao et al. (2013) reported that the low thermal stability and high crystallinity of NFC was successfully isolated from dried softwood pulp through a simple and environmentally friendly mechanical refining pre-treatment with a high shear homogenization process. A potential novel nanopaper with high optical transparency proposed for advanced applications has been fabricated. Some researchers have also used mechanical disintegration in the manufacture of nanocomposites, especially on a laboratory scale. It acts as a dispersion mechanism that provides better dispersion between the polymer matrix and the nanofibers (Hedayati et al. 2011; Sehaqui et al. 2012).

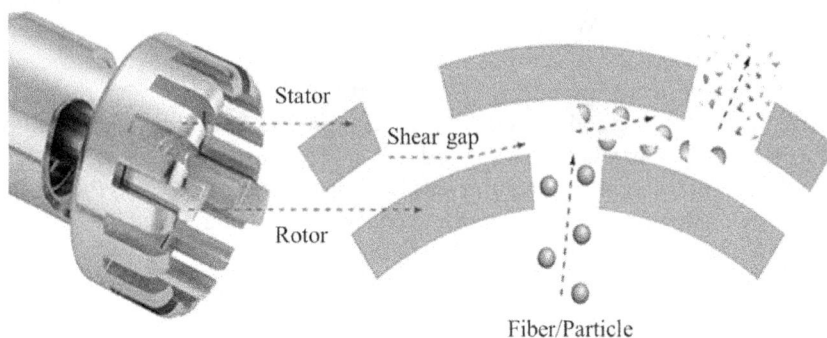

FIGURE 13.4   Fiber fibrillation by Ultra Turrax rotor-stator homogenizer (Sand & Jawaid 2019).

However, clogging problem always occurs and is time-consuming because long fibers often clog the system due to small slots or holes, which need to be disassembled and cleaned. Hence, it is crucial to reduce the size of the fibers before passing through the homogenizer to prevent small grooves from gathering. Therefore, it requires pre-treatment and a combination of additional chemical and mechanical treatments before the homogenization process. Researchers have incorporated several mechanical pre-treatments such as refining (Stelte & Sanadi 2009; Nakagaito & Yano 2004), cryocrushing (Wang & Sain 2007), enzyme (López-Rubio et al. 2007) and chemical pre-treatments (Alemdar & Sain 2008) to reduce the size of fibers before homogenization which helps to lower the energy consumption. Jonoobi et al. (2009, 2011) using a combination of pulping and bleaching processes as chemical pre-treatment while filtration cryocrushing and refining as mechanical pre-treatment before homogenization to produce NFC with a diameter of 10–90 nm from kenaf bast fibers. The authors also used milling pre-treatments on kenaf stems and cores to generate 15–80 nm and 20–25 nm wide NFCs, respectively.

Chronological Events/Related Studies

The development of nano-sized cellulose fibers and their existing use and potential in many applications cause this field to be increasingly studied by researchers. Studies have been started since the 1980s where it involves various aspects such as isolation, modification and application. Table 13.4 illustrates the chronological events related to methods of nanocellulose isolation from various lignocellulosic raw materials and description of products found in literature since year 2010.

TABLE 13.4
Methods of Nanocellulose Isolation from Various Lignocellulosic Raw Materials and Description of Products

Year	Source	Product Type	Isolation Method	References
2010	Kenaf	CNF	High-pressure homogenization	Jonoobi et al. (2010)
2010	Pineapple leaves	CNF	Steam explosion	Cherian et al. (2010)
2011	Pineapple leaves	CNC	Steam explosion, acid hydrolysis	Abraham et al. (2011)
2011	Corn husk, corn cob	MCC	Acid hydrolysis	Jantip and Suwanruji (2011)
2012	Rice straw	CNC	Acid hydrolysis	Lu and Hsieh (2012)
2013	Cotton linter waste	CNC	Acid hydrolysis	Morais et al. (2013)
2013	Rice straw, banana plant	MCC	Alkaline-acid pulping, hypochlorite bleaching	Ibrahim et al. (2013)
2014	Bamboo	MFC	High-pressure enzymatic hydrolysis	Abdul Khalil et al. (2014)
2014	Kenaf	CNF	Alkali pulping, bleaching, mechanical shearing	Karimi et al. (2014)
2015	Bamboo	MFC	High-pressure enzymatic hydrolysis	Aprilia et al. (2015)
2015	Corn stover	CNC	Acid hydrolysis, centrifugation	Costa et al. (2015)
2015	Cotton stalk	CNC	Acid hydrolysis, TEMPO-mediated oxidation, ultrasonication	Soni and Mahmoud (2015)
2016	Corn stover	CNF	TEMPO-mediated oxidation, high-pressure homogenization	Balea et al. (2016)
2016	Cotton stalks	MFC	Masuko grinder	Adel et al. (2016)
2016	Tobacco	CNF	Steam explosion, grinding	Tuzzin et al. (2016)
2017	Wood pulp	CNC	Acid hydrolysis	Jasmani and Adnan (2017)
2017	Oil palm frond leaves	CNC	Alkaline treatment, acid hydrolysis;	Elias et al. (2017)
2018	Rice straw	CNC	Acid hydrolysis	Oun and Rhim (2018)
2018	Sugarcane bagasse	CNC	Steam explosion, alkaline treatment, ultrasonication	Feng et al. (2018)

(Continued)

TABLE 13.4 (*Continued*)
Methods of Nanocellulose Isolation from Various Lignocellulosic Raw Materials and Description of Products

Year	Source	Product Type	Isolation Method	References
2018	Pineapple leaves	CNC	High-shear homogenization, ultrasonication	Mahardika (2018)
2018	Empty fruit bunch	CNC	Acid hydrolysis, ultrasonic homogenization	Hastuti et al. (2018)
2019	Pineapple leaves	MFC	Steam explosion	Tanpichai et al. (2019)
2019	Empty fruit bunch	CNC	Ni(II)-catalyzed hydrolysis, ultrasonication, centrifugation	Yahya et al. (2019)
2019	Pineapple waste	CNC	Acid hydrolysis	Prado and Spinacé (2019)
2020	Oil palm leaves, oil palm frond	CNC	Centrifugation, acid hydrolysis	Hussin et al. (2020); Azani et al. (2020)
2020	Tobacco	CNC	Centrifugation, acid hydrolysis	Shi et al. (2020)
2020	Lime residue, citrus pomace	CNF	Microwave pre-treatment, high shear homogenizer, high-pressure homogenizer	Impoolsup et al. (2020)
2021	Pineapple leaves	CNC	Acid hydrolysis	Chawalitsakunchai et al. (2021)
2021	Oil palm fronds, leaves, coir	CNC	Acid hydrolysis, centrifugation	Mehanny et al. (2021)
2022	Abaca	CNF	TEMPO-mediated oxidation	Lapuz et al. (2022)
2022	Hemp	CNC	Acid hydrolysis, ultrasonication	Barbash et al. (2022)

CONCLUSION

The development of nanocellulose-based materials has fascinated significant interest in recent decades. It is due to the potential uses and unique characteristics, i.e. abundance, renewability, environmental friendliness, high strength and stiffness and low weight. The use of lignocellulosic materials, especially from the agricultural and forestry sectors can help reduce excessive dependence on petrochemical resources and providing a sustainable waste management alternative. Lignocellulosic biomass can be fractionated into separate components with the aid of physical, mechanical, chemical and biological pre-treatment. In this chapter, the cellulose classification and extraction methods of nanocellulose from lignocellulosic materials using different approaches have been highlighted. Different methods lead to different properties of nanocellulose, which is usually depending on the targeted application. The process of extracting and isolating nanocellulose requires more studies and adjustments to improve the properties and expand the applications of this biopolymer. Following this, this chapter also addresses the recent development and provides the chronological events related to nanocellulose isolation from various lignocellulosic raw materials available in the literature since year 2010.

REFERENCES

Abdul Khalil, H.P.S., Hossain, M.S., Rosamah, E., Nik Norulaini, N.A., Peng, L.C., Asniza, M., Davoudpour, Y., & Zaidul, I.S.M. (2014). High-pressure enzymatic hydrolysis to reveal physicochemical and thermal properties of bamboo fiber using a supercritical water fermenter. *BioResources*, *9*(4), 7710–7720.

Abitbol, T., Rivkin, A., Cao, Y., Nevo, Y., Abraham, E., Ben-Shalom, T., Lapidot, S., & Shoseyov, O. (2016). Nanocellulose, a tiny fiber with huge applications. *Current Opinion in Biotechnology*, *39*, 76–88.

Abraham, E., Deepa, B., Pothan, L.A., Jacob, M., Thomas, S., Cvelbar, U., & Anandjiwala, R. (2011). Extraction of nanocellulose fibrils from lignocellulosic fibres: A novel approach. *Carbohydrate Polymers*, *86*(4), 1468–1475.

Adel, A.M., El-Gendy, A.A., Diab, M.A., Abou-Zeid, R.E., El-Zawawy, W.K., & Dufresne, A. (2016). Microfibrillated cellulose from agricultural residues. Part I: Papermaking application. *Industrial Crops and Products*, *93*, 161–174.

Alemdar, A. & Sain, M. (2008). Biocomposites from wheat straw nanofibers: Morphology, thermal and mechanical properties. *Composite Science Technology*, *68*, 557–565.

Aprilia, N.A.S., Hossain, M.S., Asniza, M., Siti Suhaily, S., Nik Noruliani, N.A., Peng, L.C., Mohd Omar, A.K., & Abdul Khalil, H.P.S. (2015). Optimizing the isolation of microfibrillated bamboo in high pressure enzymatic hydrolysis. *BioResources*, *10*(3), 5305–5316.

Azani, N.F.S.M., Haafiz, M.M., Zahari, A., Poinsignon, S., Brosse, N., & Hussin, M.H. (2020). Preparation and characterizations of oil palm fronds cellulose nanocrystal (OPF-CNC) as reinforcing filler in epoxy-Zn rich coating for mild steel corrosion protection. *International Journal of Biological Macromolecules*, *153*, 385–398.

Bacakova, L., Pajorova, J., Bacakova, M., Skogberg, A., Kallio, P., Kolarova, K. & Svorcik, V. (2019). Versatile application of nanocellulose: From industry to skin tissue engineering and wound healing. *Nanomaterials*, *9*(2), 164.

Balea, A., Merayo, N., Fuente, E., Delgado-Aguilar, M., Mutje, P., Blanco, A., & Negro, C. (2016). Valorization of corn stalk by the production of cellulose nanofibers to improve recycled paper properties. *BioResources*, *11*(2), 3416–3431.

Barbash, V.A., Yashchenko, O., Yakymenko, O., Zakharko, R., & Myshak, V. (2022). Preparation of hemp nanocellulose and its use to improve the properties of paper for food packaging. *Cellulose*, *29*, 8305–8317.

Chang, V.S., & Holtzapple, M.T. (2000). Fundamental factors affecting biomass enzymatic reactivity. *Applied Biochemical and Biotechnology*, *84*, 5–37.

Chauhan, V.S., & Chakrabarti, S.K. (2012). Use of nanotechnology for high performance cellulosic and papermaking products. *Cellulose Chemistry and Technology*, *46*(5), 389–400.

Chawalitsakunchai, W., Dittanet, P., Loykulnant, S., Sae-oui, P., Tanpichai, S., Seubsai, A., & Prapainainar, P. (2021). Properties of natural rubber reinforced with nano cellulose from pineapple leaf agricultural waste. *Materials Today Communications*, *28*, 102594.

Cherian, B.M., Leão, A.L., De Souza, S.F., Thomas, S., Pothan, L.A., & Kottaisamy, M. (2010). Isolation of nanocellulose from pineapple leaf fibres by steam explosion. *Carbohydrate Polymers*, *81*(3), 720–725.

Costa, L.A., Assis, D.D.J., Gomes, G.V., da Silva, J.B., Fonsêca, A.F., & Druzian, J.I. (2015). Extraction and characterization of nanocellulose from corn stover. *Materials Today: Proceedings*, *2*(1), 287–294.

Dahmen, N., Lewandowski, I., Zibek, S., & Weidtmann, A. (2019). Integrated lignocellulosic value chains in a growing bioeconomy: Status quo and perspectives. *Gcb Bioenergy*, *11*(1), 107–117.

Dhankhar, P. (2014). Homogenization fundamentals. *IOSR Journal of Engineering*, *4*(5), 2278–8719.

Ek, M., Gellerstedt, G., & Henriksson, G. (2009). *Pulp and Paper Chemistry and Technology, Pulping Chemistry and Technology*. Berlin: Walter de Gruyter GmbH & Co.

Elias, N., Chandren, S., Attan, N., Mahat, N.A., Razak, F.I.A., Jamalis, J., & Wahab, R.A. (2017). Structure and properties of oil palm-based nanocellulose reinforced chitosan nanocomposite for efficient synthesis of butyl butyrate. *Carbohydrate polymers*, *176*, 281–292.

Feng, Y.H., Cheng, T.Y., Yang, W.G., Ma, P.T., He, H.Z., Yin, X.C., & Yu, X.X. (2018). Characteristics and environmentally friendly extraction of cellulose nanofibrils from sugarcane bagasse. *Industrial Crops and Products*, *111*, 285–291.

Galbe, M., & Zacchi, G. (2007). Pretreatment of lignocellulosic materials for efficient bioethanol production. *Advances in Biochemical Engineering/Biotechnology*, *108*, 41–65.

Hakeem, K.R., Jawaid, M., & Alothman, O.Y. (Eds.). (2015). *Agricultural Biomass Based Potential Materials*, Springer, Cham, pp. 1–505.

Hastuti, N., Kanomata, K., & Kitaoka, T. (2018). Hydrochloric acid hydrolysis of pulps from oil palm empty fruit bunches to produce cellulose nanocrystals. *Journal of Polymers and the Environment*, *26*(9), 3698–3709.

Hedayati, M., Salehi, M., Bagheri, R., Panjepour, M., & Maghzian, A. (2011). Ball milling preparation and characterization of poly (ether ether ketone)/surface modified silica nanocomposite. *Powder Technology*, *207*, 296–303.

Hussin, F.N.M., Attan, N., & Wahab, R.A. (2020). Extraction and characterization of nanocellulose from raw oil palm leaves (*Elaeis guineensis*). *Arabian Journal for Science and Engineering*, *45*(1), 175–186.

Ibrahim, M.M., El-Zawawy, W.K., Jüttke, Y., Koschella, A., & Heinze, T. (2013). Cellulose and microcrystalline cellulose from rice straw and banana plant waste: Preparation and characterization. *Cellulose*, *20*(5), 2403–2416.

Impoolsup, T., Chiewchan, N., & Devahastin, S. (2020). On the use of microwave pre-treatment to assist zero-waste chemical-free production process of nanofibrillated cellulose from lime residue. *Carbohydrate Polymers*, *230*, 115630.

Janardhnan, S., & Sain, M. (2006). Isolation of cellulose microfibrils: An enzymatic approach. *BioResources*, *1*, 176–188.

Janardhnan, S., & Sain, M. (2011). Targeted disruption of hydroxyl chemistry and crystallinity in natural fibers for the isolation of cellulose nano-fibers via enzymatic treatment. *BioResources*, *6*(2), 1242–1250.

Jantip, S., & Suwanruji, P. (2011). Preparation and properties of microcrystalline cellulose from corn residues. *Advanced Materials Research*, *332*, 1781–1784.

Jasmani, L., & Adnan, S. (2017). Preparation and characterization of nanocrystalline cellulose from *Acacia mangium* and its reinforcement potential. *Carbohydrate Polymers*, *161*, 166–171.

Jonoobi, M., Harun, J., Mathew, A.P., & Oksman, K. (2010). Mechanical properties of cellulose nanofiber (CNF) reinforced polylactic acid (PLA) prepared by twin screw extrusion. *Composites Science and Technology*, *70*, 1742–1747.

Jonoobi, M., Harun, J., Mishra, M., & Oksman, K. (2009). Chemical composition, crystallinity and thermal degradation of bleached and unbleached kenaf bast (*Hibiscus cannabinus*) pulp and nanofiber. *BioResources*, *4*(2), 626–639.

Jonoobi, M., Harun, J., Paridah, M.T., Shakeri, A., Saiful Azry, S., & Makinejad, M.D. (2011). Physicochemical characterization of pulp and nanofibers from kenaf stem. *Material Letters*, *65*(7), 1098–1100.

Kargarzadeh, H., Mariano, M., Huang, J., Lin, N., Ahmad, I., Dufresne, A., & Thomas, S. (2017). Recent developments on nanocellulose reinforced polymer nanocomposites: A review. *Polymer*, *132*, 368–393.

Karimi, S., Paridah, M.T., Ali, K., Dufresne, A., & Abdulkhani, A. (2014). Kenaf bast cellulosic fibers hierarchy: A comprehensive approach from micro to nano. *Carbohydrate Polymer*, *101*, 878–885.

Klemm, D., Cranston, E.D., Fischer, D., Gama, M., Kedzior, S.A., Kralisch, D., Kramer, F., Kondo, T., Lindström, T., Nietzsche, S., & Petzold-Welcke, K. (2018). Nanocellulose as a natural source for groundbreaking applications in materials science: Today's state. *Materials Today*, *21*(7), 720–748.

Kumar, P., Barrett, D.M., Delwiche, M.J., & Stroeve, P. (2009). Methods for pre-treatment of lignocellulosic biomass for efficient hydrolysis and biofuel production. *Industrial and Engineering Chemistry Research*, *48*, 3713–3729.

Lapuz, A.R, Tsuchikawa, S., Inagaki, T., Ma, T., & Migo, V. (2022). Production of nanocellulose film from abaca fibers. *Crystals*, *12*(5), 601.

Lee, J.W., Gwak, K.S., Park, J.Y., Park, M.J., Choi, D.H., Kwon, M., & Choi, I.G. (2007). Biological pretreatment of softwood *Pinus densiflora* by three white rot fungi. *Journal of Microbiology*, *45*(6), 485–491.

Lee, K.Y. (Ed.). (2018). *Nanocellulose and Sustainability: Production, Properties, Applications, and Case Studies*. CRC Press, Boca Raton, pp. 1–314.

López-Rubio, A., Lagaron, J.M, Ankerfors, M., Lindström, T., Nordqvist, D., & Mattozzi, A. (2007). Enhanced film forming and film properties of amylopectin using micro-fibrillated cellulose. *Carbohydrate Polymer*, *68*, 718–727.

Lu, P., & Hsieh, Y.L. (2012). Preparation and characterization of cellulose nanocrystals from rice straw. *Carbohydrate Polymers*, *87*(1), 564–573

Mahardika, M. (2018). Production of nanocellulose from pineapple leaf fibers via high-shear homogenization and ultrasonication. *Fibers*, *6*, 28.

Mehanny, S., Abu-El Magd, E.E., Ibrahim, M., Farag, M., Gil-San-Millan, R., Navarro, J., & El-Kashif, E. (2021). Extraction and characterization of nanocellulose from three types of palm residues. *Journal of Materials Research and Technology*, *10*, 526–537.

Morais, J.P.S., de Freitas Rosa, M., Nascimento, L.D., do Nascimento, D.M., & Cassales, A.R. (2013). Extraction and characterization of nanocellulose structures from raw cotton linter. *Carbohydrate Polymers*, *91*(1), 229–235.

Mosier, N., Wyman, C., Dale, B., Elander, R., Lee, Y.Y., Holtzapple, M., & Ladisch, M. (2005). Features of promising technologies for pretreatment of lignocellulosic biomass. *Bioresource Technology*, *96*, 673–686.

Nakagaito, A.N., & Yano, H. (2004). The effect of morphological changes from pulp fiber towards nano-scale fibrillated cellulose on the mechanical properties of high strength plant fiber based composites. *Applied Physics A: Materials Science & Processing*, *78*(4), 547–552.

Okano, K., Kitagaw, M., Sasaki, Y., & Watanabe, T. (2005). Conversion of Japanese red cedar (*Cryptomeria japonica*) into a feed for ruminants by white-rot basidiomycetes. *Animal Feed Science and Technology*, *120*, 235–243.

Oun, A.A., & Rhim, J.W. (2018). Isolation of oxidized nanocellulose from rice straw using the ammonium persulfate method. *Cellulose*, *25*(4), 2143–2149.

Paakko, M., Ankerfors, M., Kosonen, H., Nykanen, A., Ahola, S., & Osterberg, M. (2007). Enzymatic hydrolysis combined with mechanical shearing and high pressure homogenization for nanoscale cellulose fibrils and strong gels. *Biomacromolecules*, *8*(6), 1934–1941.

Pérez, J.A., Gonzalez, A., Oliva, J.M., Ballesteros, I., & Manzanares, P. (2007). Effect of process variables on liquid hot water pretreatment of wheat straw for bioconversion to fuel-ethanol in a batch reactor. *Journal of Chemical Technology and Biotechnology*, *82*, 929–938.

Prado, K.S., & Spinacé, M.A. (2019). Isolation and characterization of cellulose nanocrystals from pineapple crown waste and their potential uses. *International Journal of Biological Macromolecules*, *122*, 410–416.

Qing, Y., Sabob, R., Zhub, J.Y., Cai, Z., & Wu, Y. (2013). Comparative study of cellulose nanofibrils: Disintegrated from different approaches. *Bioresource Technology*, *130*, 783–788.

Sand, C.S., & Jawaid, M. (2019). The effect of Bi-functionalized MMT on morphology, thermal stability, dynamic mechanical, and tensile properties of epoxy/organoclay nanocomposites. *Polymers*, *11*(12), 2012.

Sehaqui, H., Mushi, N.E., Morimune, S., Salajkova, M., Nishino, T., & Berglund, L.A. (2012). Cellulose nanofiber orientation in nanopaper and nanocomposites by cold drawing. *ACS Applied Materials and Interfaces*, *4*, 1043–1049.

Sharma, V., & Kuila, A. (Eds.). (2017). *Lignocellulosic Biomass Production and Industrial Applications*. John Wiley & Sons.

Shi, J., Liu, W., Jiang, X., & Liu, W. (2020). Preparation of cellulose nanocrystal from tobacco-stem and its application in ethyl cellulose film as a reinforcing agent. *Cellulose*, *27*(3), 1393–1406.

Siddiqui, N., Mills, R.H., Gardner, D.J., & Bousfield, D. (2011). Production and characterization of cellulose nanofibers from wood pulp. *Journal of Adhesive Science and Technology*, *25*(6–7), 709–721.

Siró, I., & Plackett, D. (2010). Microfibrillated cellulose and new nanocomposite materials: A review. *Cellulose*, *17*(3), 459–494.

Soni, B., & Mahmoud, B. (2015). Chemical isolation and characterization of different cellulose nanofibers from cotton stalks. *Carbohydrate Polymers*, *134*, 581–589.

Stelte, W., & Sanadi, A.R. (2009). Preparation and characterization of cellulose nanofibers from two commercial hardwood and softwood pulps. *Industrial and Engineering Chemistry Research*, *48*, 11211–11219.

Tanpichai, S., Witayakran, S., & Boonmahitthisud, A. (2019). Study on structural and thermal properties of cellulose microfibers isolated from pineapple leaves using steam explosion. *Journal of Environmental Chemical Engineering*, *7*(1), 102836.

Tuzzin, G., Godinho, M., Dettmer, A., & Zattera, A.J. (2016). Nanofibrillated cellulose from tobacco industry wastes. *Carbohydrate Polymers*, *148*, 69–77.

Wan Rosli, W.D., Leh, C.P., Zainuddin, Z., & Tanaka, R. (2003). Optimization of soda pulping variable for preparation of dissolving pulps from oil palm fiber. *Holzforschung*, *57*, 106–114.

Wang, B., & Sain, M. (2007). Dispersion of soybean stock-based nanofiber in a plastic matrix. *Polymer International*, *56*, 538–546.

Wang, S., & Cheng, Q. (2009). A novel process to isolate fibrils from cellulose fibers by high-intensity ultrasonication. Part 1. Process optimization. *Journal of Applied Polymer Science*, *113*(2), 1270–1275.

Wei, H., Rodriguez, K., Renneckar, S., & Vikesland, P.J. (2014). Environmental science and engineering applications of nanocellulose-based nanocomposites. *Environmental Science: Nano*, *1*(4), 302–316.

Yahya, M., Chen, Y.W., Lee, H.V., Hock, C.C., & Hassan, W.H.W. (2019). A new protocol for efficient and high yield preparation of nanocellulose from *Elaeis guineensis* biomass: A response surface methodology (RSM) study. *Journal of Polymers and the Environment*, *27*(4), 678–702.

Zhao, J., Zhang, W., Zhang, X., Zhang, X., Lu, C., & Deng, Y. (2013). Extraction of cellulose nanofibrils from dry softwood pulp using high shear homogenization. *Carbohydrate Polymers*, *97*, 695–702.

14 Characteristics of Cellulose Nanocrystals from Sugarcane Bagasse Isolated from Various Methods
A Review

Eti Indarti, Zalniati Fonna Rozali, Dewi Yunita,
Laila Sonia, and Marwan Mas
Universitas Syiah Kuala

CONTENTS

INTRODUCTION

Sugarcane (*Saccharum officinarum* L.) is a tropical plant widely grown in Indonesia. The area of sugarcane plantations in 2018 was 415.66 thousand hectares, based on data from the statistical center. The total sugarcane production reached 2.16 million tons and produced 2.17 million tons of sugar (BPS Statistics Indonesia, 2019). The sugar industry may produce large amounts of waste, covering an area of ±84%, according to the data on plantation areas. The stems of the sugarcane plant are ground and extracted to produce juice that is processed into sugar. The waste produced from the sugar industry is bagasse in the dry fiber residue form, which is usually used as fuel to produce steam and electricity in the sugar production industry (Mandal & Chakrabarty, 2011). In addition, bagasse contains cellulose, which can be used to produce cellulose nanocrystals (Evans et al., 2019).

Bagasse comprises 45%–55% cellulose, 25%–30% hemicellulose, 18%–24% lignin, 1%–4% ash, and <1% wax (Ferreira et al., 2018). The high cellulose content in bagasse makes it potentially used as a product with nanoparticle sizes, such as cellulose nanocrystals (CNC), cellulose nanofibers, and cellulose nanowhiskers. Nanoscale particles have advantages in their applications, including being biodegradable, derived from a renewable source, and having a good physical structure on the cellulose nanocrystals surface (Kumar et al., 2014).

Several methods can isolate cellulose nanocrystals from bagasse, including the acid hydrolysis method, mechanical methods such as high-pressure homogenization, ultrasonication, and TEMPO oxidation (2,2,6,6-*Tetramethylpiperidin*-1-*Oxy*) or a combination using mechanical methods (Zhang et al., 2016). Several researches have carried out studies on the characteristics of cellulose nanocrystals produced from bagasse using various isolation methods. The current work provides a clear review of the recent developments and advances in biofibers, biofilms, biopolymers, and biocomposites, helping the industry and engineering sector to develop advanced bio-based composites for potential applications (Murawski et al., 2019; Payal, 2019; Sanjay et al., 2018; Vinod et al., 2020).

This paper focuses on the isolation method and characteristics of nanocellulose produced from sugarcane bagasse as a raw material. All publications from the last 5–10 years of CNC production from sugarcane bagasse were selected. The characteristics discussed include the CNC morphology, examined size dimensions using transmission electron microscopy (TEM), crystallinity degree (X-ray diffraction), and the yield obtained from each method.

BAGASSE CELLULOSE AND NANOCELLULOSE

Bagasse comprises of 40%–50% cellulose and consists of tree bark, the outermost part of the trunk and pith (Mahmud & Anannya, 2021). The bagasse pulping stage is the initial step before the cellulose isolation process. Desilication, milling, depithing, soda pulping, and bleaching are the stages to obtain bagasse pulp. In the desilication stage, sand is removed by rotating to drop the remaining sand on the bagasse. Next is the milling stage to reduce the bagasse size. The next stage is depithing, a grinding process to remove the pith using a hammer mill and continued with soda pulping using NaOH with a ratio of 5.5:1. After turning it into a pulp, it is washed with water to remove excess alkaline. Finally, ClO$_2$ is used in the bleaching stage to change the pulp color from brown to white (Plengnok & Jarukumjorn, 2020).

CELLULOSE

Cellulose is a linear biopolymer consisting of amorphous and crystalline parts, where D- anhydroglucopyranose units are linked together with β-1,4-glycosidic bonds (Figure 14.1). Cellulose is the basic structural component of all plant fibers and has biodegradable and nontoxic properties (Azeh, 2017). Based on the polymerization degree and solubility in NaOH, cellulose is divided into α-cellulose, β-cellulose, and γ-cellulose. α-cellulose has a long cellulose chain, which is insoluble in the NaOH solution, and has a polymerization degree of 600–15,000. α-cellulose is a purity level indicator of cellulose with a purity degree of $\alpha > 92\%$. β-cellulose has short chains, which dissolves in a strong base of NaOH 17.5%, and has a polymerization degree of 15–90. Meanwhile, γ-cellulose has a short chain with the main hemicellulose content, which dissolves in strong alkaline solutions or the NaOH solution of 17.5%, and has a polymerization degree of less than 15 (Tamara & Sumada, 2012).

Cellulose is the main component obtained in plant cells and is synthesized by cellulose synthase enzymes (CESAs). The plant cell wall consists of three layers, namely the middle lamella, primary cell wall, and secondary cell wall. All layers in the cell wall are comprised of two phases: microfibrillar and matrix. The microfibrillar phase (crystal phase) comprises cellulose microfibrils

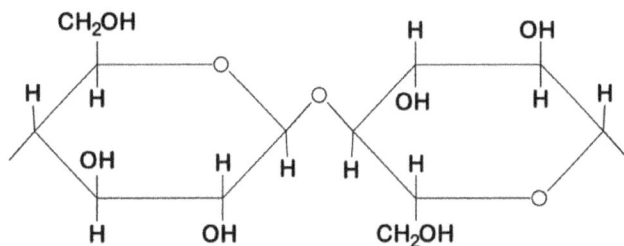

FIGURE 14.1   The chemical structure of cellulose.

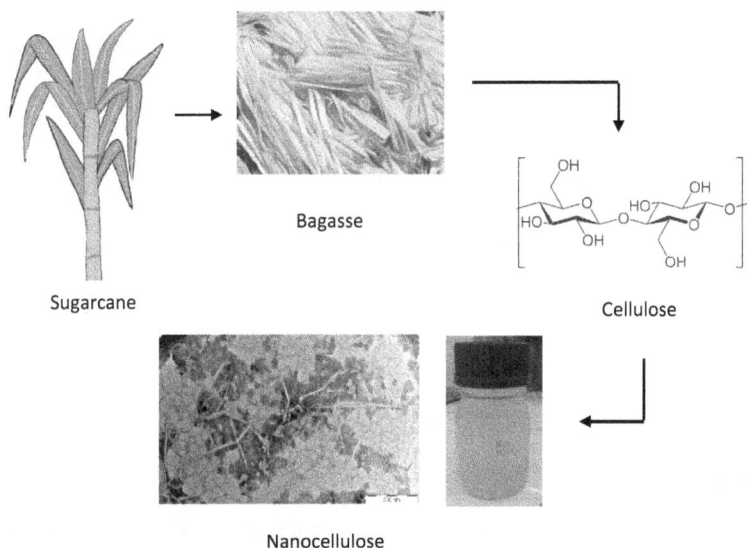

FIGURE 14.2   The cellulosic biomass source (sugarcane bagasse).

(Figure 14.2). In contrast, the matrix phase (noncrystalline) consists of various polysaccharides (pectin and hemicellulose), proteins, and phenolic compounds such as lignin and coumaric acid (Festucci-Buselli et al., 2007). Figure 14.2 in section (a) shows the cellulosic biomass source (e.g., sugarcane), section (b) displays a cross-sectional view of the sugarcane bagasse.

CELLULOSE NANOCRYSTALS

Cellulose nanocrystals (CNC), also known as cellulose nanowhiskers, are needle-like cellulose particles with diameters ranging from 2 to 20 nm and varied in lengths from 100 nm to several microns. The cellulose particles can be obtained through the hydrolysis of strong acids by removing amorphous regions in the cellulose chains (Figure 14.3). However, the crystallinity degree, dimensional diversity, and morphology depend on the material's source and the cellulose's initial preparation. Cellulose nanocrystals have a crystallinity level of less than 400, no more than 10% of the material, and a particle size of less than 5 μm (Bhat et al., 2017).

Cellulose nanocrystals have unique characteristics, including biodegradability, a low specific mass (~1.566 g/cm^3), and low costs due to abundant material availability. In addition, cellulose has nanometer scale dimensions on the surface, high elasticity modulus (~150 GPa), high tensile strength, high stiffness, and high flexibility. It also has good thermal, electrical, and optical properties (Bhat et al., 2017).

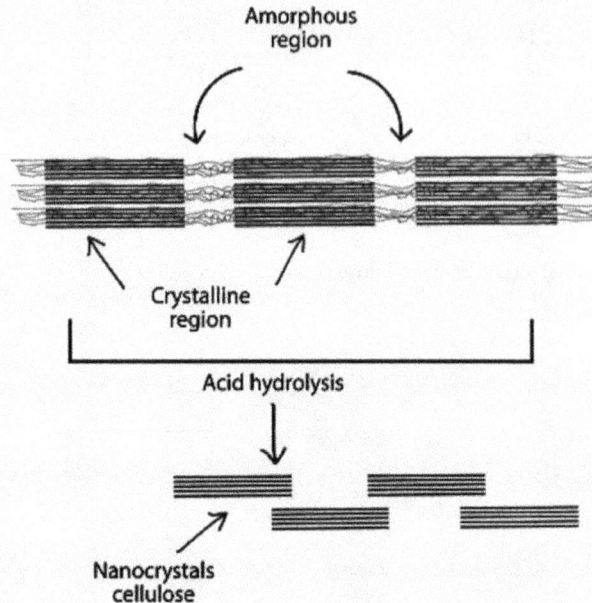

FIGURE 14.3   Cellulose nanocrystals isolation by removing the amorphous part.

CELLULOSE NANOCRYSTALS ISOLATION

PRETREATMENT METHODS ON BAGASSE

The material size reduction and pretreatment stage are the initial steps in the isolation process of cellulose nanocrystals. The pretreatment stage aims to remove unwanted compounds in lignocellulosic raw materials. Several methods, such as alkaline and biological treatments, are used at the pretreatment stage (enzymatic method) (Camassola & Dillon, 2009).

Alkaline Treatment

Alkaline pretreatment is the most traditional method for processing pulp and paper. The pretreatment stage with the alkaline method can be divided into two groups based on the chemicals used; pretreatment using sodium hydroxide (NaOH) and calcium hydroxide or lime ($Ca(OH)_2$), and pretreatment using ammonia (NH_3) (Hu & Ragauskas, 2012). The alkaline method is highly influenced by temperature, time, and the alkaline amount. In addition, this alkaline method is suitable for herbaceous plants or agricultural plants with a lower lignin content (Zheng et al., 2009). Basically, alkaline treatment is a delignification process where hemicellulose is dissolved in large quantities. This method causes swelling, decreases crystallinity, increases internal surface area, and separates lignin and cellulose (Zheng et al., 2009).

The chemicals used in the alkaline method have both advantages and disadvantages. Sodium hydroxide and lime are the main chemicals that effectively break down lignocellulosic cell walls. The disadvantages of using sodium hydroxide are the cost, safety level, and difficulty in recovering or neutralizing (Hu & Ragauskas, 2012). On the other hand, the alkaline method using lime is a low-cost pretreatment and can improve the lignocellulosic biomass digestibility (Agbor et al., 2011). The advantages of using lime as a reagent are that it is cheap, safe to use, and easy to neutralize. The disadvantages of using lime are that it is ineffective to remove lignin in plants with high lignin and require a large water volume at the washing stage (Agbor et al., 2011).

Biological Treatment (Enzymatic Method)

The enzymatic treatment method is used in the delignification process. This treatment method requires less energy, is environmentally friendly, and has higher yields and selectivity compared with chemical treatment. Microorganisms such as soft rot fungi and bacteria are used in biological pretreatment to degrade lignin and hemicellulose. In addition, enzymes obtained from microbes can also be used for biological treatment (Bhat et al., 2017). The enzymes obtained are cellulase enzymes consisting of an endoglucanase, exoglucanase, and cellobiohydrolase (Filson et al., 2009). However, the biological treatment method is not appropriate to be applied on an industrial scale. The disadvantages of using this method include the time length used for the microorganism growth (10–14 days), the need for a large space, and the loss of the carbohydrate fraction consumed by microorganisms (Agbor et al., 2011).

THE ISOLATION METHOD OF CELLULOSE NANOCRYSTALS

The main structure of natural fibers needs to be broken down to obtain nanosized cellulose molecules with high crystallinity. Several methods can be used to obtain nanocrystals from cellulose by breaking glycosidic bonds. These methods include mechanical treatment such as ultrasound, chemical treatment with acid hydrolysis, TEMPO oxidation, and a combination of these methods (Satyamurthy et al., 2011). Depending on the isolation method selection, cellulose nanocrystals with variations in the crystal structure, crystallinity degree, CNC size, and different physical and chemical properties can be produced. Here are some methods that can be used to isolate cellulose nanocrystals from bagasse.

Chemical Treatment (Acid Hydrolysis)

Due to easy operating condition, the acid hydrolysis method is the most used technique to isolate cellulose nanocrystals and produce a well-balanced suspension. The amorphous part of cellulose fibers can be destroyed during the acid hydrolysis process under controlled conditions and keeps the crystal area intact. This method has several drawbacks, such as taking a long time and the high cellulose degradation that can affect the CNC yield. However, a mechanical treatment of ultrasound combined with acid hydrolysis can be used to overcome these disadvantages. This method can improve mixing and chemical reactions in liquids (Azrina et al., 2017).

There are three steps of the acid hydrolysis mechanism in the cellulose chain. The first is the conjugate acid formation and the interaction between protons from the acid and oxygen from glycosides. In the second stage, the C-O bond side separates the conjugate acid into cyclic carbonium ions. The third stage is the liberation of protons and sugar after the adding water. After these three steps, cellulose nanocrystals can be produced. The chemical components often applied to the isolation of nanocrystals using the acid hydrolysis method are phosphoric acid (H_3PO_4), hydrogen bromide (HBr), sulfuric acid (H_2SO_4), and hydrochloric acid (HCl) (Jonoobi et al., 2015).

TEMPO (2,2,6,6-Tetramethylpiperidin-1-Oxy)

TEMPO (2,2,6,6-*tetramethylpiperidine*-1-*oxy*) is a catalyst that mediates oxidation reaction in which replaces the primary C6 hydroxyl group of cellulose with a carboxylate (COO^-) group. The TEMPO oxidation method is typical for preparing and modifying the CNC surface by increasing the carboxyl groups on the surface (Figure 14.4). This method aims to make the modified CNC particles well dispersed in the aqueous phase. This occurs due to the presence of many carboxyl groups on the CNC surface, which can produce strong electrostatic repulsion between the CNC particles (Li et al., 2015).

Cooxidants, such as NaBrO, NaCIO, or $NaCIO_2$, produce an oxidizing component ($TEMPO^+$) in this reaction. Defibrillation of nanocelluloses is induced by the negative charge of the carboxyl groups to the nanocellulose surface, which creates electrostatic repulsion between adjacent units.

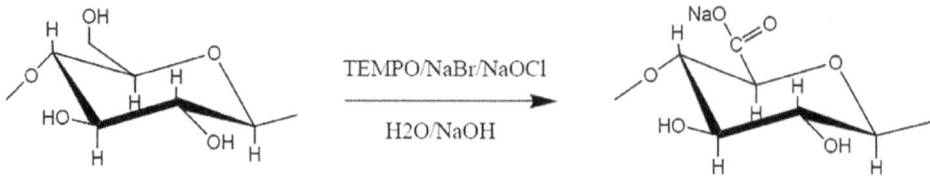

FIGURE 14.4 Oxidation reaction scheme by the TEMPO. Reproduced with permission from Mishra et al. (2011). Copyright, Bioresources.

Figure 14.4 shows the oxidation reaction of the C6 hydroxyl group to a carboxyl group by the TEMPO system (Bhat et al., 2017).

Generally, TEMPO-mediated oxidation is carried out under pH 10 and with a low temperature of −4°C. However, this oxidation reaction can also be carried out under acidic conditions with a pH value of 4.5–5.0 and a temperature ranging from 50°C to 60°C. The number of carboxyl groups that appear on the cellulose surface, and the polymerization degree depends on the NaClO amount in the reaction medium. Therefore, this method is an effective method for surface modification as an over-oxidation process that can be controlled by the NaClO amount. The results include 90% of cellulose nanocrystal particles that are insoluble in water and have a nanocrystalline size of less than 10 nm (Bhat et al., 2017).

Mechanical Treatment (High-Pressure Homogenization)

High-pressure homogenization (HPH) is a mechanical treatment using a high-pressure homogenizer to produce nanocellulose of microcrystalline cellulose (Li et al., 2012). After 30 homogenization passes, the size of the cellulose fibrils can range from 28 to 100 nm. Cellulose is insoluble in water and mostly soluble in organic solvents due to the extensive network through the many hydrogen bonds between molecules. This can cause blockage of the homogenizer valve. Therefore, prior treatments such as microfluidization, ultrafine grinding, or disk grinding are required before homogenization (Hassan et al., 2012). Combining two or more of these methods impacts the increase in the nanomaterials amount of cellulose fibers (Jonoobi et al., 2015).

Mechanical Treatment (High-Intensity Ultrasonication)

High-Intensity Ultrasonication (HIUS) is a mechanical separation method of fibrils involving large shear forces (Figure 14.5) (Khalil et al., 2014). Waves in HIUS can produce strong mechanical oscillations for cavitation, a physical phenomenon that includes the formation, expansion, and explosion of micro gas bubbles when molecules in a liquid absorb ultrasonic energy. The sound spectrum produced by the transducer ranges from 20 kHz to 10 MHz by converting mechanical energy into high-intensity energy. Ultrasonic radiation can be used in many processes, including emulsification and homogenization, and as a catalyst and dispersant (Yang et al., 2010).

Six factors affect the efficiency of HIUS, namely power, temperature, time, concentration, distance, and size. Almost all mechanical methods involve the use of high energy levels, which can reduce the length of the fibrils and the yields produced (Khalil et al., 2014). Overall, the mechanical method can provide an advantage when combined with two or more mechanical methods, thus giving better results.

CHARACTERIZATION OF BAGASSE CELLULOSE NANOCRYSTALS

Yield

The yield is the cellulose nanocrystals (CNC) amount obtained based on bagasse-bleached pulp. Yield can be measured using the following formula (Ghazy et al., 2016):

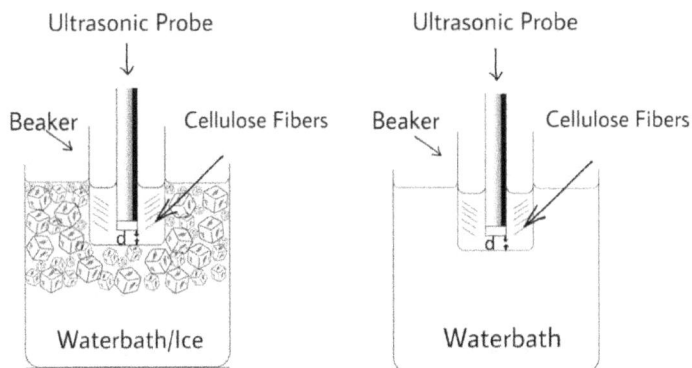

FIGURE 14.5 The scheme for the ultrasonication treatment.

$$\text{Yield } (\%) = (M1 \ / \ M2) \times 100\% \qquad\qquad (14.1)$$

where

M1 = weight of the bleached pulp material
M2 = constant dried CNC.

This discussion will explain the yield produced from the acid hydrolysis isolation method. In contrast, from the TEMPO oxidation method and mechanical isolation, no research result has been reported regarding the yield of bagasse raw materials. The data on the CNC yield can be seen in Table 14.1.

The CNC production through the acid hydrolysis process generally uses sulfuric acid with various concentrations. When sulfuric acid concentration is 64% with a 30 min reaction time, the yield is 45% (Achaby et al., 2016). However, at lower concentrations (33%) with a time of 30 and 75 min, the yields were higher, namely 58% and 50% (Teixeira et al., 2011). However, cellulose degradation will occur at higher sulfuric acid concentrations >45%, resulting in a lower yield. It can also be concluded that the yields with high sulfuric acid concentrations cause cellulose degradation of the pulp. This is due to the amorphous part and the crystal part being missing. Apart from the sulfuric acid concentration used, the yield is also influenced by time.

This is as reported by Panicker et al. (2017), using a combination of sulfuric acid and hydrochloric acid (1:1) with a concentration of 44% in 2 h reaction time. This process resulted in a lower yield of 32.6% compared with previous studies. This is presumably due to the acid combination during hydrolysis and prolonged reaction time, which led to low yields (Panicker et al., 2017). This supports the statement above that the concentration and reaction time greatly affect the yield.

MORPHOLOGY

The morphology of fiber and nanocellulose from bagasse before and after treatment can be determined by analyzing TEM (de Campos et al., 2013). TEM is an electron microscope used to visualize and analyze specimens in micro sizes (1 μm = 10^{-6} m) to nano sizes (1 nm = 10^{-9} m). CNC particle dimension data from bagasse raw material are shown in Table 14.1.

In the acid hydrolysis method, the acid concentration produces different particle dimensions. For example, for sulfuric acid concentrations of 64%–65%, the obtained particle dimensions were (L) 84–275 nm for length and (D) 4–20 nm for diameter (Bras et al., 2010). However, at 60% sulfuric acid concentration, particle dimensions were 80–400 nm in length and 3–20 nm in diameter (Lam, et al., 2017b). Higher acid concentration results in a smaller diameter because the higher the

TABLE 14.1

Yield and Dimensions of CNC Based on the Hydrolysis Method

Acid Hydrolysis Acid Concentration	Isolation Process	Yield	Particle Dimensions			References
			Diameter (nm)	Length (nm)	Aspect Ratio	
Sulfuric acid 65%	Acid hydrolysis (H_2SO_4 65%, 45°C, 45 min). Centrifuged and dialyzed Sonicated (2 min). Few drops of chloroform and stored (4°C)	-	5–12	84–102	~13	Bras et al. (2010)
Sulfuric acid 64%	Acid hydrolysis (H_2SO_4 64%, 55°C, 30 min). Centrifuged and dialyzed to neutral pH.	45%	5±1.1	275±73	~55	Achaby et al. (2016)
	Acid hydrolysis (H_2SO_4 64%, 45°C, 60 min). Centrifuged (12,000rpm, 15 min).	-	20–30	160–400	~20	Sofla et al. (2016)
	Pretreatment with alkaline (NaOH 6%, 50°C, 4 h) Filtered and dialyzed. Acid hydrolysis (H_2SO_4 64%, 40°C, 3 h).	-	10.11±3.36	247.51	~24.49	Slavutsky and Bertuzzi (2014)
Sulfuric acid 60%	Alkaline pretreatment and enzyme Washed with distilled water Acid hydrolysis (H_2SO_4 60%, 45°C, 75 min)	-	9.8±6.3	280.1±73.3	~28.58	Lam, et al. (2017a)
	Acid hydrolysis (H_2SO_4 60%, 45°C, 75 min).	-	3–20	400–500	~25	Lam, et al. (2017b)
	Alkaline pretreatment (17.5% NaOH, 5 h).Filtered and washed with distilled water Acid hydrolysis (H_2SO_4 60%, 50°C, 5 h)	-	35	170	4.86	Mandal and Chakrabarty (2011)
	Acid hydrolysis (H_2SO_4 60%, 45°C, 75 min). Dialyzed and centrifuged (15 min)	-	20–40	200–300	~15	Sukyai et al. (2018)
Sulfuric acid and hydrochloric acid 44% (1:1)	Acid hydrolysis (H_2SO_4 + HCl 44%, 50°C, 2 h) Washed and centrifuged (15 min)	32.6%	10–20	-	-	Panicker et. al. 2017
Sulfuric acid 33%	Acid hydrolysis (H_2SO_4 33%, 30 and 75 min, 45°C) Centrifuged (10 min)	58% (30 min), 50% (75 min)	4 (30 min), 8 (75 min)	255±55	64 32	Teixeira et al. (2011)
Sulfuric acid 32	Acid hydrolysis (H_2SO_4 32%, 24 h). Dialyzed and centrifuged (15 min) Sonicated (1 h)	-	38	-	-	Evans et al. (2019)

(Continued)

TABLE 14.1 (Continued)
Yield and Dimensions of CNC Based on the Hydrolysis Method

Acid Hydrolysis Acid Concentration	Isolation Process	Yield	Diameter (nm)	Length (nm)	Aspect Ratio	References
Sulfuric acid 60% (5 min) Sulfuric acid 50% (10 min)	Alkaline pretreatment (NaOCl 0,735%, 45°C, 6 h) Washed until neutral Acid hydrolysis (ratio 1:25) Dialyzed and centrifuged (30 min) Sonicated (10 min)	-	196.7 111	-	-	Wulandari et al. (2016)
TEMPO Oxidation						
TEMPO	Cellulose fiber (TEMPO, NaBr). Drop it (NaClO) and stirred for 5 h (pH 10). Sonicated (30 min) and added ethanol. Washed until neutral Centrifuged (30 min)	-	15±8	264±69	~33	Zhang et al. (2016)
Mechanical Method						
HPH (40 and 400 bar)	Homogenized cellulose fiber for 10 min on the pressure of 40 and 400 bar	-	3.5	-	-	Hassan et al. (2012)
HPH (400 bar and 1,400 bar)	Alkaline pretreatment (NaOH 1%, 90°C, 110 min) Washed with distilled water Cellulose fiber+BmimCl solution (stirred for 2 h, 130°C). Homogenized from 400 to 1,400 bar	-	10–20	-	-	Li et al. (2012)
HPH (1,035 bar)	Alkaline pretreatment and enzyme. Dialyzed for 24 h Homogenized on 1,035 bar Cooled suspension	-	5–10	-	-	Saelee et al. (2016)
HIUS	Alkaline pretreatment-hydrothermal catalyst Stirred the pretreatment fiber with 300 mL deionized water for 1 h. Ultrasonicated (40 min)	-	20–40	-	-	Feng et al. (2018)
HIUS	Alkaline pretreatment (NaOH 2%–3%, 24 h). Washed with distilled water. Ultrasonicated (60°C for 4–5 h)	-	60–100	-	-	Bansal et al. (2016)
Ultrafine grinder	Bleached pulp suspension on the ultrafine grinder (1,500 rpm). dropped chloroform and stored (4°C)	-	11.13±1	-	-	Heidarian et al. (2016)
DHPM	Dissolved cellulose with BmimCl in a microwave (130°C). Homogenized (340–1,720 bar) Centrifuged and dried with vacuum freeze dryer	-	5–12	-	-	Li et al. (2014)

Note: L, length; D, diameter; HPH, high-pressure homogenization; HIUS, high-intensity ultrasonication; DHPM, dynamic high-pressure microfluidization.

concentration, the easier the acid penetrates inner part of cellulose fibril. The reaction will alter the OH in the C6 of anhydrous cellobiose to SO_{4-} negative charge. When the sulfuric acid is combined with 44% hydrochloric acid (1:1), similar particle dimensions of 10–20 nm in diameter are produced (Panicker et al., 2017). In contrast, a lower concentration of 32% has a dimension of 38 nm. However, it is different from Teixeira et al. (2011), where the sulfuric acid concentration of 33% and the hydrolysis time of 30 and 75 min resulted in smaller dimensions; 4 and 8 nm in diameter and 255 ± 55 nm in length, and the axial ratio were 64 and 32, respectively. Therefore, it is noted that the hydrolysis time also affects the CNC dimensions.

In the mechanical method using HPH with a pressure of 400–1,400, 1,035, and 40–400 bar, the particle dimensions are 10–20 nm (Li et al., 2012), 5–10 nm (Saelee et al., 2016), and 3.5 nm (Hassan et al., 2012), respectively. In contrast, HIUS produces particle dimensions of 20–40 nm (Feng et al., 2018) and 60–100 nm (Bansal et al., 2016). Other results by Heidarian et al. (2016) showed that when they used an ultrafine grinder, the particle dimensions were 11.13 nm. When using the dynamic high-pressure microfluidization (DHPM) method, particle dimensions obtained were 5–12 nm (Li et al., 2014). The difference in the dimensions of the particles is due to the different tools used when isolating the CNC. In addition, the pretreatment also affects the nanofibrillation process. The pretreatment process can destroy the primary wall and weaken the hydrogen bonds in the cellulose fibers to become cellulose microcrystals, which can then be easily fibrillated into cellulose nanocrystals by mechanical force. It can be concluded that the applied pressure is not directly in line with the CNC dimensions and yield obtained. Like the acid hydrolysis process, many factors affect the CNC dimension, for example, the type of equipment used, pretreatment, time, and heating process.

CRYSTALLINITY INDEX

Crystallinity is the crystal fraction amount in the nanocellulose. The crystallinity index (CrI) is the parameter used in testing the crystallinity, which describes the relative number of crystals in a material (Park et al., 2010). The purpose of this analysis is to find out the effectiveness and crystallinity level of each isolation treatment. In this discussion, the CNC crystallinity degree for various methods and the concentrations of each treatment will be compared. The crystallinity degree of CNC from bagasse with various methods and concentrations can be seen in Table 14.2. The crystallinity index (CrI) of cellulose samples can be determined using X-ray diffraction (XRD) analysis, which is calculated in Equation (14.2).

$$CrI = \frac{I_{cr} - I_{am}}{I_{002}} \times 100 \qquad (14.2)$$

Overall, I_{Cr} represents a maximum intensity of crystalline material at 2θ of 22°, and I_{am} represents a minimum intensity of amorphous material at 2θ of 18°. XRD calculates the crystal mass fraction of cellulose among all the total cellulose content samples. Therefore, the cellulose crystallinity can be directly compared if it is in its pure form (Ahvenainen et al., 2016). The XRD method can determine the crystallinity degree of the crude cellulose material and its derivatives. The crystallinity degree can be influenced by the chemicals used during the isolation treatment of cellulose into nanocellulose (Jonoobi et al., 2015). Some cellulose fibers consist of more than one crystal form (a mixture of cellulose I and II) and have different diffraction patterns.

The CNC crystallinity degree ranged from 67.83% to 87.5% in the acid hydrolysis method. This result differs from the crystallinity degree obtained from the TEMPO oxidation method, which is 40%. This is related to the difference between the two processes, where the acid hydrolysis removes almost the amorphous part of the cellulose fibers and leaves the crystalline part so that the crystalline percentage is high. Meanwhile, in the TEMPO oxidation method, the oxidation process on the OH group of the C6 cellulose chain occurs in both amorphous and crystalline regions. Therefore,

TABLE 14.2

Crystallinity Index in Several Isolation Methods

Acid Hydrolysis Method		
Isolation Process	**Crystallinity Index (%)**	**References**
Sulfuric acid 65%	CNC=89	Bras et al. (2010)
Sulfuric acid 64%	Bagasse = 19 Bleached pulp=45 CNC=78	Achaby et al. (2016)
	Bagasse=45 Bleached pulp=65 CNC=73	Sofla et al. (2016)
Sulfuric acid 60%	CNC=68.54	Lam et al. (2017a)
	CNC=68.28	Sukyai et al. (2018)
Sulfuric acid and hydrochloric acid 44% (1:1)	CMC=69.66 CNC=82.65	Panicker et al. (2017)
Sulfuric acid 33%	CNC=87.5 (30 min) CNC=70.5 (75 min)	Teixeira et al. (2011)
Sulfuric acid 32%	Bagasse=40.66 Bleached pulp=67.26 CNC=76.89	Evans et al. (2019)
H_2SO_4 60% (5 min)	CNC=67.83	Wulandari et al. (2016)
H_2SO_4 50% (10 min)	CNC=76.01	
TEMPO Oxidation Method		
TEMPO	Bagasse=65 Bleached pulp=63.3 CNC=40	Zhang et al. (2016)
Mechanical Method		
HPH 400 and 1,400 bar	Bleached pulp=52 CNC=36	Li et al. (2012)
HPH 1,035 bar	Bagasse=59.52 Bleached pulp=69.72 CNC=68.10	Saelee et al. (2016)
HIUS	Intensity of bagasse (1,453.78) Intensity of CNC (2,994.04)	Bansal et al. (2016)
HIUS	Bagasse=55.1 Bleached pulp=73.6 CNC=71.2	Feng et al. (2018)
Ultrafine grinder	Bagasse=45.81 Bleached pulp=63.81 CNC=58.21	Heidarian et al. (2016)

Note: CMC, cellulose microcrystals;, HPH, high-pressure homogenization; HIUS, high-intensity ultrasonication; DHPM, dynamic high-pressure microfluidization.

the amorphous still remain in the TEMPO process, and this causes the crystallinity degree of the TEMPO process to be lower than the acid hydrolysis process. On the other hand, mechanical methods with varying pressures produce a crystallinity degree in the 36%–71.2% range.

At a sulfuric acid concentration of 65%, the CNC crystallinity was 89% (Bras et al., 2010) and higher than that at a concentration of 64% with the CNC crystallinity of 78% (Achaby et al., 2016) and 73% (Sofla et al., 2016). Further, at a lower sulfuric acid concentration (60%), resulted in a lower crystallinity degree of 68.54% (Lam, et al., 2017a) and 68.28% (Sukyai et al., 2018). It can be concluded that higher acid concentrations can result in a higher crystallinity degree. This is because the process of removing the amorphous regions occurs completely, leaving only the crystalline regions.

Panicker et al. (2017) used a combination of sulfuric acid and hydrochloric acid at 44% wt (1:1), where there was an increase in the crystallinity degree of cellulose microcrystals from 69.66% to 82.65% for cellulose nanocrystals. The crystallinity increase is due to the hydrolysis process using a combination of the two acids, which is thought to have removed the amorphous regions which work more effectively even at lower concentrations (44%).

Apart from the acid concentration, the hydrolysis time also affects the crystallinity degree. For example, Teixeira et al. (2011) produced CNC through a hydrolysis process using sulfuric acid concentration of 33% and obtained a high crystallinity degree, 87.5% and 70.5% in 30 and 75 min, respectively. This indicates that the increased time can remove amorphous parts and damage the crystalline structure. However, Evans et al. (2019) reported that a lower concentration of sulfuric acid (32%) resulted in a lower crystallinity degree (76.89%). On the other hand, different results were also obtained by Wulandari et al. (2016) using hydrochloric acid with a concentration of 60% (5 min) and 50% (10 min). The crystallinity index obtained with the difference in concentration and time was 67.83% and 76.01%, respectively.

In the TEMPO oxidation method by Zhang et al. (2016), the crystallinity degree for each step process tended to decrease. Bagasse fiber crystallinity was 65%, and the degree of crystallinity decreased to 63.3% after TEMPO-oxidation process (pulp). Furthermore, at the final stage of the TEMPO oxidation method, the crystallinity degree of the CNC obtained dropped drastically to 40%. This is due to in the ultrasonication stage, causing defibrillation to be non-selective, resulting in damage to amorphous regions and cellulose crystals (Zhang et al., 2016).

In the mechanical methods, the crystallinity index produced using HPH with a pressure of 400 and 1,400 bar (Li et al., 2012) was 36%, lower than the crystallinity index of sugarcane bagasse of 60%. In contrast, Saelee et al. (2016) reported using a pressure of 1,035 bar producing a crystallinity index of 68.10%, showing an increase compared with its initial crystallinity index of 59.52%.

It is assumed that the high pressure exerted due to the shear force on the tool causes the loss of the crystal part. However, the crystallinity index produced using an ultrafine grinder was 58.21% (Heidarian et al., 2016). Feng et al. (2018) found that using ultrasonication, the crystallinity index was 71.2%, while Bansal et al. (2016) showed an intensity increase from good crystallinity of sugarcane bagasse of 1,453 to nano cellulose of 2,994 (there is no information regarding the % crystallinity obtained).

Unlike the case with the acid hydrolysis method, the crystallinity index in the mechanical method decreased after performing the mechanical treatment. This is in accordance with Heidarian et al. (2016), who showed that the crystallinity index of bagasse fiber was 45.81%, which increased to 63.81% after the bleaching process. However, it decreased to 58.21% after being treated with an ultrafine grinder at 1,500 rpm.

Feng et al. (2018) also reported a decrease in the crystallinity index by ultrasonication. The crystallinity index was decreased after ultrasonication treatment of bleached pulp, from 73.6% to 71.2%, respectively. This is due to the crystalline region destruction in the cellulose after mechanical treatment. The vibrations in ultrasonic waves are thought to damage the amorphous and crystalline regions structures in cellulose, resulting in the crystal region loss.

CONCLUSION

Cellulose nanocrystals can be isolated from bagasse through acid hydrolysis, TEMPO oxidation, and mechanical methods such as HPH, ultrasonication, ultrafine grinder, disk grinder, and DHPM. The yield, morphology, and crystallinity of CNC obtained differed for each method, depending on the concentration of chemical used, process time, and mechanical treatment, for example, tools and pressure. The highest yield has been obtained (58%) from the acid hydrolysis method with a sulfuric acid concentration of 33% and a hydrolysis time of 30 min. Although reports on the CNC isolation from other materials stated that the TEMPO method usually produces higher yields than the acid hydrolysis method, unfortunately, in the TEMPO process obtained for this CNC sugar bagasse, no yields were reported. The acid hydrolysis process also resulted in the highest crystallinity index (89%) by the sulfuric acid concentration of 65% in 45 min. The morphological characteristics included the highest axial ratio (64) from acid hydrolysis using a sulfuric acid concentration of 33% and a hydrolysis time of 30 min, with dimensions of 2 nm in width and 255 ± 55 nm in length.

REFERENCES

Achaby, M. E., Miri, N. E., Aboulkas, A., Zahouily, M., Essaid, B., Barakat, A., & Solhy, A. (2016). Processing and properties of eco-friendly bio- nanocomposite films filled with cellulose nanocrystals from sugarcane bagasse. *International Journal of Biological Macromolecules*, 96, 340–352

Agbor, V. B., Cicek, N., Sparling, R., Berlin, A., & Levin, D. B. (2011). Biomass pretreatment: Fundamentals toward application. *Biotechnology Advances*, 29(6), 675–685.

Ahvenainen, P., Kontro, I., & Svedstr, K. (2016). Comparison of sample crystallinity determination methods by X-ray diffraction for challenging cellulose I materials. *Cellulose*, 23, 1073–1086.

Azeh, Y. (2017). Synthesis and characterization of cellulose nanoparticles and its derivatives using a combination of spectro-analytical techniques. *International Journal of Nanotechnology in Medicine & Engineering*, 2(6), 65–94.

Azrina, Z. Z., Beg, M. D. H., Rosli, M. Y., Ramli, R., Junadi, N., & Alam, A. M. (2017). Spherical nanocrystalline cellulose (NCC) from oil palm empty fruit bunch pulp via ultrasound assisted hydrolysis. *Carbohydrate polymers*, 162, 115–120.

Bansal, M., Chauhan, G. S., Kaushik, A., & Sharma, A. (2016). Extraction and functionalization of bagasse cellulose nanofibres to Schiff-base based antimicrobial membranes. *International Journal of Biological Macromolecules*, 91, 887–894.

Bhat, A. H., Dasan, Y. K., Khan, I., Soleimani, H., & Usmani, A. (2017). Application of nanocrystalline cellulose: Processing and biomedical applications. In Jawaid, M., Boufi, S., & Abdul Khalil H.P.S (eds) *Cellulose-Reinforced Nanofibre Composites: Production, Properties and Applications*. Elsevier Inc, Sawston, pp. 215–240.

Bras, J., Hassan, M. L., Bruzesse, C., Hassan, E. A., El-wakil, N. A., & Dufresne, A. (2010). Mechanical, barrier, and biodegradability properties of bagasse cellulose whiskers reinforced natural rubber nanocomposites. *Industrial Crops & Products*, 32(3), 627–633.

Camassola, M., & Dillon, A. J. (2009). Biological pretreatment of sugar cane bagasse for the production of cellulases and xylanases by Penicillium echinulatum. *Industrial Crops and Products,* 29(2–3), 642–647.

de Campos, A., Carolina, A., David, C., & Eliangela, C. (2013). Obtaining nanofibers from curaua fibers using enzymatic hydrolysis followed by sonication. *Cellulose*, 20, 1491–1500.

Evans, S. K., Wesley, O. N., Nathan, O., & Moloto, M. J. (2019). Chemically purified cellulose and its nanocrystals from sugarcane baggase: Isolation and characterization. *Heliyon*, 5(10), e02635

Feng, Y., Cheng, T., Yang, W., Ma, P., He, H., Yin, X., & Yu, X. (2018). Characteristics and environmentally friendly extraction of cellulose nanofibrils from sugarcane bagasse. *Industrial Crops & Products*, 111, 285–291.

Ferreira, F. V., Mariano, M., Rabelo, S. C., Gouveia, R. F., & Lona, L. M. F. (2018). Isolation and surface modification of cellulose nanocrystals from sugarcane bagasse waste: From a micro- to a nano-scale view. *Applied Surface Science*, 436, 1113–1122.

Festucci-Buselli, R. A., Otoni, W. C., & Joshi, C. P. (2007). Structure, organization, and functions of cellulose synthase complexes in higher plants. *Brazilian Journal of Plant Physiology*, 19, 1–13.

Filson, P. B., Dawson-Andoh, B. E., & Schwegler-Berry, D. (2009). Enzymatic-mediated production of cellulose nanocrystals from recycled pulp. *Green Chemistry*, 11, 1808–1814.

Ghazy, M. B., Esmail, F. A., El-Zawawy, W. K., Al-Maadeed, M. A., & Owda, M. E. (2016). Extraction and characterization of Nanocellulose obtained from sugarcane bagasse as agro-waste. Journal: *Journal of Advances in Chemistry*, 12(3), 4256–4264.

Hassan, M. L., Mathew, A. P., Hassan, E. A., El-Wakil, N. A., & Oksman, K. (2012). Nanofibers from bagasse and rice straw: Process optimization and properties. *Wood science and technology,* 46(1), 193–205.

Heidarian, P., Behzad, T., & Karimi, K. (2016). Isolation and characterization of bagasse cellulose nanofibrils by optimized sulfur-free chemical delignification. *Wood Science and Technology*, 50(5), 1071–1088.

Hu, F., & Ragauskas, A. (2012). Pretreatment and lignocellulosic chemistry. *Bioenergy Research*, 5(4), 1043–1066.

Jonoobi, M., Oladi, R., Davoudpour, Y., Oksman, K., Dufresne, A., Hamzeh, Y., & Davoodi, R. (2015). Different preparation methods and properties of nanostructured cellulose from various natural resources and residues: A review. *Cellulose*, 22(2), 935–969.

Khalil, H. A., Davoudpour, Y., Islam, M. N., Mustapha, A., Sudesh, K., Dungani, R., & Jawaid, M. (2014). Production and modification of nanofibrillated cellulose using various mechanical processes: A review. *Carbohydrate polymers*, 99, 649–665.

Kumar, A., Negi, Y. S., Choudhary, V., & Bhardwaj, N. K. (2014). Sugarcane bagasse: A promising source for the production of nanocellulose. *Journal Polymer and Composites*, 2(3), 23–27.

Lam, N. T., Chollakup, R., Smitthipong, W., Nimchua, T., & Sukyai, P. (2017a). Utilizing cellulose from sugarcane bagasse mixed with poly (vinyl alcohol) for tissue engineering scaffold fabrication. *Industrial Crops and Products*, 100, 183–197.

Lam, N. T., Chollakup, R., Smitthipong, W., Nimchua, T., & Sukyai, P. (2017b). Characterization of cellulose nanocrystals extracted from sugarcane bagasse for potential biomedical materials. *Sugar Tech*, 19(5), 539–552.

Li, B., Xu, W., Kronlund, D., Määttänen, A., Liu, J., Smått, J. H., Peltonen, J., Willfor, S., Mu, X., & Xu, C. (2015). Cellulose nanocrystals prepared via formic acid hydrolysis followed by TEMPO-mediated oxidation. *Carbohydrate Polymers*, 133, 605–612.

Li, J., Wang, Y., Wei, X., Wang, F., Han, D., Wang, Q., & Kong, L. (2014). Homogeneous isolation of nanocelluloses by controlling the shearing force and pressure in microenvironment. *Carbohydrate Polymers*, 113, 388–393.

Li, J., Wei, X., Wang, Q., Chen, J., Chang, G., Kong, L., Su, J., & Liu, Y. (2012). Homogeneous isolation of nanocellulose from sugarcane bagasse by high pressure homogenization. *Carbohydrate Polymers*, 90(4), 1609–1613.

Mahmud, M. A., & Anannya, F. R. (2021). Sugarcane bagasse-A source of cellulosic fiber for diverse applications. *Heliyon*, 7(8), e07771.

Mandal, A., & Chakrabarty, D. (2011). Isolation of nanocellulose from waste sugarcane bagasse (SCB) and its characterization. *Carbohydrate Polymers*, 86(3), 1291–1299.

Mishra, S. P., Thirree, J., Manent, A. S., Chabot, B., & Daneault, C. (2011). Ultrasound-catalyzed TEMPO-mediated oxidation of native cellulose for the production of nanocellulose: Effect of process variables. *BioResources*, 6(1), 121–143.

Murawski, A., Diaz, R., Inglesby, S., Delabar, K., & Quirino, R. L. (2019). *Polymer Nanocomposites in Biomedical Engineering*. Springer, Cham.

Panicker, A., Rajesh, K. A., & Varghese, T. O. (2017). Mixed morphology nanocrystalline cellulose from sugarcane bagasse fibers/poly (lactic acid) nanocomposite films: Synthesis, fabrication and characterization. *Iranian Polymer Journal*, 26(2), 125–136.

Park, S., Baker, J. O., Himmel, M. E., Parilla, P. A., & Johnson, D. K. (2010). Cellulose crystallinity index: Measurement techniques and their impact on interpreting cellulase performance. *Biotechnology for Biofuels*, 3(1), 1–10.

Payal, R. (2019). *Sustainable Polymer Composites and Nanocomposites*. Springer, Cham.

Plengnok, U., & Jarukumjorn, K. (2020). Preparation and characterization on nanocellulose from sugarcane bagasse. *Biointerface Research in Applied Chemistry*, 10(3), 5675–5678.

Saelee, K., Yingkamhaeng, N., Nimchua, T., & Sukyai, P. (2016). An environmentally friendly xylanase-assisted pretreatment for cellulose nanofibrils isolation from sugarcane bagasse by high-pressure homogenization. *Industrial Crops and Products*, 82, 149–160.

Sanjay, M. R., Madhu, P., Jawaid, M., Senthamaraikannan, P., Senthil, S., & Pradeep, S. (2018). Characterization and properties of natural fiber polymer composites: A comprehensive review. *Journal of Cleaner Production*, 172, 566–581.

Satyamurthy, P., Jain, P., Balasubramanya, R. H., & Vigneshwaran, N. (2011). Preparation and characterization of cellulose nanowhiskers from cotton fibres by controlled microbial hydrolysis. *Carbohydrate Polymers*, 83(1), 122–129.

Slavutsky, A. M., & Bertuzzi, M. A. (2014). Water barrier properties of starch films reinforced with cellulose nanocrystals obtained from sugarcane bagasse. *Carbohydrate polymers*, 110, 53–61.

Sofla, M. R. K., Brown, R. J., Tsuzuki, T., & Rainey, T. J. (2016). A comparison of cellulose nanocrystals and cellulose nanofibres extracted from bagasse using acid and ball milling methods. *Advances in Natural Sciences: Nanoscience and Nanotechnology*, 7(3), 035004.

Sukyai, P., Anongjanya, P., Bunyahwuthakul, N., Kongsin, K., Harnkarnsujarit, N., Sukatta, U., Sothornvit, R., & Chollakup, R. (2018). Effect of cellulose nanocrystals from sugarcane bagasse on whey protein isolate-based films. *Food Research International*, 107, 528–535.

Tamara, P. E., & Sumada, K. (2012). Isolation study of efficient α-cellulose from waste plant stem Manihot esculenta crantz. *Jurnal Teknik Kimia*, 5(2), 434–438.

Teixeira, E. M., Bondancia, T. J., Teodoro, K. B. R., Corrêa, A. C., Marconcini, J. M., & Mattoso, L. H. C. (2011). Sugarcane bagasse whiskers: Extraction and characterizations. *Industrial Crops and Products*, 33(1), 63–66.

Vinod, A., Sanjay, M. R., Suchart, S., & Jyotishkumar, P. (2020). Renewable and sustainable biobased materials: An assessment on biofibers, biofilms, biopolymers and biocomposites. *Journal of Cleaner Production*, 258, 120978.

Wulandari, W. T., Rochliadi, A., & Arcana, I. M. (2016). Nanocellulose prepared by acid hydrolysis of isolated cellulose from sugarcane bagasse. *IOP Conference Series: Materials Science and Engineering*, 107(1), 012045.

Yang, Z., Peng, H., Wang, W., & Liu, T. (2010). Crystallization behavior of poly (ε-caprolactone)/layered double hydroxide nanocomposites. *Journal of Applied Polymer Science*, 116(5), 2658–2667.

Zhang, K., Sun, P., Liu, H., Shang, S., Song, J., & Wang, D. (2016). Extraction and comparison of carboxylated cellulose nanocrystals from bleached sugarcane bagasse pulp using two different oxidation methods. *Carbohydrate Polymers*, 138, 237–243.

Zheng, Y., Pan, Z., & Zhang, R. (2009). Overview of biomass pretreatment for cellulosic ethanol production. *International Journal of Agricultural and Biological Engineering*, 2(3), 51–68.

15 Applications of Regenerated Cellulose Products

Kushairi Mohd Salleh
Universiti Sains Malaysia

*Nur Amira Zainul Armir, Swarna Devi Palanivelu,
Amalia Zulkifli, and Sarani Zakaria*
Universiti Kebangsaan Malaysia

CONTENTS

Introduction 217
Cellulose Insolubility, Dissolution, and Regeneration Process 218
Regenerated Cellulose Products 220
   Hydrogel, Aerogel, Xerogel, Cryogel 221
      Fibers 223
      Membrane and Thin Films 224
Applications 225
   Medical 225
   Agricultural 226
   Automotive 228
   Aerospace 229
   Textile 230
Conclusion 232
Acknowledgments 232
References 232

INTRODUCTION

Cellulose is a polymer where glucose is repeated in its linear structure via 1–4-linked β-d-glucopyranose. Anselme Payen first chemically identified it in 1838, where it consists of 44%–45% carbon, 6%–6.5% hydrogen, and oxygen making up the rest (Payen, 1838). Cellulose is mainly found in the of higher plants, where they are entangled together with lignin, hemicellulose, pectin, and proteins in the heterogeneous matrix. Cellulose is an inexhaustible material that yields billions of tons of biomass every year. This makes them an important material that plays a significant role in human development and civilization. As it has been used for numerous applications from fuel to clothing to writing materials, it can be believed to have yet more potential. Currently, an attempt to use cellulose in a more complex system and devices has attracted researchers. Because of its moldability, compatibility, and alterability, its polymer matrix is an ideal candidate for mobilizing a wide variety of materials, each of which possesses a distinct functionality and a significant capacity to retain.

The size of cellulose molecules is defined by the average degree of polymerization (DP). A higher DP is usually associated with a higher average molecular weight (Mw). The quality of cellulose is often relying on these two essential characteristics. Cellulose with a high DP and Mw is unlikely to be used for the melting process as they are required to be pretreated to reduce the DP and Mw. Pretreatments involve cost, time, and chemical waste. Pretreatments are unfavored as a solvent system to dissolve the cellulose since they involve additional laborious and time-consuming stages. Usually, cellulose with manageable DP and Mw is used for the melting process. If the DP and Mw of cellulose are not identified and controlled, the efficiency of the solvent system is disturbed, and the dissolving yield will be lower. Solubility is a vital display to measure the suitability of cellulose and the quality of solvents. The actual capability of the solvent system cannot be determined unless the DP and Mw of cellulose are controlled. Other than DP and Mw, constituents like lignin must also be noted. Lignin must be removed from the cellulose structure because it results in swelling and ballooning during the dissolution process. The reachability of the solvents to swell and break the intermolecular network of cellulose is easily agitated by these factors. Hence, pre-characterizations of cellulose must be carried out.

Cellulose can be physically and mechanically modified. Physical modifications on cellulose usually involve changes in its physical form. Meanwhile, chemical modifications altered chemical functionality and behavior. Physical enhancements of cellulose involve physical actions (e.g., plasma and corona discharges, gamma irradiation, and ultraviolet (UV) light) and might also apply chemical measures; for instance, a simple hydrolysis method could change cellulose into a micro- or nanocrystalline structure. Unlike physical modifications, chemical modifications are rather complex and less green but offer massive potential for cellulose. Under chemical modifications, a specific solvent system is required. The solvent system can then be divided into derivatizing and non-derivatizing systems. The former involves introducing new chemical functions, but the latter is not. Nevertheless, both approaches guarantee the solubility of cellulose in multicomponent mixtures, including inorganic and organic salts, under peculiar experiment conditions (e.g., at very high or very low temperatures, and pH).

Even though cellulose is repeatedly reported to be difficult to be dissolved, it is surprisingly highly hydrophilic. The hydrophilicity of cellulose is beneficial in paper making, where in wetted form, the cellulose fibers interact with each other, driven by the affinity to form hydrogen bonding. The formed hydrogen bonding then binds the cellulose fiber into a closer packing structure, forming a thin layer of paper. The affinity to self-aggregate among cellulose fibers also occurs at the molecular level. This behavior has made cellulose unique and intricate, and exploring specific solvents in preventing self-aggregation of cellulose is still on the quest. In addition, the cellulose crystal structure is notoriously resistant to being dissolved. Since cellulose is disreputably known not to be meltable, research on them will be continued until a perfect solvent system is found.

CELLULOSE INSOLUBILITY, DISSOLUTION, AND REGENERATION PROCESS

Extracted cellulose can be used as it is in many fields, including textile, pulp and papermaking, and packaging. However, dissolving the cellulose extends its potential in many areas, comprising medical, agriculture, automotive, aerospace, and functional textile. By dissolving the cellulose, regenerated cellulose products can be produced and provide a homogenous environment, and cellulose can be degraded more efficiently without affecting its bulk properties. Several factors are outlined to root the causes of the insolubility of cellulose in water and most organic solvents.

Intermolecular hydrogen bonding is usually implied in this problem but is not entirely true when most hydrogen-bonded substances can easily dissolve in water, for instance, glucose. Glucose, on the other hand, can easily dissolve in water because when they interact with water or any polar solvents, hydrogen bonding is even more potent than the compound itself. Due to the fact that cellulose is a polar molecule, its insolubility in nonpolar solvents is easy to realize; nevertheless, the aqueous insolubility of cellulose is more difficult to fathom. The most well-known and acceptable consensus on the insolubility of cellulose is the ability to self-aggregate via intermolecular hydrogen bonds. The -OH group at C6 is the leading site for intermolecular hydrogen bonding. The limited reachability of the solvent to this group is said to be the main factor in the insolubility of cellulose. This, however, remains untrue as the dissolution process might be controlled by thermodynamic and kinetic interactions. Lindman et al. (2010) compared dextran, methyl, and hydroxyethyl cellulose, which have similar hydrogen bonding with cellulose but are easily dissolved in water. Some claim that the crystallinity of cellulose is the leading cause of aqueous cellulose insolubility. This claim, however, should not be used as an indicator of one capability of being soluble. Many crystalline systems can easily dissolve in water, but the amorphous portions of cellulose are proven to be undissolved in water and common solvents. Nevertheless, modifications to cellulose tend to make it more soluble in water due to the disruption of cellulose crystals' chain packing. So, to some extent, crystallinity is acceptable to be the cause of the insolubility of cellulose. Another known cause of the insolubility of cellulose is the amphiphilic property. Many polymers are amphiphilic, a dual character of polar and nonpolar. Amphiphilic self-assembly is reported to occur with homopolymers like cellulose. The hydrophobic sides of the glucose ring in cellulose cause them to adhere to one another in an aqueous environment, preventing good solubility (Lindman et al., 2010).

During the cellulose dissolution process, the solvent molecules penetrate the cellulose molecular structure and cause the cell wall of the cellulose fiber to expand, resulting in dissolution. Swelling mostly appears as a "ballooning" in the selected zone of the fiber. When the osmotic pressure inside the balloon exceeds the breaking resistance of the membrane, "ballooning" membrane bursts, and the supramolecular structure of cellulose must completely dissolve to achieve total dissolution. In the "ballooning" mode, the fiber never reaches the stage of dissolution owing to the bursting event; instead, it remains in the "ballooning" phase indefinitely. The weaker fiber areas begin to inflate and increase until their maximum size is reached, at which point they dissolve. From this understanding, several factors must be considered to choose the right cellulose conditions and solvent quality. Factors that can accelerate cellulose dissolutions depend on the efficiency of the solvent system and external stimuli, including heating and stirring, which can significantly increase contact between the solvent and the solute and speed up the dissolution process. The achievement of total disintegration is also greatly aided by predetermined cellulose DP and Mw. Introducing charges on cellulose is also reported to be efficient in increasing the solubility of cellulose. Therefore, it is always anticipated that cellulose will be more soluble when it is charged up by the modification process, whether by cationic or anionic groups.

In the solvent arena, there are "derivatizing" solvents and "non-derivatizing" solvents to dissolve the cellulose, as seen in Figure 15.1. Recent advances are more interested in the latter as the non-derivatizing solvents covered more significant scientific findings and provided reasonable explanations for cellulose intermolecular interactions. The derivatizing solvent can easily alter cellulose behavior, where the added functional groups form a new class of material known as cellulose derivatives. The derivatizing solvent is usually associated with cellulose chemical modification. Dissolution is aided by functionalities added to the free hydroxyl groups of cellulose, such as ether, ester, and unstable derivative acetal. These groups introduced steric hindrance that disturbed intermolecular hydrogen bonding formation, leading to total dissolution. If this solvent system is used on cellulose, the functional group must adhere to a single rule; it must be able to detach itself from the cellulose macromolecular structure. The problem with derivatizing solvent is weak reproducibility and unknown and uncontrolled side reactions during the dissolution process.

As for non-derivatizing, the solvent can be branched into aqueous and non-aqueous media. This solvent system does not involve the chemical modification of cellulose structure. It is known as "non-derivatizing" as no intervention of the covalent bond occurs during the dissolution process. The weakness of this system is that self-aggregation of cellulose is highly likely to occur, forming inhomogeneity and incomplete dissolution. For the aqueous medium of a non-derivatizing solvent system, the aqueous alkaline–urea system is highlighted. This simple, green, cost- and time-efficient technique works efficiently in dissolving cellulose. Like any other solvent type, specific requirements on cellulose quality must be obeyed. Similarly, a multicomponent solvent system is used to promote better dissolution of cellulose for a non-aqueous non-derivatizing solvent system,

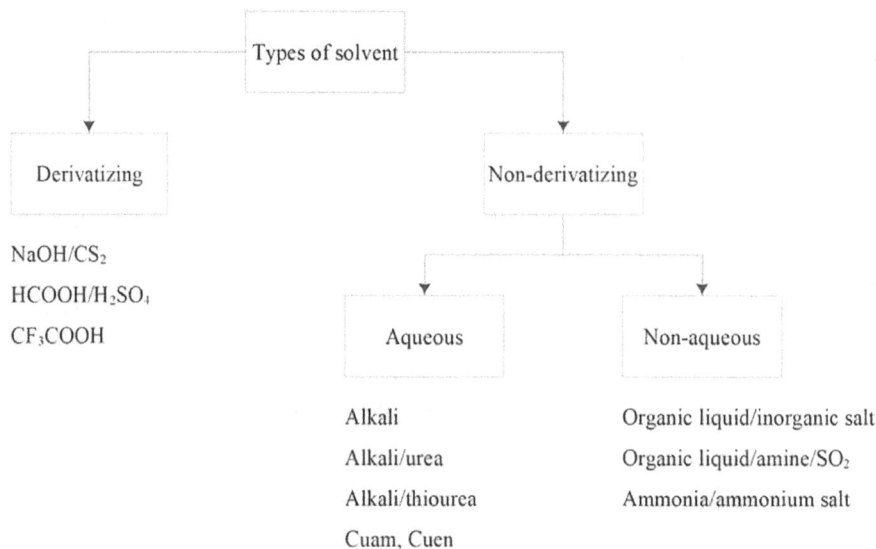

Types of solvent

Derivatizing

Non-derivatizing

NaOH/CS$_2$
HCOOH/H$_2$SO$_4$
CF$_3$COOH

Aqueous

Non-aqueous

Alkali
Alkali/urea
Alkali/thiourea
Cuam, Cuen

Organic liquid/inorganic salt
Organic liquid/amine/SO$_2$
Ammonia/ammonium salt

FIGURE 15.1   Classification of solvent for cellulose dissolution.

also known as an ionic liquid solvent system. This ionic liquid solvent system is so effective because it can dissolve cellulose on a large scale. Still, it has a restricted application because of its high cost, uncontrollable viscosity, and sensitivity to moisture. In general, the complete dissolution of cellulose is a highly uncommon cause. On the other hand, the partial dissolution of cellulose and the inhomogeneity of the cellulose solvent are not entirely undesirable as they may help promote the product's strength and stability, particularly when producing hydrogels.

When successfully dissolved, cellulose can be regenerated into many physical forms. In their liquid state, self-aggregation of cellulose is slowly taking place and accelerated when exposed to external stimuli. A control self-aggregation process is most desired, but it is seldom achieved as regeneration of cellulose occurs in a randomized action. If the regeneration process occurred an uncontrolled rate, the quality of the regenerated products is unlikely to be good. For instance, when the regeneration occurs faster, it is usually associated with a larger pore size, which denotes instability. Physical or chemical interactions could induce the regeneration of cellulose. For physical interactions, coagulation is a transition of a liquefied solution to a solid via hydrogen bonding (non-permanent linkages) interaction. The cellulose solvent is diffused from its matrix induced by osmotic pressure, and the nonsolvent penetrates the matrix, aiding self-aggregation. Even without the nonsolvent system, exposing the cellulose solution to air could also help promote regeneration. The regeneration process, structural modifications, and cellulose aggregation state mechanism are crucial for successfully manufacturing regenerated cellulose materials. Ionic liquids, H$_2$SO$_4$, Na$_2$SO$_4$, Na$_2$SO$_4$, (NH$_4$)$_2$SO$_4$/H$_2$O, and CH$_3$COOH are the most often employed coagulants for cellulose regeneration. For chemical interactions, regeneration requires the formation of covalent bonds (permanent linkages) with the aid of a specialized crosslinker. The drawback of this technique is the by-products produced after the reaction. This regeneration technique is mechanically more robust than the physical interactions as permanent linkages assure strong intermolecular interaction and good resistance from deformation.

REGENERATED CELLULOSE PRODUCTS

The dissolution–regeneration process has unlocked many potentials in turning cellulose from various sources into high-value-added products. The resultant regenerated cellulose-based products have experienced massive growth coupled with the technological advancement, which has assisted in the improved version of the respective products. Since cellulose is widely acknowledged for its water-soluble, safe, and readily available characteristics, its potential to become the predominant base material in extensive products used in agriculture, medicine, cosmetics, sensor technology, electrical devices, etc., is proliferating. Currently, the research on regenerated cellulose products such as hydrogel (Salleh et al., 2019; Zainul Armir et al., 2022), fibers (Ma et al., 2020), thin film (H. Zhang et al., 2018), and membrane (Gan et al., 2015) have entered the advance application phase. The different solvent system plays a great role in determining the final properties of regenerated cellulose products and the anti-solvent used in the regeneration process. The disturbance in the inter- and intramolecular hydrogen bonding that causes the transformation of cellulose polymorph from I to II is the uniqueness of regenerated cellulose in attracting research interests. Besides, the increase in amorphous regions and cellulose II crystallites *per se* have opened various chemical modifications with other promising materials. Thus, this section will discuss regenerated cellulose products by exploring the fundamental aspects of the regeneration process, final characteristics, and potential applications.

HYDROGEL, AEROGEL, XEROGEL, CRYOGEL

Aerogel, xerogel, and cryogel are the products that extend from hydrogel and vary in morphological structure, appearance, and properties. The differences mentioned above have been driven by the preparation method, notably how they are dried from the wet hydrogel. Thus, the different properties will also lead to various applications tailored to the fabricated products. In general, hydrogel is widely accepted as a three-dimensional crosslinked interconnected network that is able to absorb and retain water mainly subjected to the hydrophilic groups of the polymers used such as -OH, -NH$_2$, -COOH, -COO$^-$, and -SO$_3^-$ (Kabir et al., 2018). The polymers used as the hydrogel's precursor are extensively derived from natural, synthetic, or hybrid polymer type covering biodegradable and non-biodegradable aspects. It is built by using different crosslinking methods widely classified into physical and chemical crosslinking, where different crosslinking methods dictate different final properties of hydrogel (Abdeen, 2011). Chemically crosslinked hydrogel exhibits a tough and stable product because the crosslinking mechanism is permanent and irreversible, subjected to the new chemical bonding formation in the interconnected networks (Saldivar-Guerra & Vivaldo-Lima, 2013). However, on the contrary, the physically crosslinked hydrogel shows a fragile polymeric network. Still, it is opted in some usage due to reversible and "green" reaction that can be applied as a self-healing device. The use of crosslinker type relies on the type of application such as chemical hydrogel is favored for the fertilizer carrier in smart fertilizer systems and tissue replacement material for the physical hydrogel. Regardless of the interconnected polymeric network outcome, regenerated cellulose has been proven to be a flexible, excellent, tunable material and compatible for both crosslinking types. The numerous -OH groups generated from regenerated cellulose have enabled extensive flexibility toward various types of crosslinkers. Several steps are performed to produce regenerated cellulose hydrogel: (1) the dissolution sources of cellulose that can be obtained from any sources of interest in any desired compatible solvent system, (2) crosslinking process, and (3) regeneration of the crosslinked solution in any desired anti-solvent (Zainul Armir et al., 2021). After the regeneration process, the resultant regenerated cellulose hydrogel undergoes several drying methods for liquid removal to form aerogel, cryogel, and xerogel.

The hydrogel is well-known for its characteristic similar to water, where its swelling behavior is one of the most significant aspects that have attracted a wide research interest. Regenerated cellulose with a more amorphous region than native cellulose creates larger void volumes in the resultant hydrogel and produces a larger swelling result (Kassem et al., 2020). Also, a larger void enhances the cohesive force between an interconnected network wall and water, which can retain the absorbed water. Besides, the more amorphous region of regenerated cellulose also enhances the reactivity with the crosslinker by forming more intermolecular hydrogen bonding (Yang et al.,

2019). As a result, various crosslinkers affect the characteristics of regenerated cellulose hydrogel differently. For example, the most common crosslinker epichlorohydrin tends to create rigid and low swelling of regenerated cellulose hydrogel upon increasing the crosslinker content subjected to more reaction with -OH groups (Gan et al., 2017; Yang et al., 2019). In contrast to epichlorohydrin, N, N-methylenebisacrylamide is also recognized as a water-soluble and non-toxic crosslinker. A higher concentration of N, N-methylenebisacrylamide results in a bigger swelling of the hydrogel by creating a larger void in the linked polymeric network to some extent (Geng, 2018). With the mentioned circumstances, it highlights the benefit of regenerated cellulose in having more amorphous regions that could be tuned according to the nature of crosslinkers.

The transformation of cellulose I to cellulose II structure possessed by regenerated cellulose had caused the disruption of the orders structure of native cellulose and resulted in more loose hydrogen bonding, as described by Kassem et al. (2020). The advantage of having a looser structure in regenerated cellulose hydrogel makes it more flexible when mixed with other materials. For instance, the loose structure of regenerated cellulose hydrogel has facilitated the physical entanglement with other materials of polyvinyl alcohol chains that eventually create tougher hydrogel (Ding et al., 2021). Furthermore, the numerous -OH groups provide stability toward regenerated cellulose hydrogel for filtration technology such as oil filter systems whereby the hydrophilic parts that cling to the -OH parts are stabilized due to the abundant -OH group, while glucopyranose rings will attach to the oil (Jiang et al., 2019).

In terms of thermal behavior, regenerated cellulose hydrogel provides low thermal stability since low degradation temperature is recorded in various studies due to disturbance toward inter- and intramolecular hydrogen bonding during the dissolution and regeneration process (Shin et al., 2020; H. Zhang et al., 2020). However, thermally stable char was produced due to a higher amount of residue at an increasing temperature produced by regenerated cellulose hydrogel subjected to the easier decomposition of the polymeric network, as reported by Huang et al. (2019). Theoretically, the thermally stable char generated by regenerated cellulose hydrogel acts as the barrier by insulating the surface to avoid further decomposition that can enhance fire-retardant properties (Zhu et al., 2020).

The methods applied to remove liquid from a hydrogel are crucial in determining the hydrogel's dried state, which are aerogel, xerogel, and cryogel. The liquid removal via supercritical drying, freeze-drying, and ambient pressure drying resulted in the formation of lightweight aerogel, cryogel, and xerogel (Budtova, 2019). The appearance of aerogel and cryogel is similar as they are opaque and white, while xerogel is translucent and slightly yellowish color (Zainul Armir et al., 2021). The most crucial aspect to consider during the formation of aerogel, xerogel, and cryogel is the structure collapsing as these products are prone to high shrinkage when subjected to drying. Since regenerated cellulose is used, where cellulose itself is as a water-soluble polymer, it contributes to larger pores during drying. Alongside, the choice of cellulose (depending on its Mw, DP), dissolution solvent, regeneration solvent, crosslinking application, and direct coagulation for the formation of cryogel, aerogel, and xerogel plays vital roles in the outcomes of the respective products. Minimal studies are investigated on regenerated cellulose xerogel since it has a highly shrunk and compact structure that restricts its usage for broad applications. Also, the great shrinkage that occurred in xerogel is subjected to the capillary pressure alongside the densification of regenerated cellulose and causes low-density and porosity products (Buchtová & Budtova, 2016).

Regarding the drying effect, aerogel and cryogel will experience higher shrinkage upon the respective products' formations from hydrogel on the ice crystal formation before the drying process. The higher Mw and DP of cellulose have contributed to the low density or lower volume of pores due to a high viscose dissolution formed since the cellulose–cellulose interaction is greater than cellulose–solvent interaction (Tyshkunova et al., 2021). Further drying of aerogel and cryogel will result in more shrinkage, leading to reduced density and surface area, which are reported to be irreversible due to the hornification process (Navarra et al., 2015). Thus, the drying method as the critical step is continuously improved to avoid a higher shrinkage phenomenon.

Aerogel and cryogel have a similar appearance, and cryogel is sometimes called aerogel in the literature. However, cryogel production through the freeze-drying technique is usually improved using the supercritical drying technique. As an example, carbon dioxide is majorly employed as the supercritical miscible fluid for the supercritical drying method due to its excellent properties, such as non-toxic, cheap, chemically inert, and mild critical point and pressure (Budtova, 2019). Moreover, supercritical drying is also considered one of the methods that produce the lowest shrinkage toward dried hydrogel, notably in retaining pore volume (Lin & Jana, 2021). The benefit of using regenerated cellulose hydrogel as the precursor of aerogel fabrication is that it can alter the solvent exchange during the regeneration process to control the shrinkage of pores.

In regard to the pore structure, in the case of monolithic aerogel formation, the surface of aerogel is observed to be less porous due to the regenerated cellulose-rich region, while larger pores are obtained along the bottom of the sample (Lin & Jana, 2021; Xie et al., 2021). This phenomenon could be explained by the formation of the fibrillar network from an aggregated network that occurred during the slow progress of the regeneration process that moves along from the top to the bottom of the sample. On the other hand, the freeze-drying of hydrogel that leads to the formation of cryogel is often associated with the cracked structure due to the large porosity obtained from the frozen water that forms ice crystals during the freezing process. Besides, some literature recorded that the two-dimensional sheet morphology was observed in the regenerated cellulose aerogel samples (Beaumont et al., 2016; Zeng & Byrne, 2021). This morphological structure is also observed in regenerated cellulose aerogel samples but with less appearance subjected to the drying technique. The sheet is the result of the ice crystal expansion during the freezing process, and the sheet formation can be reduced by using the instant freeze technique by using liquid nitrogen (Zeng & Byrne, 2021). Instant freezing will shorten the time taken for the ice crystal to expand. Increasing the crosslinker concentration will lead to a considerable reduction in the two-dimensional sheet structure in the aerogel and cryogel that is subjected to the disruption of hydrogen bonds between microfibrils of cellulose formed during the regeneration process by the covalent bonds arise from crosslinking (Moosavi et al., 2020). The formation of aeogel, cryogel, and xerogel from regenerated cellulose hydrogel is summarized in Figure 15.2.

Fibers

Regenerated cellulose fibers are man-made fibers and have potential for many applications outside the textile industry. Two types of regenerated cellulose fibers are available to date: (1) viscose fibers – first generation and (2) lyocell fibers – second generation (Santamala et al., 2016). The Lyocell fibers are favored compared to the viscose fiber due to the higher mechanical strength widely employed for reinforcement and the sustainable approach by dissolving in a specific solvent and regenerating in water. Lyocell fiber is produced through the dry jet-wet spinning from the

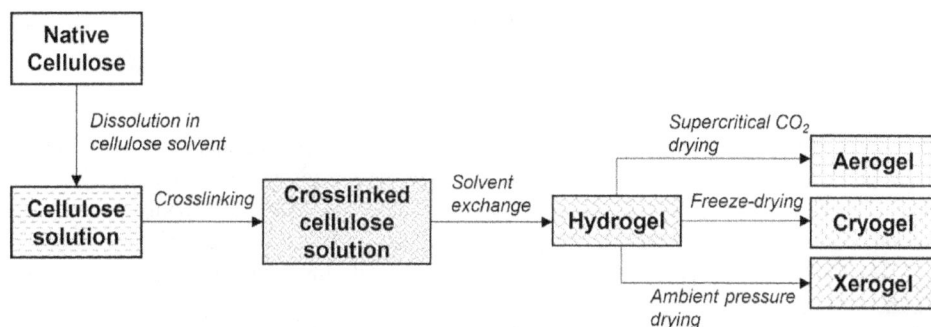

FIGURE 15.2 Schematic presentation of regenerated cellulose hydrogel in forming aerogel, cryogel, and xerogel.

N-methylmorpholine N-oxide (NMMO) cellulose solvent (Harlin & Leppa, 2019). Furthermore, the high mechanical strength of lyocell fibers is subjected to the lower elongation of cellulose microfibril orientation that is highly oriented at the fiber longitudinal axis (Ma et al., 2020). Continuous research has found various cellulose solvents, such as ionic liquids and alkali/urea, as the solvent to produce the regenerated cellulose fibers, where the alkali/urea system is favored due to its cheaper option. The oldest viscose fibers are still available and are currently being used despite the hazardous waste of hydrogen sulfide gas produced by the fibers (Li et al., 2021). Despite the toxic production of viscose fibers, it has presented fundamental knowledge about the utilization of cellulose into regenerated cellulose for the fiber formation. Along with the technological advancement, the emergences of new man-made fibers from regenerated cellulose such as the Tencel™, Super 3 cord, Ioncell, and Fortisan have shown improved mechanical strength and high fatigue and are tougher and are employed from the regenerated cellulose (Moriam et al., 2021; Sharma et al., 2021).

The regenerated cellulose fibers acknowledged for their porosity has receives wide attention since they have more applications, such as wearable medical devices, electrical devices, and smart textiles. A few parameters and conditions must be highly considered and evaluated that significantly affect the properties of regenerated fibers: first, the type of solvent and procedure for the dissolution–regeneration process. For example, regenerated cellulose fibers from alkali/urea solvent have low mechanical strength when regenerating in water. Qiu et al. (2018) regenerate fibers in phytic acid, eventually improving the mechanical strength subjected to the more oriented fibril arrangement. The authors also emphasized the balance between the dissolution and regeneration process because vigorously fast regeneration leads to fragile fiber formation.

Other than the type of cellulose solvent and anti-solvent for fiber fabrication, the drying method as the final step is relatively important in producing porous fibers. Supercritical drying and freeze-drying have provided a higher porosity of regenerated cellulose fibers, where no shrinkage is observed for the former drying; however, air-dried fibers have resulted in classic morphology of regenerated cellulose fibers (Zeng & Byrne, 2021). Moreover, spinning parameters, such as draw orientation in the air gap, coagulation bath desolvation temperature, and dope temperature, are critical in producing strong and quality regenerated cellulose fibers (J. Zhang et al., 2017, 2020). Next, the degree of cellulose orientation that involves the amorphous and crystalline region also affects the properties of the regenerated cellulose fibers. In relation to this, for the fiber morphology, regenerated cellulose fibers possess smooth, micro-protruded filament, and scaly structures that are more possibly formed due to disrupted hydrogen bonds and a shorter chain of cellulose microfibrils during the dissolution–regeneration process (Xue et al., 2021). Having said that, these abundances of the protruded and shorter chain of microfibrils enable homogenous interaction with other materials, which improve the regenerated cellulose fibers' mechanical strength. Considering this, the shorter hydrogen bond length in the regenerated cellulose fibers has caused a compact and closed packing unit that has contributed to the excellent tensile strength close to the lyocell (Xue et al., 2022). Besides, the higher amorphous regions in regenerated cellulose enable small molecule movements when stress is applied, making regenerated cellulose fibers retain a higher load (Moriam et al., 2021).

Membrane and Thin Films

The commercial membranes have structures of microporous and hydrophilic non-porous membranes, which are non-degradable and fabricated as they are made from expanded polytetrafluoroethylene and hydrophilic polyurethane, respectively. Therefore, they do not only pose hazardous environmental issues but also consume high energy during the complex fabricating process (Shi et al., 2019). On the other hand, regenerated cellulose membranes and thin films are much more environmentally friendly.

Membranes and thin films of regenerated cellulose are produced with a step-by-step process initiated with cellulose dissolution in a solvent system, followed by casting and immersion in a coagulation bath. Then, the resulting membrane is soaked in deionized water to neutralize before the drying process to obtain the desired regenerated cellulose membrane (Mazlan et al., 2019). The mechanical properties differ based on the type of solvents, crosslinking methods, and processing strategies. Chemical crosslinking offers the material good elasticity, toughness, and strong mechanical properties. A strong membrane with breaking strength and elongation at break was obtained, which displayed self-adaptive breathability and excellent biocompatibility (Tu et al., 2021). In recent times, membranes and thin films have been made to function in water purification systems and wastewater treatments to curb water pollution through a method such as electrochemical treatments, flocculation-coagulation, and photocatalysis. In water treatments, the membranes work on the adsorption principle (Azmi et al., 2021).

Films via viscose (cellophane) and cuprammonium (cuprophane) are employed in food casing, cosmetics, medical and pharmaceutical packaging, and pressure-sensitive tapes. The cellophane films display larger tensile strength and smaller elongation than the synthetic polymer films in water purification (e.g., polypropylene and polyethylene terephthalate). Besides, composite films and membranes such as fluorescent and photoluminescent cellulose films have been produced by treating the never-dried regenerated films with fluorescent dyes. The composite film was prepared from a cellulose solution comprising photoluminescent alkaline earth aluminates embedded in the matrix of the photoluminescent film. The photoluminescent film could emit visible light for more than 10h in the dark after 10min of sunlight exposure. Films constructed from hydrophobic polyaniline (PANI)/cellulose solution exhibited high homogeneity, fair miscibility, excellent mechanical properties, and conductivity (Wang et al., 2016).

APPLICATIONS

A great number of fields of industries utilize regenerated cellulose to its potential. Regenerated cellulose extends its use in medical, agriculture, automotive, aerospace, textile, and many more industries. Extensive research into regenerated cellulose adaptability and its many facets has helped create a suitable product for the above industries.

MEDICAL

In the medical field, regenerated cellulose in the form of cellulose derivatives has been supportive in biomedical, healthcare, and pharmaceutical industries as they release matrices, tablets, granules, delivery systems, stabilizers, semi-solid gelling agents, artificial wound dressing, and cell capsulation. Hydrogels are known as encapsulating agents for probacteria and continuous delivery upon association with different food matrix systems. Self-healing and remolding properties were achievable using a hydrophobic hydrogel network produced by cellulose nanowhiskers (extracted crystalline fractions of cellulose), acrylamide, and stearyl methacrylate. The unique feature of the formed hydrogel possessed good mechanical strength, making it highly suitable in different applications within the biomedical field. Hydrogel, known for its water-absorbing and -holding capacity found to have treated edema issues. Phosphor-doped cellulose hybrid hydrogel with epichlorohydrin (crosslinker) is known to assist in bioimaging applications. Its ability to fluoresce and glow for long durations prevented the use of dangerous radiations and further eased the detection under the skin (Singh et al., 2015).

Bacterial cellulose (BC) is produced by gram-negative bacterial cultures *Gluconacetobacter, Acetobacter, Agrobacterium, Achromobacter, Aerobacter, Sarcina, Azobacter, Rhizobium, Pseudomonas, Salmonella,* and *Alcaligenes* extracellularly. The BC is regenerated from its original chemical structure consisting of $(1\rightarrow4)$-D-anhydroglucopyranose chains bounded through ß-glycosidic linkages (cellulose I) to anti-parallel packing (cellulose II) formed by treatment with sodium hydroxide known as mercerization process. This BC has an immense contribution to regenerative and diagnostic medicine due to its properties, such as being highly pure, biocompatible, and versatile material. BC membranes serve as wound dressing material due to their high *in vivo* biocompatibility. The commercial BC membrane is the wound dressing device with trademarks, such as Bionext®, Membracell®, and Xcell® that impersonates the extracellular matrix to enhance epithelialization. It is also reported that BC membrane wound treatments are more efficient than traditional gauze and synthetic materials such as Tegaderm®, Cuprophan®, or Xeroform™. The BC membranes expedite the epithelialization and tissue regeneration process in wound healing and burn treatments. Not only that, BC membranes aid the removal of necrotic residues and act as a natural graft for a tympanic membrane, which suffers from local perforation. Nevertheless, the BC membrane is limited in bacterial infection and seems unsuitable for prolonged occlusion intervals, owing to fast release rates during infections or epithelialization process treatments. Consequently, BC silver-containing composites are developed to prevent bacteria growth and bacteriostatic and bactericide effects. BC works as a stabilizing agent to control particle nucleation, avoid aggregation, and produce nanosilver particles. The BC fiber surface porosity and hydrophilicity facilitate synthesizing and stabilizing the nanoparticle.

BC composites are also used as biosynthetic grafts to treat eye diseases. BC composites are employed creatively in ensuring their ability to adhere to and enhance the proliferation of retinal pigment epithelium and keratinocytes. BC grafts also lower the rejection rates of transplanted corneas and improve the treatment of eye diseases. Moreover, BC biocomposites are mentioned to adapt the material's characteristics in eye therapeutics fully. It has been reported that BC with polyvinyl alcohol raises the light transmittance and UV absorbance. Surface modification by chitosan and carboxymethylcellulose (CMC) in BC has increased BC's hydrophilicity, displaying an improved retinal pigment epithelium proliferation. BC composites have also known to be able to grow corneal stromal cells while enhancing complete vision in the patients. Furthermore, BC's bioengineering has contributed to generating convex-shaped contact lenses to rectify presbyopia, astigmatism, hyperopia, and myopia. BC-based contact lenses are great potential for wound dressing after eye surgery, replacing antibiotics eye drops, or ameliorating ocular burns recovery (Picheth et al., 2017).

BC is an excellent material for 3D-bioprinting in the medical field, such as bioprinting costal, auricle, and nasal cartilage. After 1 week of implantation of increased cellulose content (17%), the modified BC showed similarities with high compatibility with human auricular cartilage. BC plays an essential role in oral implants and bone regeneration techniques. Moreover, BC replaces collagen as a shielding membrane as it has a low biodegradation rate. In tissue engineering, BC shows up as a feasible material, especially in stem cell culture. A research study verified that the BC membrane could inhibit the differentiation of mouse embryonic stem cells and improve mouse embryonic fibroblast cultivation compared to the conventional culture media. BC can also serve as a scaffold for tissue engineering (Popa et al., 2022).

The advances in drug delivery have explored regenerated cellulose in the fabrication of two-piece hard shell capsules (dip-coating approach) as a medium for extemporaneous controlled release. The release mechanisms of active pharmaceutical ingredients (APIs) encapsulated in self-pore forming regenerated cellulose capsules were observed. The API release is bound to diffusion and osmotic mechanisms of the self-formed pores. High aqueous soluble APIs have osmotic mechanisms as the main release mode in the regenerated cellulose capsule and lowered diffusion through the hydrated capsule wall. The dominance of a flux mechanism across the capsule wall depends on the pore size variations, which are self-formed when exposed to aqueous conditions. The large pores shift the mechanism of API release from osmotic to diffusion mechanism, and the smaller pores only take part in the encapsulated API osmosis-mediated flux. The fluid permeability of the regenerated cellulose capsule showed that it is more suitable as an osmotic drug delivery ground (Bhatt & Kumar, 2017).

AGRICULTURAL

The intense land development for the growing human population has turned conventional agriculture into modern farming. However, lack of arable land, water supply to crops, and labors are

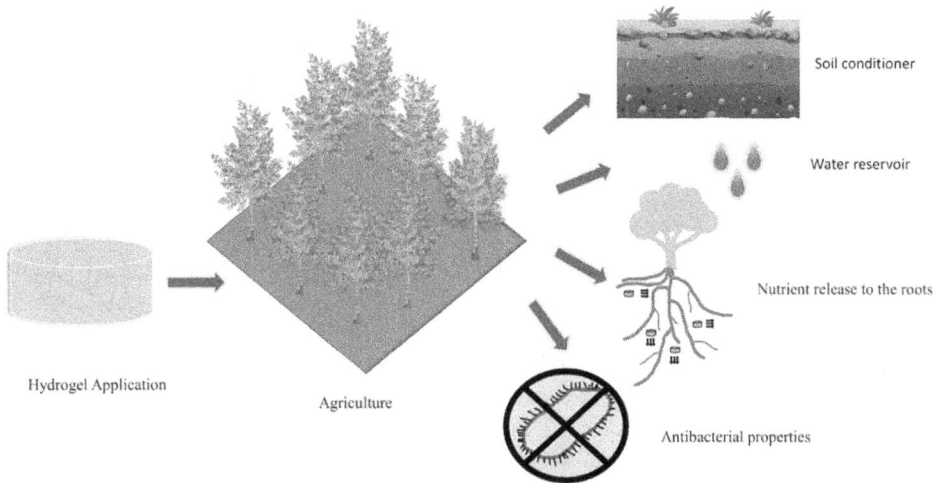

FIGURE 15.3 Hydrogel applications in agriculture sector.

concerns in conventional farming. Hydrogel has always been introduced in soil application as a soil conditioner and controlled-release fertilizer (CRF) (Winarti et al., 2021). They also influence soil permeability, density, structure, texture, evaporation, and water infiltration rates through the soil. Hydrogel application has also gathered immense scientific attention and interest in soilless agriculture. Figure 15.3 shows the use of hydrogel in the agricultural sector. Hydrogels' well-known feature of water holding capacity, as described previously, is interesting because it is so useful in the seed germination processes in agriculture. The water-holding capacity feeds the seeds with adequate supply throughout the germination process. The early stages of seed germination is seed hydration, known as seed imbibition, followed by initiating principle metabolic processes (Sarmah & Karak, 2020). Additionally, granulated nutrients with materials like CMC and chitosan have been developed to improve seed germination and increase the leaf water and chlorophyll content in arid regions. Improved root development, plant growth, minimized nutrient losses by leaching, and soil penetration can substantially decrease the adverse effects of water stress after plant transplantation.

Hydrogels are also seen as great water reservoirs in agriculture. The development of superabsorbent polymer is intended to reduce the release of water from the network (Palanivelu et al., 2022). Superabsorbent of CMC and starch crosslinked with aluminum sulfate octadecahydrate are reported to be useful even in the midst of the acute drought. The hydrogel can be synthesized to provide water, nutrient, and agrochemicals in sufficient amounts. CMC and hydroxyethylcellulose-based hydrogels can be effective water reservoirs in farming. Hydrogel based on methylcellulose, polyacrylamide, and calcium montmorillonite is an efficient nutrient carrier. The addition of urea leads to a more porous hydrogel structure, enabling absorption capabilities (Singh et al., 2015).

Hydrogel releases water and nutrients targeting a plant's root zone when it detects dryness in the soil; thus, scientists develop new techniques to increase the intervals of nutrient release and decrease fertilizer loss through rain or irrigation water. In agriculture, hydrogels are commonly utilized as slow-release fertilizer (SRF) and CRF. SRF and CRF release nutrients gradually and ensure nutrient availability for longer than standard fertilizer. These improved fertilizers reduce run-off and decrease the split application of fertilizer, which causes increased cost (Rop et al., 2018).

SRF-based synthetic polymers such as poly(acrylic acid), poly(acrylamide), and copolymer are non-biodegradable. Thus, this matter proposes natural polymers for agriculture applications due to their low cost, readily available, and biodegradability properties. SRFs are synthesized in photopolymerization, suspension polymerization, reversible addition-fragmentation chain transfer polymerization, solution polymerization or aqueous polymer solution, and free radical polymerization.

Fertilizers can be incorporated into the hydrogel using a two-step process: (1) the dry hydrogel is submerged in liquid fertilizer and then the swelled hydrogel is dried, and (2) in an *in situ* process, all components, including the fertilizer, are put into the reaction mixture during polymerization, entrapping the fertilizer within the hydrogel matrix (Ramli, 2019).

In a CFR application, the factors affecting the rate and duration of the nutrient release are known and controlled. CRF encapsulated with κ-carrageenan hydrogel was prepared and tested for mechanical properties and as an end-use product. It was found that the gel does not disrupt acidic pH and can reduce the speed at which the fertilizer's NO_3^-, PO_4^{3-}, and NH_4+ions diffuse into the medium. The NPK fertilizer encapsulated with κ-carrageenan hydrogel may potentially be a CRF (Rozo et al., 2019).

Hydrogel with antibacterial properties was developed by incorporating chitosan and chitin. Chitin is a long-chain polymer of N-acetylglucosamine, a polysaccharide that is the primary component of the cell walls of fungi and the exoskeletons of arthropods such as crustaceans and insects, the radulae of mollusks, cephalopod beaks, the scales of fish, and the skin of lissamphibians. Chitin is also found to promote rhizobacteria, which cause increased plant growth and is sometimes associated with biological control of plant pathogens, nutrient cycling, and seedling establishment and enhances the growth.

The derivative of chitin is chitosan, which is obtained by treating the chitin shells of shrimp and other crustaceans with an alkaline substance, such as sodium hydroxide. Chitosan not only displays antiviral, antifungal, and antibacterial, antimicrobial properties but is also biocompatible and biodegradable (Michalik & Wandzik, 2020). In addition, chitosan helps plants become more resistant to biotic and abiotic stresses (Sharif et al., 2018). It plays a vital role in the defensive system in plants and plant-growth promotion (Ahmed et al., 2012). Generally, chitosan is used as a coating material onto fruits, seeds, and vegetables to increase host defense. It has been reported that chitosan treatment increases the nitrogen metabolism-related key enzyme activities (protease, glutamine synthetase, and nitrate reductase) and improves the transportation of nitrogen in the functional leaves. This results in an increase in the plant's growth and development due to the improved transportation of nitrogen. In a separate study, chitosan-treated adventitious root cultures elicitor individually or in chitosan/pectin combinations showed enriched biosynthesis of secondary metabolites at an optimal rate of 0.2 mg/mL chitosan concentration. This concentration resulted in 103.16 mg/g of anthraquinones, 48.57 mg/g of phenolics, and 75.32 mg/g of flavonoids dry weight (Baque et al., 2012).

Automotive

Nowadays, automobile makers struggle to manufacture green composite with lightweight materials for hybrid electric car and electric vehicle batteries. In addition, using a composite can minimize carbon emissions and enhance fuel efficiency. All materials used in the automotive industry are required to fulfil the basic requirements such as lightweight, cost, safety, strength, crashworthiness and recycling, and life cycle considerations (Baque et al., 2012). Consequently, regenerated cellulose products are the suitable reinforcing agents in the composite for this application. Among the regenerated cellulose products, regenerated cellulose fibers such as lyocell and viscose fibers have a great potential to be utilized in semi-structural applications such as automotive parts. Nonetheless, it acts as a reinforcement material in the thermoplastic and thermoset composites as an alternative to natural and synthetic fibers. Based on the literature studies, regenerated cellulose fiber-based composites are considered mediocre compared to glass fibers due to their mechanical strength. However, glass fiber has its limitations, such as being difficult to handle, having high fiber density, poor adhesion to the matrix, not being environmentally friendly, and unexpectedly, hazardous to human health. For instance, it can cause lung diseases. On the other hand, the innovation of using natural fibers as reinforcement in making composites received great attention. Nevertheless, the diverse quality, high water absorption, poor adhesion, harvesting technique, and maturity of plants become the limitations of utilizing them for automotive applications, primarily indoor automotive parts, as they can give an unpleasant odor and discomfort to consumers.

There are various benefits of using regenerated cellulose fiber-reinforced composites: their reduction of overall cost, lightweight, durability, and flexibility. Additionally, low density and non-abrasiveness are the best properties of regenerated cellulose for use in the automotive industry. Interestingly, composite behavior with high corrosion resistance and extreme weather are the best attributes for automotive use. Hence, interior parts like a dashboard, door panels, seat cushions, underbody panels, and external parts like trunk lids and bumper can be manufactured. Regenerated cellulose fibers are commonly reinforced in the matrix, such as polypropylene, polylactic acid, and polyamide. In addition, diverse techniques can be carried out for the molding process on the composites to obtain the quality products such as compression molding, injection, extrusion, sheet, and resin transfer molding. These methods are commonly used to make regenerated cellulose fiber-reinforced composites. However, the favored method employed is compression molding as this method satisfies all the requirements for composite properties.

Mechanical, thermal, and viscoelasticity of composites are the properties of composites that should be focused on producing remarkable automotive parts. Various factors could affect the performance of composites, such as orientation, fiber content, aspect ratio, and fiber–matrix interaction (Ranganathan et al., 2015). A higher amount of fiber content in the composites increases the impact resistance of composites. Evaluating the energy absorption and toughness of composites is significant to identifying the crashworthiness of automotive structures. Impact testing like Izod impact strength is required to carry out to determine the energy absorption of products. Fiber length could affect the end products, such as short and long fibers. A long fiber is preferable to a short fiber because it increases energy dissipation, which causes higher impact strength and improves energy absorption. Hence, it is advisable to incorporate the longer fiber than critical fiber length in the composites as it enhances the fiber–matrix interaction, leading to fewer pull-outs and reduced impact strength and fracture toughness. Likewise, the presence of regenerated cellulose fiber can bridge the cracks in the matrix and resist crack propagation and crack opening to have occurred. As a result, regenerated cellulose fiber can be used as an impact modifier in any composites.

AEROSPACE

The effort by researchers to study the development of polymer composites for aerospace systems is rising. Lightweight, high stiffness, and excellent resistance toward thermal and high strength are the basic characteristics of composites to be used in the aerospace sector. Aerospace is a field that produces vehicles commonly related to the atmosphere and outer space, such as military airplanes or helicopters, commercial airliners, aircraft, and spacecraft. The production of any components in the aerospace sector should satisfy security assurances and safety levels. Hence, the making of polymer composites should meet the desired requirements.

Like automotive applications, cellulose-based products like regenerated cellulose fiber have been used extensively in the aerospace industry. The rationales for utilizing regenerated cellulose products are low density, non-toxicity, abundance, and renewable. For that reason, regenerated cellulose fiber-reinforced composites are appropriate for any metal components in the aerospace industry due to their heavy weight and low corrosion resistance. Commonly, regenerated cellulose fiber like viscose and lyocell fiber reinforced in the matrix to form composites provide minimum airframe weight, leading to low fuel consumption and inexpensive operating costs. The continuous, long fibers with dimensional variations are the desirable characteristics of regenerated cellulose fibers that should be incorporated into the matrix. In addition, the optimum amount of the regenerated cellulose content in the composites provides excellent mechanical properties. The high resistance to heat and flame made the regenerated cellulose fiber-reinforced thermoplastics and thermosets capable of being manufactured as parts of the aircraft body for the interior panel (Muhammad et al., 2021). Common aerospace applications of regenerated cellulose fiber-reinforced composites are rotor blades, propellers, seats, and aircraft wings.

During the fabrication of aerospace materials, the orientation, fiber length, composition, dispersion, aspect ratio, chemistry, and matrix selection have to be prioritized to achieve excellent performance and functionalities of end products. Several aspects should be determined to ensure the materials are suitable and safe for aerospace systems. The mechanical properties of composites reinforced by regenerated cellulose fiber should be appropriately examined as the safety of materials is significant for the end-user.

It is necessary to conduct specific analyses, such as static, dynamic, and impact testing, to assess the rigidity and strength of material structures. Other than that, the strength of materials concerning fatigue and vibrations, impact strength, and fracture properties should be determined to ensure the materials can withstand long periods of flying. Hence, the production of any components in aerospace systems should fulfilled the excellent fatigue, impact, and fracture properties. Next, the structural stability of aerospace materials should have high resistance to damage or, in other words, strong. Aerospace materials tend to be exposed to conditions like high temperatures, lightning strikes, and freezing, including corrosive liquids. Thus, the structures used should not easily oxidize, crack, or corrode when encountering such conditions. Determining materials' strength is significant to assure a long life cycle. Therefore, these factors should meet aerospace systems' requirements to avoid accidents.

TEXTILE

In the new era of globalization, sustainable and green products are demanded for the production of textiles. Textile can be defined as fabric made from yarn, the second basic human need. The exploitation of textiles has been thriving, causing the necessity of fiber production to increase. The utilization of regenerated cellulose fibers via the green spinning process caught researchers' attention, which is biodegradable and safe for the environment (Salleh et al., 2021). Regenerated cellulose fibers, considered natural/man-made fibers, have been used extensively in textile applications, similar properties to commercial synthetic fibers like nylon and polyester. Viscose and lyocell fibers are the two types of regenerated cellulose fibers utilized widely in textile products. These fibers act as a substitution for cotton or silk fabric.

The oldest commercial fiber, viscose rayon fiber, which can be known as "artificial silk," is a soft fiber. It has high drapability, which is commonly used in making blouses, dresses, lingerie, linings, slack, skirts, jackets, and sportswear. Besides, the fabric based on viscose fiber is breathable, and the moisture-absorbent properties of the fiber make the consumers feel comfortable wearing the clothes. The steps for making viscose rayon are as follows: (1) steeping (mercerization process), (2) shredding, (3) aging, (4) xanthation, (5) ripening, (6) preparation of the spinning solution, (7) filtration, and (8) wet spinning. However, viscose fiber has limitations such as low abrasion resistance, poor elastic recovery, low wrinkle resistance, and wet shrinkage.

To date, lyocell fibers with the brand name "Tencel" are considered preferable to replace viscose fibers. This is because the processing of viscose fiber does not comply with the environmental requirement and has low resiliency (Jiang et al., 2020). The usage of sensitive and hazardous chemicals like carbon disulfide is one of the reasons viscose fiber is undesirable in the green textile industry. Later, Lenzing discovered the lyocell method in which lyocell fibers can be formed using a low-cost solvent, NMMO. This solvent is safe to be used in the environment compared to the viscose method. However, the production and improvement of lyocell fiber in the textile industry are still developing.

Unlike viscose fiber, the manufacturing process of lyocell fiber, is more straightforward than viscose fiber due to its physical dissolution. The differences of methodology for production viscose and lyocell fiber can be seen in Figure 15.4. The methods of production are as follows: (1) dissolution, (2) filtration, (3) spinning regeneration, (4) washing, and (5) finishing. In addition, the spinning

Viscose fiber   Lyocell fiber

Mercerization

↓

Shredding

↓

Aging

↓

Xanthation

↓

Ripening

↓

Preparation of spinning solution

↓

Filtration

↓

Wet spinning

Dissolution

↓

Filtration

↓

Spinning regeneration

↓

Washing

↓

Finishing

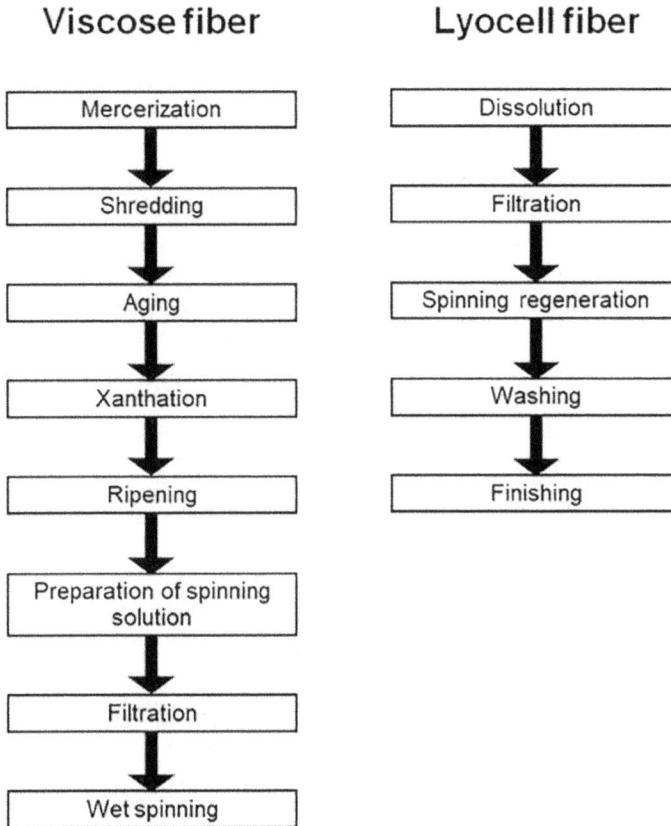

FIGURE 15.4 Steps in the production of viscose and lyocell fiber

process of lyocell fiber is dry-jet wet spinning. Higher viscosity will be obtained from dopes forming for lyocell fiber than viscose fiber (S. Zhang et al., 2018). Lyocell fiber is used to produce home-textile products like mattresses, bed covers, and a few apparels such as towels, underwear, denim, and dress. In addition, it is proven that the literature studies show the improvement of lyocell fiber compared to viscose fiber in mechanical strength (Jiang et al., 2020). Lyocell fiber has excellent resilience as it does not easy to wrinkle compared to cotton and silk. The anti-odor, breathable, and absorbent of the lyocell fabric are the remarkable characteristics that could attract the consumers' attention to buy this type of material in the market industry.

Several factors should be considered in regenerating cellulose fibers to produce great clothes. Clothing covers human bodies and protects them from extreme weather like heat and cold. Accordingly, diverse analysis methods can be employed to evaluate the best fabric for balancing the heat transfer between the human body and the surrounding environment. As for thermal behavior, porosity, perspiration, and air permeability are vital to assure the fabric can attain the wearer at the normal body temperature. On the other hand, the issue of ozone depletion has worsened, which causes UV radiation increases in the present climate. UV radiation is the most dangerous radiation due to the highest energy per photon, surpassing carbon–carbon single bond energy. Consequently, humans need high UV protection to avoid chronic effects on the skin, like photoaging, photosensitivity disorders, and skin cancer. Hence, the textile engineering approach to producing clothing-based regenerated cellulose fibers with photoprotective properties can minimize these harmful diseases. The introduction of a UV absorber in the regenerated cellulose fiber can improve the level of UV protection. Type of fiber, thickness, porosity and density, coloration, and chemical additives are the parameters that should be optimized as they affect the effectiveness of UV protection (Kocić et al., 2019).

Notably, to ensure the excellent properties of fibers used in textile applications with outstanding quality, several crucial analyses are required to be carried out. For instance, tensile and shear properties, abrasion resistance, dimensional stability, and wrinkle resistance testing. Tensile and shear properties are vital analyses for examining the mechanical properties of fabric based on standard ASTM 5035. The purpose of mechanical tests on the fabric is to measure the maximum extension of the fabric when force is applied. Next, abrasion resistance is defined as the ability of a fabric to withstand abrasion. Some standard methods for determining abrasion resistance are ASTM D 3884, ASTM D 3885, ASTM D 4966, and AATCC 93. The fabric with high dimensional stability has an excellent quality of clothing. Hence, monitoring the dimensions changes upon laundering should be implemented because wet shrinkage might affect the fabric. Thus, the standard method AATCC 135 can be used to evaluate dimensional stability. Other than that, the most significant aspect of clothing is the level of wrinkle resistance. These days, people search and buy the least wrinkled clothes before or after the laundering process. Hence, the fabric's recovery angle and wrinkle recovery appearance must be carried out according to the standard methods AATCC 66 and AATCC 128.

CONCLUSION

The most applauding applications of cellulose-based materials are their melted form. Bountiful applications with the most complex structures can be achieved from the dissolved state via control and efficient regeneration mechanisms. Interchangeable size, shape, and texture have made cellulose-based materials alluring in vast applications. The physical, mechanical, thermal, and biological qualities can be pre-designed and manufactured in line with the intended application. The compatibility with various materials significantly enhances the quality of regenerated cellulose products to unbounded sectors. Cellulose is no longer associated with typical applications as it is emerging as the most flexible material. Cellulose is indeed remarkable when its properties are tunable in the right direction.

ACKNOWLEDGMENTS

The Malaysian Ministry of Higher Education is acknowledged for the research grant LRGS/1/2019/UKM-UKM/5/1. We want to express our heartfelt gratitude to Universiti Sains Malaysia and Universiti Kebangsaan Malaysia for their unfailing cooperation and resources throughout the writing process.

REFERENCES

Abdeen, Z. (2011). Swelling and reswelling characteristics of cross-linked poly(vinyl alcohol)/chitosan hydrogel film. *Journal of Dispersion Science and Technology*, 32, 1337–1344.

Ahmed, S., Kanchi, S., Kumar, G., Kumar, V., Sangeetha, K., Ajitha, P., Aisverya, S., Sashikala, S., & Sudha, P. N. (2012). *Chitin and Chitosan: The Defense Booster in Agricultural Field*. In Ahmed,S., Kanchi, S., & Kumar, G (eds) *Handbook of Biopolymers: Advances and Multifaceted Applications* (pp. 1–322), Jenny Stanford Publishing, New York..

Azmi, A., Lau, K. S., Chin, S. X., Khiew, P. S., Zakaria, S., & Chia, C. H. (2021). Zinc oxide-filled polyvinyl alcohol–cellulose nanofibril aerogel nanocomposites for catalytic decomposition of an organic dye in aqueous solution. *Cellulose*, 28, 2241–2253.

Baque, M. A., Shiragi, M. H. K., Lee, E. J., & Paek, K. Y. (2012). Elicitor effect of chitosan and pectin on the biosynthesis of anthraquinones, phenolics and flavonoids in adventitious root suspension cultures of *Morinda citrifolia* (L.). *Australian Journal of Crop Science*, 6, 1349–1355.

Beaumont, M., Rennhofer, H., Opietnik, M., Lichtenegger, H. C., Potthast, A., & Rosenau, T. (2016). Nanostructured cellulose II gel consisting of spherical particles. *ACS Sustainable Chemistry and Engineering*, 4, 4424–4432.

Bhatt, B., & Kumar, V. (2017). Regenerated cellulose capsules for controlled drug delivery: Part IV. In-vitro evaluation of novel self-pore forming regenerated cellulose capsules. *European Journal of Pharmaceutical Sciences*, 97, 227–236.

Buchtová, N., & Budtova, T. (2016). Cellulose aero-, cryo- and xerogels: Towards understanding of morphology control. *Cellulose*, 23, 2585–2595.

Budtova, T. (2019). Cellulose II aerogels: A review. *Cellulose*, 26, 81–121.

Ding, L., Chen, L., Hu, L., Feng, X., Mao, Z., Xu, H., Wang, B., & Sui, X. (2021). Self-healing and acidochromic polyninyl alcohol hydrogel reinforced by regenerated cellulose. *Carbohydrate Polymers*, 255, 117331.

Gan, S., Padzil, F. N. M., Zakaria, S., Chia, C. H., Jaafar, S. N. S., & Chen, R. S. (2015). Synthesis of liquid hot water cotton linter to prepare cellulose membrane using NaOH/urea or LiOH/urea. *BioResources*, 10, 2244–2255.

Gan, S., Zakaria, S., Chia, C. H., Chen, R. S., Ellis, A. V., & Kaco, H. (2017). Highly porous regenerated cellulose hydrogel and aerogel prepared from hydrothermal synthesized cellulose carbamate. *PLoS ONE*, 12, 1–13.

Geng, H. (2018). Preparation and characterization of cellulose/N,N'-methylene bisacrylamide/graphene oxide hybrid hydrogels and aerogels. *Carbohydrate Polymers*, 196, 289–298.

Harlin, A., & Leppa, A. H. I. (2019). Optical cellulose fiber made from regenerated cellulose and cellulose acetate for water sensor applications. *Cellulose*, 27, 1543–1553.

Huang, S., Wu, L., Li, T., Xu, D., Lin, X., & Wu, C. (2019). Facile preparation of biomass lignin-based hydroxyethyl cellulose super-absorbent hydrogel for dye pollutant removal. *International Journal of Biological Macromolecules*, 137, 939–947.

Jiang, X., Bai, Y., Chen, X., & Liu, W. (2020). A review on raw materials, commercial production and properties of lyocell fiber. *Journal of Bioresources and Bioproducts*, 5, 16–25.

Jiang, Y., Liu, L., Wang, B., Yang, X., Chen, Z., Zhong, Y., Zhang, L., Mao, Z., Xu, H., & Sui, X. (2019). Polysaccharide-based edible emulsion gel stabilized by regenerated cellulose. *Food Hydrocolloids*, 91, 232–237.

Kabir, S. M. F., Sikdar, P. P., Haque, B., Bhuiyan, M. A. R., Ali, A., & Islam, M. N. (2018). Cellulose-based hydrogel materials: Chemistry, properties and their prospective applications. *Progress in Biomaterials*, 7, 153–174.

Kassem, I., Kassab, Z., Khouloud, M., & Achaby, M. E. L. (2020). Phosphoric acid-mediated green preparation of regenerated cellulose spheres and their use for all-cellulose cross-linked superabsorbent hydrogels. *International Journal of Biological Macromolecules*, 162, 136–149.

Kocić, A., Bizjak, M., Popović, D., Poparić, G. B., & Stanković, S. B. (2019). UV protection afforded by textile fabrics made of natural and regenerated cellulose fibres. *Journal of Cleaner Production*, 228, 1229–1237.

Li, J., Lu, S., Liu, F., Qiao, Q., Na, H., & Zhu, J. (2021). Structure and properties of regenerated cellulose fibers based on dissolution of cellulose in a CO_2 switchable solvent. *ACS Sustainable Chemistry and Engineering*, 9, 4744–4754.

Lin, W., & Jana, S. C. (2021). Analysis of porous structures of cellulose aerogel monoliths and microparticles. *Microporous and Mesoporous Materials*, 310, 110625.

Lindman, B., Karlström, G., & Stigsson, L. (2010). On the mechanism of dissolution of cellulose. *Journal of Molecular Liquids*, 156(1), 76–81.

Ma, Y., Nasri-Nasrabadi, B., You, X., Wang, X., Rainey, T. J., & Byrne, N. (2020). Regenerated cellulose fibers wetspun from different waste cellulose types. *Journal of Natural Fibers*, 19, 2338–2350.

Mazlan, N. S. N., Zakaria, S., Gan, S., Hua, C. C., & Baharin, K. W. (2019). Comparison of regenerated cellulose membrane coagulated in sulphate based coagulant. *Cerne*, 25, 18–24.

Michalik, R., & Wandzik, I. (2020). A mini-review on chitosan-based hydrogels with potential for sustainable agricultural applications. *Polymers*, 12, 1–16.

Moosavi, S., Gan, S., Chia, C. H., & Zakaria, S. (2020). Evaluation of crosslinking effect on thermo-mechanical, acoustic insulation and water absorption performance of biomass-derived cellulose cryogels. *Journal of Polymers and the Environment*, 28, 1180–1189.

Moriam, K., Sawada, D., Nieminen, K., Hummel, M., Ma, Y., Rissanen, M., & Sixta, H. (2021). Towards regenerated cellulose fibers with high toughness. *Cellulose*, 9, 9547–9566.

Muhammad, A., Rahman, M. R., Baini, R., & Bin Bakri, M. K. (2021). Applications of sustainable polymer composites in automobile and aerospace industry. In *Advances in Sustainable Polymer Composites* (pp. 185–207). Elsevier.

Navarra, M. A., Dal Bosco, C., Moreno, J. S., Vitucci, F. M., Paolone, A., & Panero, S. (2015). Synthesis and characterization of cellulose-based hydrogels to be used as gel electrolytes. *Membranes*, 5, 810–823.

Palanivelu, S. D., Armir, N. A. Z., Zulkifli, A., Hafiza, A., Ainul Hafiza, A. H., Salleh, K. M., Lindsey, K., Che-othman, M. H., & Zakaria, S. (2022). Hydrogel application in urban farming : Potentials and limitations — A review. *Polymers*, 14, 2590.

Payen, M. (1838). Mémoire sur la composition du tissu propre des plantes et du ligneux. *Comptes-Rendus de l'académie Des Sciences*, 7, 1052–1057.

Picheth, G. F., Pirich, C. L., Sierakowski, M. R., Woehl, M. A., Sakakibara, C. N., de Souza, C. F., Martin, A. A., da Silva, R., & de Freitas, R. A. (2017). Bacterial cellulose in biomedical applications: A review. *International Journal of Biological Macromolecules*, 104, 97–106.

Popa, L., Ghica, M. V., Tudoroiu, E. E., Ionescu, D. G., & Dinu-Pîrvu, C. E. (2022). Bacterial cellulose—a remarkable polymer as a source for biomaterials tailoring. *Materials (Basel)*, 15, 1054.

Qiu, C., Zhu, K., Zhou, X., Luo, L., Zeng, J., Huang, R., Lu, A., Liu, X., Chen, F., Zhang, L., & Fu, Q. (2018). Influences of coagulation conditions on the structure and properties of regenerated cellulose filaments via wet-spinning in LiOH/urea solvent. *ACS Sustainable Chemistry and Engineering*, 6, 4056–4067.

Ramli, R. A. (2019). Slow release fertilizer hydrogels: A review. *Polymer Chemistry*, 10, 6073–6090.

Ranganathan, N., Oksman, K., Nayak, S. K., & Sain, M. (2015). Regenerated cellulose fibers as impact modifier in long jute fiber reinforced polypropylene composites: Effect on mechanical properties, morphology, and fiber breakage. *Journal of Applied Polymer Science*, 132, 1–10.

Rop, K., Karuku, G. N., Mbui, D., Michira, I., & Njomo, N. (2018). Formulation of slow release NPK fertilizer (cellulose-graft-poly(acrylamide)/nano-hydroxyapatite/soluble fertilizer) composite and evaluating its N mineralization potential. *Annals of Agricultural Sciences*, 63, 163–172.

Rozo, G., Bohorques, L., & Santamaría, J. (2019). Controlled release fertilizer encapsulated by a κ-carrageenan hydrogel. *Polimeros*, 29, 1–12.

Saldivar-Guerra, E., & Vivaldo-Lima, E. (2013). *Handbook of Polymer Synthesis, Characterization, and Processing*. John Wiley & Sons, New Jersey.

Salleh, K.M., Armir, N. A. Z., Mazlan, N. S. N., Wang, C., & Zakaria, S. (2021). Cellulose and its derivatives in textiles: Primitive application to current trend. In M. I. H. Mondal (Ed.), *Fundamentals of Natural Fibres and Textiles* (pp. 33–63). Woodhead Publishing, Sawston.

Salleh, K. M., Zakaria, S., Sajab, M. S., Gan, S., & Kaco, H. (2019). Superabsorbent hydrogel from oil palm empty fruit bunch cellulose and sodium carboxymethylcellulose. *International Journal of Biological Macromolecules*, 131, 50–59.

Santamala, H., Livingston, R., Sixta, H., Hummel, M., Skrifvars, M., & Saarela, O. (2016). Advantages of regenerated cellulose fibres as compared to flax fibres in the processability and mechanical performance of thermoset composites. *Composites Part A: Applied Science and Manufacturing*, 84, 377–385.

Sarmah, D., & Karak, N. (2020). Biodegradable superabsorbent hydrogel for water holding in soil and controlled-release fertilizer. *Journal of Applied Polymer Science*, 137, 1–12.

Sharif, R., Mujtaba, M., Rahman, M. U., Shalmani, A., Ahmad, H., Anwar, T., Tianchan, D., & Wang, X. (2018). The multifunctional role of chitosan in horticultural crops; a review. *Molecules*, 23, 1–20.

Sharma, A., Wankhede, P., Samant, R., Nagarkar, S., Thakre, S., & Kumaraswamy, G. (2021). Process-induced microstructure in viscose and lyocell regenerated cellulose fibers revealed by SAXS and SEM of acid-etched samples. *ACS Applied Polymer Materials*, 3, 2598–2607.

Shi, S., Zhu, K., Chen, X., Hu, J., & Zhang, L. (2019). Cross-linked cellulose membranes with robust mechanical property, self-adaptive breathability, and excellent biocompatibility. *ACS Sustainable Chemistry & Engineering*, 7, 19799–19806.

Shin, I., Postnova, I., Shchipunov, Y., & Ha, C. (2020). Transparent regenerated cellulose bionanocomposite film reinforced by exfoliated montmorillonite with polyhedral oligomeric silsesquioxane bearing amino groups. *Composite Interfaces*, 28, 653–669.

Singh, P., Duarte, H., Alves, L., Antunes, F., Le Moigne, N., Dormanns, J., Duchemin, B., Staiger, M. P., & Medronho, B. (2015). From Cellulose Dissolution and Regeneration to Added Value Applications — Synergism between Molecular Understanding and Material Development. In Poletto, M (ed) *Cellulose - Fundamental Aspects and Current Trends* (pp. 1–284). Intech Open, London.

Tu, H., Zhu, M., Duan, B., & Zhang, L. (2021). Recent progress in high-strength and robust regenerated cellulose materials. *Advanced Materials*, 33, 1–22.

Tyshkunova, I. V., Chukhchin, D. G., Gofman, I. V., Poshina, D. N., & Skorik, Y. A. (2021). Cellulose cryogels prepared by regeneration from phosphoric acid solutions. *Cellulose*, 28, 4975–4989.

Wang, S., Lu, A., & Zhang, L. (2016). Recent advances in regenerated cellulose materials. *Progress in Polymer Science*, 53, 169–206.

Winarti, C., Sasmitaloka, K. S., & Arif, A. B. (2021). Effect of NPK fertilizer incorporation on the characteristics of nanocellulose-based hydrogel. *IOP Conference Series: Earth and Environmental Science*, 648, 1–8.

Xie, C., Liu, S., Zhang, Q., Yang, S., Guo, Z., Qiu, T., & Tuo, X. (2021). Macroscopic-scale preparation of aramid. *ACS Nano*, 15, 10000–10009.

Xue, Y., Li, W., Yang, G., Lin, Z., Qi, L., Zhu, P., Yu, J., & Chen, J. (2022). Strength enhancement of regenerated cellulose fibers by adjustment of hydrogen bond distribution in ionic liquid. *Polymers*, 14, 2030.

Xue, Y., Qi, L., Lin, Z., Yang, G., & He, M. (2021). High-strength regenerated cellulose fiber reinforced with cellulose nanofibril and nanosilica. *Nanomaterials*, 11, 2664.

Yang, B., Hua, W., Li, L., Zhou, Z., Xu, L., Bian, F., Ji, X., Zhong, G., & Li, Z. (2019). Robust hydrogel of regenerated cellulose by chemical crosslinking coupled with polyacrylamide network. *Journal of Applied Polymer Science*, 47811, 1–10.

Zainul Armir, N. A., Salleh, K. M., Zulkifli, A., & Zakaria, S. (2022). pH-responsive ampholytic regenerated cellulose hydrogel integrated with carrageenan and chitosan. *Industrial Crops and Products*, 178, 114588.

Zainul Armir, N. A., Zulkifli, A., Gunaseelan, S., Palanivelu, S. D., Salleh, K. M., Hafiz, M., Othman, C., & Zakaria, S. (2021). Regenerated cellulose products for agricultural and their potential: A review. *Polymers*, 13, 3586.

Zeng, B., & Byrne, N. (2021). The effect of drying method on the porosity of regenerated cellulose fibres. *Cellulose*, 28, 8333–8342.

Zhang, H., Chen, K., Gao, X., Han, Q., & Peng, L. (2018). Improved thermal stability of regenerated cellulose films from corn (Zea mays) stalk pith using facile preparation with low-concentration zinc chloride dissolving. *Carbohydrate Polymers*, 217, 190–198.

Zhang, H., Lang, J., Lan, P., Yang, H., Lu, J., & Wang, Z. (2020). Study on the dissolution mechanism of cellulose by ChCl-based deep eutectic solvents. *Materials*, 13, 278.

Zhang, J., Tominaga, K., Yamagishi, N., & Gotoh, Y. (2020). Comparison of regenerated cellulose fibers spun from ionic liquid solutions with lyocell fiber. *Journal of Fiber Science and Technology*, 76, 257–266.

Zhang, J., Yamagishi, N., Tominaga, K., & Gotoh, Y. (2017). High-strength regenerated cellulose fibers spun from 1-butyl-3-methylimidazolium chloride solutions. *Journal of Applied Polymer Science*, 45551, 1–9.

Zhang, S., Chen, C., Duan, C., Hu, H., Li, H., Li, J., Liu, Y., Ma, X., Stavik, J., & Ni, Y. (2018). Regenerated cellulose by the Lyocell process, a brief review of the process and properties. *BioResources*, 13, 4577–4592.

Zhu, F. L., Feng, X. L., & Feng, Q. (2020). Thermal decomposed behavior and kinetic study for untreated and flame retardant treated regenerated cellulose fibers using thermogravimetric analysis. *Journal of Thermal Analysis and Calorimetry*, 145, 423–435.

16 Developments and Applications of Nanocellulose-Based Hydrogels in the Biomedical Field

Junidah Lamaming
Universiti Malaysia Sabah

Sofie Zarina Lamaming
Universiti Sains Malaysia

Mohd Hazim Mohamad Amini
Universiti Malaysia Kelantan

Abu Zahrim Yaser
Universiti Malaysia Sabah

CONTENTS

INTRODUCTION

In recent years, cellulose-based applications have been burgeoning in various applications. Researchers and scholars all around the world reviewed, explored, and discovered more about the use of cellulose, the most abundant renewable biopolymer with an annual yield of approximately 1.5×10^{12} tons (Acharya et al., 2021). Cellulose is known as a linear polymer made up of repeating units of two cellobiose (disaccharide D glucose) linked by β-1,4 glycosidic bonds. Strong intra-molecular and intermolecular hydrogen bonds as well as van der Waals forces between similar or neighboring hydroxyl groups and oxygens are used to assemble cellulose chains to create elementary fibrils, which are subsequently held together to create microfibrils with diameters between 5 and 50 nm and lengths up to several micrometers (Lamaming et al., 2017; Sharma et al., 2019; Baghaei & Skrifvars, 2020). Cellulose has the potential to serve as a raw material for the creation of a variety of green products due to its superior mechanical qualities, chemical stability,

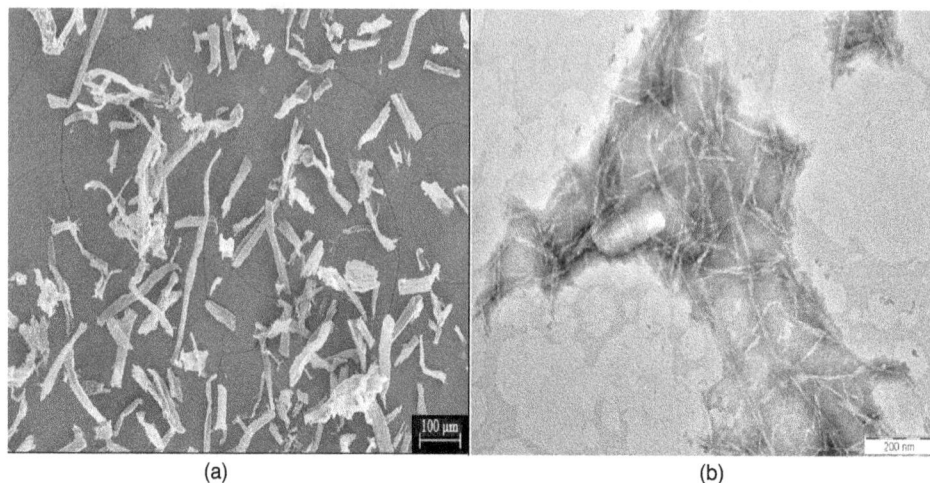

(a)   (b)

FIGURE 16.1 Micrographs of nanocellulose from oil palm trunk as viewed under (a) Scanning Electron Microscope (SEM) and (b) Transmission Electron Microscope (TEM). Photograph by author.

biocompatibility, renewability, biodegradability, and adaptability (Acharya et al., 2014; Joseph et al., 2020; Debnath et al., 2021; Tanpichai, 2022).

Nanocellulose (NC) (Figure 16.1) has been extensively used in many industries including biomedical, papermaker, and packaging and as a novel composite, biosensor, and so on (Ghaderi et al., 2014; Zhao et al., 2014; Hickey & Pelling, 2019). NC can be divided into three main groups, namely cellulose nanocrystals (CNC), cellulose nanofibrils (CNF), and bacterial cellulose (BC). CNC is known for its high crystallinity and is a needle-like form of NC that can be obtained through acid hydrolysis of wood pulp, cotton, cellulose agricultural residue, or microcrystalline cellulose (Lamaming et al., 2015). Compared to cellulose fibers, CNC is more likely to be utilized in a wide range of applications since it can offer high strength, high specific modulus, high specific surface area, and special optical properties and is in nanoscale size. On the other hand, CNF is a nanoscale cellulose fiber made up of long and flexible cellulose chains, forming a structure that is linked together by hydrogen bonds (Ciolacu & Popa, 2020). BC is basically a cellulose product from bacterial synthesis (Abeer et al., 2014). Compared to cellulose plants, BC is different in its macromolecular structures and physical properties, especially due to its high surface area, degree of polymerization, wet tensile strength, purity, crystallinity, and nanostructured fibers (Chen et al., 2013). BC appears in microfibrillar structures and nanostructures that cause it to have enhanced water retention and a high degree of biocompatibility (Aravamudhan et al., 2014).

NANOCELLULOSE-BASED HYDROGELS: STRUCTURE AND PROPERTIES

Hydrogel (Figure 16.2) is a three-dimensional cross-linked material made up of hydrophilic polymers that can absorb or hold a significant amount of water (up to 99.9%), saline, or other solutions (Parhi, 2017; Curvello et al., 2019). The structure of hydrogels is created by hydrated polymer networks or hydrophilic groups ($-OH$, $-COOH$, $-NH_2$, $-CONH_2$, and $-SO_3H$) under aqueous conditions (Sagar et al., 2018). They take on a soft, rubbery structure when inflated, mimicking the behavior of extracellular matrix (ECM) in actual tissues (Shojaeiarani et al., 2019). According to Nicu et al. (2021) and Jacob et al. (2021), hydrogels can be shaped to suit a variety of surfaces. These characteristics make hydrogels viable candidates for biological applications, along with their mucoadhesiveness, elasticity, swelling, and deswelling responses to environmental stimuli.

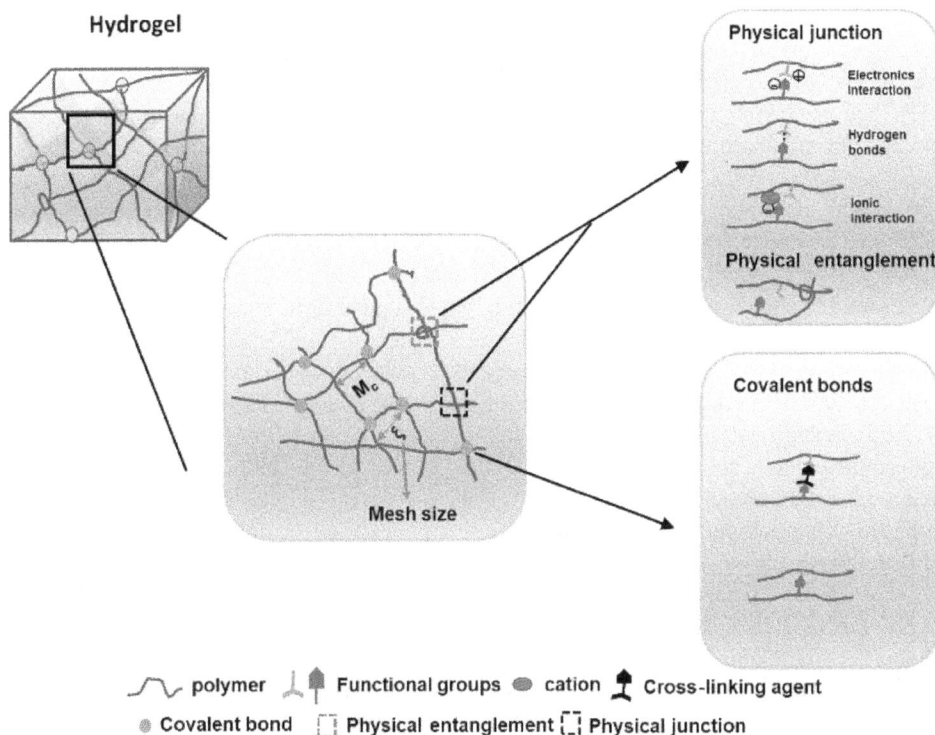

FIGURE 16.2 Structure of hydrogels produced through chemical linking and physical junctions. Reproduced from Ho et al. (2022). Copyright, CC BY license.

One-step hydrogel production processes use the polymerization method and parallel cross-linking of many monomers, while multistep processes also involve the creation of a polymer molecule's reactive group (Huber et al., 2019). Hydrogel can be classified based on its material, production techniques, and its application (Kaco et al., 2014; Yom-Tov et al., 2016). However, producing it from a natural polymer is more appealing due to its low cost, improved biocompatibility, and biodegradability (Yin et al., 2015). Based on its characteristics stated earlier, cellulose from biomass resources has been shown to be an excellent material for manufacturing hydrogels. The usage of cellulose-based hydrogels for pharmaceutical/medical applications has been the subject of multiple thorough review articles in recent years (Luo et al., 2019; Du et al., 2019; Saddique and Cheong, 2021; Pradeep et al., 2022).

Despite being a very attractive and promising polymer that may be used in a variety of applications, NC has several limitations, such as a low solubility that restricts its application (Sezer et al. 2019; Rusu et al., 2019). Obtaining cellulose derivatives through various chemical modification processes, such as esterification or oxidations, can address this limitation (Sezer et al., 2019). CNF has been employed in a number of clinical applications such as in drug delivery and as hemo-dialysis membranes and an antimicrobial nanomaterial and is claimed to be cytocompatible and highly tolerogenic (Čolić et al., 2015; Del Valle et al., 2017). There are two main problems that are associated with CNF's physical properties which are the high hydrophilicity of cellulose and the high capacity of nanofibrils to interact and bind through a large number of hydrogen bridges. This problem will cause an alteration of the desired structure in the drying process and limitations in its application. However, by modifying the loading density of the fibers employed, selecting the ideal condition of the swelling environment, and using the ideal working approach, this material can be used to produce hydrogels with improved porosity and specific surface area as well as a controllable degree of swelling (Del Valle et al., 2017).

The hydrogel must possess the necessary biological and mechanical characteristics, and these characteristics must be comparable to those of the tissue it will replace. One of the key considerations in the engineering design of hydrogels for use in the medical field is their mechanical qualities. The mechanical properties of scaffolds, both on a macroscopic and microscopic scale, are essential for regulating cell behavior in tissue engineering (TE). The mechanical strength of hydrogels can be increased by including NC in their formulation. Incorporating polymers into the hydrogel system to develop interpenetrating networks or semi-interpenetrating networks is an effective technique for enhancing the mechanical strength of hydrogels (Ho et al., 2022). The hydrogel's structure must be compatible with cells, tissues, and bodily fluids. It must also be nontoxic and noncancerous and should not induce long-lasting physiological or inflammatory effects after being degraded (Chamkouri and Chamkouri, 2021).

The production of hydrogels from cellulose and its derivatives can be either by the interaction of physical nature or chemical cross-linking. Physical natural processes, according to Ciolacu and Suflet (2018), include electronic associations, hydrogen bridges, hydrophobic interactions, mechanical chain entanglement, and van der Waals interactions. Due to the tendency for cellulose–cellulose contacts over cellulose–solvent interactions, physical gelation, which is commonly followed by micro-phase separation, includes the self-association of cellulose chains (Ciolacu et al., 2016). TE has utilized physically networked hydrogels as a substrate for a number of medicinal purposes, such as the release of bioactive substances and the confinement of cells. The principal method by which pharmacological compounds are released from hydrogels is via swelling of the structure (Chamkouri & Chamkouri, 2021). The use of cross-linkers is necessary for chemical cross-linking (Ciolacu & Suflet, 2018). Cross-linker is a material that can interconnect the molecules and improve the properties of the hydrogel in order to increase the molecular weight and enhance the mechanical properties and stability as well as the physical properties of the final hydrogel (Sirajuddin et al., 2014; Reddy et al., 2015). Cross-linking techniques can be physical, chemical, or polymerization. Table 16.1 below shows the cross-linking techniques used in the production of hydrogels.

Cellulose derivatives normally come from the chemical reactions of cellulose. For example, the active hydroxyl groups of cellulose will react with organic compounds (methyl and ethyl units) during esterification and produce water-soluble cellulose derivatives, also known as cellulose ether derivatives (Tosh, 2015). The degree of substitution was determined by the average number of

---

TABLE 16.1

Some of the Cross-linking Techniques Related to Producing Hydrogel from Cellulose and its Derivatives

Technique	Type	Reference
Physical cross-linking	Freeze-thawing technique	Guan et al. (2014), Butylina et al. (2016), Dai et al. (2017), Zhao et al. (2018)
	Photoinitiator technique	Zhao et al. (2016), Lu et al. (2018), Qi et al. (2018), Yuan et al. (2018)
	Radiation-induced technique	Saiki et al. (2011), Wach et al. (2014) Singh and Bala (2014), González-Torres et al. (2018)
Chemical cross-linking	Citric acid	Ghorpade et al. (2016), Mali et al. (2018), Dharnakingam et al. (2019), Kanafi et al. (2019), Uyanga et al. (2020), Manikandan and Lens (2022)
	Epichlorohydrin	Hasanah et al. (2015) and Almeida et al. (2022)
	Glutaraldehyde	Yu et al. (2017)
Polymerization	Bulk polymerization	Rasoulzadeh et al. (2017)
	Copolymerization	Mandal and Ray (2016)
	Polymerization by irradiation	Khoylou and Naimian (2009) and Chu et al. (2020)

etherified hydroxyl groups in the end product of cellulose esterification. This degree of substitution was controlled so that the cellulose derivatives could have the desired solubility and viscosity in a water solution (Mischnick & Momcilovic, 2010). Cellulose ether is a hydrophilic compound and is preferable in certain areas of application since it is able to form a gel in the presence of a high amount of water (Ribeiro et al., 2018). Cellulose ether derivatives include hydroxyprophyl methyl-cellulose, carboxymethyl cellulose (CMC), and hydroxypropyl cellulose. Aside from biodegradabil-ity and biocompatibility, cellulose ether hydrogels can target, sustain, and deliver drugs over time in the presence of specific environmental stimuli (Ribeiro et al., 2018).

NANOCELLULOSE-BASED HYDROGELS IN BIOMEDICAL APPLICATIONS

Drug Delivery

Cellulose and its derivatives are used as excipients that are responsible for controlling the rate of drug release and achieving the correct drug concentration (Debele et al., 2016; Pradeep et al., 2022). Due to its hydrophilicity, bioadhesivity, pH sensitivity, and nontoxicity, CMC has become the most uti-lized cellulose ether used in drug administration and pharmaceutical applications. The existence of carboxylic groups in hydrogels triggers swelling and deswelling behaviors of hydrogels when recog-nizing pH stimuli, so they are considered to be of great potential application in controlled drug deliv-ery systems (Zhang et al., 2004). In this area, CMC can prevent crystallization or degradation of the drug while at the same time increasing the frequency of drug release (Javanbakht & Shaabani, 2019).

A novel anticancer drug carrier system with controlled and targeted release capabilities was developed by Rao et al. (2018) using graphene oxide (GO) functionalized with acid dihydrazide (ADH) and CMC. The GO-CMC/DOX drug loading system was made by bonding the small mol-ecule of the anticancer drug doxorubicin hydrochloride (DOX) to GO-CMC with π-π bonds and hydrogen bonds. The developed GO-CMC/DOX has pH sensitivity, good drug release, certain tar-geting, and good biocompatibility and is safe and proven to have great antitumor activity. A similar good finding was also presented by Javanbakht and Namazi (2018) when fabricating CMC with graphene quantum dots loaded with DOX as a drug model. Gholamali and Yadollahi (2020) also described a new biocompatibility carrier for controlled anticancer drug delivery.

Daneshmoghanlou et al. (2022) fabricated a pH-responsive magnetic nanocarrier based on car-boxymethyl cellulose-aminated graphene oxide (GO-ADH-CMC) for loading and in vitro release of curcumin. The pH-responsive system was prepared through the cross-linking of cobalt ferrite $(CoFe_2O_4)$/GO to CMC. The study demonstrated that the maximum adsorption capacity of curcumin was 23.64 mg/g for $CoFe_2O_4$/GO-ADH-CMC and that under neutral conditions, the total amount of curcumin released was roughly 86%, which was relatively higher than the nonfunctionalized $CoFe_2O_4$/GO (15.69 mg/g). Moreover, the in vitro release of curcumin from the resulting CMC polymer-based carrier was pH-dependent and exhibited a non-Fickian transport behavior. The study suggests that the curcumin-loaded $CoFe_2O_4$/GO-ADH-CMC may be a good candidate for the con-tinuous and delayed release of curcumin in cancer treatment.

CMC/mesoporous magnetic GO (MG@SiO_2) was formulated by Pooresmaeil et al. (2020) as a safe and long-lasting ibuprofen delivery system. In their study, MG@SiO_2 was loaded with CMC as a coating agent to encapsulate the drug to overcome the burst release of ibuprofen. The pre-pared hydrogel showed a sustained release of ibuprofen for 8 h with a first-order kinetic model. The information provided indicated that the generated CMC might be recommended as a reli-able colon administration for prolonged ibuprofen delivery. Hossieni-Aghdam et al. (2018) designed pH-sensitive bionanocomposite hydrogel beads based on CMC hydrogel and halloysite nanotube (HNT). Using the co-precipitation technique, the hydrogel was loaded with atenolol (AT) as a model drug into the lumen of the nanotubes. The new pH-sensitive CMC/HNT-AT bionanocomposite beads were made by mixing HNT-AT nanohybrids into a CMC matrix and then using Fe^{3+} ions to cross-link the molecules. Based on how the beads swelled and how the drugs were released, the study found that the CMC/HNT-AT beads could be used as a new oral drug carrier for regulated and gradual drug release.

Both CNC and CNF-based hydrogels are promising materials for drug delivery systems due to their biocompatibility, adaptable surface chemistry, high surface area and open structure, and biodegradability. In recent years, Md Abu et al. (2020) prepared NC films incorporated with honey and discovered that the films followed the first-order kinetic model for drug release control. The drug release system was successfully utilized in treating chronic wounds. Ongoing research aims to improve the pharmacokinetics and optimize the substance distribution and inward variation in pH, ionic strength, and temperature around the target sites (organs and tissues) (Subhedar et al., 2021).

Wang et al. (2022) formulated self-healable NC composite hydrogels via simple mixing, which combines multiple dynamic bonds. Sodium tripolyphosphate was incorporated into hydrogels to develop the physical network and CNC/CNF was used to improve the mechanical properties using the nano-doping technique, providing a hydrogen bond as the triple network in complex hydrogels. Triple-cross-linked hydrogels were very resistant to fatigue and could heal themselves quickly. They also released drugs slowly and steadily, with less burst release and more release in acidic environments, and they were biocompatible when tested with sheep blood in an in vitro hemolysis assay.

TISSUE ENGINEERING

The utilization of cellulose hydrogels and its discoveries in TE has evolved in bone tissue engineering, cartilage tissue engineering, as well as 3D cell culture (Subhedar et al., 2021). TE involves isolating and inoculating specific cells in scaffolds with the regulation of biochemical factors to construct functional substitutes. The application of TE involves the repair and regeneration of skin, blood vessels, muscles, intervertebral discs, bones, ligaments, and others. The biomaterial template, known as ECM, is important since it is responsible for providing the conditions for the survival and proliferation of all different types of cells. ECM has intrinsic biochemical and mechanical cues that regulate cell phenotype and function in development and homeostasis, in response to injury (Hussey et al., 2018). For the 3D cell culture to work, the properties of the biomaterials need to be like those of the ECM.

In TE, NC-based hydrogels (CNCs and CNFs) were studied because of their highly hydrated three-dimensional porous structure that looks like a biological tissue and their good mechanical properties. Cheng et al. (2019) fabricated tunable self-healing biodegradable chitosan (CS)–CNF-based hydrogels that were successfully produced. CNFs were dispersed in DF-PEG before being mixed with glycol CS to create the composite hydrogels. By adding a low amount of CNFs (0.06–0.15 wt%) to the pristine CS self-healing hydrogel, the reversible dynamic Schiff bonding, strain sensitivity, and self-healing of the hydrogel are obviously affected. The study also found that the CS–CNF hydrogels could be injected easily through a 160 µm diameter syringe needle without clogging. Neural stem cells embedded in the CS–CNF hydrogel with better self-healing properties reveal significantly enhanced oxygen metabolism as well as neural differentiation. In vivo therapeutic function evaluation by a zebrafish brain injury model suggested that hydrogel with 0.09 wt% of CNFs and the best self-healing properties showed a 50% improvement over the pristine CS hydrogel.

The advanced technology in 3D bioprinting allows the development of highly structured scaffolds for TE. The 3D-shaped structures hold great promise for therapies aimed at tissue regeneration and organ repair and replacement (Athukoralalage et al., 2019). Cells are encapsulated within the printed material gel in a homogenous density and quantity (Curvello et al., 2019). The bioprinting device constructs the tissue using a specific printing method such as inkjet 3D bioprinting, micro-extrusion 3D bioprinting, laser-assisted 3D bioprinting, and stereolithography by employing a combination of printing materials such as scaffold, bio-ink, and other additive factors (Figure 16.3). Patel et al. (2021) fabricated alginate/gelatin/CNC hydrogels through a physical cross-linking process as a

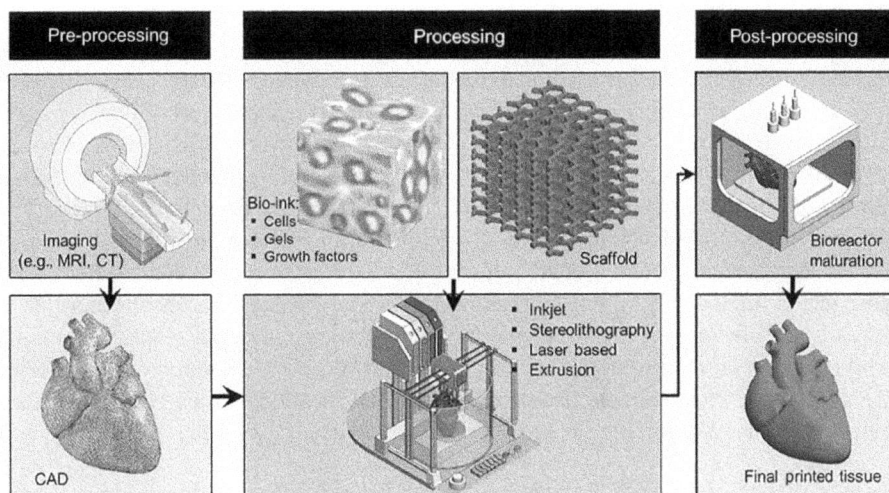

FIGURE 16.3 Schematic view of bioprinting processes. Reproduced from Ramadan and Zourob (2021). Copyright, CC BY License.

printable bio-ink and printed them using a CELLINK® BIOX 3D bioprinter. The fabricated scaffold exhibits superior swelling potential and enhanced mineral deposition. When the biocompatibility of the scaffold was tested in human bone marrow-derived mesenchymal stem cells using the WST-1 assay, better cell viability was observed, demonstrated by healthy cells and the cells adhering appropriately to the surface of the scaffold. In addition, the scaffold also exhibited enhanced mineralization and expression of osteogenic-related genes that occurred in the scaffold-treated media.

In other studies, Maturavongsadit et al. (2021) developed novel CNC-incorporated CS-based formulations as bio-inks for 3D extrusion-based bioprinting of cell-laden scaffolds to promote osteogenic differentiation and bone regeneration. The study found that bio-inks were biocompatible and printable at an optimized range of printing pressures (12–20 kPa) that did not compromise cell viability. Also, all different formulations of bio-inks were biocompatible and could encapsulate cells at a high density of 5 million cells/mL.

WOUND DRESSING AND WOUND HEALING

Functional hydrogels rapidly contribute to wound healing due to their bioactive properties in healing processing (Ho et al., 2022). The use of hydrogels in wound dressing management is attractive due to their good porosity, offering the debriding and desloughing capacity on fibrotic and necrotic tissues and providing moisture, biocompatibility, the ability to load and release various drugs, high water content, and flexibility (Jagur-Grodzinski, 2011). Apart from their biocompatibility, hydrogels from cellulose and its derivatives provide good moisture content for the wound healing process and give protection against infection by bacteria (Alven and Aderibigbe, 2020). Cellulose has the capability to accelerate the wound healing process by maintaining and releasing several growth factors (basic fibroblast, phosphodiesterase, and epidermal) at the site. It promotes the migration and proliferation of dermal fibroblasts and inhibits the proliferation of bacteria in the wound (Cullen et al., 2010).

A few studies have successfully investigated the use of cellulose-based hydrogels in wound dressing and wound healing management. Gupta et al. (2019) used BC hydrogels encapsulated with curcumin for the purpose of wound dressing. Curcumin-loaded hydrogels provide a moist environment, effectively controlling bacteria-associated infection and reducing oxidative stress at the wound site, according to the researchers. In other studies, the oxidized cellulose nanofiber-PVA hydrogels with curcumin revealed the sustained release of curcumin and significant wound closure (Shefa et al., 2020). Fan et al. (2019) discovered that pH-sensitive dual drug-loaded cellulose-based hydrogels have good results and promote a high percentage of wound closure.

Jiang et al. (2022) developed a hydrogel with a synergistic therapeutic effect by a one-pot method using hydroxyethyl cellulose as a framework and epichlorohydrin as a cross-linking agent. Menthol coated with cationic-β-cyclodextrin (C-β-CD) was embedded in the voids of the 3D system of the hydrogel. Zinc oxide nanoparticles were synthesized in situ on the hydrogel matrix and were uniformly distributed without agglomeration. The study found that the hydrogel had sufficient mechanical strength, swelling ability, water retention, and oxygen transmission rate. Hydrogels exhibited efficient bactericidal effects and no toxic effects on normal human epithelial cells. In vitro release showed that the hydrogel system could release menthol stably over a long period of time. The performance of the hydrogel formulation with a 7% addition of $Zn(NO_3)_2 \cdot 6H_2O$ is the most suitable for wound healing applications. Other studies also reported good biocompatibility when using PVA/ BC/silver nanoparticle hydrogels in promoting wound healing (Song et al., 2021).

Zhang et al. (2022) developed a CNF-reinforced hydrogel with pH sensitivity and enhanced mechanical stability for wound dressing purposes. The hydrogel was constructed by mixing dopamine, CMC, and methacrylic acid and loaded with levofloxacin as an antibacterial drug. When in vivo testing is performed on burned patients, hydrogels exhibit ideal therapeutic efficacy and accelerate the healing process. Hydrogels exhibited favorable cytocompatibility after being cultured for 24 h, and CNFs were nontoxic to L929 fibroblast cells. The inflammatory response test also showed a significant decrease of interleukin 1β in the produced hydrogels.

A novel pH-responsive CNF-reinforced PVA/borax hydrogel was formulated by Yang et al. (2022). CNFs were added to enhance the mechanical properties and grafted with natural antibiotic resveratrol to equip the hydrogel with potent antibacterial and antioxidant properties. CNF-reinforced Prussian blue (PB) hydrogel exhibited increased mechanical strength that reached 150 kPa and displayed a pH-responsive drug release behavior (at pH 5.4) and excellent biocompatibility and antioxidant effect. Moreover, in vitro and in vivo results revealed that such reinforced PEG/chitosan/PB hydrogel had excellent antibacterial properties, skin tissue regeneration, and wound closure.

Currently, there are some cellulose-based hydrogels that were effective in treating wounds that underwent clinical trials or were marketed, for example, nanofibrillated cellulose wound dressing for skin graft donor site treatment that was run for a clinical trial by Koivuniemi et al. (2020). The outcomes of the trial show efficient wound healing and it self-detaches after re-epithelialization, reduces pain, and requires no changes in dressing. Other marketed hydrogels that can be found are CelMat®, Nanoskin®, Nanoderm Ag, Biofill, and EpiProtect®, among others (Portela et al., 2019; Alven & Aderibigbe, 2020). However, more related studies should be further conducted for real clinical applications to exploit the use of tailored hydrogels for wound dressing and wound healing.

BIOMONITORING AND BIOSENSING

Biomonitoring and biosensing devices can be fabricated using a combination of conductive fillers with different types of polymer matrices. A conductive polymer hydrogel was synthesized for the development of capacitive sensors for applications in monitoring human activities. The capacitive sensor is the main functional material in wearable strain sensors for healthcare and medical diagnosis. Wearable devices with a strain sensor can be attached to clothes or garments or directly placed on the human skin by adhesive or elastic straps for monitoring human's physical and biochemical signals and motions (Lu et al., 2019). In a medical situation, these strain sensors can be used to detect finger movements, throat vibrations due to speaking, or wrist pulses (Horta-Velázquez & Morales-Narváez, 2022). NC is a dispersant, binding agent, or building block in the hydrogel network, producing highly flexible and stretchy electronic materials suitable for pressure sensing (Misenan et al., 2022).

TABLE 16.2

Some Examples of Research on Nanocellulose-Based Hydrogels for Biomedical Applications in recent years

Nanocellulose/Cellulose Derivatives/Polymers	Application	Model Drug	Findings	Reference
Graphene oxide (GO) functionalized with acid dihydrazide (ADH) and CMC	Drug delivery	Doxorubicin hydrochloride (DOX)	• pH sensitivity, good drug release, and certain targeting, safe, good biocompatibility, and proven to have a great antitumor activity. • The cumulative release rate of drugs can reach 65.2% at pH 5.0 • No obvious cytotoxicity and good biocompatibility	Rao et al. (2018)
Carboxymethyl cellulose/graphene quantum dot (GQD)	Drug delivery	Doxorubicin (DOX)	• CMC/GQDs showed improvement in in vitro swelling, degradation, water vapor permeability, and mechanical properties • pH-sensitive drug delivery properties along with not significant toxicity against blood cancer cells (K562).	Javanbakht and Namazi (2018)
Carboxymethyl cellulose /halloysite nanotube (HNT)	Drug delivery	Atenolol (AT)	• pH-sensitive swelling behavior with its maximum content at pH 6.8. • More sustained and controlled drug releases were observed for CMC/HNT-AT beads • The release kinetics of AT from the CMC/HNT-AT bionanocomposite beads followed the Higuchi kinetic model	Hossieni-Aghdam et al. (2018)
Mesoporous silica/magnetic graphene oxide/carboxymethyl cellulose (CMC/MG@mSiO2-IBU)	Drug delivery	Ibuprofen (IBU)	• CMC-coated MG@mSiO2-IBU sustainably released IBU in 8 h with a first-order kinetic model • VSM analysis displayed the super-paramagnetic properties • A safe carrier with controlled release of the poorly water-soluble IBU drug	Pooresmaeil et al. (2020)
Carboxymethyl cellulose (CMC)/starch/ZnO nanoparticles (ZnO-NPs)	Drug carrier	Doxorubicin (DOX)	• Enhanced swelling ratio of CMC/starch/ZnO nanocomposite • Higher amount of ZnO-NPs, drug release was reduced because of strong interactions between DOX amine groups and CMC/starch carboxyl groups • Cytotoxicity using SW480 human colon cancer cells showed that the DOX-loaded CMC/starch/ZnO nanocomposite beads resulted in a considerable decline in cell viability	Gholamali and Yadollahi (2020)
Carboxymethyl cellulose/aminated magnetic graphene oxide	Drug carrier	Curcumin (CUR)	• The highest amount of drug loading was achieved (23.7 mg/g at 323 K, pH 6.0, and 120 mg L^{-1} CUR concentration) • Maximum cumulative releases were found to be 86% and 38% at pH 5.6 and 7.4 • Following Peppas–Sahlin kinetic model • Exhibited good biocompatibility	Daneshmoghanlou et al. (2022)
Cellulose nanocrystals or cellulose nanofibrils/polysaccharide/sodium tripolyphosphate (TPP)	Drug delivery	5-fluorouracil (5-FU)	• The maximum tensile strength of the complex hydrogels with CNC and CNF additions could reach 890 KPa and 910 KPa • In vitro drug release experiments confirmed that 5-FU-loaded TPP-nanocellulose hydrogels were capable of releasing drugs sustainably • The hemolysis rate of all samples is less than 2%, showing good biocompatibility	Wang et al. (2022)

(Continued)

TABLE 16.2 (*Continued*)

Some Examples of Research on Nanocellulose-Based Hydrogels for Biomedical Applications in recent years

Nanocellulose/Cellulose Derivatives/Polymers	Application	Model Drug	Findings	Reference
Cellulose nanocrystals/alginate/gelatin	Tissue engineering	-	• Exhibited superior swelling potential • Better cell viability was observed in hBMSCs • The cells showed an elongated and flattened morphology, and the fluorescence intensity was higher • Enhanced mineralization and enhanced expression of osteogenic-related genes that occurred in the scaffold-treated media	Patel et al. (2021)
Methacrylated gellan-gum (GGMA)/cellulose nanocrystals	Tissue regeneration	-	• GGMA reinforcement with CNC and thus matrix entanglements with higher stiffness Upon ionic cross-linking • Mechanical tests demonstrate values close to those of the human annulus fibrosus (AF) tissue • Gel loading with bovine AF cells indicated that the construct promoted cell viability and a physiologically relevant morphology for up to 14 days in vitro	Pereira et al. (2018)
Chitosan-CNF/telechelic difunctional poly(ethylene glycol) PEG	Tissue regeneration	-	• Favorable self-healing properties when it contained 0.09 wt% of CNFs. • Passes through a 160 μm needle and promotes oxygen metabolism and differentiation of neural stem cells • Better thermal stability and prolonged degradation of the hydrogel • Similar rigidity (approximately 2 kPa)	Cheng et al. (2019)
Chitosan/glycerophosphate/hydroxyethyl cellulose/cellulose nanocrystals (CNCs)	Tissue engineering	-	• The addition of CNCs and cells (5 million cells/mL) significantly improved the viscosity of bio-inks and the mechanical properties of chitosan scaffolds post-fabrication • Biocompatible and printable at an optimized range of printing pressures (12–20 kPa) • CNCs promoted greater osteogenesis of MC3T3-E1 cells in chitosan scaffolds	Maturavongsadit et al. (2021)
Bacterial cellulose (BC)/-hydroxypropyl-β-cyclodextrin (HPβCD)	Wound dressing	Curcumin (CUR)	• IC75 emerged to be the best preparation method with the highest encapsulation efficacy • HPβCD enhanced the aqueous solubility of CUR and allowed loading into BC hydrogels • High moisture content, biocompatibility (cytocompatibility and hemocompatibility), antimicrobial, and antioxidant properties	Gupta et al. (2019)
Cellulose nanocrystals/honey/polyvinylpyrrolidone (PVP)	Wound dressing	-	• A sustained release of the maximum amount of active ingredients after 48 h • Exhibited significant antimicrobial activity on both gram-positive and gram-negative bacteria • Significant in vitro release kinetic and good antimicrobial efficacy against the test microbes primarily offering the nanocellulose film to be acted as a good wound dressing	Md Abu et al. (2020)

(*Continued*)

TABLE 16.2 (*Continued*)

Some Examples of Research on Nanocellulose-Based Hydrogels for Biomedical Applications in recent years

Nanocellulose/Cellulose Derivatives/Polymers	Application	Model Drug	Findings	Reference
Cellulose nanofibrils/poly(-vinyl alcohol)/borax	Wound dressing	Resveratrol (RSV)	• Robust mechanical properties (fracture strength of 149.6 kPa), high self-healing efficiency (>90%), and excellent adhesion performance (tissue shear stress of 54.2 kPa) • pH-responsive drug release behavior • Excellent biocompatibility and antioxidant effect • In vitro and in vivo results revealed excellent antibacterial effects, skin tissue regeneration, and wound closure capabilities	Yang et al. (2022)
Hydroxyethyl cellulose (HEC)/epichlorohydrin (EPI)/zinc oxide nanoparticles (ZnO NPs) (ZnO NPs@ HEC/C-β-CD/menthol)	Wound dressing	–	• Sufficient mechanical strength, swelling ability, water retention, and oxygen transmission rate • Hydrogels exhibited efficient bactericidal effects and no toxic effects on normal human epithelial cells • In vitro release showed that the hydrogel system could release menthol stably over a long period of time • The performance of the hydrogel formulation with a 7% addition of $Zn(NO_3)_2 \cdot 6H_2O$ is the most suitable for wound healing applications	Jiang et al. (2022)
Polyvinyl alcohol (PVA)/BC/silver nanoparticles (AgNPs)	Wound healing		• Hydrogels contained a porous three-dimensional reticulum structure and had high mechanical properties • Hydrogels possessed outstanding antibacterial properties and good biocompatibility • Effectively repaired wound defects in mice models and wound healing reached 97.89% within 15 days	Song et al (2021)

Many studies had been done on NC hydrogel-based strain sensors. Heidarian et al. (2022) produced a hydrogel by imine formation of carboxyl methyl chitosan, oxidized cellulose nanofibers, and chitin nanofibers followed by two subsequent cross-linking stages which are immersion in tannic acid solution to create hydrogen bonds and soaking in Fe(III) solution to give electrical conductivity property to the hydrogel.

Meanwhile, Zheng et al. (2019) prepared strain sensors with a combination of CNFs and graphene (GN) co-incorporated poly(vinyl alcohol)-borax (GN-CNF@PVA) hydrogel. CNFs act as a bio-template and dispersant to support GN to create a homogeneous GN–CNF aqueous dispersion, increasing the mechanical flexibility and strength while providing good conductivity. Results showed high stretchability, excellent viscoelasticity with storage modulus up to 3.7 kPa, rapid self-healing ability of 20 s, and high healing efficiency of around 97.7% (Table 16.2).

CONCLUSION

Cellulose has become an interesting material for various applications due to its high mechanical strength, biocompatibility, biodegradability, and eco-friendliness. NC as hydrogels, either from native cellulose or derivatives of cellulose, appears to be an exciting and attractive material, not only in the biomedical fields but also in other applications and future applications. In terms of advanced technology and engineering progress, there is still room for more discoveries and fundamental knowledge in cellulose-based research fields. For example, 3D bioprinting technology, super-hydrophobic membranes, and superabsorbents are some of the many emerging and growing fields that will benefit many industries. Many significant works on the use of cellulose-based materials together with their related technologies have been demonstrated and reported. In applying the products beyond the laboratory scale, comprehensive and systematic studies are still needed to enable high-throughput production.

REFERENCES

Abeer, M.M., Mohd Amin, M.C.I., & Martin, C. (2014). A review of bacterial cellulose-based drug delivery systems: Their biochemistry, current approaches and future prospects. *Journal of Pharmacy and Pharmacology*, 66(8), 1047–1061.

Acharya, S., Abidi, N., Rajbhandari, R., & Meulewaeter, F. (2014). Chemical cationization of cotton fabric for improved dye uptake. *Cellulose*, 21(6), 4693–4706.

Acharya, S., Hu, Y., & Abidi, N. (2021). Cellulose dissolution in ionic liquid under mild conditions: Effect of a hydrolysis and temperature. *Fibers*, 9(1), 5.

Almeida, A.P., Saraiva, J.N., Cavaco, G., Portela, R.P., Leal, C.R., Sobral, R.G., & Almeida, P.L. (2022). Crosslinked bacterial cellulose hydrogels for biomedical applications. *European Polymer Journal*, 177, 111438.

Alven, S., Aderibigbe, B.A. (2020). Chitosan and cellulose-based hydrogels for wound management. *International Journal of Molecule Science*, 21, 1–30.

Aravamudhan, A., Ramos, D.M., Nada, A.A., & Kumbar, S.G. (2014). Natural polymers: Polysaccharides and their derivatives for biomedical applications. In Kumbar, S.G., Laurencin, C.T., & Deng, M. (eds) *Natural and Synthetic Biomedical Polymers* (pp. 67–89). Elsevier, Amsterdam.

Athukoralalage, S.S., Balu, R., Dutta, N.K., & Choudhury, N.R. (2019). 3D bioprinted nanocellulose-based hydrogels for tissue engineering applications: A brief review. *Polymers*, 11, 898.

Baghaei, B., & Skrifvars, M. (2020). All-cellulose composites: A review of recent studies on structure, properties and applications. *Molecules*, 25(12), 2836.

Butylina, S., Geng, S., & Oksman, K. (2016). Properties of as-prepared and freeze-dried hydrogels made from poly (vinyl alcohol) and cellulose nanocrystals using freeze-thaw technique. *European Polymer Journal*, 81, 386–396.

Chamkouri, H., & Mahyodin Chamkouri, M. (2021). A review of hydrogels, their properties and applications in medicine. *American Journal of Biomedical Science & Research*, 11(6), 485–493.

Chen, L., Hong, F., Yang, X.X., & Han, S.F. (2013). Biotransformation of wheat straw to bacterial cellulose and its mechanism. *Bioresource Technology*, 135, 464–468.

Cheng, K.C., Huang, C.F., Wei, Y., & Hsu, S.H. (2019). Novel chitosan–cellulose nanofiber selfhealing hydrogels to correlate self-healing properties of hydrogels with neural regeneration effects. *NPG Asia Materials*, 11(25), 1–17.

Chu, S., Maples, M. M., & Bryant, S. J. (2020). Cell encapsulation spatially alters crosslink density of poly (ethylene glycol) hydrogels formed from free-radical polymerizations. *Acta Biomaterialia*, 109, 37–50.

Ciolacu, D., & Popa, V.I. (2020). Nanocelluloses: Preparations, properties, and applications in medicine. In Popa, V.I. (Ed.), *Pulp Production and Processing* (2nd ed., pp. 317–340). Walter de Gruyter, Berlin.

Ciolacu, D., Rudaz, C., Vasilescu, M., & Budtova, T. (2016). Physically and chemically cross-linked cellulose cryogels: Structure, properties and application for controlled release. *Carbohydrate Polymers*, 151, 392–400.

Ciolacu, D. E., & Suflet, D.M. (2018). Cellulose-based hydrogels for medical/pharmaceutical applications. In Popa, V., & Volf, I. (ed) *Biomass as Renewable Raw Material to Obtain Bioproducts of High-Tech Value* (pp. 401–439). Elsevier, Amsterdam.

Čolić, M., Mihajlović, D., Mathew, A., Naseri, N., & Kokol, V. (2015). Cytocompatibility and immunomodulatory properties of wood based nanofibrillated cellulose. *Cellulose*, 22(1), 763–778.

Cullen, M.B., Silcock, D.W., Boyle, C. (2010). Wound dressing comprising oxidized cellulose and human recombinat collagen. U.S. Patent 7833790 B2.

Curvello, R., Raghuwanshi, S., & Garnier, G. (2019). Engineering nanocellulose hydrogels for biomedical applications. *Advances in Colloid and Interface Science*, 267, 47–61.

Dai, H., Ou, S., Liu, Z., & Huang, H. (2017). Pineapple peel carboxymethyl cellulose/polyvinyl alcohol/ mesoporous silica SBA-15 hydrogel composites for papain immobilization. *Carbohydrate Polymers*, 169, 504–514.

Daneshmoghanlou, E., Miralinaghi, M., Moniri, E., & Sadjady, S.K. (2022). Fabrication of a pH-responsive magnetic nanocarrier based on carboxymethyl cellulose-aminated graphene oxide for loading and in-vitro release of curcumin. *Journal of Polymers and the Environment*, 30, 3718–3736.

Debele, T.A., Mekuria, S.L., & Tsai, H.C. (2016). Polysaccharide based nanogels in the drug delivery system: Application as the carrier of pharmaceutical agents. *Materials Science and Engineering: C*, 68, 964–981.

Debnath, B., Haldar, D., & Purkait, M. K. (2021). A critical review on the techniques used for the synthesis and applications of crystalline cellulose derived from agricultural wastes and forest residues. *Carbohydrate Polymers*, 273, 118537.

Del Valle, L. J., Díaz, A., & Puiggalí, J. (2017). Hydrogels for biomedical applications: Cellulose, chitosan, and protein/peptide derivatives. *Gels*, 3(3), 27.

Du, H., Liu, W., Zhang, M., Si, C., Zhang, X., & Li, B. (2019). Cellulose nanocrystals and cellulose nanofibrils based hydrogels for biomedical applications. *Carbohydrate Polymers*, 209, 130–144.

Fan, X., Yang, L., Wang, T., Sun, T., & Lu, S. (2019). pH-responsive cellulose-based dual drug-loaded hydrogel for wound dressing. *European Polymer Journal*, 121, 109290

Ghaderi, M., Mousavi, M., Yousefi, H., & Labbafi, M. (2014). All-cellulose nanocomposite film made from bagasse cellulose nanofibers for food packaging application. *Carbohydrate Polymers*, 104, 59–65.

Gholamali, I., & Yadollahi, M. (2020). Doxorubicin-loaded carboxymethyl cellulose/Starch/ZnO nanocomposite hydrogel beads as an anticancer drug carrier agent. *International Journal of Biological Macromolecules*, 160, 724–735.

Ghorpade, V. S., Yadav, A. V., & Dias, R. J. (2016). Citric acid crosslinked cyclodextrin/ hydroxypropylmethylcellulose hydrogel films for hydrophobic drug delivery. *International Journal of Biological Macromolecules*, 93, 75–86.

González-Torres, M., Leyva-Gómez, G., Rivera, M., Krötzsch, E., Rodríguez-Talavera, R., Rivera, A. L., & Cabrera-Wrooman, A. (2018). Biological activity of radiation-induced collagen–polyvinylpyrrolidone–PEG hydrogels. *MAterials Letters*, 214, 224–227.

Guan, Y., Bian, J., Peng, F., Zhang, X. M., & Sun, R.C. (2014). High strength of hemicelluloses based hydrogels by freeze/thaw technique. *Carbohydrate Polymers*, 101, 272–280.

Gupta, A., Keddie, D.J., Kannappan, V., Gibson, H., Khalil, I.R., Kowalczuk, M., Martin, C., Shuai, X., & Radecka, I. (2019). Production and characterisation of bacterial cellulose hydrogels loaded with curcumin encapsulated in cyclodextrins as wound dressings. *European Polymer Journal*, 118, 437–450.

Hasanah, A. N., Muhtadi, A., Elyani, I., & Musfiroh, I. (2015). Epichlorohydrin as crosslinking agent for synthesis of carboxymethyl cellulose sodium (Na-CMC) as pharmaceutical excipient from water hyacinth (Eichorrnia crassipes L.). *International Journal of Chemical Science*, 13(3), 1227–37.

Heidarian, P., Gharaie, S., Yousefi, H., Paulino, M., Kaynak, A., Varley, R., & Kouzani, A. Z. (2022). A 3D printable dynamic nanocellulose/nanochitin self-healing hydrogel and soft strain sensor. *Carbohydrate Polymers*, 291, 119545.

Hickey, R. J., & Pelling, A. E. (2019). Cellulose biomaterials for tissue engineering. *Frontiers in Bioengineering and Biotechnology*, 7, 45.

Ho, T.C., Chang, C.C., Chan, H.P., Chung, T.W., Shu, C.W., Chuang, K.P., Duh, T.H., Yang, M.H., & Tyan, Y.C. (2022). Hydrogels: Properties and applications in biomedicine. *Molecules*, 27, 2902.

Horta-Velázquez, A., & Morales-Narváez, E. (2022). Nanocellulose in wearable sensors. *Green Analytical Chemistry*, 1, 100009.

Hossieni-Aghdam, S.J., Foroughi-Nia, B., Zare-Akbari, Z., Mojarad-Jabali, S., Motasadizadeh, H., & Farhadnejad, H. (2018). Facile fabrication and characterization of a novel oral pH-sensitive drug delivery system based on CMC hydrogel and HNT-AT nanohybrid. *International Journal of Biological Macromolecules*, 107, 2436–2449.

Huber, T., Feast, S., Dimartino, S., Cen, W., & Fee, C. (2019). Analysis of the effect of processing conditions on physical properties of thermally set cellulose hydrogels. *Materials*, 12(7), 1066.

Hussey, G.S., Dziki, J.L., & Badylak, S.F. (2018). Extracellular matrix-based materials for regenerative medicine. *Nature Review Materials*, 3, 159–173.

Jacob, S., Nair, A., Shah, J., Sreeharsha, N., Gupta, S., & Shinu, P. (2021). Emerging role of hydrogels in drug delivery systems, tissue engineering and wound management. *Pharmaceutics*, 13, 357.

Jagur-Grodzinski, J. (2011). Polymeric gels and hydrogels for biomedical and pharmaceutical applications. *Polymers for Advanced Technologies*, 21, 27–47.

Javanbakht, S., & Namazi, H. (2018). Doxorubicin loaded carboxymethyl cellulose/graphene quantum dot nanocomposite hydrogel films as a potential anticancer drug delivery system. *Materials Science and Engineering: C*, 87, 50–59.

Javanbakht, S., & Shaabani, A. (2019). Carboxymethyl cellulose-based oral delivery systems. *International Journal of Biological Macromolecules*, 133, 21–29.

Jiang, L., Han, Y., Xu, J., & Wang, T. (2022). Preparation and study of cellulose-based ZnO NPs@HEC/C-β-CD/Menthol hydrogel as wound dressing. *Biochemical Engineering Journal*, 184, 108488.

Joseph, B., Sagarika, V.K., Sabu, C., Kalarikkal, N., & Thomas, S. (2020). Cellulose nanocomposites: Fabrication and biomedical applications. *Journal of Bioresources and Bioproducts*, 5(4), 223–237.

Kaco, H.A.T.I.K.A., Zakaria, S., Razali, N.F., Chia, C.H., Zhang, L., & Jani, S.M. (2014). Properties of cellulose hydrogel from kenaf core prepared via pre-cooled dissolving method. *Sains Malaysiana*, 43(8), 1221–1229.

Kanafi, N. M., Rahman, N. A., & Rosdi, N. H. (2019). Citric acid cross-linking of highly porous carboxymethyl cellulose/poly (ethylene oxide) composite hydrogel films for controlled release applications. *Materials Today: Proceedings*, 7, 721–731.

Khoylou, F., & Naimian, F. (2009). Radiation synthesis of superabsorbent polyethylene oxide/tragacanth hydrogel. *Radiation Physics and Chemistry*, 78(3), 195–198.

Koivuniemi, R., Hakkarainen, T., Kiiskinen, J., Kosonen, M., Vuola, J., Valtonen, J., Luukko, K., Kavola, H., & Yliperttula, M. (2020). Clinical study of nanofibrillar cellulose hydrogel dressing for skin graft donor site treatment. *Advanced in Wound Care*, 9, 199–210.

Lamaming, J., Hashim, R., Sulaiman O., & Leh, C.P. (2017). Properties of cellulose nanocrystal from oil palm trunk isolated by total chlorine free method. *Carbohydrate Polymers*, 156, 409–416.

Lamaming, J., Hashim, R., Sulaiman O., Leh, C P., Sugimoto, T., & Nordin, N.A. (2015). Cellulose nanocrystal isolated from oil palm trunk. *Carbohydrate Polymers*, 127, 202–208.

Lu, M., Liu, Y., Huang, Y.C., Huang, C.J., & Tsai, W.B. (2018). Fabrication of photo-crosslinkable glycol chitosan hydrogel as a tissue adhesive. *Carbohydrate Polymers*, 181, 668–674.

Lu, Y., Biswas, M.C., Guo, Z., Jeon, J.W., & Wujcik, E.K. (2019). Recent developments in bio-monitoring via advanced polymer nanocomposite-based wearable strain sensors. *Biosensors and Bioelectronics*, 123, 167–177.

Luo, H., Cha, R., Li, J., Hao, W., Zhang, Y., & Zhou F. (2019). Advances in tissue engineering of nanocellulose-based scaffolds: A review. *Carbohydrate Polymers*, 224, 115144.

Mali, K. K., Dhawale, S. C., Dias, R. J., Dhane, N. S., & Ghorpade, V. S. (2018). Citric acid crosslinked carboxymethyl cellulose-based composite hydrogel films for drug delivery. *Indian Journal of Pharmaceutical Sciences*, 80(4), 657–667.

Manikandan, N. A., & Lens, P. N. (2022). Green extraction and esterification of marine polysaccharide (ulvan) from green macroalgae Ulva sp. using citric acid for hydrogel preparation. *Journal of Cleaner Production*, 366, 132952.

Mandal, B., & Ray, S. K. (2016). Removal of safranine T and brilliant cresyl blue dyes from water by carboxy methyl cellulose incorporated acrylic hydrogels: Isotherms, kinetics and thermodynamic study. *Journal of the Taiwan Institute of Chemical Engineers*, 60, 313–327.

Maturavongsadit, P., Narayanan, L. K., Chansoria, P., Shirwaiker, P., & Benhabbou, S.R. (2021). Cell-laden nanocellulose/chitosan-based bioinks for 3D bioprinting and enhanced osteogenic cell differentiation. *ACS Applied Bio Materials*, 4(3), 2342–2353.

Md Abu, T., Zahan, K.A., Rajaie, M.A., Leong, C.R., Ab Rashid, S., Mohd Nor Hamin, N.S., Tan, W.N., & Tong, W.Y. (2020). Nanocellulose as drug delivery system for honey as antimicrobial wound dressing. *Materials Today Proceedings*, 31, 14–17.

Mischnick, P., & Momcilovic, D. (2010). Chemical structure analysis of starch and cellulose derivatives. *Advances in Carbohydrate Chemistry and Biochemistry*, 64, 117–210.

Misenan, M.S.M., Akhlisah, Z.N., Shaffie, A.H., Saad, M.A.M., & Norrrahim, M.N.F. (2022). 8- Nanocellulose in sensors. In Sapuan, S. M., Norrrahim, M. N. F., Ilyas, R. A., & Soutis, C. (Eds.), *Industrial Applications of Nanocellulose and Its Nanocomposites* (pp. 213–243). Woodhead Publishing, Sawston.

Nicu, R., Ciolacu, F., & Ciolacu, D.E. (2021). Advanced functional materials based on nanocellulose for pharmaceutical/medical applications. *Pharmaceutics*, 13(8), 1125.

Parhi, R. (2017). Cross-linked hydrogel for pharmaceutical applications: A review. *Advanced Pharmaceutical Bulletin*, 7(4), 515–530.

Patel, D.K., Dutta, S.D., Shin, W.C., Ganguly, K., & Lim, K.T. (2021). Fabrication and characterization of 3D printable nanocellulose-based hydrogels for tissue engineering. *RSC Advances*, 11, 7466–7478.

Pereira, D.R., Silva-Correia, J., Oliveira, J.M., Reis, R.L., Pandit, A., & Biggs, M.J. (2018). Nanocellulose reinforced gellan-gum hydrogels as potential biological substitutes for annulus fibrosus tissue regeneration. *Nanomedicine: Nanotechnology, Biology and Medicine*, 14(3), 897–908.

Pooresmaeil, M., Javanbakht, S., Nia, S.B., & Hamazi, H. (2020). Carboxymethyl cellulose/mesoporous magnetic graphene oxide as a safe and sustained ibuprofen delivery bio-system: Synthesis, characterization, and study of drug release kinetic. *Colloids and Surfaces A: Physicochemical and Engineering Aspects*, 594, 124662

Portela, R., Leal, C.R., Almeida, P.L., & Sobral, R.G. (2019). Bacterial cellulose: A versatile biopolymer for wound dressing applications. *Microbial Biotechnology*, 12(4), 586–610.

Pradeep, H.K., Patel, D.H., Onkarappa, H.S., Pratiksha, C.C., & Prasanna, G.D. (2022). Role of nanocellulose in industrial and pharmaceutical sectors - A review. *International Journal of Biological Macromolecules*, 207, 1038–1047.

Qi, C., Liu, J., Jin, Y., Xu, L., Wang, G., Wang, Z., & Wang, L. (2018). Photo-crosslinkable, injectable sericin hydrogel as 3D biomimetic extracellular matrix for minimally invasive repairing cartilage. *Biomaterials*, 163, 89–104.

Rao, Z., Ge, H., Liu, L., Zhu, C., Min, L., Liu, M., Fan, L., & Li, D. (2018). Carboxymethyl cellulose modified graphene oxide as pH-sensitive drug delivery system. *International Journal of Biological Macromolecules,* 107, 1184–1192.

Ramadan, Q., & Zourob, M. (2021). 3D bioprinting at the frontier of regenerative medicine, pharmaceutical, and food industries. *Frontiers in Medical Technology*, 2, 607648.

Ribeiro, A.M., Magalhães, M., Veiga, F., & Figueiras, A. (2018). Cellulose-Based Hydrogels in Topical Drug Delivery: A challenge in medical devices. In Mondal, M.I.H. (Ed.), *Cellulose-Based Superabsorbent Hydrogels* (1st ed., pp. 1–29). Springer Nature, Switzerland.

Reddy, N., Reddy, R., & Jiang, Q. (2015). Crosslinking biopolymers for biomedical applications. *Trends in Biotechnology*, 33(6), 362–369.

Rusu, D., Ciolacu, D., & Simionescu, B.C. (2019). Cellulose-based hydrogels in tissue engineering applications. *Cellulose Chemical and Technology*, 53, 907–923.

Saddique, A., & Cheong, I.W. (2021). Recent advances in three-dimensional bioprinted nanocellulose-based hydrogel scaffolds for biomedical applications. *Korean Journal of Chemical Engineering*, 38, 2171–2194.

Sagar, S., Mandaloju, S., Lavanya, S., Nanjwade, B., & Reddy, B. (2018). Hydrogels as advanced bio-materials for drug delivery system. *Asian Journal of Science and Technology*, 9(5), 8117–8125.

Saiki, S., Nagasawa, N., Hiroki, A., Morishita, N., Tamada, M., Kudo, H., & Katsumura, Y. (2011). ESR study on radiation-induced radicals in carboxymethyl cellulose aqueous solution. *Radiation Physics and Chemistry*, 80(2), 149–152.

Sezer, S., Şahin, İ., Öztürk, K., Şanko, V., Koçer, Z., & Sezer, Ü.A. (2019). Cellulose-based hydrogels as biomaterials. In Mondal, M. H. (ed) *Cellulose-Based Superabsorbent Hydrogels* (pp. 1177–1203). Springer, Cham.

Sharma, A., Thakur, M., Bhattacharya, M., Mandal, T., & Goswami, S. (2019). Commercial application of cellulose nano-composites–A review. *Biotechnology Reports*, 21, e00316.

Shefa, A.A., Sultana, T., Park, M.K., Lee, Y.S., Gwon, J., & Lee, B. (2020). Curcumin incorporation into an oxidized cellulose nanofiber-polyvinyl alcohol hydrogel system promotes wound healing. *Materials and Design*, 186, 108313

Shojaeiarani, J., Bajwa, D., & Shirzadifar, A. (2019). A review on cellulose nanocrystals as promising biocompounds for the synthesis of nanocomposite hydrogels. *Carbohydrate Polymers*, 216, 247–259.

Singh, B., & Bala, R. (2014). Development of hydrogels by radiation induced polymerization for use in slow drug delivery. *Radiation Physics and Chemistry*, 103, 178–187.

Sirajuddin, N.A., Jamil, M.S.M., & Lazim, M.A.S.M. (2014). Effect of cross-link density and the healing efficiency of self-healing poly (2-hydroxyethyl methacrylate) hydrogel. *e-Polymers*, 14(4), 289–294.

Song, S., Liu, Z., Abubaker, M.A., Ding, L., Zhang, J., Yang, S., & Fan Z. (2021). Antibacterial polyvinyl alcohol/bacterial cellulose/nano-silver hydrogels that effectively promote wound healing. *Materials Science & Engineering: C*, 126, 112171.

Subhedar, A., Bhadauria, S., Ahankari, S., & Kargarzadeh, H. (2021). Nanocellulose in biomedical and biosenesing applications: A review. *International Journal of Biological Macromolecules*, 166, 587–600.

Tanpichai, S. (2022). Recent development of plant-derived nanocellulose in polymer nanocomposite foams and multifunctional applications: A mini-review. *Express Polymer Letters*, 16(1), 52–74.

Tosh, B. (2015). Esterification and etherification of cellulose: Synthesis and application of cellulose derivatives. In Mondal, I. H. (ed) *Cellulose and Cellulose Derivatives: Synthesis, Modification and Applications*, (pp. 259–298). Nova Science Publishers Inc, New York.

Uyanga, K.A., Okpozo, O.P., Onyekwere, O.S., & Daoud, W.A. (2020). Citric acid crosslinked natural bi-polymer-based composite hydrogels: Effect of polymer ratio and beta-cyclodextrin on hydrogel microstructure. *Reactive and Functional Polymers*, 154, 104682.

Wach, R. A., Rokita, B., Bartoszek, N., Katsumura, Y., Ulanski, P., & Rosiak, J.M. (2014). Hydroxyl radical-induced crosslinking and radiation-initiated hydrogel formation in dilute aqueous solutions of carboxymethyl cellulose. *Carbohydrate Polymers*, 112, 412–415.

Wang, F., Huang, K., Xu, Z., Shi, F., & Chen, C. (2022). Self-healable nanocellulose composite hydrogels combining multiple dynamic bonds for drug delivery. *International Journal of Biological Macromolecules*, 203,143–152.

Yang, G., Zhang, Z., Liu, K., Ji, X., Fatehi, P., & Chen, J. (2022). A cellulose nanofibril-reinforced hydrogel with robust mechanical, self-healing, pH-responsive and antibacterial characteristics for wound dressing applications. *Journal of Nanobiotechnology*, 20, 312.

Yin, O.S., Ahmad, I., & Amin, M.C.I.M. (2015). Effect of cellulose nanocrystals content and pH on swelling behaviour of gelatin based hydrogel. *Sains Malaysiana*, 44(6), 793–799.

Yom-Tov, O., Seliktar, D., & Bianco-Peled, H. (2016). PEG-Thiol based hydrogels with controllable properties. *European Polymer Journal*, 74, 1–12.

Yu, S., Zhang, X., Tan, G., Tian, L., Liu, D., Liu, Y., Yang, X., & Pan, W. (2017). A novel pH-induced thermosensitive hydrogel composed of carboxymethyl chitosan and poloxamer cross-linked by glutaraldehyde for ophthalmic drug delivery. *Carbohydrate Polymers*, 155, 208–217.

Yuan, M., Bi, B., Huang, J., Zhuo, R., & Jiang, X. (2018). Thermosensitive and photocrosslinkable hydroxypropyl chitin-based hydrogels for biomedical applications. *Carbohydrate Polymers*, 192, 10–18.

Zhang, Q., Zhu, S., Zheng, Y., Gao, W., Wu, D., Yu, J., & Dai, Z. (2022). Cellulose-nanofibril-reinforced hydrogels with pH sensitivity and mechanical stability for wound healing. *Materials Letters*, 323, 132596.

Zhang, X., Wu, D., & Chu, C.C. (2004). Synthesis and characterization of partially biodegradable, temperature and pH sensitive Dex–MA/PNIPAAm hydrogels. *Biomaterials*, 25(19), 4719–4730.

Zhao, J., He, X., Wang, Y., Zhang, W., Zhang, X., Zhang, X., & Lu, C. (2014). Reinforcement of all-cellulose nanocomposite films using native cellulose nanofibrils. *Carbohydrate Polymers*, 104, 143–150.

Zhao, Y., He, M., Jin, H., Zhao, L., Du, Q., Deng, H., Tian, W., Li, Y., Lv, X., & Chen, Y. (2018). Construction of highly biocompatible hydroxyethyl cellulose/soy protein isolate composite sponges for tissue engineering. *Chemical Engineering Journal*, 341, 402–413.

Zhao, Y., Shen, W., Chen, Z., & Wu, T. (2016). Freeze-thaw induced gelation of alginates. *Carbohydrate Polymers*, 148, 45–51.

Zheng, C., Yue, Y., Gan, L., Xu, X., Mei, C., & Han, J. (2019). Highly stretchable and self-healing strain sensors based on nanocellulose-supported graphene dispersed in electro-conductive hydrogels. *Nanomaterials*, 9(7), 937.

17 Overview of Cellulose Fiber as Materials for Paper Production

Nurul Syuhada Sulaiman
Universiti Sains Malaysia

CONTENTS

INTRODUCTION

Paper is a thin sheet of fiber network (frequently manufactured from cellulose fiber) bound to each other on a fine screen upon dewatering. The modern word "paper" was initiated from the papyrus plant, a grass-like aquatic plant with a woody and bluntly triangle stem. The papyrus paper was first invented in Egypt as this plant was abundantly available along the riverside of the Nile River. The formation of papyrus paper was started by removing the outer rind of its long stem. Then, the stem was lengthwise cut into thin strips of about 40 cm long, exposing its sticky fibrous inner pith. The strips were then placed side by side with their edges slightly overlapping, and the other layer was put on the top at right angles. The layers were dampened, hammered into a single sheet, and dried under pressure. The sticky fibrous inner pith will cement the layers together, acting as an adhesive upon drying. Papyrus is highly stable in a dry climate like Egypt due to its highly rot-resistant cellulose

DOI: 10.1201/9781003358084-17

fiber. Cellulose fiber is paper's primary and essential structural element, which contributes about 90%–99% of its content, significantly impacting end product properties.

CELLULOSE FIBER SOURCES

Cellulose fiber is a long, linear polysaccharide that serves as the plant cell wall's primary load-bearing component. Technically, the term cellulose fiber is referring to as fiber that is composed of cellulose as a major constituent, followed by hemicellulose and lignin. Meanwhile, cellulose is a linear polysaccharide composed of glucose derivatives that makes up the cellulose fiber. The cellulose fiber is very high in strength and durability and has excellent flexibility. The colors can be white to colorless after the removal of impurities. Cellulose fiber is water-insoluble, despite being hydrophilic. It is chemically stable and can form physical and chemical bonding with other fibers when the condition changes from wet to dry. These combinations of cellulose fiber properties make cellulose a very suitable raw material for papermaking. Additionally, various sources of fiber have different morphological and chemical characteristics. Therefore, cellulose fibers can be suitable for producing different grades of final products.

WOOD

Wood is the predominant and preferable source of cellulose fiber in paper production. Cellulose fiber from trees or wood falls into two categories: hardwood (deciduous tree) and softwood (coniferous tree). Poplars, maple, birch, sweetgum, and hickory are examples of hardwood trees that are suitable for producing sturdier printing paper and magazines. Meanwhile, softwood trees such as hemlocks, pines, firs, and spruces serve as excellent cellulose fiber sources for papermaking. The properties of paper and paper-based products are highly influenced by cellulose fiber's physical properties, such as breadth (broad), cell wall thickness, and length. Due to this, vigilant choices of cellulose fiber sources need to be made.

Cellulose fiber properties vary depending on the types of wood, species, the percentage of juvenile wood and mature wood, the ratio of earlywood (springwood) to latewood (summerwood), and the type of pulping process (Cameron, 2004). The length of softwood cellulose fibers ranges from about 2 to 6 mm and 20–60 μm in width, while the length of hardwood fibers measures from about 0.7 to 1.7 mm and 14–40 μm in width. The other difference in morphology between softwood and hardwood is that softwood generates fiber with less cell wall thickness than hardwood. The thick cell wall fiber has poor collapsibility, leading to poor bonding between the fibers and a low rigidity coefficient. The rigidity coefficient is calculated by dividing cell wall thickness with diameter and multiplying by a hundred. Rigid fibers, as in hardwood, are less elastic, negatively affecting paper resistance properties. Thus, it is not suitable for paper production (Akgül & Tozluoğlu, 2009). Another factor that needs to be considered is the tree's ratio of latewood and earlywood. Latewood is composed of fiber with a thick cell wall compared to earlywood. Thus, wood species with a high proportion of latewood than softwood tend to produce paper with low resistance properties. Juvenile wood is short fiber xylem at the core with high microfibril angles (Shmulsky & Jones, 2019). In comparing the percentage of juvenile and mature wood, wood species containing a greater proportion of juvenile wood will produce paper with excellent properties due to its lower cell wall thickness.

Softwood has gained much more interest for papermaking than hardwood due to ease and less energy required to convert them into paper, in consort with its good fiber properties that give rise to good quality of the end product. A long fiber associated with softwood is advantageous in providing high felting power during paper production. High felting power positively affects the physical properties of paper, such as tear resistance, double folding resistance, burst, breaking off, and strength. However, hardwood fibers have an advantage in providing paper with high opacity and smooth surface properties, which is good for sturdier paper products. Therefore, some paper products must be made from a mixture of softwood and hardwood fibers to gain specifically targeted properties of the end products.

The depletion of forest resources due to the rapid and vast increasing consumption of wood for pulp has gained attention. Therefore, a new interest in utilizing other cellulose fiber sources has been developed. Cellulose fiber sources from non-wood such as kenaf, bamboo, hemp, and cotton and waste from wood mills such as sawdust, recycled paper, and paperboard have been investigated. Some of them are readily commercialized.

Non-Wood

Cellulose fiber from non-wood sources is one of the good alternatives to reduce wood consumption in the papermaking industry, especially in a less developed country where few forest resources are available, and paper consumption is comparatively low. Paper production using non-wood is frequently produced on a small scale, and low capital investment with its sources comes from local annual fiber that is suited for local markets. The essential characteristic of non-wood is its vascular bundle. Tracheid, fiber, and vessels are embedded in this vascular bundle, surrounded by parenchyma tissue. There are four categories of non-wood: (1) annual fiber crops (hemp, kenaf, sisal, jute), (2) agricultural residue (wheat, corn, wheat straw, rice straw, bagasse, empty fruit bunch, oil palm frond, oil palm trunk), (3) wild plants (grasses, bamboo, seaweed), and (4) well-defined wastes (cow dung, kangaroo poo).

Straw is a standard non-wood used in the pulp and papermaking industry, followed by bagasse and bamboo. Straw has a short fiber similar to hardwood fiber and contains high silica content, which is typical for non-wood resources. Compared to wood, non-wood plant fiber is thinner and has less cellulose and lignin but is high in hemicellulose, silica, and ash. These characteristics result in a low pulp yield that is high in purity. By contrast, high pulp yield produced from non-woody sources contain a high number of extractives. Therefore, paper made from only non-wood cellulose fiber tends to be stiff and dense, with low opacity and tear resistance. However, pulp and paper produced from non-wood like kenaf, hemp, and flax will exhibit high quality and are frequently used for unique papers such as banknotes and cigarette papers.

The non-wood plant contains many nonfibrous cells (parenchyma cells), which differ entirely from coniferous wood (softwood). Nonfibrous cells are less desirable in paper manufacturing than fibers. Still, the mixture of these two results in paper with high opacity due to the filling of the void by nonfibrous cells during the papermaking process. This morphology is similar to deciduous wood (hardwood) as hardwood frequently comprises many nonfibrous cells. A milder pulping process is essential when using non-wood as a pulp precursor. Moreover, alkaline chemicals such as soda ash, lime, caustic soda, and kraft liquor are favorable to be used in the pulping process.

Many researchers are still finding and working on non-wood as a source for pulp and papermaking. They have utilized non-wood sources such as sugarcane bagasse (Varghese et al., 2020), jute (Day et al., 2006), kenaf (Shakhes et al., 2011), bamboo (Dwiky et al., 2019), rice straw (Kaur et al., 2018), wheat straw (Malik et al., 2020), rapeseed straw (Mousavi et al., 2013), hemp (Barberà et al., 2011; Danielewicz & Surma-Ślusarska, 2010; Lee et al., 2011; Miao et al., 2014), and flax (Sain & Fortier, 2002). Researchers are focusing on finding new and best ways in the pulping process or are studying many aspects and factors that will affect paper properties.

Recycled Fiber

Paper recycling is a process where waste paper is reprocessed to get the recycled fiber or secondary fiber for reuse. Utilizing recycled fiber helps reduce the need for virgin cellulose fiber in addition to reducing the problem of solid waste disposal. Wastepaper is obtained from scattered sources like paper mill, discarded paper materials such as corrugated waste from retail shops and manufacturing plants, and materials discarded after consumer use such as old newspapers and magazines. Some wastepaper like packaging, wrapping papers, and corrugated boxes are typically checked for suitability before the recycling process.

Paper recovery systems are categorized into two types: the recovery with the de-inking process and the recovery without a de-inking process. The de-inking recovery process was carried out by feeding wastepaper into a cylindrical tank pulper equipped with agitator blades, which agitates the wastepaper stock. This process separates and disperses the fibers using hot water at a temperature range from 65°C to 90°C and chemicals to disperse and dissolve the ink. Caustic soda, accompanied by silicate of soda, surfactants, or wetting agents, soda ash, and phosphates, is the most commonly used chemical. After that, the stock undergoes a screening process to remove fine foreign particles and dirt before washing the chemicals and dispersed ink. In some cases, the stock was subjected to a bleaching process employing hypochlorite to improve its whiteness.

Meanwhile, the recovery process without needing de-inking was carried out the same way described above. The only difference is that hot water is used alone without adding chemicals during the pulping process. The printing grade or other white papers are frequently produced using the stock prepared with the de-inking process. Meanwhile, paper products such as coarse paper and boxboards are commonly prepared from the stock without de-inking.

RAGS

Used cloth, also known as rags, is one of the sources of cellulose fiber, which has successfully been used as a primary precursor in producing paper. The paper that originated from this cotton linter or cotton from rags is called cotton paper, rag paper, or rag stock paper. Cotton paper is extensively used for important documents such as banknotes, security certificates, and archival copies of theses. Its high durability can last hundreds of years without deterioration, fading, or discoloration. Cotton paper can be produced with different grades and is typically graded as 25%, 50%, or 100% cotton.

At the paper mill, rags are first sorted by hand to remove any foreign materials such as metal and rubber, disregarding rags comprising synthetic fibers and coatings that are tough to remove. Subsequently, those clean and selected rags are cut and dusted to remove small particles of foreign materials before the removal of iron by passing those rags over magnetic rolls. After that, the rags were cooked in a large cylindrical boiler for 3–10 h under pressure to remove any oils, natural waxes, grease, and fillers. Each part of the rags was cooked with approximately three parts of cooking liquor, soda ash or caustic soda with detergents or wetting agents, and a dilute alkaline solution of lime. Next, the rags are washed before being mechanically beaten. The beating action will optimally modify the fibers by (1) shortening the fiber on an excellent paper formation, (2) making them swell for more conformable and stronger paper, and (3) fibrillates the fiber to increase its surface area and improving bonding ability of the fibers, for a strong, smooth, and good printing paper.

PULP AND PAPER MANUFACTURING PROCESS IN PAPER INDUSTRY

The pulp and paper industry are huge, operating with a wide variety of pulping and papermaking processes that utilize a wide variety of fiber sources. The primary substance in pulp and paper manufacturing is cellulose fiber. Meanwhile, the main process is divided into two: stock preparation and papermaking. The primary objectives of stock preparation are to take the required fibrous raw materials (pulps) and nonfibrous components (additives) and modify each furnish constituent as needed. The second objective is to combine all the ingredients continuously and uniformly into the papermaking stock. Stock is a mixture (slurry) of pulp, fillers, other papermaking materials, and

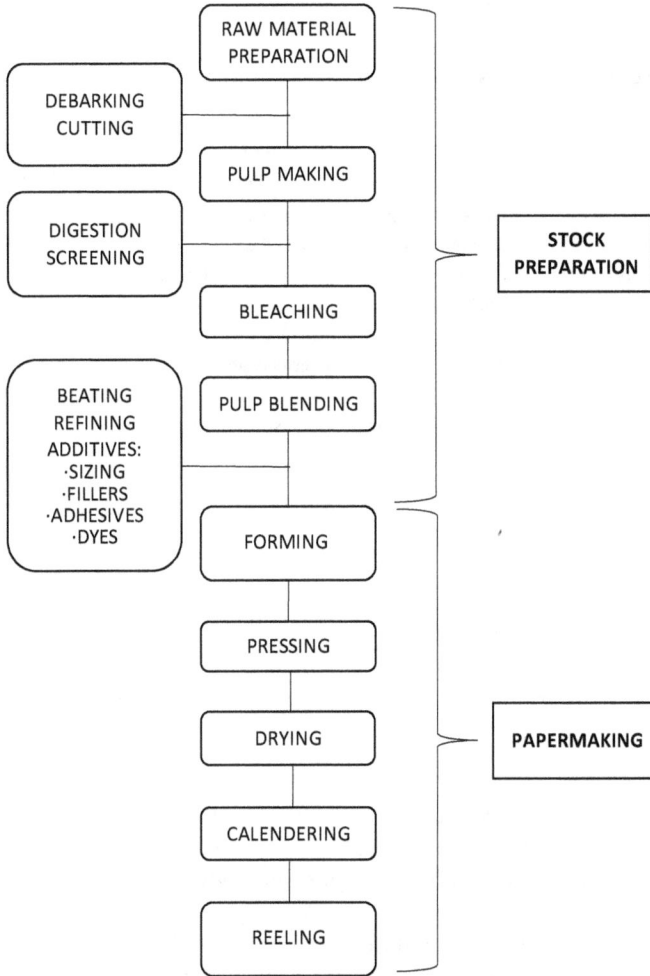

FIGURE 17.1 Basic processes in pulp and papermaking.

water. Meanwhile, furnish is the combination of all of the materials used to make paper. The basic processes in typical paper manufacturing are depicted in Figure 17.1.

There are numerous types and grades of paper that can be produced, and the differences between them are determined by several factors:

1. Types of fiber
2. Beating or refining degree of the stock
3. Addition of additives to the stock
4. Formation condition of the sheet, including grammage (basic weight)
5. Application of physical or chemical treatment to the sheet

PULPING METHODOLOGY

Generally, the pulping process is a pulp-making process that is carried out by separating and treating cellulose fibers. Pulping processes are categorized into five types: mechanical pulping, chemical pulping, chemi-mechanical pulping (CMP), semi-chemical pulping, and biopulping. The overview of all of these pulping processes is presented in Table 17.1.

TABLE 17.1

Comparison Between Five Types of Pulping Methodology

	Mechanical Pulping	Chemi-Mechanical Pulping	Semi-Chemical Pulping	Chemical Pulping	Biopulping
Chemicals	No chemical, grind-stones for logs; disc refiner for chip	Modest chemical impregnation, CTMP-NaOH, NaHSO$_3$	Partially softened or cooked with chemicals and mild disc refining, NSSP	Only chemicals and heat. Kraft, soda, sulfite	No chemicals, fungi, enzymes
Yield (%)	92–96	85–95	55–85	40–55	43–63
Pulp properties	Short, impure fibers	Intermediate	Intermediate	Long, strong fibers	Intermediate
Paper properties	High opacity, softness, bulk, low strength, and brightness	High opacity, bulk, and moderate strength	Good stiffness and moldability	High strength, light brown to dark brown pulps (depend on pulping chemicals)	High strength and high brightness
End uses	Newsprint, books, magazines	Newsprint, books, magazines	Corrugating medium	Wrapping, linerboard, white paper, tissue	White paper

Mechanical Pulping

Mechanical pulping uses mechanical action to defibrillate the cellulose fibers into fiber bundles without removing other contents such as lignin, hemicellulose, and extractives. Mechanical pulping can further be categorized into two subgroups: stone groundwood (SGW) and refiner mechanical pulping (RMP).

SGW pulping uses a wet log as its cellulose fiber source. SGW is carried out by exposing wood to a grinding action and applying pressure to ensure contact and friction between the wood and the spinning grinding stone. The rough surface of the grinder is made of aluminum oxide or silicon carbide (SiC), usually lasting for 2 years. Mechanical pulping is performed by cutting the log to a shorter length before being debarked and fed into the grinder. The log moisture content is a crucial factor that must be dealt with to ease the grinding process and ensure the pulp's quality. The minimum moisture content of the log should be at least 30% and most preferably at 45%–50%. The pulp formed enters the pit that acts as a pulp pool before getting through to a series of rifflers and screens that separate the heavy foreign material such as shives, knots, and bark. The resulting pulp is in the form of a cluster of fibers with considerable amounts of debris. In addition, other chemical constituents of wood like lignin, hemicellulose, and extractives are still present in the resulting pulp. Thus, paper produced using mechanical wood pulp tends to become yellowish over time and after exposure to heat and light such as newspaper. Due to the shorter fiber length and moderate ability of the fiber to bond to each other, paper manufactured from mechanical pulp resulted in low strength properties. However, the positive effect of paper containing mechanical pulp is that it has good opacity, is bulky, and has good printing qualities.

Another mechanical pulping type is RMP. Instead of the log, RMP uses a precursor in the form of a chip. The refiners have two discs with controllable gaps, and the discs are further classified into two types according to their spinning nature. The first type of disc was spinning in the opposite direction. Meanwhile, the other type has one stationary disc and one spinning disc. Two refining stages occur in RMP, with the first stage having a more significant breaking zone but a smaller refining zone and the other way around for the second stage. The first refining stage has a disc with three zones bar with valleys of varying sizes. The first zone is the coarse zone (breaking zone) with a large bar, deep valley, and function to break the wood chip into particles. The second zone is the middle zone with a smaller bar and shallow valley and plays a role in refining the particle further. Meanwhile, the last zone is the fine zone with a smaller bar and shallow valley that plays a role as a particle converter to the pulp. The pulp formed is then screened and washed before the papermaking process. The pulp formed using RMP is much longer, more robust, and darker than SGW.

Chemi-Mechanical Pulping

CMP is the pulping method that combines chemical and mechanical actions. The chemical action softens the lignin, while the mechanical action liberates the fibers. Neutral sulfite semi-chemical and chemi-thermo-mechanical pulping (CTMP) are pulping methodologies that are categorized under CMP pulping. Neutral sulfite semi-chemical is famous for pulping mixed hardwood species. The chemical used during the pulping process is sodium sulfite (Na_2SO_3) and sodium carbonate (Na_2CO_3). Once done, refining action takes place to separate the cellulose fiber further. Meanwhile, chemi-thermo-mechanical pulping utilized heat to pre-stim the fiber sources, followed by chemical impregnation and refining.

Semi-Chemical Pulping

Semi-chemical pulping is started by steeping and impregnating the cellulose fiber sources in the chip form, with inorganic chemical solutions like sodium sulfite, in a smaller amount and under less severe conditions. Once the impregnation is done, the chips are transferred into the disc refiners. The refining process is responsible for converting the softened chips to the pulp. The pulp produced has physical and chemical strength properties that are intermediate between the pulp produced from mechanical and chemical pulping. The resulting semi-chemical pulp has found its uses in a wide range of papers and boards, where stiffness and strength are the vital properties—for example, corrugating medium is used as the interior layer of corrugated boxboards in a heavy-duty container.

Chemical Pulping

Chemical pulping is a pulping methodology in which the pulp is obtained through only chemical action. The chemical serves as the agent to break down the lignin that binds the cellulose fiber together. The first step in chemical pulping is feeding the cellulose fiber source into a large vessel called a digester to cook the source with chemicals under high pressure and temperature. Once cooked, the pulp produced is washed before screening, bleaching (if needed), and papermaking. Three principal processes are involved in chemical pulping: soda, sulfite, and kraft processes. The soda process was developed in 1851 by Burgess and Watt. The soda process makes use of only sodium hydroxide as the chemical agent. However, it tend to produce sodium hydroxide by mixing sodium carbonate (Na_2CO_3) with calcium oxide (CaO).

The sulfite process is the following pulping process in the evolution of chemical pulping. Sulfurous acid is the cooking liquor that was used in this process. Sulfurous acid was prepared by mixing sulfur dioxide with either alkali metals (elements in group 1 in the periodic table, e.g., lithium, sodium, and potassium) or alkaline earth hydroxide (elements in group 2 in the periodic table, e.g., calcium and magnesium). This combination gives a highly acidic and aggressive system that dissolves most of the lignin that binds the cellulose fiber together. However, this very aggressive approach destroys the fibers' primary wall, leading to the destruction of other components called hemicelluloses.

Later, the kraft process (sulfate process) is developed by modifying a soda process but is less aggressive (not very high in pH). In the kraft process, the hemicellulose content is retained. Hemicellulose is essential in reinforcing the fiber–fiber bonds in the sheet of paper. Therefore, paper manufactured from kraft pulp is durable and has greater strength than paper manufactured from soda and sulfite pulp. The kraft process involves a mixture of sodium hydroxide (sodium carbonate (Na_2CO_3)+calcium oxide (CaO)) and sodium sulfide. The other advantage of using the kraft process is that the cooking liquor is recyclable, which is the most economically viable.

Bio Pulping

Biopulping was invented in Sweden and United States at the early 1970s as a pretreatment before the pulping process (Viikari et al., 2009). This method was first developed to cope the drawbacks facing by mechanical pulping, which are weaker pulp strength and high electricity consumption (Atalla et al., 2004). The wood chip was pretreated with lignin, degrading fungi prior to pulping by mechanical action. The softened wood chip reduces the time and energy required during the mechanical pulping. Minor mechanical action subjected to the cellulose fiber sources results in high pulp strength, thus increasing the paper strength. Different groups of fungi cause different forms of degradation. White rot fungi, brown rot fungi, and soft rot fungi are types of fungi that have been applied in the pulping process. However, white rot fungus is the predominant type used due to its ability to degrade all of the cell wall components including lignin. Later, more microorganisms have been screened, and lignin-degrading enzymes are implemented. Several enzymes that play a role in the pulping process are laccase, hemicellulase (xylanase), pectinase, and cellulase. They can be used individually or in combination of two or more. However, the combination of enzymes is more preferable as they can supporting each other's activities. For examples, the combination of hemicellulase and cellulase gives a great impact on de-inking, especially during the conversion of recycled cellulose fiber to pulp. Furthermore, processing conditions such as pH, incubation period, moisture content, oxygen availability, and temperature, and the presence of metal ions must be taken into account to ensure that the efficiency of the enzymes are not decreasing. Biopulping can serve as a cost-effective and green environmental process as well. This can reduce a huge amount of toxic wastewater, particularly from cooking chemicals during the chemical pulping process. Moreover, a simple bleaching method accompanied with a superior result can be achieved by biopulping, which can benefit particularly in kraft pulping.

Bleaching

Bleaching is a chemical process of making pulp lighter and whiter in color. However, the bleaching process is unnecessary depending on the end products' target properties. The cause of color in the pulp is the carbon chain (-C=C-C=C-C=C-) with a conjugated double bond system in lignin. Therefore, the objective of bleaching can be achieved using two strategies: altering or removing the lignin. The selection of the strategy depends on the type of pulp. Bleaching by altering the lignin structure is preferable applied to mechanical pulp. Mechanical pulp has lignin still in the fiber. Lignin can prevent degradation of the produced paper due to the polyphenol content that acts as radical scavengers (Małachowska et al., 2020). Therefore, bleaching by altering the conjugated double bond system of the lignin is suitable as the lignin is still preserved, while the bleaching objective can still be achieved. Meanwhile, lignin removal is the best strategy for chemical pulp. In chemical pulp, a small amount of lignin might remain on the cellulose fiber. Therefore, the pure pulp can be obtained by completely removing the lignin.

Both altering and removal strategies use chlorine gas (Cl_2). However, using chlorine elements (as the initial bleaching agent) in pulp bleaching will produce nasty by-products, including dioxin, furan, and polychlorinated biphenyls. Therefore, the pulp and paper industry has replaced chlorine elements with chlorine compounds like hypochlorous or hyperchloric acid. Two types of chlorine-free bleaching are elementally chlorine-free (ECF) and totally chlorine-free (TCF). Elementally chlorine-free bleaching involves replacing all chlorine elements in a bleaching sequence with chlorine dioxide. Other chlorine-containing compounds such as hypochlorite and chlorine dioxide have remained. Meanwhile, TCF bleaching avoids using chemicals with chlorine elements or compounds. The common bleaching agents used are oxygen, ozone, hydrogen peroxide, and peracid. TCF bleaching guarantees no chlorinated compounds are produced from the bleaching process, and the effluent produced is non-toxic.

PULP BLENDING

The operations involve pulp blending, including beating or refining and adding wet-end additives.

Pulp Beating/Refining

Pulp beating or refining is a mechanical treatment of pulp fibers using a beater or refiner. The objective of beating and refining is to modify the fibers optimally for the demands of the particular papermaking furnish. So, the fibers can be formed into papers or paperboards of the desired properties. The term beating refers to the early days of papermaking when the beating was performed manually beating the pulp with a stick. Meanwhile, refining is used to describe the mechanical action on fibers accomplished by using a machine called a refiner. Beating and refining of cellulose fiber allow water to penetrate its structure, cause the fiber swell, making it flexible, and thus improve the bonding ability of the fibers. These properties develop strong, smooth printing papers and shorten too long fibers for a good paper formation. Moreover, this mechanical treatment causes the bundle of fibers (especially in mechanical pulp) to separate and resulted in reduce drainage rate, producing high-density paper with high tensile strength and low porosity.

Mechanical pulps are usually not further refined during stock preparation, but light postrefining is sometimes carried out for drainage control. Chemical pulps are beaten for up to several hours, depending on the application, before the beater is discharged. The unbeaten or unrefined pulp contributes to several problems during the papermaking process, including complex drainage rate, bad paper formation, and poor wet-web strength, and the paper produced has low power due to high porosity and high bulk. Bad paper formation will also happen if the pulp (especially virgin pulp from softwood) is used without beating or refining. The pulp with long fiber will be tangled between itself and forming flocs before the pulps reach the papermaking wire machine. Meanwhile, wet-web produced from unbeaten or unrefined pulp is broken easily due to the structure of the unfebrile pulp that is slippery, inflexible, and cylindrical, and contributes to the less specific surface area for hydrogen bonding between the pulp fibers.

Additives

Additives are considered the primary precursor in papermaking other than cellulose fibers. In wet-end additives processes, a wide variety of mineral and chemical agents are added to the pulp stock, either to impart specific properties to the paper product (functional additives) or to facilitate the papermaking process (control additives). Batch-wise operation is frequently applied during the wet-end additives process. After adding additives, blending and metering processes are applied to form homogenous papermaking stock. Table 17.2 presents various types of additives and its application.

Paper is seldom made only from fiber, except for blotting paper, filter paper, and special tissue paper. A particular chemical is used to control and help the stock flow, decrease foaming, and increase the retention of the filler and fine. Nonfibrous materials can be added to the stock as solid and liquid. Solid additives are commercially packed in a certain weight to avoid mistakes during weighting, but the price is relatively high. The addition of solid additives is suitable for "batch-wise" stocks (during disintegration and beating). However, more paper mills use liquid additives. Liquid additives appear as solutions or suspensions depending on the economy or convenience. Storage and dispersion of liquid additives are more straightforward than solid additives. However, extra help (like water supply systems) is needed for additive dilution from the suspension form. Addition of additives must be performed correctly to avoid any unexpected reaction. Not all additives are blended into the stock preparation stage. For additives that are being added to the dried paper, they

TABLE 17.2

Various Types of Additives and Its Application

Additive	Application
Acids and bases	Control pH
Alum	Control pH, fix additives onto cellulose fibers, enhance retention
Sizing agents (rosin)	Control penetration of liquids
Dry strength adhesives (starches, gums)	Enhance burst and tensile, add stiffness and pick resistance
Wet-strength resins	Impart wet strength of the paper
Fillers (clay, tale, tio_2)	Enhance optical and surface properties
Coloring materials (dyes and pigments)	Impact desired color
Retention aids	Enhance retention of fines and fillers
Fiber flocculants	Enhance sheet formation
Defoamers	Enhance drainage and sheet formation
Drainage aids	Enhance water removal on wire
Optical brighteners	Enhance apparent brightness
Pitch control chemicals	Prevent deposition or accumulation of pitch
Simicides	Control microorganisms and slime growths
Specialty chemicals	Corrosion inhibitors, flame-proofing, anti-tarnish chemicals, etc.

are called dry-end additives. It is estimated that 10% of paper mills' total operation cost comes from adding additives.

Additives that affect paper properties are called functional additives. Sizing agents are one the functional additives that enable paper to resist paper penetration by fluids. Sizing can be carried out during stock preparation (as wet-end additives) or to the surface of the dried paper (as coatings). The rate of liquid penetration (wetting) on paper depends on the contact angle pictured. There are two types of sizing: internal sizing and surface sizing. Internal sizing such as rosin will develop resistance to penetrate aqueous liquid, such as ink, throughout the paper. Water penetration is retarded by the nonpolar part of the sizing molecule. The polar part of the sizing molecule anchors to the fiber surface. It is added to the stock before the stock goes into the head box.

Meanwhile, in surface sizing, starch is applied using different mechanisms from internal sizing and added at the size press part. Starch is sprayed at the paper surface, filling up paper capillaries and making water penetration much more difficult. Native starch is rarely used as an additive. In surface sizing, modified starch such as acid hydrolyzed (thinned) starches, oxidized starches, cationic starches, and starch ethers are used. Other than surface sizing, starch-based additives are often used as flocculants and retention aids. These are regularly carried out by using cationic starches produced by a nucleophilic substitution reaction with tertiary or quaternary amines. Additionally, starch-based additives also can serve as a pigment retention enhancer by using cationic waxy starches as an additive. Starch-based additives are commonly used due to its low cost and its ability in increasing fiber bonding, which increase paper strength in burst and tensile strength.

Fillers such as clay (aluminum silicate), talcum powder (magnesium silicate), calcium sulfate, zinc oxide, hydrated silica, hydrated alumina, zinc sulfide, barium sulfate, asbestos, and titanium dioxide are white minerals that fill spaces between fibers. Consequently, fillers improve paper density, opacity, optical properties (brightness), and surface smoothness, as well as receptivity. These properties are suitable for writing or printing paper. It can also cut costs because the cost of a filler like clay (also known as kaolin or china clay) is cheaper than fibers. Paperweight consists of 5%–15% of filler. Too much filler reduces the paper's strength. Alum is added together with the filler to hold the filler in the formed sheet as most fillers have no affinity to fibers. Calcium carbonate ($CaCO_3$) is a type of filler that cannot be used together with alum due to its acid reactivity. It is primarily used to improve brightness, opacity, and ink receptivity. Calcium carbonate also contributes to good burning properties, and this makes paper with calcium carbonate as fillers find its specialty in cigarette paper.

The absorption of dye by cellulose fiber gives color to paper. The absorption of the dye depends on the chemical nature of the dye, capillary pore of the fiber, and polarity of the fiber surface. Three principal types of water-soluble dyes are acid, basic, and direct. Chemically, direct and acid dyes are the same as they are both sodium salts of colored acids. Meanwhile, direct dyes have a good affinity to cellulose fiber, high molecular weight, and excellent fastness to light than acid dyes. The fiber readily absorbs direct dyes. An acid dye can only be retained by adding rosin size and alum to fix the dye to the cellulose fiber surface. Basic dye is salts of color bases of low cost and excellent brilliance. It has a strong affinity to lignin and has poor fastness to light. Also, retention aid is needed to fix on the surface of the cellulose fiber.

Dry strength additives such as polyacrylamides and starch are used to increase the dry strength of paper through hydrogen bonding. They also act as a retention aid. Dry strength additives such as secondary and hardwood cellulose fiber are frequently used on low-quality pulp. Wet strength resins like thermoset resins (formaldehyde, urea formaldehyde) are added to the stock to impart the wet strength of the paper. The wet paper's tensile strength can increase from 0%–5% to 15%–50% compared to dry paper. Therefore, the addition of wet strength resin helps reduce the possibility of a wet paper break during the papermaking process. Additionally, the curing process of resin leads to the formation of covalent fiber–fiber bonds. Special chemicals are some of the other additives that are added for special purpose paper, such as flame retardants and anti-tarnish chemicals, which are frequently used in the manufacturing of tissue paper to wrap silverware.

Control additives are additives that are added to the stock to improve the papermaking process. These additives do not directly affect the paper properties and do not necessarily retain paper products. Retention aid additives such as polyamines and alum improve the retention of fine, filler, and internal sizing agents. Retention aid molecules are attracted to the fiber surface and are large enough to attract several fibers, fines, and fillers. This results in increased retention and drainage. Drainage aid additives are materials that can increase the drainage rate of water from the pulp stock on the wire. Almost all retention aids are considered to improve the drainage rate as fines and fillers are removed from the white water (decrease the solid content of white water). They also influence the moisture content (reduce) of the wet paper (web) in the press section. Moisture content increases of 1% can reduce the web strength by over 10%. Formation aids like anionic polyacrylamides (linear polyelectrolytes) promote fiber dispersion, improve web formation, and may allow higher headbox consistency.

Defoamer and anti-foamer are used to control foaming. Foam exists when air or some other insoluble gas is mixed with water containing surfactants (surface active agents) such as soap and detergent. Defoamer is used to break apart the existing foams, while anti-foamer is used to prevent foam formation. Bloides control the growth of microorganisms (bacteria and fungi) around the paper machine, which produce slime (consisting of proteins and polysaccharides). Slime may break off in pieces and lead to the pitting of paper, holes, and even the web, which leads to costly downtime.

PAPERMAKING

There are several types of papermaking machines including cylinder machine, twin-wire machine, and Fourdrinier machine. The cylinder machine composes of wire covering the cylinder that is partially submerged in a pulp stock. As the cylinder revolves, the pulp stock is transferred onto the wire and forms a mat. The mat is then detached at the top of the cylinder prior to the pressing process. In the twin-wire machine, the pulp stock was brought into the gap of the two moving wires. Pressure is applied onto both bottom and top surfaces to remove water. The twin-wire machine is the less used machine in the papermaking industry (Young et al., 2003). One of the popular papermaking machines is the Fourdrinier machine. The following process in papermaking describes continuous procedures inside the Fourdrinier machine. First, the ready pulp stock is transferred from the pipe-line flow and uniformly distributed across the width of the Fourdrinier machine (headbox) through a connection component known as flow spreader. The headbox is a pressurized device that delivers a uniform jet flow on a forming wire through the slice. A uniform stock distribution is crucial to get good formation and evenly paper. Therefore, the headbox operation is critical for a successful papermaking system. The specific operational objectives of the headbox are as follows:

1. Spreading the stock evenly across the width of the paper machine
2. Leveling out cross-currents consistency variations
3. Leveling out machine direction velocity gradients
4. Creating controlled turbulence to eliminate flocculation
5. Discharging evenly from the slice opening and impinging the jet onto the forming wire at an appropriate angle and place.

Forming and Dewatering

Forming wire is an endless and finely woven belt. The wire acts as a drainage medium and allows fibers and additives to accumulate on it during web forming. The wire travels between two large rolls: the breast roll (near the headbox) and the couch roll (near the pressing section). Until the late 1960s, only forming wire made of metal (usually bronze) was used. Nowadays, plastic fabrics (polyesters or polyamide filaments) are used because they provide a much longer service life (up to ten times longer) than metal wire.

Various types of forming wire fabrics are produced, and the choice depends on the grades of paper needed. A balance between the filament web (to reduce wire effect on paper) and drainage rate is necessary for a good web forming. The paper side of the fabric should be smooth, but the machine side is wear-resistant. A breast roll is a firm roll covered with rubber, which supports the forming wire. The other component in the forming section is the forming board, which retards the initial drainage. Drainage retardation helps the formation of a layer of the flat web before drainage.

Table rolls are one of the dewatering components in the drainage section located at the lower part of the forming wire before the vacuum box. At high speed, rotating table rolls touching the moving forming wire produce a vacuum at the "outgoing nip." This vacuum is responsible for removing water, which helps drainage. However, the vacuum created by the table roll at high speed can ruin the web formation. Therefore, a hydrofoil is introduced below the forming wire to drain some water before the web reaches the table roll.

After the table rolls, the web passes through a vacuum box component, which contributes to a more dewatering process from the web. Two types of vacuum boxes are wet and dry vacuum boxes. The dry vacuum box has a higher vacuum strength than the wet vacuum box. The number of vacuum boxes depends on the speed of the paper machine. A dandy roll is another dewatering component used to impart a watermark, help offset flocculation, and smooth out the top side of the web. The dandy roll is on top of the forming wire between the wet and dry vacuum boxes. In cases when a watermark is needed, the design is sewed or welded on the surface roll. It presses into the web on the forming wire. The pressed part in the web will become compact and thin. A couch roll is the last dewatering component in the forming wire part of the papermaking machine. Water is sucked into the couch roll through the suction box inside the roll. It operates at a high vacuum strength of 40–63.5 mmHg. Sometimes, a lump breaker roll is used together with the couch roll to consolidate the web to reduce the probability of breaks during couching and to impact the web and press out any lumps, which might cause breaks in the paper at the press section.

Pressing

The pressing operation continues the water removal process, which starts from the forming wire part. At this level, the web is relatively permeable to air; therefore, further water removal by vacuum is impractical. The main objectives of pressing are removing water and consolidating the sheet (compressing the sheet to bring fibers into close contact). The other goals depend on the product's requirement, like providing surface smoothness, reducing bulk, and promoting higher sheet strength for good runnability in the dryer section. It is more economical to remove water by mechanical means than by evaporation. Water removal should be uniform across the machine, so the pressed sheet has a level moisture profile entering the dryer section. Sheet consolidation is a crucial phase where the fibers are forced into intimate contact, so fiber-to-fiber bonding develops during drying. If not, the sheet will be bulky. Pressing involves contact between the sheet and the felt in a two-roll press nip. Nip is a contact point between the two rolls where the sheet and felt are pressed.

The pressing mechanism inside a nip can be divided into four phases. In phase 1, compression of the sheet and felt begins, which causes air flow out of both structures (sheet and felt) until the sheet is saturated. No hydraulic pressure is built up and therefore not many changes in dewatering. In phase 2, the sheet is soaked, and hydraulic pressure is built. This causes water to move from the sheet into the felt. As the felt is saturated, water moves out of it through the action of the capillary and the pores between the felt filaments. This phase continues up to the mid-nip, where total pressure reaches maximum. Water is removed from the sheet, and felt is forced out to the back from the nip through hydraulic pressure. High hydraulic pressure is built in two conditions: if the felt cannot absorb moisture or there is no other way to remove felt from the sheet before reaching the mid-nip. This will cause crushing to the sheet, where fibers on the surface will stick to the felt. In phase 3, the nip pressure becomes less, and the hydraulic pressure becomes zero as the sheet and felt leave the mid-nip. This is the maximum point of paper dryness in the nip. In phase 4, both sheet and felt re-expand, and the sheet becomes unsaturated. A negative hydraulic pressure is created. Since the vacuum in the sheet is higher than the felt, re-absorption from the felt to the sheet occurs. The re-absorption of water in this phase interferes with the pressing process.

There are two methods for web transfer from forming wire to press section: open draw or vacuum/suction pick-up. In the open draw, the web is transferred onto felt without contact with the felt from the pressing area. Web tension is needed to pull the web off the forming wire. For a successful web transfer, web tension should be higher than the work required to detach the web from the forming wire. The air blast also separates the web from the forming wire. The open draw is always used for paper machines operating at a speed below 600 m/min or for the production of heavyweight paper. For the vacuum/suction pick-up method, the web is picked up off the forming wire by felt, which wraps a vacuum or suction roll at the point of contact.

Drying

After pressing, the sheet is conveyed through the drying section, where residual water is removed by evaporation. Water remaining in the web cannot be removed by vacuum or pressing at the drying stage. Fiber–fiber bonding happens at the drying stage. The dryer section received the wet sheet from the press section at a moisture content between 50% and 65%. The water removal from the sheet continues until the sheet reaches the moisture level required for finishing and converting operations (4%–9%). The drying section is the most expensive part of the papermaking operation (about 60% of the paper mill energy requirement and 80% of the heat requirement). The sheet is dried by wrapping, including passing through a series of rotating stem-heated cylinders (1.5–2.2 m in diameter). A porous fabric (dryer felt) holds the sheet firmly against the hot surface. The felt also helps control the cross-machine direction (CD) shrinkage and keeps the sheet flat.

The drying objectives are first to evaporate water as much as possible using a minimum machine. The machine should be designed to reduce energy usage. Second, drying is used to spread the evaporation evenly as possible. Evaporation in CD is the most critical. The variation in moisture at the end of the drying process will affect the sheet quality, such as creeping. The drying ability depends on two factors: the evaporation rate and the steam economy. A high evaporation rate (water in $kg/h.m^2$) is needed, but it must be suitable for the grade of the paper wanted. Meanwhile, low steam usage is more economical in terms of steam economy (energy used per evaporated water, kJ/kg). The drying process consists of two fundamental theories: heat transfer and water evaporation.

Heat is transferred from steam to the sheet through the cylinder surface of the dryer. Initially, the sheet touches the dryer shell, and the pressure by the felt increases the touching. Water evaporation occurs as the heat transferred to the sheet during connection with the drying cylinder changes the water in the sheet to vapor.

Calendering

After drying, most paper grades go through arrangements of rolls in calendering section. Calendering is a process where paper passes through nips pressed between rolls under high pressure. The rolls may or may not be of equal hardness. The time taken for pressing the nip is short. The main objectives for calendering are reducing the thickness and obtaining a smooth surface of the paper. It also controls paper density and smoothens the paper watermark. The level of calendering depends on the finishing grade desired. Low-grade finishing needs the paper to pass through only one or two calendering nips. For high grades, paper may pass through a series of calendering up to 3 sets of rolls, with each set having 7–9 rolls. Calendering operations are carried out on-machine or off-machine. For economic reasons and current technology expansions, on-machine calendering is used. Calendering under high temperature makes the paper more pliable, and calendaring can be carried out under low pressure.

Usually, the first two rolls in the intermediate rolls stack are heated through an inlet of hot water into the rolls. There are two types of calenders: which are machine calenders and supercalenders. The machine calender, also known as a hard nip machine calender, is the most common type used. It is an on-machine calendering and operates before the reeling process. Paper is pressed between a pair of hard metal rolls and produces paper with the same thickness. Using the recent technology, the soft-nip machine calender can make paper of uniform density. Meanwhile, a supercalender is an off-machine with low speed. Calendering is carried out after reeling or winding, or after the paper is coated. A supercalender is a multi-roller calender composed of alternating hard and soft rolls. Paper is pressed between hard and soft rolls. Soft rolls allow good smoothness without severe blackening of the paper.

Reeling and Winding

After drying and calendering, paper must be collected at the end of the machine by wrapping it at a metal cylinder called a reel spool crossing the paper machine. This process is called reeling. Reeling is a way of collecting paper from the paper machine. It produces a big paper loop (full reel). The reeling is performed until it reaches the desired diameter (full reel). The typical reeling defect is wrinkled and broken due to the air pocket development, especially on dense and low porosity paper grades.

After reeling, paper needs to go through the final process called winding. The purpose of winding is to cut and wind the full paper reel into suitable-sized rolls. During the winding process, the edges of the paper are trimmed off at the slitter section. The broken is conveyed back to the stock preparation section to be recycled. Meanwhile, the middle part of the paper is cut depending on the width desired by customers. Paper of the desired width is rolled on a core made from paper or plastic on the winding drum. Usually, the core is 76 mm in diameter with a wall thickness of about 10 mm. Finally, paper rolls are wrapped with plastic and labeled. Important details often written on the labels include grammage, thickness, type, length, and width of the paper. Then, paper rolls are sent to customers.

CONCLUSION

Cellulose is an important component in the papermaking industry that results in greater economic and technical impacts. The abundance of availability and renewable nature of its source make cellulose a suitable precursor for being applied in the papermaking industry. Meanwhile, the recyclable and biodegradable characteristics of cellulose provide added value to the economy.

The diverse properties of cellulose fiber from different sources provide the prospect of developing various types of paper and paper products. Future research on increasing the efficiency of cellulose resource utilization for papermaking is still needed to ensure the sustainability of the raw material.

REFERENCES

Akgül, M., & Tozluoğlu, A. (2009). Some chemical and morphological properties of juvenile woods from beech (*Fagus Orientalis L.*) and pine (*Pinus Nigra A.*) plantations. *Trends in Applied Sciences Research*, *4*(2), 116–125.

Atalla, R. H., Reiner, R. S., Houtman, C. J., & Springer, E. L. (2004). PULPING I New technology in pulping and bleaching. In J. Burley (Ed.), *Encyclopedia of Forest Sciences* (pp. 918–924). Oxford: Elsevier.

Barberà, L., Pèlach, M. A., Pérez, I., Puig, J., & Mutjé, P. (2011). Upgrading of hemp core for papermaking purposes by means of organosolv process. *Industrial Crops and Products*, *34*(1), 865–872.

Cameron, J. H. (2004). PULPING I Mechanical pulping. In B. Jeffery (Ed.), *Encyclopedia of Forest Sciences* (pp. 899–904). Massachusetts: Academic Press.

Danielewicz, D., & Surma-Ślusarska, B. (2010). Processing of industrial hemp into papermaking pulps intended for bleaching. *Fibres & Textiles in Eastern Europe*, *18*(6), 110–115.

Day, A., Chattopadhyay, S. N., & Ghosh, I. N. (2006). White and coloured handmade paper from jute waste by ambient temperature process. *Journal of Indian Pulp and Paper Technical Association*, *18*(2), 55–58.

Dwiky, M. I., Kumala, S. N., Fety, I. R., & Khaliq, F. N. (2019). The effect of NaOH Concentration variation in the process of paper making from bamboo fiber. *IOP Conference Series: Materials Science and Engineering*, *535*, 012008.

Kaur, D., Bhardwaj, N. K., & Lohchab, R. K. (2018). A study on pulping of rice straw and impact of incorporation of chlorine dioxide during bleaching on pulp properties and effluents characteristics. *Journal of Cleaner Production*, *170*, 174–182.

Lee, M.-K., Kim, J.-S., & Yoon, S.-L. (2011). Effective utilization of hemp fiber for pulp and papermaking (ii)-characteristics of hemp-wood paper made of hemp fiber cooked at low temperature. *Journal of Korea Technical Association of The Pulp Paper Industry*, *43*(5), 27–33.

Małachowska, E., Dubowik, M., Boruszewski, P., Łojewska, J., & Przybysz, P. (2020). Influence of lignin content in cellulose pulp on paper durability. *Scientific Reports*, *10*(1), 19998. doi:10.1038/s41598-020-77101-2.

Malik, S., Rana, V., Joshi, G., Gupta, P. K., & Sharma, A. (2020). Valorization of wheat straw for the paper industry: pre-extraction of reducing sugars and its effect on pulping and papermaking properties. *ACS Omega*, *5*(47), 30704–30715.

Miao, C., Hui, L.-F., Liu, Z., & Tang, X. (2014). Evaluation of hemp root bast as a new material for papermaking. *BioResources*, *9*(1), 132–142.

Mousavi, S. M. M., Hosseini, S. Z., Resalati, H., Mahdavi, S., & Garmaroody, E. R. (2013). Papermaking potential of rapeseed straw, a new agricultural-based fiber source. *Journal of Cleaner Production*, *52*, 420–424.

Sain, M., & Fortier, D. (2002). Flax shives refining, chemical modification and hydrophobisation for paper production. *Industrial Crops Products*, *15*(1), 1–13.

Shakhes, J., Zeinaly, F., Marandi, M. A. B., & Saghafi, T. (2011). The effects of processing variables on the soda and soda-aq pulping of kenaf bast fiber. *BioResources*, *6*(4), 4626–4639.

Shmulsky, R., & Jones, P. D. (2019). *Juvenile Wood, Reaction Wood, and Wood of Branches*. In Shmulsky, R., & Jones, P. D. (eds) *Forest Products and Wood Science: An Introduction* (pp. 107–139). New Jersey: John Wiley & Sons Ltd.

Varghese, L. M., Nagpal, R., Singh, A., Mishra, O. P., Bhardwaj, N. K., & Mahajan, R. J. (2020). Ultrafiltered biopulping strategy for the production of good quality pulp and paper from sugarcane bagasse. *Environmental Science Pollution Research*, *27*(35), 44614–44622.

Viikari, L., Suurnäkki, A., Grönqvist, S., Raaska, L., & Ragauskas, A. (2009). Forest products: biotechnology in pulp and paper processing. In M. Schaechter (Ed.), *Encyclopedia of Microbiology* (3rd ed., pp. 80–94). Oxford: Academic Press.

Young, R. A., Kundrot, R., & Tillman, D. A. (2003). Pulp and paper. In R. A. Meyers (Ed.), *Encyclopedia of Physical Science and Technology (Third Edition)* (pp. 249–265). New York: Academic Press.

18 Challenges and State of the Art of *Allium* Pulp Development for Papermaking
A Brief Review

Mohammad Harris M. Yahya and Noor Azrimi Umor
Universiti Teknologi MARA Cawangan Negeri Sembilan

CONTENTS

INTRODUCTION

Allium is a genus of onion, scallion, garlic, shallot, leek and chives. It has been used in food preparation for centuries for its aroma and flavor. In addition, due to its extensive nutritional, health and healing properties, *Allium* is available in the market in many consumable forms, including fresh *Allium*, *Allium* extracts, *Allium* oil, dehydrated macerated oil, black garlic and *Allium* powder (Subramanian et al., 2020, Bontempo et al., 2021). The increasing demand and supply of *Alliums* around the world generate large amounts of waste associated with *Allium* harvesting and processing, such as damaged cloves, straws, flowers, petioles, stems and leaves (Fatma and Semia, 2017). *Allium* peels are one of the collected wastes that have been tested for various purposes (Nigel et al., 2009; Pereira et al., 2017; Kiassos et al., 2009; Sharma et al., 2016; Thivya et al., 2021; Sha et al., 2013; Gawish et al., 2016; Sara et al., 2021; Poushali et al., 2019; Michalak-Majewska et al., 2020; Poh et al., 2019; Gomaa et al., 2021; Ekemini et al., 2021; Manoj et al., 2022).

The paper and board industry is a major contributor to environmental problems, including deforestation and global warming, as virgin wood is the main lignocellulosic material used in paper production. To address this, various non-wood fibers such as bagasse, wheat and rice stalks, bamboo and kenaf are being explored and currently used in manufacturing worldwide as a lignocellulosic source for papermaking (Swarnima et al., 2010; Wael, 2018; Zicheng et al., 2019; Jalal et al., 2010). The non-wood fibers for papermaking can also be obtained from waste, either from the agriculture or food industry (Moriam et al., 2021).

Although Allium peels are biodegradable wastes, it is still a challenge to manage the wastes without burning or landfilling them. So, it is time to find sustainable ways to use these wastes, and papermaking can be a good option. In this paper, the state of the art for papermaking from Allium peels is discussed. It also discusses the challenges and prospects of the substrate for the paper industry.

DOI: 10.1201/9781003358084-18

ALLIUM PEELS AS RAW PULP FOR PAPER

Allium peels can be considered inedible waste from vegetables that are usually discarded or improperly disposed (Heena and Farida, 2020). In the European Union, more than half a million tons of allium waste are discarded every year (Vanesa et al., 2011). While in Asian countries such as Japan, more than 144,000 tons of Allium waste are produced annually (Feridoun et al., 2013).

In previous work, *Allium cepa* peel waste was studied for papermaking, and it was found that the cooking time plays an important role in determining the strength of the paper sheet, with an increase in the cooking time from 120 to 180 min increasing the tensile and tear strength of the paper sheet (Mohammad et al., 2020). Moreover, *Allium sativum* (garlic) peel alone has 37.2% of cellulose, which is comparable to wheat straw, jute stick (Taslima et al., 2021), dried durian peel fiber (Shaiful et al., 2016), cassava peel (Aripin et al., 2013), red lentil stalks (Taslima et al., 2021) and rapeseed straw (Hosseinpour et al., 2010) where they were 37%, 37.7%, 35.6%, 37.9%, 36.5% and 36.6%, respectively (see Table 18.1). Shakles (2011) stated that the suitable cellulose composition in plant materials for pulp production in papermaking is 34% and above. Although the amount of cellulose is lower than that of the commercially used hardwood cellulose (42.5%), the lignin content of 9.96% is still the lowest among other non-wood fibers and hardwood fibers. Lignin is an undesirable polymer and can affect the reaction times of delignification in the digester or the consumption of pulp chemicals. The lower the lignin content in the component, the shorter the digestion times and chemical consumption.

The use of *Allium* peels as paper pulp has some additional advantages and some disadvantages. For example:

i. *Allium* peels can be obtained as abundant agricultural or food industry waste. Their volume is so large and the cost of obtaining the peels is low to non-existent that they contribute to supporting SDG12 (Luis et al., 2020).

ii. They can be easily recycled because a simple cleaning process does not harm the existing flora, thus preserving a large portion of the perennial biomass.

iii. Since *Allium* peels play a major role in paper production, it helps reduce the problem of open burning or landfills that cause air pollution, thus supporting SDG15 (Luis et al., 2020).

iv. Focusing on papermaking, the difficult points for *Allium* peels are the removal of pigments and water: bleaching and drying; production and energy costs related to pilot or large scale production (not yet tested), financing and market value. Paper made from pure *Allium* peel fibers could suffer from weak bursting, tearing and folding strength. Therefore, blending *Allium* peel pulp with other raw materials such as softwood fibers must be used to improve paper properties (comparable to kraft paper).

PAPERMAKING TECHNOLOGY

In papermaking, a dilute suspension consisting largely of separate and individual cellulose fibers in water is dewatered through a sieve-like screen to produce a mat of randomly interwoven fibers. Water is removed from this sheet by pressing, sometimes with the aid of suction or vacuum, or by heating. A traditional handmade papermaking process can be applied to hardwood or non-wood cellulose, such as *Allium* husks, to produce paper sheets. This technique uses maximum water as the main medium and tends to create a hydrogen bond between the fibers in the paper sheet.

Figure 18.1 provides an overview of the papermaking process from *Allium* (*Allium sativum*) peels. The papermaking process begins with the preparation of fibers suitable for papermaking. The fibers are cooked in a digester at a controlled temperature, pressure and time. The fibers are added to the digester with chemicals such as sodium hydroxide or organic solvents. This is done to help the delignification process of the fibers. It takes about 3 h (180 min) and 170°C for the fibers to be fully

TABLE 18.1

Comparison of the Chemical Composition of Various Sources of Pulp

Fiber/Components (%)	Allium sativum L. Peels (Present Study)	Dried Durian Peel Fiber (Shaiful et al., 2015)	Cassava Peels (Aripin et al., 2013)	Canola Straw (Hosseinpour et al., 2010)	Wheat Straw (Taslima et al., 2021)	Jute Stick (Taslima et al., 2021)	Red Lentil Stalks (Taslima et al., 2021)	Hardwooods (Mazhari Mousavi et al., 2013)
Cellulose	37.22	35.60	37.90	36.60	37	37.7	36.5	38–49
Hemicellulose	35.21	18.60	37.00	40.90	28.6	31.5	22.7	Not available
Lignin	9.96	10.70	7.50	20.00	25.1	27.1	23.8	23–30

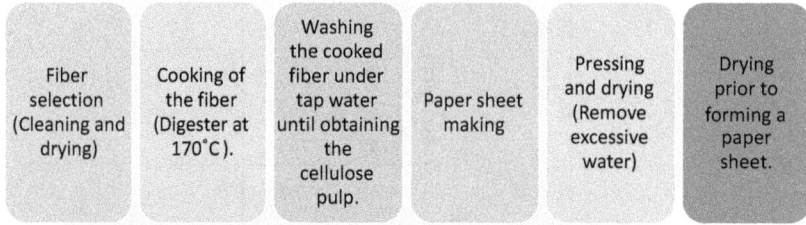

FIGURE 18.1   Papermaking process using *Allium* peels (*Allium sativum*).

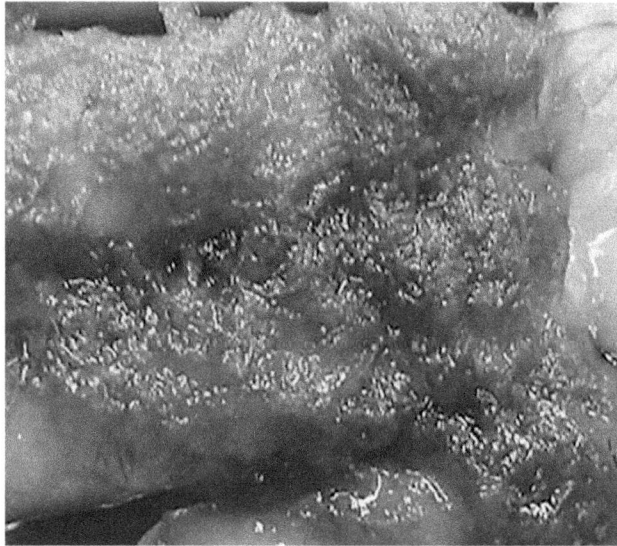

FIGURE 18.2   *Allium sativum* fiber after undergoing fractionation.

cooked. The stove can be in a rotating form or the liquid in the stove is circulated with extreme air to ensure even distribution of the delignification chemicals in the fibers in the stove. After complete cooking, the black liquor in the digester must be removed, and the cellulose, which was already turned into a slurry form, is then washed under tap water until it is completely free of the black liquor. The clean slurried fibers are then fed to a fractionator. The fibers are separated into a uniform size of about 0.05 mm using a water medium, mechanical vibration and a standard slot size to filter the standard size of fibers. The fine fibers with a size of 0.05 mm or less pass through the slot and are collected over a single jersey mesh. Figure 18.2 shows the filtered pulp after passing through the fractionator. The pulp is then weighed based on the moisture content of the fibers and shredded

FIGURE 18.3 *Allium sativum L.* paper sheet.

in the disintegrator at 3,000 rpm for 3 min to break the fiber bundle into individual fibers. Then, the broken fibers together with water are poured into the semi-automatic sheet machine to turn the pulp into a sheet of paper. The *Allium* paper sheet is then pressed on both sides, front and back, for 5 min and 3 min respectively. In this way, the excess water is squeezed out of the sheet while confirming the uniform thickness of the paper sheet. Finally, the paper sheet is dried at room temperature with a standard weight on the edge to prevent the paper sheet from curling or rippling during the drying process. Figure 18.3 shows the *Allium sativum* paper sheet produced.

PROSPECTIVE OF *ALLIUM* PEELS AS A PAPER PULP

Although the peels are easy and plentiful to obtain, some aspects of the challenges of processing them into a paper sheet must be considered. The first challenge in using *Allium sativum* is cleaning the peels before they are used in the cooking process. The peels are very light and thin (they just float in water and are difficult to submerge for a good cleaning process) and have spots of rotten color, so lots of clean water and a special technique (not yet discovered) are needed to ensure that the peels are white in color and free of impurities. The adhering impurities could interfere with the reaction between the fibers and the chemical in the digester. This might be due to the foreign elements existing that appeared as a colored spot in the solution bath. In addition, the impurities are likely to vary the chemical composition results, making it difficult to determine the exact range of chemical composition in the end.

The fiber yield of *Allium sativum* is another challenge because a large number of peels are needed to obtain a large amount of yield for papermaking. An optimum chemical concentration of about 20% sodium hydroxide is required for chemical pulping in the digester to achieve an overall pulp yield of 37.22%. A lower or higher sodium concentration results in a lower pulp yield because the fibers are more prominent or completely dissolved in the solvent chemical. On the other hand, consumer perception and behavior to properly store valuable wastes is another challenge to ensure that the wastes can still be recovered under fresh conditions before they become contaminated. Systematic storage of the waste could increase the volume of fresh waste peel fibers for the following process (Al Seadi and Holm-Nielsen, 2004; Kondusamy et al., 2022). Therefore, another challenge in the use of *Allium sativum* peel waste for papermaking is the specific storage of this waste to ensure the quality of the waste, or it can be used for other types of recyclables to increase economic efficiency (Richa et al., 2022; Sharma and Garg, 2019).

The use of pure *Allium sativum* for papermaking affects the physical and mechanical performance for some time. It is not able to achieve excellent tensile, tear and burst strength alone. To improve the mechanical properties of *Allium sativum* paper, blending with other fibers, such as other wood-free filament fibers or recycled pulp from waste paper, is suggested. Other additional chemical treatments of *Allium sativum* peels, such as organic solvents, are proposed to increase the strength of the individual fibers to strengthen the bond between the fibers in the paper sheet. Another challenge in treating the fibers is the selection of appropriate chemical solvents and their suitable concentration to avoid overtreating the fibers and damaging the chemical and physical structure of the fibers (has not been researched yet). In addition, the machine for plucking the peels of *Allium sativum* is available in many designs and methods (Umesha et al., 2011), but still, a small amount of flesh of *Allium sativum* can be obtained in the container with the separated waste peels, and it must be completely filtered and separated from the peels to avoid problems during the cooking process. The pulp contains a different chemical composition than the peels and could affect the pulp quality and the physical and mechanical properties of the paper.

All the predicted problems and challenges are very important to establish proper management of processing waste, especially *Allium sativum* peels, into paper sheets. Proper management, starting with waste storage, sorting out impurities, cleaning and drying, a special cooking process including the use of chemical solvents and the development of special treatment or other additional fibers with the *Allium sativum* peels, could contribute to another source of income for the small food industry by introducing other product innovations instead of food.

CONCLUSION

The *Allium* peels, especially *Allium sativum*, have a cellulose content of 37.22% suitable for papermaking. Although the cellulose content is still lower compared to commercial hardwood cellulose for papermaking, the lignin content of 9.96% is still much lower than hardwood and most non-wood fibers such as canola straw, cassava peels and dried durian peels. A simple papermaking process can be applied to produce paper from these wastes, thus maximizing the use of waste to achieve zero-waste production in the industry. Proper waste management or storage system, especially for the *Allium* peels, is one of the challenges really needed to ensure fresh waste in abundant volume available for papermaking or other applications. It is proposed that in commercial papermaking, a blended fiber composed of *Allium* peels and other fibers such as recycled paper can be mass produce. The use of blended fiber for pulp production can ensure better physical and mechanical properties of the paper sheet.

REFERENCES

Al Seadi, T., & Holm-Nielsen, J.B. (2004). III.2- Agricultural wastes. *Waste Management Series*, 4, 207–215.
Aripin A. M., Mohd Kassim A. S., Daud Z., & Mohd Hatta M. Z. (2013). Cassava peels for alternative fibre in pulp and paper industry: Chemical properties and morphology characterization. *International Journal of Integrated Engineering*, 5(1), 30–33.
Bontempo, P., Stiuso, P., Lama, S., Napolitano, A., Piacente, S., Altucci, L., Molinari, A.M., De Masi, L., & Rigano, D. (2021). Metabolite profile and in vitro beneficial effects of black garlic (*allium sativum* L.) polar extract. *Nutrients*, 13, 2771.

Ekemini I., Lin Y., Ambrish S., & Ruiyun L. (2021). Chemical modification of waste Allium cepa peels to Cu-complex composite and application as eco environmental oilfield anticorrosion additive. *Journal of King Saud University - Engineering Sciences*, 33(6), 375–385.

Fatma, K., & Semia, E. C. (2017). Perspective of garlic processing wastes as low-cost substrates for production of high-added value products: A review. *Environmental Progress & Sustainable Energy*, 36(6), 1765–1777.

Feridoun, S., Somayeh, D., Jalal, A., & Koji F. (2013). Adding value to onion (Allium cepa L.) waste by sub-critical water treatment. *Fuel Processing Technology*, 112, 86–92.

Gawish, S. M., Helmy, H. M., Ramadan, A. N., Farouk, R., & Mashaly, H. M. (2016). UV Protection properties of cotton, wool, silk and nylon fabrics dyed with red onion peel, madder and chamomile extracts. *Journal of Textile Science & Engineering*, 6(4), 1–13.

Gomaa, A. M. A., Supriya, S., Kwok, F. C., Essam R. S., Algarni, H., & Gurumurthy Hegde T. M. (2021). Superior supercapacitance behavior of oxygen self-doped carbon nanospheres: A conversion of Allium cepa peel to energy storage system. *Biomass Conversion and Biorefinery*, 11, 1311–1323.

Heena, S. K., & Farida, P. M. (2020). ADMET analysis of phyto-components of Syzygium cumini seeds and Allium cepa peels. *Future Journal of Pharmaceutical Sciences*, 6(117), 1–9.

Hosseinpour, R., Fatehi, P., Ahmad Jahan, L., Yonghao, N., & Javad, S.S. (2010). Canola straw hemimechenical pulping for pulp and paper production. *Bioresource Technology*, 101, 4193–4197.

Jalal, S., Mohamma, R. D. F., Pejman, R. C., & Farhad, Z. (2010). Evaluation of harvesting time effects and cultivars of Kenaf on papermaking. *Bioresources*, 5(2), 1268–1280.

Kiassos, E., Mylonaki, S., Makris, D. P., & Kefalas, P. (2009). Implementation of response surface methodology to optimise extraction of onion (Allium cepa) solid waste phenolics. *Innovative Food Science & Emerging Technologies*, 10(2), 246–252.

Kondusamy, D., Tharun, K., Mani, K., & Sandeep, K. M. (2022). Chapter 11- Techno-economic feasibility and hurdles on agricultural waste management. In Hussain, C. M., Singh, S., & Goswami, L (eds) *Emerging Trends to Approaching Zero Waste Environmental and Social Perspectives*, 243–264. Elsevier, Amsterdam.

Luis, M. F., José, P. D., & Alina, M. D. (2020). Mapping the sustainable development goals relationships. *Sustainability*, 12(3359), 1–16.

Manoj, K., Mrunal, D. B., Muzaffa, r H., Sneh, P., Sangram, D., Radha, Nadeem R., Deepak C., R. Pandiselvam, Anjineyulu, K., Maharishi, T., Varsha S., Marisennayya, S., T. Anitha, Abhijit, D., Ali A.S.Sayed Farouk, M., Gadallah Ryszard, A., & Mohamed, M. (2022). Onion (Allium cepa L.) peels: A review on bioactive compounds and biomedical activities, *Biomedicine & Pharmacotherapy*, 146, 112498.

Mazhari Mousavi, S. M., Hosseini, S. Z., Resalati, H., Mahdavi, S., & Garmaroody, E. R. (2013). Papermaking potential of rapeseed straw, a new agricultural-based fiber source. *Journal of Cleaner Production*, 52, 420–424

Michalak-Majewska, M., Teterycz, D., Muszyński, S., Radzki, W., Sykut, & Domańska, E. (2020). Influence of onion skin powder on nutritional and quality attributes of wheat pasta. *PLoS ONE*, 15(1), e0227942.

Mohammad, H. M. Y., Mohd R. A., & Normala H. (2020). Characteristics and mechanical properties of thin paper sheets made from onion peels (allium cepa). *Journal of Academia*, 8(2), 15–21.

Moriam, D. A., Abdulazeez, T. L., Abibat, O. J., Alabi, K. A., Owolabi, O. O., Nelly A. N., Taofeek S., & Sheriff A. (2021). Fascinating physical-chemical properties and fiber morphology of selected waste plant leaves as potential pulp and paper making agents. *Biomass Conversion and Biorefinery*, 11, 3061–3070.

Nigel, C. S., Partington, S. M., & Towns, A. D. (2009). Industrial organic photochromic dyes, society of dyers and colourists. *Coloration Technology*, 125, 249–261.

Pereira, G. S., Cipriani, M., Wisbeck, E., Souza, O., Strapazzon, J. O., & Gern R. M. M. (2017). Onion juice waste for production of Pleurotus sajor-caju and pectinases. *Food and Bioproducts Processing*, 106, 11–18.

Poh, W. C., Poh, S. C., Mohd, H. A., Siti Aisha, M. R., Fu, S. J. Y., & Su-Yin K. (2019). Water Extract of onion peel ash: An efficient green catalytic system for the synthesis of isoindoline-1,3-dione derivatives. *Malaysian Journal of Analytical Sciences*, 23(1) 23–30.

Poushali, D., Sayan, G., Priti, P. M., Hemant, K. S., Madhuparna, B., Santanu, D., Sharba, B., Amit, K. D., Susanta, B., & Narayan C. D. (2019). Converting waste Allium sativum peel to nitrogen and sulphur co-doped photoluminescence carbon dots for solar conversion, cell labeling, and photobleaching diligences: A path from discarded waste to value-added products. *Journal of Photochemistry and Photobiology B: Biology*, 197, 111545.

Richa, G., Anamika, K., Dushyant, D., & Niva, R. M. (2022). Chapter 10- Waste management in fashion and textile industry: Recent advances and trends, life-cycle assessment, and circular economy. In Hussain, C. M., Singh, S., & Goswami, L (eds) *Emerging Trends to Approaching Zero Waste Environmental and Social Perspectives*, 215–242. Elsevier, Amsterdam.

Sara, P.-R., David, S., Cinthia, A., Tanya, T., Nartzislav, P., Daniela, P., & María, J. L. (2021). Biomass waste-derived nitrogen and iron co-doped nanoporous carbons as electrocatalysts for the oxygen reduction reaction. *Electrochimica Acta*, 387, 138490.

Sha, L., Xueyi, G., & Qinghua, T. (2013). Adsorption of Pb2+, Cu2+ and Ni2+ from aqueous solutions by novel garlic peel adsorbent. *Desalination and Water Treatment*, 51, 7166–7171.

Shaiful, R. M., Mohd Halim, I. I., Sharmiza, A., Muhammad Syauqi, A., Ahmad, T., Radhi, A. R., Siti Nurul Aqma, A. R., & Siti Nur Faeza M. Z. (2016). Characteristics of linerboard and corrugated medium paper made from durian rinds chemi-mechanical pulp. *MATEC Web of Conferences*, 51(02007), 1–10.

Shakles, J., Marandi, M.A.B., Zeinaly, F., Saraian, A., & Saghafi, T. (2011). Tobacco residuals as promising lignocellulosic materials for pulp and paper industry. *Bioresources*, 6(4), 4481–4493.

Sharma, K., & Garg, V. K. (2019). Chapter 10- Vermicomposting of waste: A zero-waste approach for waste management. In Taherzadeh, M., Bolton, K., Wong, J., & Pandey, A (eds) *Sustainable Resource Recovery and Zero Waste Approaches*, 133–164. Elsevier, Amsterdam.

Sharma, K., Mahato, N., Nile, S. H., Lee, E. T., & Lee, Y. R. (2016). Economical and environmentally-friendly approaches for usage of onion (Allium cepa L.) waste. *Food & Function*, 7, 3354–3369.

Subramanian, M.S., Nandagopal, G.M.S., Nordin, S.A., Thilakavathy, K., & Joseph, N. (2020). Prevailing knowledge on the bioavailability and biological activities of Sulphur compounds from Alliums: A potential drug candidate. *Molecules*, 25, 4111.

Swarnima, A., Dharm, D., & Tyagi, C. H. (2010). Complete Characterization of bagasse of early species of saccharum officinerum-co 89003 for pulp and paper making, *BioResources*, 5(2), 1–18.

Taslima, F., Yonghao, N., Mohammad, A. Q., Mohammad, N. U., & Md Sarwar, J. (2021). Non-wood fibers: Relationships of fiber properties with pulp properties. *ACS OMEGA, ACS Publications*, 6, 21613–21622.

Thivya, P., Bhosale, Y. K., Anandakumar, S., Hema, V., & Sinija, V. R. (2021). Exploring the effective utilization of shallot stalk waste and tamarind seed for packaging film preparation. *Waste and Biomass Valorization*, 12, 5779–5794.

Umesha, Prakasha, T. L., Mandhar, S. C., Rajasekharppa, K. S., & Venkatachalapathy, K. (2011). Evaluation of garlic (Allium sativum L.) peeling methods. *Environment and Ecology*, 29(1A), 316–318.

Vanesa, B., Esperanza, M., María, A. M.-C., Yolanda, A., Francisco, J. L.-A., Katherine, C., Leon, A. T., & Rosa, M. E. (2011). Characterization of industrial onion wastes (allium cepa L.): Dietary fibre and bioactive compounds. *Plant Foods for Human Nutrition*, 66, 48–57.

Wael, A. E. (2018). Rice straw as a raw material for pulp and paper production. *Encyclopedia of Renewable and Sustainable Materials*, 296–304. Elsevier, Amsterdam.

Zicheng, C., Huiwen, Z., Zhibin, H., Lanhe, Z., Xiaopeng, Y. (2019). Bamboo as an emerging resource for worldwide pulping and papermaking. *Bamboo for Pulp & Paper*, 14(1), 3–5.

19 Application of Cellulose in Leather

Victória Vieira Kopp, Vânia Queiroz, Mariliz Gutterres,
and João Henrique Zimnoch dos Santos
Federal University of Rio Grande do Sul

CONTENTS

INTRODUCTION

Leather is a product with a high commercial value. There is a great demand for leather goods, including footwear, clothing, furniture, automotive upholstery (Winter et al., 2015), as well as membranes and shielding materials (Jiang et al., 2020). In 2020, the global leather goods market was valued at USD 394.12 billion and is expected to grow from 2021 to 2028 with a compound annual growth rate of 5.9%. It is expected that the increasing demand for comfortable, trendy, and fancy leather apparel, footwear, and accessories, along with growing brand awareness, has a positive impact on the market, according to Grand View Research (2020). Hides or skin are converted into leather by the leather industry. The classic processing of leather production uses limited and nonrenewable resources, such as chrome (Ding et al., 2019). Cellulose in turn is one of the most abundant sustainable and renewable biomasses on the planet. It is a tough, fibrous, water-insoluble polysaccharide that is important in maintaining the structure of plant cell walls. It is usually synthesized by plants (Brigham, 2018; Jiang et al., 2020). It has several applications, including the potential to be used in the processing of leather in the tanning, retanning, and finishing stages.

The leather industry generates tons of solid waste every year, which is mainly produced during the operation of adjusting the thickness of the leather or lowering it. For each ton of rawhide processed, approximately 100 to 150 kg of shaving is produced. It is estimated that approximately 0.8 million tons of chromium shavings are generated globally each year (Steffanello Piccin and

DOI: 10.1201/9781003358084-19

Gutterres, 2019; Agustini and Gutterres, 2017). Due to their composition, these wastes can be a source of raw materials for several products. When cellulose is added, composites are developed.

LEATHER PROCESSING

The conversion of skin or hide into leather requires several processing steps that involve complex chemical reactions and mechanical processes (Gutterres and Mella, 2015). The aim is to achieve a final leather product with the desired stability, appearance, water and temperature resistance, elasticity, perspiration, and air permeability (Selvaraju et al., 2019). The fabrication of leather requires various chemicals, such as chromium, vegetable tannins, and polymer resins, including cellulose, and involves many steps. The three main phases of leather processing are beamhouse, tanning, and finishing (Hansen et al., 2020). The leather processing steps and their purposes are as follows (Gutterres and Mella, 2015; Kopp et al. 2021):

BEAMHOUSE

- Reception: receiving the natural or salted hide (preserved to avoid its decomposition);
- Salt shake-off: mechanically removes salt from the hide;
- Presoaking: replenishing part of the water content to the hides that has been removed (dehydration) during conservation;
- Prefleshing: removing the subcutaneous tissue of the hide using a stripping machine;
- Soaking: reporting the original water content of the hides;
- Dehairing and liming: removing the epidermis along with hair and other keratinous hide materials. Lime is added to swell, loosen, and open the fibrous structure, and sulfide is usually applied to degrade the hair;
- Fleshing: eliminating the remaining subcutaneous waste;
- Splitting: dividing the hide into two layers;
- Deliming: eliminating lime, reverse swelling, and adjusting pH for bating;
- Bating: an enzymatic step that removes the residues of the epidermis and keratin;
- Pickling: interrupting enzymatic activity, acidifying the hide in the presence of salt, conditioning it for tanning.

TANNING

- Tanning: designation of the hide treatment used to obtain leather resistant to microbial attack with chemical, thermal, and physical transformation: wet-blue leather (chrome-tanned) and vegetable-tanned leather (vegetable tannin);

FINISHING

- Samming and shaving: adjusting and standardizing the thickness of the part;
- Deacidification: adjusting the pH to neutralize the acidity present;
- Retanning: giving specific texture and physical–mechanical characteristics;
- Dyeing: giving color;
- Fatliquoring: lubricating the fibers and giving softness to the leather;
- Drying: removing water;
- Prefinishing: preparing the leather and surface for finishing;
- Finishing: treating the surface to obtain the desired properties and appearance of the final leather.

Cellulose can be applied to the leather during the following three different processing steps: tanning, retanning, and finishing. These steps will be better explained in the following sections.

Tanning

Tanning is the process that bonds tanning substances to the collagen fiber structure of hide/skin, mostly by collagen protein cross-linkage. The tanning process provides putrefaction stability (microbial and enzymatic), chemical stability, and hydrothermal stability to the hide (Gutterres and Mella, 2015). Reactions of cross-linking occur between the tanning agent and the collagen matrix, in the active groups $-NH_2$ and $-COOH$. The hydrothermal stability of leather is measured by the shrinkage temperature (Ts), which denotes the cross-linking degree of collagen fibers (Li et al., 2019). The reaction between the tanning agent and hide collagen fiber contributes to the enhancement of Ts, and a higher Ts represents a better tanning effect. Chrome-tanned leather has a Ts above 100°C (Ding et al., 2019).

Mineral salts, such as chromium, aluminum, titanium, iron, and zirconium salts, are used as tanning agents because they have a favorable tanning effect and are widely available (Jiang et al., 2020). Chrome salt is currently the most prevalent tanning agent used in leather production because of the favorable properties of the resulting leather. However, as a limited and nonrenewable resource, the permanent use of chrome salt in tanning, in addition to its traditional use as a cleaning agent, may lead to resource depletion and environmental pollution issues. This means that developing other tanning materials that have a favorable tanning effect and are versatile is essential and will be good for the sustainable development of the leather industry (Ding et al., 2019). Examples of alternative tanning agents have been proposed in the literature, such as tetrakis(hydroxymethyl)phosphonium sulfate and commercial Laponite clay (Shi et al., 2019) and graphene oxide (Lv et al., 2016).

Retanning

Retanning determines the leather's desired level of fullness, softness, and grain properties. These properties can be attained by using phenol–formaldehyde resins, acrylic resins, and protein fillers (Selvaraju et al., 2019). Post-tanning, the most significant inorganic pollution load (due mainly to chlorides and sulfates) in wastewater comes from natural and synthetic retanning agents (Hansen et al., 2020). In addition, they are difficult to degrade naturally or treat in wastewater treatment plants (Selvaraju et al., 2019).

There is a new class of retanning agents that uses biopolymers to create better-quality leather products. This new range of materials consists of products that are compatible with these traditional retanning agents based on renewable raw materials such as carbohydrates, proteins, protein hydrolysates (PHs), and biopolymers. Such biobased materials are expected to increase the economic feasibility of retanning agents (Selvaraju et al., 2019), such as cellulose.

Finishing

Leather finishing is the final manufacturing process and comprises a set of treatment steps that are performed to give the appropriate/desired appearance and properties for the final leather article. During the finishing process, surface defects are corrected, thus improving the quality of the leather and its color, fullness, elasticity, shine, solidity, and stability (Winter et al., 2018; Tamilselvi et al., 2019).

Finishing involves successive stages of applying various chemical product formulations, followed or interspersed by drying operations and mechanical operations such as pressing, polishing, and drumming. The successive applications are called impregnation, stucco, prebottom, bottom, topcoat, middle topcoat, final topcoat, and feel modifiers. Depending on the required characteristics of the final product, these steps may or may not all be present. The binder products used are normally softer in the lower layers and increase in hardness until reaching the top layers. The quality and characteristics of the finish depend on the intermediate drying of the layers that must be performed after each layer is applied (Winter, 2014).

To know if the finishing process is effective, various physical tests for flexing endurance, adhesion of finish, light fastness, dry and wet rubbing fastness, and heat fastness properties can be performed (Gumel and Dambatta, 2013; Tamilselvi et al., 2019).

Currently, the most common leather finishing agents include polyacrylates, polyurethanes, nitro-cellulose, and caseins (Maina et al., 2019; Gumel and Dambatta, 2013; Zhang, 2021). Nitrocellulose is widely used in leather topcoats because it can produce a good clear finishing film on leather through a simple process (without any fixation), and the film is waterproof, washable, easy to clean, very fast to rub, and mechanically resistant to stress, with anti-fouling characteristics (Gumel and Dambatta, 2013).

Leather Solid Wastes

During leather processing, a substantial amount of solid waste is generated that impacts the environment. However, these wastes can be utilized as raw materials to produce valuable products. Many wastes come from the beamhouse, such as hide residues, from tanning, such as chromed leather waste, and from finishing, such as buffing dust. Solid leather waste is a protein source that can be used to develop composites with cellulose.

APPLICATION OF CELLULOSE IN LEATHER PROCESSING

Cellulose in Tanning

Cellulose can be used as a tanning agent, as a masking ligand to aid in the penetration of metals into the hide matrix, or can be modified to produce aldehyde groups (Jiang et al., 2020). The strong reaction between the metal, such as aluminum, and the collagen fibers leads to metal overcharge on the leather surface and then a weak tanning effect. To overcome this problem, Jiang et al. (2020) developed a tanning agent with Al-oligosaccharide complexes synthesized via $AlCl_3$-catalyzed cellulose depolymerization. $AlCl_3$ is the catalyst for cellulose degradation, and it also plays a role in tanning by coordinating metal ions. The Al_3^+ reacts with the hydroxyl groups of the oligosaccharides and promotes the decolorization of the leather, also enabling the penetration of Al into the collagen matrix of the skin. Thereafter, the Al species are free from the Al-oligosaccharide and coordinate with the $-NH_2$ group of collagen fibers of the leather, contributing to the stabilization of the fiber bundle and to a realization of satisfactory tanning performance. High-purity tanning agents were produced by liquid–liquid extraction with tetrahydrofuran as the solvent to remove small, oxygenated substances and high-molecular-weight oligosaccharides decomposed by cellulose. The novel tanning agent achieved a leather Ts close to 80°C (Jiang et al., 2020). However, the tanning ability of this Al-oligosaccharide tanning agent is not sufficient for commercial use, limiting its potential due to the inferior tanning performance of Al species. The general Ts requirement for commercial use is that the temperature must be above 80°C (Jiang et al., 2021).

Jiang et al. (2021) developed an Al–Zr-oligosaccharide complex. The conversion of cellulose used a biphasic solvent system, using $AlCl_3$-$NaCl$-H_2O as a catalytic reaction phase and γ-valerolactone as an in situ extraction phase, through a stepwise degradation and oxidation process to produce a high-quality oligosaccharide-based masking agent. The oligosaccharides were oxidized to give active coordination groups, which would help mask the penetration of metal ions. Zr and Al are two different types of metals that can form strong cross-links when combined. This makes them ideal for improving the performance of tanned leather. This product is appropriate for commercial applications at 85.2°C (Jiang et al., 2021).

Ding et al. (2019) prepared a polysaccharide-based tanning agent with peroxide–periodate com-modification of carboxymethyl cellulose (CMC). CMC is a commercially available biomass derived from cellulose. It is a popular choice for energy sources due to its high energy density and low cost. Sodium carboxymethyl cellulose (Na-CMC) was first predegraded with H_2O_2 assisted by a Cu–Fe catalyst to reduce the viscosity of CMC. This was then proceeded by periodate oxidation to produce high solid dialdehyde carboxymethyl cellulose (DCMC), where the aldehyde groups give the chemical structure of DCMC for leather tanning, reaching a Ts of 85°C. The fiber dispersion of DCMC-tanned leather was comparable to that of Cr-tanned leather. DCMC-tanned leather had physical and organoleptic properties, such as tensile strength, tear strength, burst strength, compressive resilience performance, and softness, comparable to those of chrome-tanned leather.

However, formaldehyde was detected in DCMC products. The World Health Organization has identified formaldehyde as a carcinogenic and teratogenic substance. Therefore, the formaldehyde tenor in leather is severely controlled by regulations and standards to maintain the safety of leather products. DCMC is supposed to be formaldehyde free due the aldehyde groups located on the polysaccharide chain of DCMC. However, it is possible to reduce the formaldehyde content in DCMC by lowering the degradation rate and raising the rate of substitution of CMC (Yi et al., 2020).

CELLULOSE IN RETANNING

After tanning, leather is often treated with various auxiliary substances to improve its physical and aesthetic properties. Hence, Selvaraju et al. (2019) used biocomposites made from natural and biological sources developed to replace these unfavorable adjuvants. The biocomposite was prepared with cellulose (acting as a synthetic tanning agent), soybean oil, and a bioemulsifier termed emulsan (acting as a fatliquor) for the leather post-tanning process, which served as a simultaneous retanning–fatliquoring agent. Biocomposites were prepared through ultrasonication by blending different amounts of microcrystalline cellulose with soybean oil and a constant concentration of emulsan prepared from the bacterial strain *Acinetobacter calcoaceticus*. The value of an example composite for use in leather after tanning was evaluated by using it as a retanning agent. Composites containing mostly encapsulated cellulose and soybean oil have been found to penetrate the fiber structure of the leather matrix due to their excellent emulsifying properties. A composite material of emulsified soybean oil containing cellulose was well dispersed in an emulsifier, which serves to fill the pores of the leather and give a great deal of elongation. Fuller leather was found to achieve better retanning than the control leather, demonstrating a better restoration effect. Leathers treated with a biopolymeric composite showed improved strength and physical properties, such as tensile strength, elongation at break, grain crack strength, and tear strength, compared to control leathers. The leather treated with a biopolymeric composite had better properties than the leather made with standard agents.

CELLULOSE IN FINISHING

Many changes have been made to the leather finishing process as a result of the search for better quality and more efficient use of the finish applied. The changes in technology and development have led to the use of water-based lacquers (which use less solvent) instead of organic lacquers (which use more solvent). Some types of paints, sealants, adhesives, and plastics are made from acrylics, urethanes, butadiene rubber, vinyl resins, and so on. The ultimate purpose of these resins is to bind pigment molecules, adhere to the skin, impart elasticity so that the final layer of leather can be tightened, and protect the leather surface (Gumel and Dambatta, 2013).

Tamilselvi and collaborators (2019) extracted cellulose from peanut husks and sugarcane bagasse for use in leather finishing. The composition of peanut husk and bagasse is mostly cellulose, hemicellulose, which makes these wastes viable raw materials for cellulose. Extraction was performed by modified acid hydrolysis. The spray cellulose gave fullness to the final leather without affecting its aesthetic properties.

To improve the solubility of cellulose acetate and increase its use in leather finishing, a self-emulsifying aqueous emulsion was synthesized using cellulose diacetate, dimethylolbutyric acid, and hexamethylene diisocyanate. Cellulose acetate is produced by esterifying natural cellulose and acetic acid. This material has some advantages, including being nontoxic and reproducible and having good film-forming ability (Zhang et al., 2021).

Gumel and Dambatta (2013) compared aqueous solutions of polyvinyl alcohol (PVA), nitrocellulose, and a PVA/nitrocellulose mixture as finishes on dyed goat leather. They found that the

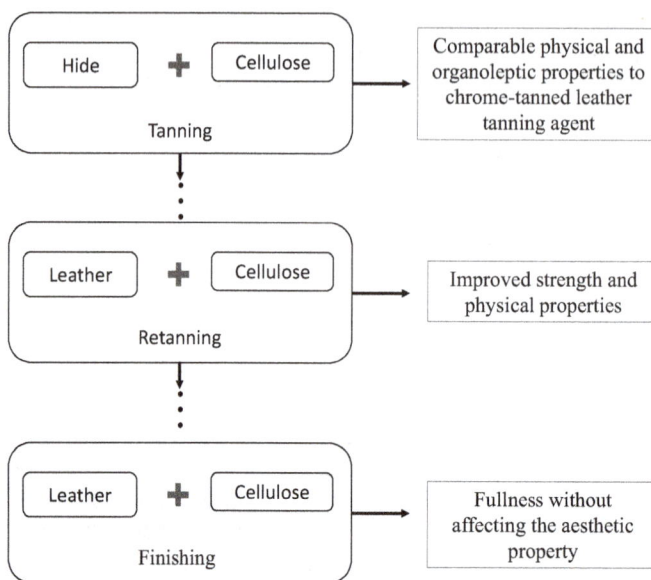

FIGURE 19.1   Function of cellulose in leather processing.

nitrocellulose coatings exhibit superior coating properties in almost every aspect. Products for processing leather containing nitrocellulose are currently found on the market (Tanquimica, 2022; Nitroquímica, 2021). Figure 19.1 shows a diagram with the function of cellulose in each step of leather processing based on the examples cited in the literature.

COMPOSITES OF CELLULOSE AND SOLID LEATHER WASTE

Leather wastes can be turned into valuable materials with the addition of cellulose. These wastes are used to develop composites for use in many areas. Zainescu et al. (2014) used tanned and finished leather wastes to obtain polymeric compositions. The production process involved defibering chrome leather waste and mixing it with cellulose. During defibering, 6% of hydrochloric acid and boric acid were used and neutralized with 8% of sodium carbonate solution and 1% glycerol and a blend of natural rubber latex and synthetic latex based on acrylonitrile butadiene with tannin, soda ash, fish oil, and anti-foaming agent added. Then, the latex was precipitated with aluminum sulfate. The product obtained can be used in automotive, footwear, handbag, and bookbinding applications.

Waste leather buff (WLB) with cellulose was utilized to make biocomposites for packaging applications. Cellulose was dissolved in the ionic liquid 1-allyl-3-methylimidazolium chloride and added 5%–25% of WLB. The cellulose and cellulose–WLB composite films were prepared by regenerating the cast solutions in a water coagulation bath followed by washing and drying. The composite films showed a higher percentage of elongation at break and thermal stability when compared to the matrix. The thermal stability produced was attributed to cross-linked collagen protein leather fibers in WLB. The composite films showed good interfacial bonding between the cellulose and the leather fibers of WLB. The product may be considered for use in packaging and wrapping applications, such as wrappers (Xia et al., 2015).

Sartore et al. (2016) developed a blend based on PH and poly(ethylene-co-vinyl acetate) (EVA). PH is a chromium-free product of chemical hydrolysis from the waste of leather manufacturing. PH–EVA blends were prepared with different percentages and the EVA has different vinyl acetate contents. The addition of PH promotes a regular stiffening effect for all the materials, with the elongation at break over 900% for a PH of 35% and at approximately 600% for the biggest PH. The product obtained mostly from renewable sources represents a biodegradable material that appears

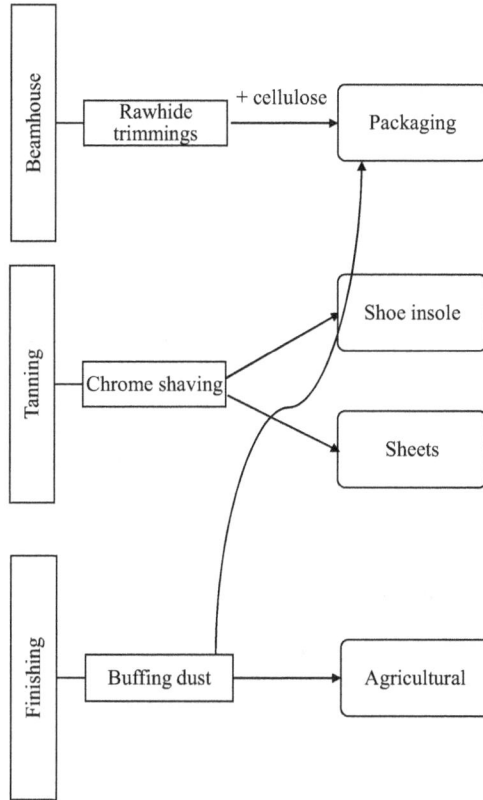

FIGURE 19.2 Composites made from leather solid wastes and cellulose.

promising for several applications, such as in packaging or in agriculture as transplanting or mulching films with the additional fertilizing action of PH.

Ashokkumar et al. (2011) prepared flexible composite sheets made up of chromium-containing collagenous wastes (CS) and derivatives of cellulose. The leather wastes were partially hydrolyzed and converted into composite sheets under microwaves with the addition of 2.5–20 wt% 2-hydroxyethyl cellulose (HEC). With 20% of HEC in the composite sheets, a strength of 3.14 MPa was achieved. The reduction in pores exhibits a better interfacial adhesion of HEC in CS. In addition, the thermal stability was higher in the sheets with more HEC. The developed CS–HEC composite sheets are suitable for leather product applications, due to their flexibility and thermo-mechanical properties. The composites made from leather solid wastes and cellulose are shown in Figure 19.2.

CONCLUSION

Hide is a byproduct of the slaughterhouse that is turned into leather, a valuable commercial product. As this chapter demonstrates, cellulose is generally applied in the following stages of leather processing: tanning, retanning, and finishing. Cellulose is a renewable raw material that can replace many chemicals in the leather industry. Although solid leather wastes are generated during the processing of leather, they are turned into a valuable raw material that combined with cellulose produces composites for several applications, ranging from agricultural to packaging. In this way, cellulose can be one solution used to minimize the environmental issues of the leather industry.

REFERENCES

Agustini, C., & Gutterres, M. (2017). *Biogas Production from Solid Tannery Wastes.*In Vico. A. & Artemio, N. (eds) *Biogas: Production, Applications and Global Developments* (pp. 1–340). Nova Science Publishers, New York.

Ashokkumar, M., Thanikaivelan, P., Krishnaraj, K., & Chandrasekaran, B. (2011). Transforming chromium-containing collagen wastes into flexible composite sheets using cellulose derivatives: Structural, thermal, and mechanical investigations. *Polymer Composites*, 32(6), 1010–1017.

Brigham, C. (2018). Biopolymers: Biodegradable Alternatives to Traditional Plastics. In Török, B. & Dransfield, T (eds) *Green Chemistry An Inclusive Approach* (pp. 753–770). Elsevier, Amsterdam.

Ding, W., Yi, Y., Wang, Y. N., Zhou, J., & Shi, B. (2019). Peroxide-periodate co-modification of carboxymethyl-cellulose to prepare polysaccharide-based tanning agent with high solid content. *Carbohydrate Polymers*, 224, 115169.

Grand View Research. (2020). *Leather Goods Market Size, Share & Trends Analysis Report*. Grand View Research. https://www.grandviewresearch.com/industry-analysis/leather-goods-market.

Gumel, S.M., & Dambatta, B. (2013). Application and evaluation of the performance of poly (vinyl alcohol) and its blend with nitrocelulose in leather top coating. *International Journal of Chemical Engineering and Applications*, 4(4), 249.

Gutterres, M., & Mella, B. (2015). Chromium in Tannery Wastewater. In Sharma, S. (ed) *Heavy Metals in Water: Presence, Removal and Safety* (pp. 315–344). Royal Society of Chemistry, Cambridge.

Hansen, É., de Aquim, P.M., Hansen, A.W., Cardoso, J.K., Ziulkoski, A.L., & Gutterres, M. (2020). Impact of post-tanning chemicals on the pollution load of tannery wastewater. *Journal of Environmental Management*, 269, 110787.

Jiang, Z., Ding, W., Xu, S., Remón, J., Shi, B., Hu, C., & Clark, J.H. (2020). A 'Trojan horse strategy'for the development of a renewable leather tanning agent produced via an AlCl 3-catalyzed cellulose depolymerization. *Green Chemistry*, 22(2), 316–321.

Jiang, Z., Xu, S., Ding, W., Gao, M., Fan, J., Hu, C., & Clark, J. H. (2021). Advanced masking agent for leather tanning from stepwise degradation and oxidation of cellulose. *Green Chemistry*, 23(11), 4044–4050.

Kopp, V.V., Agustini, C.B., Gutterres, M., & Dos Santos, J.H.Z. (2021). Nanomaterials to help eco-friendly leather processing. *Environmental Science and Pollution Research*, 28(40), 55905–55914.

Li, K., Yu, R., Zhu, R., Liang, R., Liu, G., & Peng, B. (2019). pH-sensitive and chromium-loaded mineralized nanoparticles as a tanning agent for cleaner leather production. *ACS Sustainable Chemistry & Engineering*, 7(9), 8660–8669.

Lv, S., Zhou, Q., Li, Y., He, Y., Zhao, H., & Sun, S. (2016). Tanning performance and environmental effects of nanosized graphene oxide tanning agent. *Clean Technologies and Environmental Policy*, 18(6), 1997–2006.

Maina, P., Ollengo, M.A., & Nthiga, E.W. (2019). Trends in leather processing: A review. *International Journal of Scientific Research*, 7(8), 212–223.

Nitroquímica. (2021). *Nitrocelulose para acabamento em couro*. Nitroquímica. https://nitro.com.br/mercados/couro

Sartore, L., Bignotti, F., Pandini, S., D'Amore, A., & Landro, L. (2016). Green composites and blends from leather industry waste. *Polymer Composites*, 37(12), 3416–3422.

Selvaraju, S., Ramalingam, S., & Rao, J.R. (2019). Polyanionic bio-emulsifier: A heteropolysaccharide based bio-composite for leather post tanning process. *Journal of the American Leather Chemists Association*, 114(3), 72–79.

Shi, J., Wang, C., Hu, Xiao, Y., & Lin, W. (2019). Novel wet-white tanning approach based on laponite clay nanoparticles for reduced formaldehyde release and improved physical performances. *ACS Sustainable Chemistry & Engineering*, 7(1), 1195–1201.

Steffanello Piccin, J.S., & Gutterres, M. (2019). Otimização de parâmetros de transferência de massa e capacidade de adsorção de corante por resíduos de couro. *Revista CIATEC-UPF*, 11(3), 50–61.

Tamilselvi, A., Jayakumar, G.C., Charan, K.S., Sahu, B., Deepa, P.R., Kanth, S.V. & Kanagaraj, J., (2019). Extraction of cellulose from renewable resources and its application in leather finishing. *Journal of Cleaner Production*, 230, 694–699.

Tanquimica. (2022). *Acabamento*. http://www.tanquimica.com.br/categoria/112,acabamento.

Winter, C. (2014). *Caracterização de filmes poliméricos utilizados em acabamento de couros*. Master Dissertation. Universidade Federal do Rio Grande do Sul,Porto Alegre, Brazil.

Winter, C., Agustini, C. B., RS, M. E., & Gutterres, M. (2018). Behavior of polymer films and its blends for leather finishing. *Trends in Textile Engineering & Fashion Technology*, 1(4), 93–102.

Winter, C., Schultz, M.E.R., & Gutterres, M. (2015). Evaluation of polymer resins and films formed by leather finishing. *Latin American Applied Research*, 45, 213–217.

Xia, G., Sadanand, V., Ashok, B., Obi Reddy, K., Zhang, J., & Rajulu, A. (2015). Preparation and properties of cellulose/waste leather buff biocomposites. *International Journal of Polymer Analysis and Characterization*, 20(8), 693–703.

Yi, Y., Jiang, Z., Yang, S., Ding, W., Wang, Y.N., & Shi, B. (2020). Formaldehyde formation during the preparation of dialdehyde carboxymethyl cellulose tanning agent. *Carbohydrate Polymers*, 239, 116217.

Zainescu, G., Albu, L., Deselnicu, D., Constatinescu, R., Vasilescu, A., Nichita, P., & Sirbu, C. (2014). A new concept of complex valorization of leather wastes. *Materiale Plastice*, 51(1), 90–93.

Zhang, J., Bao, Y., Guo, M., & Peng, Z. (2021). Preparation and properties of nano SiO2 modified cellulose acetate aqueous polymer emulsion for leather finishing. *Cellulose*, 28(11), 7213–7225.

20 Utilization of Cellulose in Wastewater Treatment

Nur Syazwani Abd Rahman and Norhafizah Saari
Universiti Sains Malaysia

CONTENTS

INTRODUCTION

Water pollution issues have become a serious concern nowadays. Due to improper handling and management, the water quality has deteriorated tremendously, which has had a big impact, especially on the living organisms (Sjahro et al., 2021). So, what are the consequences of this phenomenon? According to the 2018 edition of the United Nations (UN) World Water Development Report (WWDR), researchers have predicted that by 2050, water scarcity will affect about 57% of the global population for at least 1 month per year in some areas (Boretti & Rosa, 2019; World Water Assessment Programme (United Nations), 2018). In order to address the issues of water demand and quality, researchers are now eager to find a new alternative in the development of sustainable treatment technologies (Jamshaid et al., 2017).

In the process of treating wastewater, the fundamental physical, chemical, and biological basis of the contaminant should be considered. Wastewater comes from either a point source (singular or identified source) or a non-point source (combination of pollutants in larger areas). Commonly, water contaminants include organic and inorganic pollutants, synthetic organic chemicals, heavy metals, and nutrients (Jamshaid et al., 2017). Thus, a proper and specific approach is required depending on the properties and characteristics of the effluent removed (Sjahro et al., 2021). Various methods have been used in solving the issues of water pollution, such as reverse osmosis, chemical precipitation, coagulation/flocculation, ion exchange, and electrochemical technologies (Zamora-Ledezma et al., 2021). However, most of these conventional methods require a large number of chemicals, a high operational cost, poor regeneration, and high electric power consumption. Due to this reason, researchers have recently developed a treatment method called adsorption technology. The efficiency of using this method to remove organic pollutants was reported at about 99.9% (Zamora-Ledezma et al., 2021; Ali et al., 2012).

The selection of materials also plays an important role in this application. A few aspects need to be considered, including the availability, cost, and performance of the material. Materials with

DOI: 10.1201/9781003358084-20

a high specific area, low toxicity, good mechanical strength, and thermal stability would be among the best selections since they will help in increasing the adsorption capacity and efficiency of the effluent removal and, at the same time, do not chemically react with the effluent and produce a new substance that might worsen the treated water condition. Since biomass and biomaterials have become an alternative and have good potential in wastewater treatment, their demand and extended focus have been studied among researchers. This chapter presents a review of cellulose as the main material for wastewater treatment. It includes the form of cellulose commonly used, modification of cellulose, effluent removal using modified cellulose, and common challenges that occur during the application.

FORMS OF CELLULOSE

The physical form of adsorbent material is an important aspect in removing various pollutants that have contaminated fresh water sources. For adsorbents that use natural fibers as their main raw material, the most common physical forms of cellulose absorbent material that have been widely used include powder or granules, fine particles, fiber, hydrogel, film, or membranes (Liu et al., 2002).

The powder form of cellulose is commonly used in the production of adsorbents such as activated carbon. Generally, the process to produce activated carbon adsorbent is divided into two steps: carbonization and physical (carbon dioxide or water steam) or chemical (acid or base) activation (Rodríguez Correa et al., 2017). The decomposition and formation of char during the process releases oxygen and hydrogen compounds. Furthermore, the condensable volatile compound is formed by cleaving weak bonds such as the C–O bond and the -O-4 structure (Rodríguez Correa et al., 2017; Shen et al., 2017). This reaction results in the formation of single- and multi-phenolic compounds (Faravelli et al., 2010; Rodríguez Correa et al., 2017). Dorrestijn et al. (2000) asserted that during the secondary phase of the reaction, the solid structure begins to rearrange due to the crosslinking of C–C bonds, resulting in the formation of char. Moreover, at temperatures greater than 500°C, the aromatic structure is observed to grow and also rearrange into a turbostratic graphene sheet (Kercher & Nagle, 2003).

The advantage of using cellulose in the fiber form is that all the impurities can be removed during the pretreatment process, which also helps in exposing the OH functional group of the cellulose. Then, the hydroxyl groups of the cellulose are modified to improve sorption and selectivity (Wang et al., 2016). As shown in Figure 20.1, pretreatment helps in increasing the roughness of the surface of the fiber and exposes more cavities, thus increasing the rate of modification whereby more OH reactive groups are exposed during the process.

In the case of cellulose in the fiber form, the fiber typically goes through a fabrication process before being treated to produce a high-sorption and high-selectivity water adsorbent by modifying the hydroxyl groups of the cellulose (Wang et al., 2016). Cellulose in the fiber form is used to produce

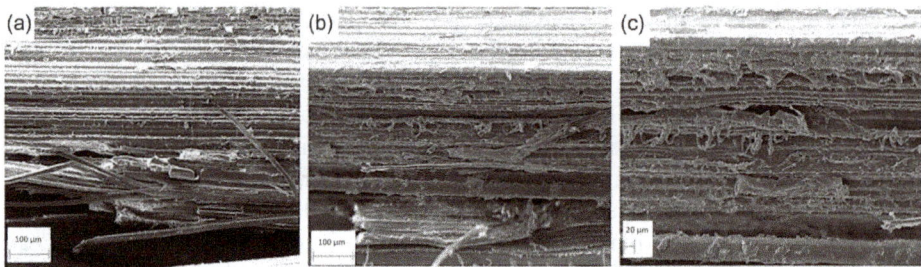

FIGURE 20.1   The effect of (a) before and (b), (c) after treatment process. Reproduced with permission from Adel Salih et al. (2020). Copyright, CC BY License.

cellulose acetate/zeolite (CA/Z) composite fibers using the technique of wet spinning. Zeolite particles embedded in the cellulose acetate act as a polymer matrix. The porous structure of the fiber, with pore sizes ranging from 300 to 500 nm, allows heavy metal ions such as Cu (II) and Ni (II) to pass through and diffuse quickly into internal pores. The zeolite particles then act as adsorptive sites with a size of about 1 μm. The maximum adsorption is found to be between 28.57 and 16.95 mg/g. According to Alila and Boufi (2009), aromatic organic compounds such as 2-naphtol, nitrobenzene, chlorobenzene, dichlorobenzene, trichlorobenzene, and chlorophenol can be removed by modifying cellulose fibers in a heterogeneous environment by grafting long hydrocarbon chains. The reagents responsible for grafting the various hydrocarbons in this study are 4-4'-methylenebis (phenyl isocyanate) (MDI) and N, N'-carbodiimidazole. During the adsorption process, the organic pollutants will self-assemble at the polar active sites on the modified fiber surface. The findings showed that the adsorption capacity had improved after the chemical modification of the fibers' surface, increased from 400 to 1,000 mol/g for the modified substrates compared with virgin fibers, which increased from 20 to 50 mol/g.

The membrane form is one of the physical forms of cellulose adsorbent materials, which are mostly used to remove trapped particles and contaminants, especially oil, from wastewater. The filtration process by membrane is normally carried out in a semi-batch process through continuous withdrawal of permeates and recycling of oil-enriched non-permeates (Madaeni et al., 2013). The benefits of using this membrane technology include less pollution and lower energy consumption, which reduces total processing costs when compared to other conventional methods. In addition, absorbents in the membrane form in processes such as microfiltration, nanofiltration, ultrafiltration, and reverse osmosis were found to be highly effective in removing contaminants in wastewater treatment (Gebru & Das, 2017a, 2017b, 2018). In a study by Juang and Shiau (2000), the membrane synthesis involved the utilization of a commercial membrane solution of cellulose triacetate and cellulose nitrate. In another study, chitosan membranes were used with the aim of removing metal ions such as Cu (II) and Zn (II) from synthetic wastewater (Chen et al., 2009). Ultrafiltration membranes with high permeability and excellent antifouling properties for oil/water emulsion separation were created using cellulose acetate and polyacrylonitrile (PAN).

Researchers have recently discussed four membrane preparation methods for wastewater filtration: direct use of biosynthesized bacterial cellulose (BC) membranes, electrospun mats impregnated with cellulose nanomaterial, vacuum filtration, and composite membranes (Dufresne, 2017). BC membranes were synthesized by bacteria and were formed after 2 days of incubation. Wanichapichart and team produced BC membranes using *Acetobacter xylinum* as a cellulose-producing bacterium in a culture medium. The produced cellulose membrane sheet was tested for filtration of bovine serum albumin and Chlorella sp. The low porosity of this membranes was reported to be around 1.4%–2.4% with an average pore size of 0.08 μm (Wanichapichart et al., 2003). The hydraulic permeability coefficient of the BC membrane depended on cell density and time during the forming process. An aqueous oxidation system based on TEMPO/NaBr/NaClO impregnated into an electrospun PAN nanofibrous scaffold supported by a poly(ethylene terephthalate) (PET) nonwoven substrate used for cellulose nanocrystals (CNCs) preparation was used to prepare electrospun mats (Ma et al., 2012). It had the capability of full retention against bacteria such as *E. coli* and *B. diminuta*. Vacuum filtration is an accessible method to produce the layered structure of nanocellulose membranes. CNC membranes prepared from tunicin CNC via vacuum-assisted filtration onto nylon filter membranes were found to be highly efficient in the separation of oily water, which includes oil-in-water and water-in-oil emulsions (Cheng et al., 2017). Composite membranes that consist of TEMPO-oxidized cellulose nanofiber (CNF) and cellulose triacetate can be prepared by casting 1-methyl-2-pyrrolidinone (NMP) mixtures (Kong et al., 2014). This technique improved the permeation flux and anti-fouling performance due to the hydrophilicity of the surface, which had been imparted by TEMPO-oxidized CNF.

Hydrogel is the most advanced form of adsorbent type that is mainly made from polysaccharide-based material and involves crosslinking-polysaccharide-based materials with other functional groups or coupling agents in order to produce a water-insoluble crosslinked network, where the nanocellulose acts as an additive, binder, or reinforcing agent (Carpenter et al., 2015).

MODIFICATION OF CELLULOSE

Overview

Over the years, cellulose has been widely studied and applied in water treatment. The unique and attractive characteristics have given it a promising potential to be applied as a membrane and adsorbent due to its excellent mechanical, physical, and chemical properties (Jamshaid et al., 2017). Moreover, cellulose is also more stable, nontoxic, and eco-friendly, which makes it a good option for raw material selection in wastewater treatment (Yang et al., 2020).

Availability of cellulose materials is also abundant since they can be obtained from many sources such as wood and non-wood plants, bacteria such as algae, and cellulose-containing animals such as tunicates (Seddiqi et al., 2021), among which, wood and non-wood have become the most preferred sources since they are easily obtained. According to Felgueiras et al. (2021), cotton is the most often used natural fabric for apparel, home furnishings, and industrial products, accounting for almost 90% of all natural fibers. Table 20.1 shows the different types of natural fibers and their cellulose composition as a percentage.

Cellulose is commonly found in the fiber wall along with other lignocellulosic compounds. It is a linear polymer with a -1,4 linkage and a glucose-repeating unit joint. Despite the abundance of hydroxyl groups, the string intermolecular hydrogen bonding and compact structure of cellulose make it difficult to penetrate and replace or dissolve in water (Phanthong et al., 2018; Felgueiras et al., 2021). Therefore, cellulose modification was required to meet the desired industrial requirements. The modification may involve a physical or chemical approach.

Types of Cellulose Modification

Cellulose modification can be carried out through several methods, such as dissolution, allomorphic modifications (chemical reagents, mercerization, and thermal treatment), and derivatization via chemical modification (Park et al., 2010; Sjahro et al., 2021). Each type of modification may have a different effect on the final product.

Dissolution of Cellulose

The linear structure of cellulose, combined with strong intermolecular hydrogen bonding, has made it difficult to penetrate or replace, which makes cellulose insoluble in water despite the abundance of

TABLE 20.1

Types of Natural Fiber and its Cellulose Percentage Composition (Hernandez & Rosa, 2016)

Types of Fiber	Origin	Percentage Cellulose (%)
Banana	Leaf	60.0–65.0
Coir	Fruit	32.0–43.0
Cork bark	Leaf	12.0–25.0
Corn cob	Stalk	33.7–41.2
Cotton	Seed	82.7–95.0
Flax	Stem	64.0–84.0
Hemp	Stem	67.0–78.0
Jute	Bast	51.0–78.0
Sisal	Leaf	60.0–73.0
Wheat stalk	Stalk	30.0–35.0

hydroxyl groups. Thus, dissolution of cellulose is one approach that can be considered, which commonly involves the utilization of organic solvents. However, only selected organic solvents can be used to dissolve cellulose without prior chemical modification, such as ionic liquids, amine oxides, aqueous alkali solutions, inorganic salts, organic solvents, and inorganic molten salt hydrates (Chen et al., 2009).

According to Acharya et al. (2021), theoretically, in order for dissolution of cellulose to take place, the native hydrogen-bonded network, especially in the crystalline region, should be broken. However, by using non-derivatizing solvent systems, the hydrogen bonding between cellulose is disrupted by establishing new hydrogen bonds with the hydroxyl group of the cellulose, which at the same time destroys the crystalline structure. Regenerated cellulose, such as rayon, is commonly produced by dissolving cellulose in a specific solvent and regenerating it via precipitation (Sjahro et al., 2021). Viscous rayon can be used in treating water-soluble zinc (Mamyachenkov et al., 2017).

Allomorphic Modification

In allomorphic modification, the process involves the conversion of cellulose I to cellulose II, III, and IV. Among all forms of cellulose, cellulose I is most commonly found in its native form, which is converted into cellulose II using alkaline treatment. Cellulose III is produced via chemical treatment of cellulose I or II using certain amines or liquid ammonia. However, cellulose IV cannot be produced directly from cellulose I. Cellulose IV can be obtained by treating cellulose II and III at high temperatures using glycerol or by directly converting cellulose III to IV via thermal treatment (Park et al., 2010; Sjahro et al., 2021).

Derivation of Cellulose

The derivation of cellulose, also known as cellulose modification, involves many types of modifications, such as polymer grafting, esterification, etherification, hydrolysis, and others. Different types of modifications may have a different effect on the properties of the cellulose. Depending on the preference, it entails changing or substituting a new functional group at the hydroxyl cellulose. Table 20.2 summarizes the types of modifications.

TABLE 20.2
Types of Fiber Modification

Types of Modification	Attached/Alter Group	Perform Using	Effect of Modification
Esterification	O-C=O	Carboxylic acid, alkyd, alkyl ketene dimer, acid chloride	Increase hydrophobicity of cellulose surface
Oxidation	Oxidant media	• Superoxide anion (O_2^-) • Hydrogen peroxide (H_2O_2)	Act as an excellent coagulation–flocculation agent.
Alkaline	Use as an activating agent in production of activated carbon	NaOH	Increase micropores and diameter of cavities
Silylation	Substitute silyl group	Common silane type • gamma-aminopropyl triethoxy silane (APS) • gamma-diethylenetriaminopropyl trimethoxy silane (TAS) • gamma- methacryloxypropyl trimethoxy silane (MPS)	Normally used to capture CO_2 from air
Grafting	Combination two or more polymers	• RAFT • NMF • ATRP	Grafted cellulose absorbs better than common sorbate.

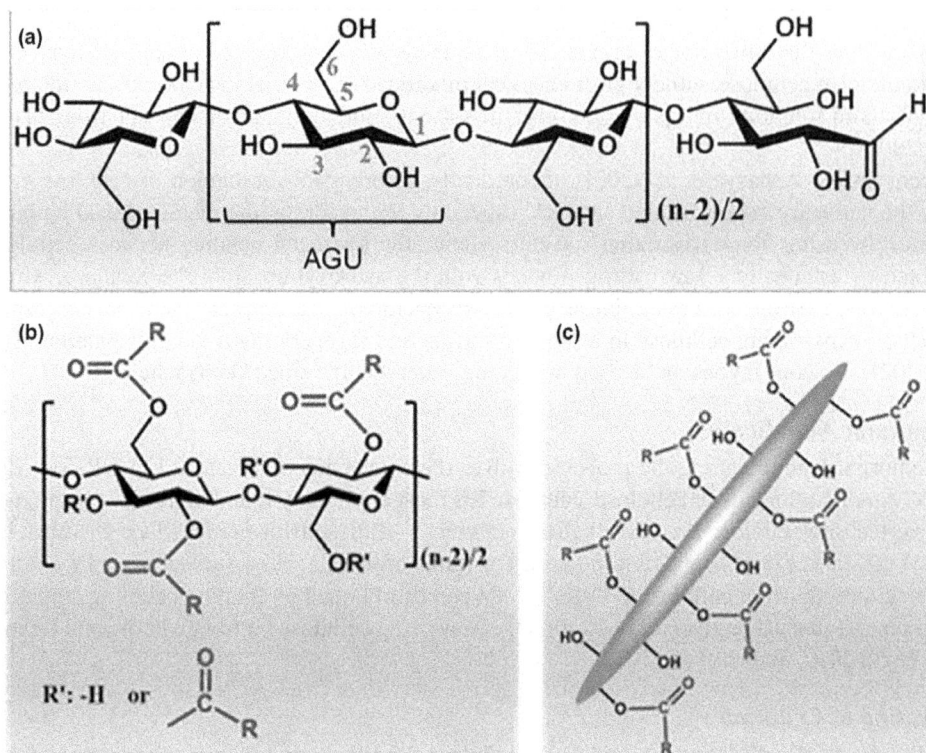

FIGURE 20.2   (a) The location of the OH functional group on the surface of the fiber. (b) and (c) The effect of esterification on the molecular structure of the cellulose. Reproduced with permission from Wang et al. (2018). Copyright, CC BY License.

As shown in Figure 20.2, the mechanism of the process for esterification involves acylation of cellulose using carboxylic acid under acidic conditions or anhydride reaction in base or Lewis acid concentration. The OH groups of cellulose (Figure 20.2a), especially at C2, C3, and C6, are substituted by the ester group (Figure 20.2b and c). This type of modification helps increase the hydrophobicity of cellulose.

Meanwhile, modification by oxidation of cellulose involves oxidation media such as sodium periodate or hydrogen periodate, nitroxyl radicals, strong acids, or ozone (Sang et al., 2017). Among these, the periodate media (sodium periodate or hydrogen periodate) and nitroxyl radicals such as TEMPO (2,2,6,6-tetramethylpiperidine-1-oxyl) are the most preferred media used since they can selectively oxidize the OH groups in cellulose. The difference between the periodate and nitroxyl media is that the periodate can cleave the carbon bond between C2 and C3 of the cellulose and oxidize the C2 and C3 of the hydroxyl groups. On the other hand, the nitroxyl radicals can only oxidize the primary OH groups, as shown in Figure 20.3.

Alkaline modification has been considered one of the simplest methods to improve the interfacial bonding between the fibers (Williams et al., 2011). During the modification process, the hydroxide ions from NaOH will destroy the hydrogen bonding between the cellulose molecules (Figure 20.4). Thus, it helps to increase the active OH groups on the surface of the fiber and, at the same time, removes the cementing materials such as wax and impurities. As a result, the surface area of the fiber wall increases, as do the number of micropores and the diameter of the cavities.

Modification via silylation involves the utilization of silane coupling agents such as gamma-diethylenetriaminopropyl trimethoxy silane (TAS), gamma-methacryloxypropyl trimethoxy silane (MPS), and many more. The OH groups of cellulose are substituted by the silanol group

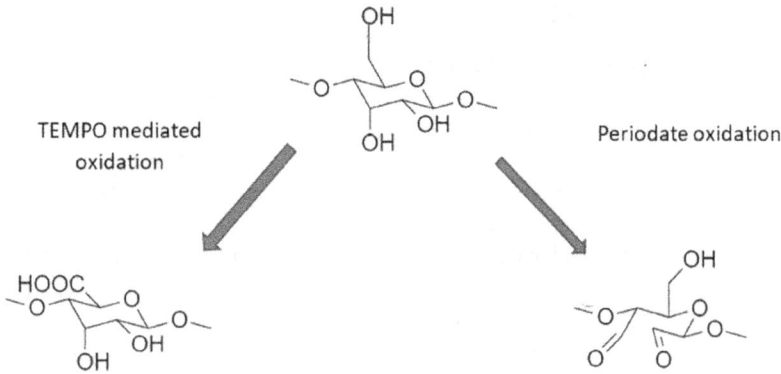

FIGURE 20.3 Effect of cellulose oxidation using nitroxyl radicals (TEMPO-mediated oxidation) and periodate oxidation.

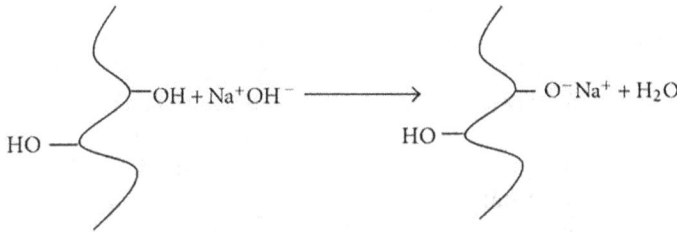

FIGURE 20.4 The alkaline modification of cellulose. Reproduced with permission from Williams et al. (2011). Copyright, CC BY License.

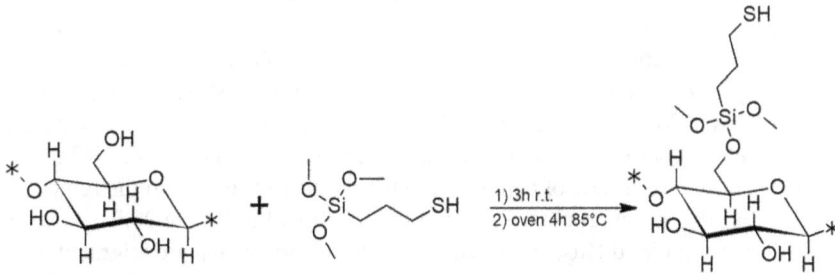

FIGURE 20.5 The silylation process of cellulose. Reproduced with permission from Coelho Braga de Carvalho et al. (2021). Copyright, CC BY License.

on the surface of the cellulose and thus increase the hydrophobicity of the surface (Figure 20.5) (Jankauskaitė et al., 2020).

Cellulose grafting is defined as combining two or more polymers with the backbone of the cellulose chain. A few techniques can be used in the grafting process, such as reversible addition fragmentation transfer (RAFT), radical polymerization, ring-opening polymerization, and many more. Commonly, most of the modification takes place at the hydroxyl group of C2, C3, and C6 of the anhydroglucose unit since it is easily accessible. The effect of grafting may alter the physical and chemical properties of the fiber.

MODIFIED CELLULOSE FOR EFFLUENT TREATMENT

Water contaminants are generally generated from the agricultural sector, domestic activities, and industrial activities, which have been constantly increased to fulfill the demands of food production and the sustenance of the human population globally (Hokkanen et al., 2016b; Gupta et al., 2015). The contaminants that have been detected polluting the fresh water include heavy metal ions, dyes, oil spills, microbes, organic and inorganic micropollutants, metalloids, nutrients, and synthetic organic chemicals (Abouzeid et al., 2019; Hokkanen et al., 2016a; Jamshaid et al., 2017). Toxic heavy metals, for example, have a high potential for causing cancer and other risks to human health by bioaccumulating in the human body through water consumption. Besides, biological pollutants have been found to exist in large quantities in our water resources, such as worms, algae, bacterial species, and viruses, which results from the degradation of organic matter, animal waste, and human activity (Abouzeid et al., 2019).

Oil spills are one of the main pollutants in fresh water that normally occur in offshore oil exploration or during transportation of the oil sources (Munirasu et al., 2016). Oil spill accidents have become a major concern because it is estimated that 30% of the world's oil is spilled (Fakhru'l-Razi et al., 2009). In addition to oil spills, organic matter, also known as total organic carbon (TOC), is one of the pollutants that contaminate fresh water. These hydrocarbons include benzene, toluene, ethylbenzene, and xylenes (BTEX), naphthalene, phenanthrene, dibenzothiophene (NPD), polyaromatic hydrocarbons (PAHs), phenols, carboxylic acids, and low molecular weight aromatic compounds. The low solubility of this hydrocarbon has caused the presence of small oil droplets in water (Peng et al., 2020). Furthermore, another pollutant in produced or processed water is inorganic matter, which includes cations (Na, K, Ca, Mg, Ba, Sr, Fe), anions (Cl, SO, CO, HO), and heavy metals (cadmium, chromium, copper, lead, mercury, nickel, silver, and zinc). Moreover, radioactive materials (radium-226 and radium-228) can also be detected in produced or processed water (Igunnu & Chen, 2014). Suspended solid matters such as solids, sand, clay, corrosion, wax, bacteria, and asphaltenes have been found in processed water (Munirasu et al., 2016).

Modification of cellulose biomaterials with respect to wastewater treatment is being widely studied due to their excellent physical, chemical and mechanical properties. Modified cellulose such as cellulose gels, cellulose composites, cellulose derivatives, functionalized cellulose and nanocrystalline cellulose has been found to be effective in removing pollutants present in wastewater. As discussed under a previous subtopic, cellulose can be chemically modified through many methods such as esterifications, halogenations, oxidation, etherification, grafting, and others. Furthermore, it also can be modified by combining with other materials to form composites beads.

Table 20.3 presents an overview of cellulose modification with respect to the removal of wastewater pollutants via adsorption process. Generally, as illustrated in Figure 20.5, adsorption occurs when the adsorbate molecules diffuse to the surface of the adsorbent and are migrated into the pores of the adsorbent. Finally, the molecules will be adsorbed on the surface of the adsorbent by monolayer or multilayer adsorption (Bharathi & Ramesh, 2013) (Figure 20.6).

THE CHALLENGE AND CONCERN OF USING CELLULOSE IN WASTEWATER TREATMENT

Wastewater treatment is among the main concerns in water management. It should be taken seriously since the uncontrollable discharge of wastewater will have serious effects on the health and water balance of the ecosystem (Nguyen et al., 2021). As a result, researchers have found cellulose as a potential raw material for wastewater treatment.

In the utilization of nanocellulose, the biggest drawback is limited permeance. Most of the nanopapers produced, as reported by researchers, has narrow pore sizes and thick active layers,

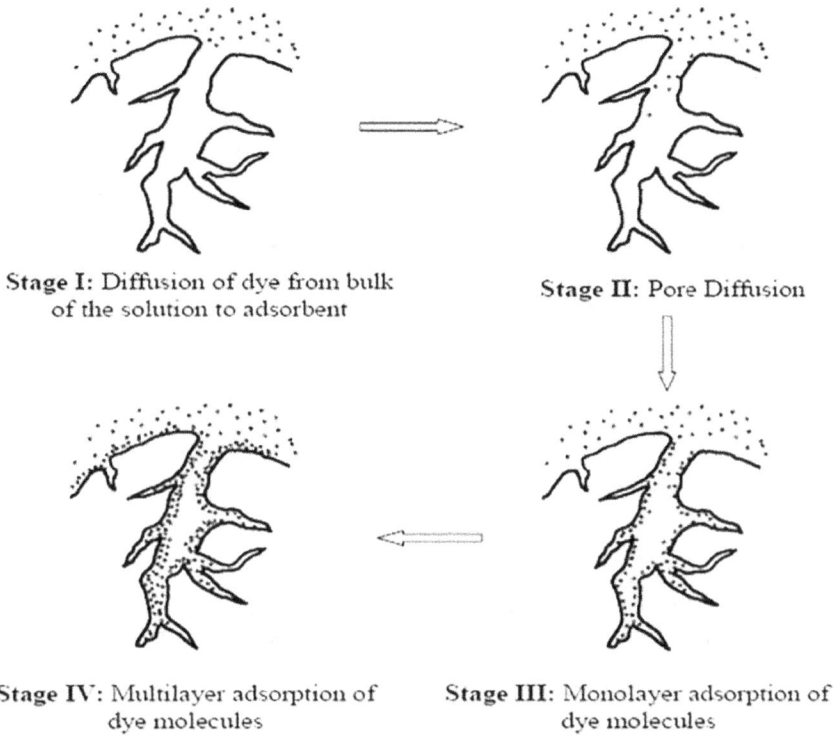

Stage I: Diffusion of dye from bulk
of the solution to adsorbent

Stage II: Pore Diffusion

Stage IV: Multilayer adsorption of
dye molecules

Stage III: Monolayer adsorption of
dye molecules

FIGURE 20.6   The four-stage mechanism of adsorption as proposed by Sivakumar and Palanisamy (2010). Reproduced with permission from Bharathi and Ramesh (2013). Copyright, CC BY License.

TABLE 20.3
Cellulose Modification with Respect to Pollutants Removed From Wastewater

Cellulose Modification	The Process Involved	Contaminants Removed	References
Organic/inorganic treatment	Mercerized cellulose with NaOH, followed by washing with distilled water and acetone. Then reacted with succinic anhydride under pyridine reflux	Cu^{2+}, Cd^{2+} and Pb^{2+}	Gurgel et al. (2008)
	Using carbonated hydroxyapatite (CHA) to produce CHA-modified microfibrillated cellulose	Cd^{2+}, Ni^{2+}	Hokkanen et al. (2014)
	Modified with thionyl chloride followed by ethylene sulfide	Pb, Zn, Co, Ni, Cu and Cd	Silva Filho et al. (2013)
	Modified with maleic anhydride	Hg (III)	Zhou et al. (2012)
	Cellulose is etherified, oxidized, and then modified to a Schiff base using lysine	Mercury ions	Kumari and Chauhan (2014)
Acid treatment	Using iron oxyhydroxide to form cellulose beads	Arsenate and arsenite	Guo et al. (2013)
Cellulose gel	Carboxymethyl cellulose crosslinked with epichlorohydrin (ECH)	PBb (II), Ni (II) and Cu (II)	Yang et al. (2011)
	Using cellulose graft polyacrylamide and hydroxyapatite	Cu (II)	Zwain et al. (2014)

(Continued)

TABLE 20.3 (*Continued*)
Cellulose Modification with Respect to Pollutants Removed From Wastewater

Cellulose Modification	The Process Involved	Contaminants Removed	References
Cellulose composites	Using sodium montmorillonite (NaMMT)	Cr (VI)	Kumar et al. (2012)
	Combined with chitosan (CTS)	Heavy metals	Wu et al. (2010)
	Using ionic liquids to form cellulose and chitin composites	Cu (II), Zn (II), Cr (VI), Ni (II) and Pb (II)	Sun et al. (2009)
	Combined TiO_2 with cellulose to form nanocomposites	Pb	Li et al. (2015)
Cellulose derivatives	Synthesizing graft copolymers from cellulose biopolymers	Pb^{2+}, Zn^{2+}, Cu^{2+}, Cd^{2+}	Singha and Guleria (2014)
	Nanoparticles of zirconium dioxide were grown on a cellulose matrix	Co^{2+}, Cd^{2+}, Cr^{3+}, Fe^{3+}, Cu^{2+}, Zn^{2+}, Ni^{2+}, Zr^{4+}	Khan et al. (2013)
Functionalized cellulose	Cellulose acetate combined with tetraethoxysilane	Cr (VI)	Khan et al. (2013)
	Functionalized a hybrid material of cellulose using glycidylmethacrylate and tetraethylenepentamine	Ag (I), Cu (II), Hg (II)	Donia et al. (2012)
Nanocellulose	Cellulose nanocrystals were chemically modified using succinic anhydride and sodium bicarbonate	Pb^{2+}, Cd^{2+}	Yu et al. (2013)
	Using xanthated nano banana cellulose	Cd (II)	Pillai et al. (2013)

especially in the ultrafiltration range, which may lead to low permeance and limit the efficiency of the filter (Nguyen et al., 2021).

Another concern that should be considered is the possibility of poor durability and leakage of the embedded effluent during the purification process (Nguyen et al., 2021). Furthermore, the filter paper's low mechanical strength may allow some fine fiber particles to enter the treated water. The exposure to the fine particles may also have a short-term effect since it is considered a carcinogenic material. As a result, more research is needed to overcome the challenges of contamination and post-treatment issues (Wang et al., 2020).

REFERENCES

Abouzeid, R. E., Khiari, R., El-Wakil, N., & Dufresne, A. (2019). Current state and new trends in the use of cellulose nanomaterials for wastewater treatment. *Biomacromolecules*, 20(2), 573–597.

Acharya, S., Liyanage, S., Parajuli, P., Rumi, S. S., Shamshina, J. L., & Abidi, N. (2021). Utilization of cellulose to its full potential: A review on cellulose dissolution, regeneration, and applications. *Polymers*, 13(24), 4344.

Adel Salih, A., Zulkifli, R., & Azhari, C. H. (2020). Tensile properties and microstructure of single-cellulosic bamboo fiber strips after alkali treatment. *Fibers*, 8(5), 26.

Ali, I., Asim, M., & Khan, T. A. (2012). Low cost adsorbents for the removal of organic pollutants from wastewater. *Journal of Environmental Management*, 113, 170–183.

Alila, S., & Boufi, S. (2009). Removal of organic pollutants from water by modified cellulose fibres. *Industrial Crops and Products*, 30(1), 93–104.

Bharathi, K. S., & Ramesh, S. T. (2013). Removal of dyes using agricultural waste as low-cost adsorbents: A review. *Applied Water Science*, 3(4), 773–790.

Boretti, A., & Rosa, L. (2019). Reassessing the projections of the world water development report. *NPJ Clean Water*, 2(1), 15.

Carpenter, A. W., de Lannoy, C. F., & Wiesner, M. R. (2015). Cellulose nanomaterials in water treatment technologies. *Environmental Science & Technology*, 49(9), 5277–5287.

Chen, W., Su, Y., Zheng, L., Wang, L., & Jiang, Z. (2009). The improved oil/water separation performance of cellulose acetate-graft-polyacrylonitrile membranes. *Journal of Membrane Science*, 337(1), 98–105.

Cheng, Q., Ye, D., Chang, C., & Zhang, L. (2017). Facile fabrication of superhydrophilic membranes consisted of fibrous tunicate cellulose nanocrystals for highly efficient oil/water separation. *Journal of Membrane Science*, 525, 1–8.

Coelho Braga de Carvalho, A. L., Ludovici, F., Goldmann, D., Silva, A. C., & Liimatainen, H. (2021). Silylated thiol-containing cellulose nanofibers as a bio-based flocculation agent for ultrafine mineral particles of chalcopyrite and pyrite. *Journal of Sustainable Metallurgy*, 7(4), 1506–1522.

Donia, A. M., Atia, A. A., & Abouzayed, F. I. (2012). Preparation and characterization of nano-magnetic cellulose with fast kinetic properties towards the adsorption of some metal ions. *Chemical Engineering Journal*, 191, 22–30.

Dorrestijn, E., Laarhoven, L. J. J., Arends, I. W. C. E., & Mulder, P. (2000). The occurrence and reactivity of phenoxyl linkages in lignin and low rank coal. *Journal of Analytical and Applied Pyrolysis*, 54(1), 153–192.

Dufresne, A. (2017). Nanocellulose. *From Nature to High Performance Tailored Materials* (2nd ed., p. 650). De Gruyter, Berlin, Boston.

Fakhru'l-Razi, A., Pendashteh, A., Abdullah, L. C., Biak, D. R. A., Madaeni, S. S., & Abidin, Z. Z. (2009). Review of technologies for oil and gas produced water treatment. *Journal of Hazardous Materials*, 170 (2), 530–551.

Faravelli, T., Frassoldati, A., Migliavacca, G., & Ranzi, E. (2010). Detailed kinetic modeling of the thermal degradation of lignins. *Biomass and Bioenergy*, 34(3), 290–301.

Felgueiras, C., Azoia, N. G., Gonçalves, C., Gama, M., & Dourado, F. (2021). Trends on the Cellulose-based textiles: Raw materials and technologies. *Frontiers in Bioengineering and Biotechnology*, 29, 608826.

Gebru, K. A., & Das, C. (2017a). Removal of bovine serum albumin from wastewater using fouling resistant ultrafiltration membranes based on the blends of cellulose acetate, and PVP-TiO2 nanoparticles. *Journal of Environmental Management*, 200, 283–294.

Gebru, K. A., & Das, C. (2017b). Removal of Pb (II) and Cu (II) ions from wastewater using composite electrospun cellulose acetate/titanium oxide (TiO2) adsorbent. *Journal of Water Process Engineering*, 16, 1–13.

Gebru, K. A., & Das, C. (2018). Removal of chromium (VI) ions from aqueous solutions using amine-impregnated TiO2 nanoparticles modified cellulose acetate membranes. *Chemosphere*, 191, 673–684.

Guo, C., Zhou, L., & Lv, J. (2013). Effects of expandable graphite and modified ammonium polyphosphate on the flame-retardant and mechanical properties of wood flour-polypropylene composites. *Polymers and Polymer Composites*, 21(7), 449–456.

Gupta, V. K., Nayak, A., & Agarwal, S. (2015). Bioadsorbents for remediation of heavy metals: Current status and their future prospects. *Environmental Engineering Research*, 20(1), 1–18.

Gurgel, L. V. A., Júnior, O. K., Gil, R. P. de, F., & Gil, L. F. (2008). Adsorption of Cu(II), Cd(II), and Pb(II) from aqueous single metal solutions by cellulose and mercerized cellulose chemically modified with succinic anhydride. *Bioresource Technology*, 99(8), 3077–3083.

Hernandez, C., & Rosa, D. D. (2016). Extraction of cellulose nanowhiskers: Natural fibers source, methodology and application. In A. Méndez-Vilas & A. Solano-Martín (Eds.), *Polymer Science: Research Advances, Practical Applications and Educational Aspects* (pp. 232–242). Formatex Research Center, Pennsylvania.

Hokkanen, S., Bhatnagar, A., Repo, E., Lou, S., & Sillanpää, M. (2016a). Calcium hydroxyapatite microfibrillated cellulose composite as a potential adsorbent for the removal of Cr(VI) from aqueous solution. *Chemical Engineering Journal*, 283, 445–452.

Hokkanen, S., Bhatnagar, A., & Sillanpää, M. (2016b). A review on modification methods to cellulose-based adsorbents to improve adsorption capacity. *Water Research*, 91, 156–173.

Hokkanen, S., Repo, E., Westholm, L. J., Lou, S., Sainio, T., & Sillanpää, M. (2014). Adsorption of Ni2+, Cd2+, PO43– and NO3– from aqueous solutions by nanostructured microfibrillated cellulose modified with carbonated hydroxyapatite. *Chemical Engineering Journal*, 252, 64–74.

Igunnu, E. T., & Chen, G. Z. (2014). Produced water treatment technologies. *International Journal of Low-Carbon Technologies*, 9(3), 157–177.

Jamshaid, A., Hamid, A., Muhammad, N., Naseer, A., Ghauri, M., Iqbal, J., Rafiq, S., & Shah, N. S. (2017). Cellulose-based materials for the removal of heavy metals from wastewater – An overview. *ChemBioEng Reviews*, 4(4), 240–256.

Jankauskaitė, V., Balčiūnaitienė, A., Alexandrova, R., Buškuvienė, N., & Žukienė, K. (2020). Effect of cellulose microfiber silylation procedures on the properties and antibacterial activity of polydimethylsiloxane. *Coatings*, 10 (567), 1–21.

Juang, R.-S., & Shiau, R.-C. (2000). Metal removal from aqueous solutions using chitosan-enhanced membrane filtration. *Journal of Membrane Science*, 165(2), 159–167.

Kercher, A. K., & Nagle, D. C. (2003). Microstructural evolution during charcoal carbonization by X-ray diffraction analysis. *Carbon*, 41(1), 15–27.

Khan, S. B., Alamry, K. A., Marwani, H. M., Asiri, A. M., & Rahman, M. M. (2013). Synthesis and environmental applications of cellulose/ZrO2 nanohybrid as a selective adsorbent for nickel ion. *Composites Part B: Engineering*, 50, 253–258.

Kong, L., Zhang, D., Shao, Z., Han, B., Lv, Y., Gao, K., & Peng, X. (2014). Superior effect of TEMPO-oxidized cellulose nanofibrils (TOCNs) on the performance of cellulose triacetate (CTA) ultrafiltration membrane. *Desalination*, 332(1), 117–125.

Kumar, A. S. K., Kalidhasan, S., Rajesh, V., & Rajesh, N. (2012). Application of cellulose-clay composite biosorbent toward the effective adsorption and removal of chromium from industrial wastewater. *Industrial & Engineering Chemistry Research*, 51(1), 58–69.

Kumari, S., & Chauhan, G. S. (2014). New cellulose–Lysine Schiff-base-based sensor–adsorbent for mercury ions. *ACS Applied Materials & Interfaces*, 6(8), 5908–5917.

Li, Y., Cao, L., Li, L., & Yang, C. (2015). In situ growing directional spindle TiO2 nanocrystals on cellulose fibers for enhanced Pb2+ adsorption from water. *Journal of Hazardous Materials*, 289, 140–148.

Liu, M., Deng, Y., Zhan, H., & Zhang, X. (2002). Adsorption and desorption of copper(II) from solutions on new spherical cellulose adsorbent. *Journal of Applied Polymer Science*, 84(3), 478–485.

Ma, H., Burger, C., Hsiao, B. S., & Chu, B. (2012). Nanofibrous microfiltration membrane based on cellulose nanowhiskers. *Biomacromolecules*, 13(1), 180–186.

Madaeni, S. S., Gheshlaghi, A., & Rekabdar, F. (2013). Membrane treatment of oily wastewater from refinery processes. *Asia-Pacific Journal of Chemical Engineering*, 8(1), 45–53.

Mamyachenkov, S. V., Anisimova, O. S., & Egorov, V. V. (2017). Investigation of viscose rayon manufacturing sludges, considered as a raw material for zinc production. *KnE Materials Science*, 2, 182–187.

Munirasu, S., Haija, M. A., & Banat, F. (2016). Use of membrane technology for oil field and refinery produced water treatment—A review. *Process Safety and Environmental Protection*, 100, 183–202.

Nguyen, P. X. T., Ho, K. H., Nguyen, C. T. X., Do, N. H. N., Pham, A. P. N., Do, T. C., Le, K. A., & Le, P. K. (2021). Recent developments in water treatment by cellulose aerogels from agricultural waste. *IOP Conference Series: Earth and Environmental Science*, 947(1), 12011.

Park, S., Baker, J. O., Himmel, M. E., Parilla, P. A., & Johnson, D. K. (2010). Cellulose crystallinity index: Measurement techniques and their impact on interpreting cellulase performance. *Biotechnology for Biofuels*, 3(1), 10.

Peng, B., Yao, Z., Wang, X., Crombeen, M., Sweeney, D. G., & Tam, K. C. (2020). Cellulose-based materials in wastewater treatment of petroleum industry. *Green Energy & Environment*, 5(1), 37–49.

Phanthong, P., Reubroycharoen, P., Hao, X., Xu, G., Abudula, A., & Guan, G. (2018). Nanocellulose: Extraction and application. *Carbon Resources Conversion*, 1(1), 32–43.

Pillai, S. S., Deepa, B., Abraham, E., Girija, N., Geetha, P., Jacob, L., & Koshy, M. (2013). Biosorption of Cd(II) from aqueous solution using xanthated nano banana cellulose: Equilibrium and kinetic studies. *Ecotoxicology and Environmental Safety*, 98, 352–360.

Rodríguez Correa, C., Stollovsky, M., Hehr, T., Rauscher, Y., Rolli, B., & Kruse, A. (2017). Influence of the carbonization process on activated carbon properties from lignin and lignin-rich biomasses. *ACS Sustainable Chemistry & Engineering*, 5(9), 8222–8233.

Sang, X., Qin, C., Tong, Z., Kong, S., Jia, Z., Wan, G., & Liu, X. (2017). Mechanism and kinetics studies of carboxyl group formation on the surface of cellulose fiber in a TEMPO-mediated system. *Cellulose*, 24(6), 2415–2425.

Seddiqi, H., Oliaei, E., Honarkar, H., Jin, J., Geonzon, L. C., Bacabac, R. G., & Klein-Nulend, J. (2021). Cellulose and its derivatives: Towards biomedical applications. *Cellulose*, 28(4), 1893–1931.

Shen, Y., Ma, D., & Ge, X. (2017). CO2-looping in biomass pyrolysis or gasification. *Sustainable Energy & Fuels*, 1(8), 1700–1729.

Silva Filho, E. C., Lima, L. C. B., Silva, F. C., Sousa, K. S., Fonseca, M. G., & Santana, S. A. A. (2013). Immobilization of ethylene sulfide in aminated cellulose for removal of the divalent cations. *Carbohydrate Polymers*, 92(2), 1203–1210.

Singha, A. S., & Guleria, A. (2014). Chemical modification of cellulosic biopolymer and its use in removal of heavy metal ions from wastewater. *International Journal of Biological Macromolecules*, 67, 409–417.

Sivakumar, P., & Palanisamy, N. (2010). Mechanistic study of dye adsorption on to a novel non-conventional low-cost adsorbent. *Advances in Applied Science Research*, 1(1), 58–65.

Sjahro, N., Yunus, R., Abdullah, L. C., Rashid, S. A., Asis, A. J., & Akhlisah, Z. N. (2021). Recent advances in the application of cellulose derivatives for removal of contaminants from aquatic environments. *Cellulose*, 28(12), 7521–7557.

Sun, X., Peng, B., Ji, Y., Chen, J., & Li, D. (2009). Chitosan(chitin)/cellulose composite biosorbents prepared using ionic liquid for heavy metal ions adsorption. *AIChE Journal*, 55(8), 2062–2069.

Wang, J., Geng, G., Liu, X., Han, F., & Xu, J. (2016). Magnetically superhydrophobic kapok fiber for selective sorption and continuous separation of oil from water. *Chemical Engineering Research and Design*, 115, 122–130.

Wang, Y., Liu, S., Wang, J., & Tang, F. (2020). Polymer network strengthened filter paper for durable water disinfection. *Colloids and Surfaces A: Physicochemical and Engineering Aspects*, 591, 124548.

Wang, Y., Wang, X., Xie, Y., & Zhang, K. (2018). Functional nanomaterials through esterification of cellulose: A review of chemistry and application. *Cellulose*, 25(7), 3703–3731.

Wanichapichart, P., Kaewnopparat, S., Buaking, K., & Puthai, W. (2003). Characterization of cellulose membranes produced by acetobacter xyllinum. *Songklanakarin Journal of Science and Technology*, 24, 855–862.

Williams, T., Hosur, M., Theodore, M., Netravali, A., Rangari, V., & Jeelani, S. (2011). Time effects on morphology and bonding ability in mercerized natural fibers for composite reinforcement. *International Journal of Polymer Science*, 2011, 192865.

World Water Assessment Programme (United Nations), (2018). *The United Nations World Water Development Report 2018*. United Nations Educational, Scientific and Cultural Organization, New York, USA. www.unwater.org/publications/world-water-development-report-2018/.

Wu, F.-C., Tseng, R.-L., & Juang, R.-S. (2010). A review and experimental verification of using chitosan and its derivatives as adsorbents for selected heavy metals. *Journal of Environmental Management*, 91(4), 798–806.

Yang, S., Fu, S., Liu, H., Zhou, Y., & Li, X. (2011). Hydrogel beads based on carboxymethyl cellulose for removal heavy metal ions. *Journal of Applied Polymer Science*, 119(2), 1204–1210.

Yang, X., Reid, M. S., Olsén, P., & Berglund, L. A. (2020). Eco-friendly cellulose nanofibrils designed by nature: Effects from preserving native state. *ACS Nano*, 14(1), 724–735.

Yu, X., Tong, S., Ge, M., Wu, L., Zuo, J., Cao, C., & Song, W. (2013). Adsorption of heavy metal ions from aqueous solution by carboxylated cellulose nanocrystals. *Journal of Environmental Sciences*, 25(5), 933–943.

Zamora-Ledezma, C., Negrete-Bolagay, D., Figueroa, F., Zamora-Ledezma, E., Ni, M., Alexis, F., & Guerrero, V. H. (2021). Heavy metal water pollution: A fresh look about hazards, novel and conventional remediation methods. *Environmental Technology & Innovation*, 22, 101504.

Zhou, Y., Jin, Q., Hu, X., Zhang, Q., & Ma, T. (2012). Heavy metal ions and organic dyes removal from water by cellulose modified with maleic anhydride. *Journal of Materials Science*, 47(12), 5019–5029.

Zwain, H. M., Vakili, M., & Dahlan, I. (2014). Waste material adsorbents for zinc removal from wastewater: A comprehensive review. *International Journal of Chemical Engineering*, 2014, 347912.

21 Comparing Properties and Potential of Pinewood, Dried Tofu, and Oil Palm Empty Fruit Bunch (EFB) Pellet as Cat Litter

Noor Azrimi Umor, Nurul Hidayah Adenan,
Nadya Hajar, Nurul Ain Mat Akil, and Nor Haniah A. Malik
Universiti Teknologi MARA Cawangan Negeri Sembilan

Shahrul Ismail
Universiti Malaysia Terengganu

Zaim Hadi Meskam
Usaha Strategik Sdn. Bhd

CONTENTS

INTRODUCTION

In Malaysia, about 5 million cats were recorded in 2019, and 658,000 of them were adopted as pet (Petfair, 2022).This is also contributed by the fact that 70% of the Malaysian population are pet owners (Mordor Intelligence, 2022). Pets such as cats are reported to have a positive impact on various aspects of mental health and well-being, thus creating a better condition for coping with adverse situations such as movement control due to the COVID 19 pandemic (Grajfoner et al., 2021). As a result, pet services have become one of the fastest growing business sectors, contributing to the increasing number of pet owners. It is a high-end industry that includes various services such as veterinarian, pet store, pet hotel, and others. Following this trend, the industry will continue to flourish in the coming years.

It is reported that an average cat will secrete about 40 g of fecal waste every day (Dabritz and Conrad, 2006). These feces and urine can be a nuisance due to their strong natural odor. Therefore, cat litter is commonly used for indoor or outdoor cat excretion to solve the problem (Neilson, 2009). Cat litter must have the ability to absorb liquid and odor, making the waste easy to manage (Saikeaw et al., 2021). Most commercial cat litters are imported and are either biodegradable or non-biodegradable. Non-biodegradable litter types include clay and crystal litter, while biodegradable litter types are based on cellulose and lignocellulosic materials such as pine, corn, paper cat litter, olive oil waste, and tofu (Yarnell, 2004; Wendland, 2011; Wilde, 2022). It is estimated that for indoor cats, the amount of cat litter used per cat per month is in the range of 4–6 kg. The amount of waste generated from cat litter utilization in Malaysia may reached up to 1.316 million tons, given 50% of adopted cat are kept indoor. The use of biodegradable cat litter is more beneficial in terms of environmental risk while protecting cat and owner's health. There is a possibility that cats will eat cat litter, and if it is not biodegradable, it may be toxic. When it comes to waste disposal, non-biodegradable cat litter is considered to be highly polluting and requires a special process compared to biodegradable types that are more environmentally friendly. For example, a clumping clay litter does not naturally decompose and cannot be flushed into either sewage or septic systems (Vaughn et al., 2011).

The use of empty fruit bunches (EFB) pellets of oil palm as cat litter is new and has never been reported. EFB pellets are generally used as fuel pellets and have been found to be suitable for use as cat litter. EFB pellet is a lignocellulosic material mainly composed of plant cell walls consisted of cellulose, hemicellulose, and lignin. The ability to absorb water is determined by the structure of hydrophilic functional groups (Wisetrat et al., 2012). In addition, cellulose has a structure and ion exchange with organic materials such as resin, which are also suitable as absorbents (Saueprasearsit et al., 2010). Therefore, the use of lignocellulose as cat litter continues to spark interest. To produce pellets, EFB is mechanically and thermally treated to obtain a small, uniform size and a shorter, dry form compared to EFB fibers (Umor et al., 2021). In terms of appearance, it has the normal shape of a commercial cat litter pellet. In addition, it is biodegradable, and the waste can even be safely disposed of on the ground or near plants. Although the material is marketed as cat litter and reportedly performs similarly to a commercial product, there is no scientific evidence to support this claim.

In this study, the EFB pellet and EFB pellet + wood were evaluated for their properties as cat litter. Two types of commercial biodegradable cat litter, namely, pinewood and dried tofu, were also tested for comparison of performance. It is expected that the EFB pellet and EFB pellet + wood will provide a result comparable to the commercial product.

MATERIALS

SOURCE OF BIODEGRADABLE CAT LITTER

Tofu and pine were purchased from a local pet shop in Kuala Pilah, Negeri Sembilan, while the EFB pellet was provided by EcoBed Sdn. Bhd., a local cat litter producer and supplier. All the samples were weighed prior to analysis. Table 21.1 shows the description of samples used in this study.

METHODS

Physical property characterization

The physical properties of cat litter including size, length, and diameter were measured using a digital caliper following the standard procedure. Moisture was calculated from differences in weight before and after the oven drying process, which was carried out for 24 h at 80°C. The pH of the samples were analyzed using a pH meter (Merck). The bulk density was measured using the formula shown below:-

$$BD = (m_2 - m_1)/V_c$$

TABLE 21.1

Description of Samples

Sample	Description
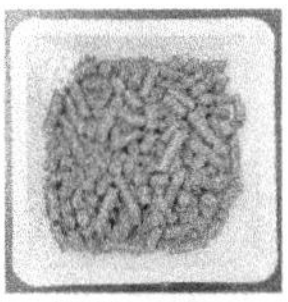 EFB pellet	100% EFB (6 mm)
 EFB pellet + wood	Mix 80% EFB pellet and 20% wood
Dried tofu	100% soya bean (tofu)
 Wood	100% natural pine wood

---

where BD is the bulk density (g/m^{-3}), m_2 is the weight of the filled container (g), m_1 is the weight of the empty container (g), and V is the container volume (m^3) (Brunerová et al., 2018).

Absorption rate, water adsorption capacity, and hydration capacity

The absorption rate of cat litter was calculated by measuring the time taken to absorb 100 mL of distilled water in a 100-mL beaker using 50 g of the substrate. The setup of equipment and apparatus used are shown in Figure 21.1. For water adsorption capacity, 50 g of the substrate was placed in the beaker and a certain amount of water was added at one time until leaking. Hydration capacity was measured by dividing the total amount of water adsorption with the 50 g of samples. All tests were performed in triplicate.

ODOR RECOGNITION TEST

Four samples were used for this study: EFB pallet, EFB pallet+wood, wood, and dried tofu. The samples were stored in their original packaging at ambient temperature in a dry and odorless place until the sensory evaluation. Fifty panelists voluntarily participated in this study. The panelists were among undergraduate, postgraduate students, and UiTM staff. The panelists were compensated with a small gift for their participation in the study. The sensory evaluation was conducted in the sensory analysis laboratory, UiTM Cawangan Negeri Sembilan, Kampus Kuala Pilah. It is an individual sensory booth with white light. Before receiving the samples, the panelists were explained about the task and questionnaire. Then, they received the samples, which were coded with 3-digit random numbers. The panelists rated their preferences using a scoring indicator, with "1=unable to identify," "2=difficult to identify," "3=easy to identify," and "4=difficult

FIGURE 21.1   Water absorption rate test.

Score Sheet
Odor Recognition

Date:
Name:
Instructions:
1. You are given six (6) samples.
2. Examine one sample at a time. Start from left to right.
3. Bring the sample to your nose and take three short sniffs.
4. Score the sample according to the indicator provided. You can repeat the test, if needed. Do not sniff too frequent or too long as it might cause the olfactory system to become fatigued. If this situation happens, take five (5) minutes rest before continuing.
5. Allow to rest after each sample.

Score	Indicator
1	Unable to identify the odor of fabric softener
2	Difficult to identify the odor of fabric softener
3	Easy to identify the odor of fabric softener
4	Very easy to identify the odor of fabric softener

Sample	Score
936	
147	
258	
387	
752	
531	

FIGURE 21.2   Sample of questionnaire given to the respondent.

to identify." The panelists answered the questionnaire, choosing the sensory attributes that characterized their ideal sample. In this test, 5% of softener solution was prepared as odor reference solution. The sample is prepared by adding 10 mL of the reference solution to 10 g of samples. The respondents were required to recognize the smell of the softener from the sample using the scale sheet provided (Figure 21.2). Then, they were required to score each samples. One-way analysis of variance (ANOVA) was used to analyze differences among the treatments. Comparison between different treatments was made using differences of least squares means, when significant F-test values from the ANOVA were obtained at $p \leq 0.05$. All statistical analyses were performed using IBM SPSS Statistics version 19.

RESULTS AND DISCUSSION

PHYSICAL PROPERTIES

Table 21.2 shows the physical properties of the biomass sample. Dried tofu has the highest moisture content compared to the other samples. However, there is not much difference between all samples in terms of moisture content. Dried tofu recorded the highest reading as 8%, followed by EFB pellet, EFB pellet+wood, and wood, which reported 6%, 4%, and 4%, respectively. It was expected that the pH of cat litter is generally almost neutral, but in this study, dried tofu was found to have an acidic pH of 4.53, while the other samples had a pH value of more than 6. The length of the samples showed that dried tofu and wood were longer, while the diameter of the wood and EFB pellet+wood samples was larger. Bulk density is defined as the ratio of the mass of biomass to its volume. The disadvantages of raw biomass are high moisture content, low bulk density, and thus lower heating value. Low bulk density leads to difficulties in handling, storing, and transporting the material (Tang et al., 2014). From the results, the bulk density of dried tofu has a higher density than the other samples. This is expected as each material consists of different compositions and properties.

ADSORPTION RATE, TOTAL VOLUME OF WATER ADSORPTION, AND HYDRATION CAPACITIES

The absorption rate and total volume of water absorption of the samples ranged from 28.5 to 100 (mL/min) and 100–131 mL, respectively, as shown in Table 21.3. In addition, the hydration capacity of the tested samples ranged from 2.00 to 2.62 (g water/g litter). The hydration capacity recorded in this study was higher than previously reported. For example, only 0.52–1.17 (g water/g litter) was measured for bentonite, while dried corn distiller can hold up to 2.17 (g water/g litter). In terms of total water retention volume for 50 g samples, wood was found to have the highest volume of

TABLE 21.2
Physical Properties of Pet Litter and Biochar

Sample	Moisture (%)	Length (mm)	Diameter (mm)	pH	Bulk Density (g m^{-3})
EFB pellet	6	24.30	5.63	6.29	0.79
EFB pellet+ wood	4	26.10	6.10	6.09	0.75
Dried tofu	8	36.70	1.50	4.53	0.84
Wood	4	35.20	6.80	6.40	0.76

TABLE 21. 3
Absorption Rate, Total Volume of Water Adsorption, and Hydration Capacity of Cat Litter

Sample	Absorption Rate (mL/min)	Total Water Adsorption (mL)	Hydration Capacity (g water /g litter)
EFB pellet	28.5	100.00	2.00
EFB pellet+ wood	100.0	113.33	2.27
Dried tofu	40.0	110.00	2.20
Wood	70.42	131.00	2.62
Bentonite (Cliff and Heymann, 1991)	–	26–58	0.52–1.17
Flour pellet (Saikeaw et al., 2021)	–	27–39	0.53–0.78
Corn dried distiller grains (Vaughn et al., 2011)	–	108	2.17

131 mL, followed by EFB pellet+wood with 113.33 mL. These were better than those in previous reports. It also shows that the tested biodegradable cat litter, especially EFB pellet, has sufficient quality compared to other types of commercial cat litter. For absorption rate, it was observed that EFB pellet+wood was the fastest absorbent compared to other samples. Combining wood with EFB pellet proves to be a good strategy to boost the absorption rate as EFB pellet alone only recorded 28.5 (mL/min) compared to the latter, 100 (mL/min). When these result were compared with the physical properties, it was observed that the lower bulk density did improve the absorption rate, and the lower moisture content of cat litter had increased the total water absorption.

ODOR SENSORY TEST

Odor can be evaluated by chemical–analytical measurements or experiments and subjective odor analysis (human perception). Subjective evaluation is a common tool for assessing the efficiency of odor reduction, e.g., in gas cleaning systems, to measure the ability to reduce the intensity and improve the hedonic odor of a gaseous effluent (VDI, 1992). The result of the odor sensory test is shown in Table 21.4. In general, the score obtained by most respondents for all samples ranged from 1.80 to 2.06. This result indicates that all samples have good adsorption capacity for softener solution as the scores were classified as difficult to detect. Dried tofu was the best odor adsorbent, followed by wood and EFB pellet. The natural odor of the samples and the associated adsorption process contribute to the evaluation. Based on the physical properties, the acidic pH of tofu contribute to better adsorption of odor. Lower pH contribute to changes in substrate towards isoelectric point (PI) thus possibly creating larger site of hydrophobic area to stretch and expose for better odor adsorption (Xue et al., 2021). For example, when dried tofu was mixed with water only, it produced a natural odor that adsorb off-odors from detergent. When it reacted with the reference solution in the experiment, it surpassed the odor of the softener. Lignocellulosic materials such as wood and EFB pellets are considered low-cost biosorbents with natural affinity for inorganic and organic pollutants (Ioannidov and Zabaniotou, 2007).

CONCLUSION

In summary, the EFB pellet and EFB pellet+wood were found to be cat litter compared to commercial products, namely, wood and dried tofu. The overall result showed that both samples have suitable properties and comparable quality to dried tofu and wood. In some parameters, the samples outperform the commercial product. In terms of hydration capacity and the adsorption rate, the EFB pellet+wood was better than dried tofu and wood. The use of EFB pellets and EFB pellets+wood

TABLE 21.4
Mean of Score for Odor Recognition Test

Sample	Odor Test Score
EFB pellet	$1.96c \pm 0.83$
EFB pellet + wood	$2.06b \pm 0.84$
Dried tofu	$1.80a \pm 0.97$
Wood	$1.92c \pm 0.90$

a,b,c Different letters show significantly different among treatments ($p < 0.05$).

as cat litter is indeed a better choice for Malaysian people and could provide new income for the oil palm industry.

REFERENCES

Brunerová, A., Roubík, H., & Brožek, M. (2018). Bamboo fibre and sugarcane skin as a bio-briquette fuel. *Energies*, 11(9), 2186.

Cliff, M., & Heymann, H. (1991). Descriptive analysis of oral pungency. *Journal of Sensory Studies*, 7(4), 279–290.

Dabritz, H. A & Conrad, P.A. (2010). Cats and toxoplasma: Implications for public health. *Zoonoses Public Health*, 57, 34–52.

Grajfoner, D., Ke, G. N., & Wong, R. (2021). The effect of pets on human mental health and wellbeing during COVID-19 lockdown in Malaysia. *Animals*, 11(9), 2689.

Ioannidou, O., & Zabaniotou, A. (2007). Agricultural residues as precursors for activated carbon production-A review. *Renewable and Sustainable Energy Reviews*, 11(9), 1966–2005.

Mordor Intelligence. (2022). Malaysia pet food market - Growth, trends, COVID-19 impact, and forecasts (2022–2027). Assessed on 20 June 2022 at https://www.mordorintelligence.com/industry-reports/malaysia-petfood-market.

Neilson, J. C. (2009). The latest scoop on litter. *Journal of Veterinary Medicine*, 104, 140–144.

Petfair. (2022). Malaysia pet market insights. Accessed on 28 June 2022 at https://www.petfair-sea.com/asia-markets/southeast-asia-pet-market/malaysia-pet-market/.

Saikeaw, N., Rungsardthong, V., Pornwongthong, P., Vatanyoopaisarn, S., Thumthanaruk, B., Pattharaprachayakul, N., Wongsa, J., Mussatto, S. I., & Uttapap, D. (2021). Preparation and properties of biodegradable cat litter produced from cassava (Manihot esculenta L. Crantz) trunk. *E3S Web of Conferences*, 302, 02017.

Saueprasearsit P., Nuanjarae, M. & Chinlapa M. (2010). Biosorption of lead (Pb2+) by Luffa cylindrical fibre. *Environmental Research Journal*, 4(1), 157–166.

Tang, J. P., Lam, H. L., Aziz, M. K. A., & Morad, N. A. (2014), Biomass characteristics index: A numerical approach in palm bio-energy estimation. *Computer Aided Chemical Engineering*, 33, 1093–1098.

Umor, N. A., Abdullah, S., Mohamad, A., Ismail, S. Bin, Ismail, S. I., & Misran, A. (2021). Energy potential of oil palm empty fruit bunch (Efb) fibre from subsequent cultivation of *Volvariella volvacea* (bull.) singer. *Sustainability*, 13(23), 1–15.

Vaughn, S. F., Berhow, M.A., Winkler-Moser, J.K., & Lee, E. (2011). Formulation of a biodegradable, odour-reducing cat litter from solvent-extracted corn dried distillers grains. *Industrial Crops and Products*, 34(1), 999–1002.

VDI. (1992). *VDI 3882 Part 2. Olfactometry - Determination of Hedonic Odour Tone*. Düsseldorf: Verein Deutscher Ingenieure (VDI).

Wendland. (2011). A thesis: Development of a novel cat litter from olive oil waste products. Master of Animal Science, Graduate College, Massy University.

Wilde, L. (2022). The 6 best eco-friendly cat litters of 2022. Accessed on 22 June 2022 at https://www.treehugger.com/best-eco-friendly-cat-litter-5180604.

Wisetrat, O., Ngamsombat, R., Saueprasearsit, P., & Prasara, J. (2012). Adsorption of suspended oil using bagasse and modified bagasse. *Journal of Science and Technology Mahasarakham University*, 31, 4.

Xue, C., You, J., Zhang, H., Xiong, S., Yin, T., & Huang, Q. (2021). Capacity of myofibrillar protein to adsorb characteristic fishy-odor compounds: Effects of concentration, temperature, ionic strength, pH and yeast glucan addition. *Food Chemistry*, 363, 130304.

Yarnell, A. (2004). Kitty litter. *Chemical & Engineering News*, 82, 26.

22 Challenges and Future Perspectives

Junidah Lamaming, Abu Zahrim Yaser and Mohd Sani Sarjadi
Universiti Malaysia Sabah

CONTENT

A molecule called cellulose is made up of hundreds, and occasionally even thousands, of carbon, hydrogen, and oxygen atoms. Cellulose is no longer associated with typical applications as it is emerging as the most flexible and smart material. Considering its excellent mechanical strength, biocompatibility, biodegradability, and environmental friendliness, cellulose has evolved as an intriguing material for a variety of applications (Figure 22.1). Cellulose is indeed remarkable when its properties are tuned in the right direction. As research studies have deepened and expanded with the advancement of technology and innovation, some of the challenges encountered and addressed for the sustainability of the cellulose in the developments, processing, or applications. Some of reviews has highlighted the concerns regarding the cellulose-based materials in various field such as in packaging (Nadeem et al., 2022; Shi et al., 2023), medical (Thomas et al., 2020; Ji et al., 2022), water treatments (Aoudi et al., 2022), composites (Trache et al., 2022) and as an adsorbents (Zhang et al., 2023), among others.

One strategy to secure future energy that is ecologically beneficial is to produce biofuel from a variety of renewable biomass feedstocks in place of fossil fuels. Two potential conversion pathways—thermochemical and biochemical conversion pathways—have been emphasized in order to ensure the biofuel production's success. Different conversion pathways will result in different products depending on the application. A wide variety of chemical distribution in various feedstocks is also obviously capable of influencing the final biofuel product in terms of both quality and quantity. Therefore, choosing the best feedstock for the intended use is crucial. The stability of the product is another factor that must be taken into account before being utilized as fuel. To solve this problem, more advancements in bio-oil and new technology are needed. Currently, the biochemical conversion route used to create biofuels like ethanol and butanol requires numerous steps. Further research into using strong microorganisms to convert syngas or hydrogen from a thermochemical pathway may offer a substitute to reduce the number of steps required in the process. On the other hand, creating single-step processes that combine fermentation and hydrolysis or do both at once could help guarantee the creation of biofuel. To assure the viability of future biofuel production, further research is needed to attempt the maximum biofuel production from a variety of feedstocks using low-energy and affordable technology.

Lignocellulosic biomass, which includes waste from oil palm trees and bamboo, is a plentiful source of energy that enables the production of a wide range of useful products, including pulp, paper, biodiesel, and palm composite. It offers great adaptability in terms of feedstock that can be sampled and can be operated in a wide range of temperatures and weather situations. Bamboo biomass can be processed by a number of methods, including pyrolysis, gasification, and thermal conversion, to produce fuel. The commercially feasible byproducts of the processes are ethanol, oil, syngas, charcoal, and syngas. The fuel characteristics of bamboo biomass are superior to those of most energy crops. It might be able to grow on ruined soil with less upkeep and less competition

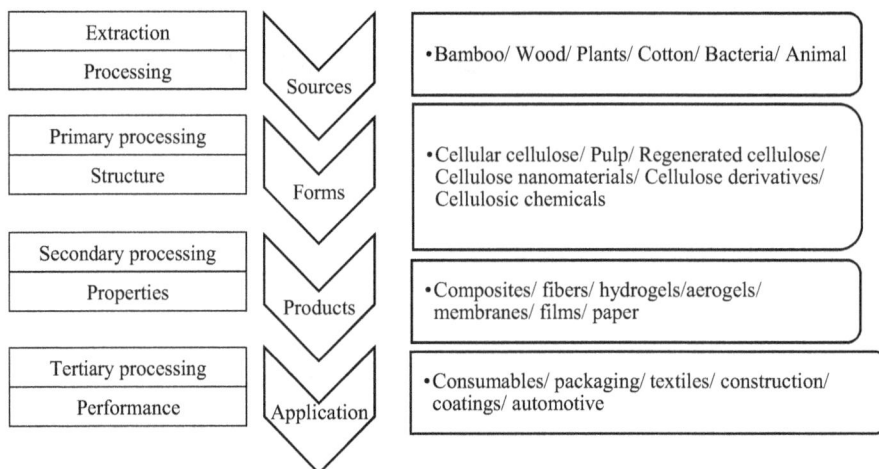

FIGURE 22.1 Diagram show the processing, structure, and discoveries about the characteristics and uses of cellulose.

from food crops for space. Despite the fact that palm oil has a wide range of applications, the palm oil industry has recently come under heavy criticism due to a number of environmental and social problems. Although the oil palm industry waste is surrounded by a number of difficulties, it also offers a number of opportunities in terms of products that can be made from it, particularly as it can be used as a renewable energy to aid in the growth of a country. Large amounts of nutrients are present in oil palm biomass, and the nutrient profiles from earlier studies have revealed that this biomass could be used as bio-compost and organic fertilizer, helping to condition the soil and lowering the use of inorganic fertilizer in agriculture sectors while also lowering the environmental impact. The importance of re-cultivation using reused compost still needs to be further investigated (Umor et al., 2021). Depending on the required ratio, mixing palm oil waste with other organic wastes could hasten the composting process and boost the macronutrients and micronutrients in the compost for use as a growing medium, particularly for horticulture plants. Different types of industrial wastewater utilized as nutrient enhancers and moisturizers demonstrated varying performances in the production of compost. Utilizing household wastewater and other types of industrial wastewater as fertilizer boosters and moisturizers in composting should be the subject of further study. Therefore, bioconversion of palm oil wastes needs to be a strong plan and policy to be followed in future in order to achieve optimal palm oil production in a more sustainable manner.

Unquestionably, advanced biomaterials have a greater impact on the world economy by enabling the production of energy-saving products that improve living standards. Study in this field is beneficial because there are so many issues that still need to be resolved. Undoubtedly, with the right design and choice of production technique, biocomposites will dominate the structural materials market in the engineering sector. Combining natural fibers with polymers generated from renewable resources can help solve several environmental issues. To promote worldwide growth in this cutting-edge class of materials for positive societal, environmental, and economic advantages, industry leaders and senior government officials must work together. In addition to the benefits of composites made from renewable and sustainable materials in terms of economics and functionality, these leaders' leadership is necessary. Apart from that, the increasing number of research conducted and published in replacing vehicle parts with natural fiber composites makes a significant contribution to the automotive industry. However, great effort must be placed in order to convince the consumers about the reliability of this natural fiber-reinforced composites in automotive sectors since many are still having doubts about its performance in comparison to conventional materials. With all factors being taken into account, the breakthrough of excellent potential of these natural fiber composites for automotive application would finally contribute to improvements in societal well-being and eventually meet the requirement of sustainable development goal.

Can cellulose-based composites compete in terms of characteristics with classic synthetic materials? The use of nanomaterials in cement composite materials, notably the use of cellulose nanocrystals (CNCs) as a strengthening agent, has attracted considerable interest in the construction industry. Despite that, only a few researchers have explored the impact of CNCs on the strength of cement composites (compressive, flexural, and tensile). The reasons behind how CNCs respond and enhance the cement matrix structure, however, are still being investigated. Research on sustainable construction materials has recently received a lot of attention. Additionally, there has been an increase in demand for "green" materials with desirable thermal properties, particularly in tropical nations. Currently, research on CNCs' thermal performance indicates that they have a strong chance of serving as the cooling agent in cement composites. One of the significant research gaps that have to be filled involves CNCs as the cooling agent in the cement matrix. In addition, the hybrid composites created by blending nanocellulose with other nanomaterials like graphene offer a variety of opportunities to create high-quality composites for next research.

The most applauding applications of cellulose-based materials are their melted form. With the right control and effective regeneration mechanisms, the dissolved state can be used for a plethora of applications with even the most complicated structures. Numerous surface alterations are possible, thanks to the presence of hydroxyl functional groups on nanocellulose surfaces, creating advantageous nanocomposites with adjustable properties. Materials composed of cellulose are appealing in a wide range of applications due to their interchangeable size, form, and texture. According to the desired use, the physical, mechanical, thermal, and biological characteristics can be pre-designed and fabricated. The compatibility with different materials vastly improves the quality of items made from regenerated cellulose. In cellulose-based research and fields, there is still potential for further discoveries and fundamental information in terms of advanced technology and engineering development. Superhydrophobic membranes, 3D bioprinting technologies, and superabsorbents are some of the numerous newly developing and growing fields that will help a wide range of industries.

One of the most discussed challenges in the cellulose research is the cost of producing cellulose from raw materials to commercialization. Large-scale cellulose isolation can be an energy- and time-intensive procedure. Recent technological advancements have made it possible to manufacture nanocellulose mechanically or chemically, and only a select few businesses and research facilities are currently able to do so in considerable quantities on a pilot scale. Nanocellulose has the tendency to irrevocably agglomerate upon drying, making it difficult to handle and store it while keeping its nanoscale structure. Agglomeration has an impact on the cellulose fibril size distribution, which results in a loss of impacts on the material's nanoscale properties. Obtaining sufficient amounts of dry cellulose nanofibrils in a non-agglomerated condition is challenging. Both the manufacturing procedure utilized to remove water from the nanocellulose suspensions and the specific nanocellulose manufacturing technique can have an impact on fibril morphology. The surface modification and redispersibility of nanocellulose may be important variables for better transportation and long-term application. Comprehensive and systematic investigations are still required to utilize products beyond the laboratory scale and for commercialization in order to enhance understanding and enable high-throughput production (Aoudi et al., 2022; Barhoum et al., 2022).

REFERENCES

Aoudi, B., Boluk, Y., & El-Din, M.G. (2022). Recent advances and future perspective on nanocellulose-based materials in diverse water treatment applications. *Science of the Total Environment*, 843, 156903.

Barhoum, A., Rastogi, V.K., Mahur, B.K., Rastogi, A., Abdel-Haleem, F. M., & Samyn, P. (2022). Nanocelluloses as new generation materials: Natural resources, structure-related properties, engineering nanostructures, and technical challenges. *Materials Today Chemistry*, 26, 101247.

Ji, F., Sun, Z., Hang, T., Zheng, J., Li, X., Duan, G., Zhang, C., & Chen, Y. (2022). Flexible piezoresistive pressure sensors based on nanocellulose aerogels for human motion monitoring: A review. *Composites Communications*, 35, 101351.

Nadeem, H., Athar, M., Dehghani, M., Garnier, G., & Batchelor, W. (2022). Recent advancements, trends, fundamental challenges and opportunities in spray deposited cellulose nanofibril films for packaging applications. *Science of the Total Environment*, 836, 155654.

Shi, J., Zhang, R., Liu, X., Zhang, Y., Du, Y., Dong, H., Ma, Y., Li, X., Cheung, P. C.K., & Chen, F. (2023). Advances in multifunctional biomass-derived nanocomposite films for active and sustainable food packaging. *Carbohydrate Polymers*, 301(Part B), 120323.

Thomas, P., Duolikun, T., Rumjit, N.P., Moosavi, S., Lai, C. W., Johan, M. F., & Fen, L.B. (2020). Comprehensive review on nanocellulose: Recent developments, challenges and future prospects. *Journal of the Mechanical Behavior of Biomedical Materials*, 110, 103884.

Trache, D., Tarchoun, A. F., Abdelaziz, A., Bessa, W., Hussin, M.H., Brossed, O.C., & Thakur, V.K. (2022). Cellulose nanofibrils–graphene hybrids: Recent advances in fabrication, properties, and applications. *Nanoscale*, 14, 12515–12546.

Umor, N.A., Ismail, S., Abdullah, S. Huzaifah, M.H. R., Huzir, N. M., Mahmood, N.A.N., & Zahrim, A.Y. (2021). Zero waste management of spent mushroom compost. *Journal of Material Cycles and Waste Management*, 23, 1726–1736.

Zhang, Z., Abidi, N., Lucia, L., Chabi, S., Denny, C. T., Parajuli, P., & Rumi, S.S. (2023). Cellulose/nanocellulose superabsorbent hydrogels as a sustainable platform for materials applications: A mini-review and perspective. *Carbohydrate Polymers*, 299, 120140.

Index